Frank Evers Beddard

The Structure and Classification of Birds

Frank Evers Beddard

The Structure and Classification of Birds

ISBN/EAN: 9783742810717

Manufactured in Europe, USA, Canada, Australia, Japa

Cover: Foto ©Klaus-Uwe Gerhardt /pixelio.de

Manufactured and distributed by brebook publishing software
(www.brebook.com)

Frank Evers Beddard

The Structure and Classification of Birds

THE

STRUCTURE AND CLASSIFICATION

OF

BIRDS

BY

FRANK E. BEDDARD, M.A., F.R.S.

PROSECTOR AND VICE-SECRETARY OF THE ZOOLOGICAL SOCIETY OF LONDON

LONGMANS, GREEN, AND CO.

39 PATERNOSTER ROW, LONDON

NEW YORK AND BOMBAY

1898

PREFACE

— • • —

IT was the intention of my predecessor in the office of Prosector to the Zoological Society, the late Professor Garrod, F.R.S., to write a treatise upon bird anatomy. This intention was so far realised that a nearly complete account of the anatomy of the fowl, with the appropriate illustrations, was actually drawn up ; it was proposed that this should be followed by a second part, in which the general anatomical characters of the different groups of birds were to be stated. Of this second part I have beside me some thirty sheets of MS. Professor Garrod's successor in the post of Prosector to the Zoological Society, the late Mr. W. A. Forbes, had every intention of finishing this work commenced ; but unfortunately death took place before any actual additions had been made to the MS. left by his predecessor. I have, on the kind encouragement of Mr. Sclater, determined to make an attempt to carry out this plan of my two forerunners, and the present volume is the result.

It must be admitted that a handbook upon bird anatomy was more wanted at the time that it was first conceived by Mr. Garrod than it is at the present day. Zoologists had then nothing of a general character save the incomplete fragment of Bronn's 'Thierreich' and the sections devoted to bird anatomy in such comprehensive works as those of Meckel and Cuvier. We have now two treatises of first-rate merit, that of Fürbringer and Dr. Gadow's completion of the section 'Aves' in Bronn's 'Thierreich.' Professor

Garrod's intended work differed from either of these in that he meant to preface it with a detailed account of Gallus. I have not thought it useful to follow him in this ; for we have an excellent treatise dealing with one particular bird type in Dr. Shufeldt's book upon the ' Raven.' Instead of this I commence with a general sketch of bird structure, purposely avoiding histological detail and the elaborate description of anatomical facts which are not, in the present state of our knowledge, of great use in classification. The main part of this book is the account of the structure of the different groups of birds. It was upon this aspect of the subject that Mr. Garrod intended to dwell most fully. Dr. Gadow has also treated bird anatomy from this point of view ; the concluding section of his contribution to Bronn's ' Thierreich ' is devoted to an enumeration of the distinguishing characters of the groups of birds. I have, however, treated of this matter more fully, and have incorporated more facts (some of them recorded for the first time) in the systematic part of my book, than did Dr. Gadow. I have felt it to be useless to attempt to vie with Professor Fürbringer's magnificent treatise upon birds. To deal with all the organs of the body as fully as he has done would require more space than it would be probable that any publisher would be disposed to allow me. I believe, however, that I have been able to note the principal facts in the anatomy of the different orders of birds, and that nothing of first-rate importance has been omitted.

Moreover under each section I have referred to the majority of the memoirs already published, so that the reader can supplement, where it is necessary, the facts which I myself detail. These references, I may remark, have been (with a few exceptions) carefully verified ; and although my bibliography of the subject is not complete it is, I hope, without important deficiencies.

The facts of bird structure contained in this book are to

some extent drawn from published memoirs; but the
majority of them, especially those relating to osteology and
muscular anatomy, have been verified by myself; and I have
also laid under heavy contribution the note books of my two
predecessors already mentioned. Some of the more important of these facts are illustrated by woodcuts and 'process'
blocks; for these I am indebted to the liberality of the
Publication Committee of the Zoological Society of London
and to the editors of the ' Ibis.'

Finally, I regret to have to acknowledge that in several
instances I have used two names for the same bird, a fact
which I did not discover in every case until it was too late
for alteration in the text. I hope, however, that any difficulties arising from this error on my part will be obviated by
the cross references in the Index.

FRANK E. BEDDARD.

July 1898.

CONTENTS

59484

CONTENTS

LIST OF ILLUSTRATIONS

———•:•———

THE

STRUCTURE AND CLASSIFICATION

OF

BIRDS

—•◦•—

THE GENERAL STRUCTURE OF BIRDS

As the aim of the present work is to detail the characteristics
of the various groups of birds, I do not propose in this
section to do more than give a general account of bird
anatomy. A fuller description will be found in the part
relating to birds by Dr. GADOW, in Bronn's 'Klassen und
Ordnungen des Thierreichs,' where much that will be found
here in the accounts of the several families of birds is
treated of in the introductory chapters. The greater part of
this I deliberately omit, to save repetition.

The Foot

The feet of birds show a large amount of variation,
which is not for the most part of great value in the determination of affinities. That the older naturalists paid great
attention to these facts is evident from the names Palmipedes,
Cursores, &c. No bird has, save for abnormalities, such as
the Dorking fowl, more than four toes. The opposite extreme is reached by the ostrich, which has only two. That
the three-toed and, *a fortiori*, the two-toed condition has
been arrived at by a reduction from four toes seems to be
shown by the condition of the feet in certain petrels, where,

B

as shown by FORBES, a rudiment of the missing toe is present in the shape of a nodule, or of two nodules, of bone, hidden in extreme cases beneath the skin, and only appearing externally as a small wart with no claw. In *Phœbetria* there is an advance upon this, since there is externally a minute claw, and beneath the skin two minute nodules of bone. In the woodpeckers, *Picoides* and *Tiga*, commonly spoken of as three-toed birds, there is a similar vestige of the fourth toe. On the other hand this does not apply to all three-toed birds. In *Rhea, Tetrax,* and *Pelecanoides* FORBES searched in vain for a trace of the missing toe. The toes of the bird's foot are arranged in different fashions, giving rise to more than one form of foot. When there are only three toes they are all directed forwards, except in *Picoides,* where the last is directed backwards; but then, as already stated, this bird has a rudimentary hallux, and it conforms therefore to the type seen in other woodpeckers. When there are four toes they are rarely all turned forwards; this is the case, however, with the swifts. Most commonly the hallux is turned more or less completely backwards; this is so with the passerines and with many other birds. In what is termed the zygodactyle foot, *e.g.* the woodpeckers, both the first and the fourth toes are turned backwards, and thus an effective grasping organ is produced.

An anomalous form of zygodactylism, termed heterodactylism by some, is offered by the trogons, where the second toe is turned back. Syndactylism is an expression used to describe toes which are united together for a longer or shorter distance, such as, for example, the todies and kingfishers. Further details in the form of foot will be found under the descriptions of the several families.

In many birds the toes are perfectly free from each other up to their attachment to the metatarsals. In others there is a condition known as webbing, where a scale-covered skin is stretched between the toes. This may be feebly developed, as in many wading birds, or complete, as in swimming birds, such as the duck. The extreme state of webbing is seen in the pelican tribe, where all the four toes are united by webs—in the ducks only three being thus united, and the hallux free. The coots have a form of webbing which is characterised by applying the term ' lobate ' to the foot. Each toe in the case of a lobate foot is bordered by a flat expansion of skin, but there is no connection between the borders of adjacent toes. The foot is covered to a varying degree with a horny integument, which is arranged as larger or smaller flat scales, or as granules of various shapes and sizes. This scutellation sometimes extends on to the tibia ; it generally occupies the tarsus as well as the foot proper ; the other extreme is shown by *Syrrhaptes*, where feathering extends down to the last digits of the foot. The form of these is sometimes useful in classification, as in the **Passeres** (*q.v.*) The digits are armed with claws, which are straighter in wading birds, and very curved in the birds of prey. Their relative lengths vary ; in the larks, for example, and in the cuckoo (*Centropus*) that of the hind toe is enormously long. The middle toe is often serrated, as in herons, owls, and goatsuckers, &c.

The order of the toes may be almost always settled by counting the number of the phalanges. This is progressive, the first toe having two phalanges, the second three, the third four, and the fourth five. To this general rule there are a few exceptions. I have described in the owl (*Photodilus*) only four phalanges in the last (fourth toe), a state of affairs, however, which is plainly due to a fusion between the two first phalanges. The goatsuckers (of genus *Caprimulgus* and allied forms) have a digital formula of 2, 3, 4, 4. In many **Tubinares** the formula is 1, 3, 4, 5. In the swifts the toes are still further reduced, for we have in that group the digital formula 2, 3, 3, 3. *Pterocles* has a digital formula agreeing with that of *Caprimulgus*. ZEHNTNER [1] has lately shown •that the digital formula of the swifts is due to reduction ; he has found, in fact, in *Cypselus melba*, in a certain stage of development, four phalanges in each of digits 3 and 4, in which stage, therefore, the bird is only one digit short of the normal.

[1] ' Beiträge zur Entwicklung von *Cypselus melba*,' *Arch. f. Naturg.* 1890.

A comparison of the varied forms of feet among birds has an important bearing upon the origin of the foot. From this follows some insight into the nature of the life led by the possessors of the most primitive form of foot. The matter has been put forward in a clear fashion by FINN,[1] and in a correspondence which his paper elicited. As all the evidence at our disposal seems to show that the four-toed foot is the more primitive, we have to decide whether the palmate foot of the pelicans or the grasping foot of the passerine is the earlier, or whether some modification of these, such as the zygodactyle, or four-toed foot with a rudimentary hallux, is the more primitive. The four-toed foot of the **Steganopodes** is often figured as if all the toes were directed forwards, but this is really not the case; they are, as in terrestrial birds with a more or less rudimentary hallux, directed at least sideways. This seems to argue that the original form of the foot was as it is now in the Passeres, a fact which is still further enforced by the foot of *Archæopteryx* (see below). The more purely terrestrial the birds are the more rudimentary is the hallux, until in the purely terrestrial bustards the hallux has disappeared altogether, and in the ostrich, most terrestrial of birds, the second toe has vanished also. Among the gallinaceous birds, moreover, the more arboreal forms, the Cracidæ and Megapodidæ, the hallux is better developed than in those that do not roost in trees. That the zygodactyle foot is a further modification of the anisodactyle seems to be shown by the transitional state of the owls, which 'always perch in the zygodactyle position,' the fourth toe being capable of reversion.[2]

Beak

In all existing birds the upper and lower mandibles are invested with a horny sheath, the beak. The form of this

---

[1] 'The Significance of the Bird's Foot,' *Natural Science*, June 1894; see also July and September Nos. of the same periodical for further notes and correspondence on the matter.

[2] A. REICHENOW, 'Die Fussbildungen der Vögel,' *J. f. O.* 1871, p. 401.

beak has been used by ornithologists for systematic purposes, and whole groups of birds have received their names from this shape, *e.g.* dentirostres, lamellirostres, &c. The bill, however, varies so greatly in admittedly allied birds that its use for classificatory purposes is not great. As a striking instance of this may be mentioned the **Limicolæ** ; we see there the spatulate bill of *Euonyrhynchus*, like a diminutive spoonbill, the upturned bill of *Recurvirostra*, the sideways-turned bill of *Anarhynchus*, the longer lower mandible of *Rhynchops*, and the ibis-like bill of *Numenius*. GADOW has used for classificatory purposes the complex or simple condition of the beak. In some birds, *e.g.* Ratitæ, the horny sheath is composed of several pieces ; in others, the majority, this is not the case. In birds of prey and in parrots there is present a structure which has been termed the cere : this is simply the basal part of the beak, which has remained soft. Its occurrence in those two groups of birds does not appear to be significant of any close affinity.

The lamellirostres afford another example of how dangerous it is to attempt any decision as to affinities from the form of this organ. It has been insisted that one reason for regarding the flamingo as a long-legged duck is the existence of lamellæ along the beak ; but this feature is also met with in the stork, *Anastomus*, to which group moreover the bird is now more generally believed to be related. The puffin is nearly exceptional [1] in the periodical moulting of a portion of the bill ; but the pelican (*P. trachyrhynchus*) casts annually an excrescence upon the top of the upper beak. Sexual dimorphism in the bill is rare, but is exhibited in a marked way in *Heteralocha*, where the female has a long and downwardly curved bill, while that of the male is shorter and straighter.[2]

[1] Several auks (*q.r.*) do so, and it has been asserted of the penguins.
[2] See in this matter and for variations EHLERS, *Zool. Miscell.* i. (Göttingen, 1894).

Feathers [1]

A bird may be known by its feathers; to define a bird it is only necessary to refer to its covering of feathers. No other animal has any structures comparable to a well-developed feather. It is true that the filo-plumes are really little more than hairs. But the processes of development serve to place a fundamental barrier between the two kinds of structures.

A hair commences as a thickening of the stratum Malpighii, which grows downwards into the dermis; a feather is from the first a slight papilla involving the outer layers of the epidermis as well as the stratum Malpighii, a papilla which is surrounded by a circular depression. This papilla gradually sinks down into the skin and assumes a cylindrical form. The cells of the Malpighian layer commence to proliferate vigorously, and form a series of thickened folds disposed radially to the longitudinal axis of the feather papilla, and towards the central pulpa. These radially arranged masses of cells undergo a process of cornification, free themselves from the overlying cells of the horny layer of the epidermis, and produce a bundle of horny fibres—the embryonic down. The feathers may retain this embryonic character throughout life, or further changes may take place. This consists in the formation below the first feather follicle of a second in continuity with it; in this a feather is developed, which may be a down feather, like the first formed, or may grow into one of the stronger varieties of feathers to be described presently. In either case the growing feather pushes the down before it, and the latter is ultimately thrown off.

The structure of feathers has been described at length by

[1] H. R. Davies, 'Beitrag zur Entwicklungsgeschichte der Feder,' *Morph. J.B.* xiv. 1888, p. 368, and 'Die Entwicklungsgesch. d. Feder,' &c. *ibid.* xv. 1889. p. 560; C. R. Hennicke, 'Die Entwicklung d. Feder,' *Monatsschr. deutsch. Ver. Vogelsch.* xiv. 1889, p. 223; K. Klee, 'Bau und Entwicklung der Feder,' *Zeitschr. f. d. ges. Naturw.* lix. p. 110; see also Gadow, article 'Feather' in Newton's *Dict. of Birds.*

NITZSCH, and among recent writers more especially by
WRAY.[1] A typical feather consists of the stem or rhachis,
of which the lower 'quill' region is termed the calamus.
From the rhachis above the calamus spring a series of lateral
branches, the rami or barbs, which in turn give rise to
barbules, and they to minute, often hooked, processes, the
barbicels. At the junction of the calamus with the barb-
bearing rhachis arises in many feathers an aftershaft (fig.
4), which has the character of a second smaller feather
arising from the shaft of the first : but in the cassowary,
emu, and the extinct *Dinornis* this aftershaft is as large as
the main feather from which it arises. The barbicels with
their terminal hamuli give the stiffness to the feather which
is caused by the interlocking of these processes. The barbules are of two sorts, those nearest to the root of the barb
being different from those nearest to its tip. The former
are shaped something like a knife blade ; they are thickened
above and bent in the middle, gradually tapering away to
a fine point ; just in the middle, where the bend is, are
two or three small teeth on the upper margin. It is by
means of these teeth that successive barbules are locked
together. The remaining set of barbules are frayed out
towards the end into a series of branchlets which are
hooked at first, but the more distal set are merely fine-
pointed branchlets ; these arise obliquely, so that a given
barbule comes into relation with four or five other barbules.

All feathers, however, have not so complicated a structure.
The strong wing feathers of the cassowary consist of the
stem alone. Filoplumes have but few radii, consisting
almost alone of the calamus and rhachis.

Down feathers are as a rule without the hamuli ; often
the radii spring at once from the calamus, there being
no rhachis.[2] A peculiar form of these feathers, called

---

[1] 'On the Structure of the Barbs, Barbules, and Barbicels of a Typical
Pennaceous Feather,' *Ibis*, 1887, p. 420.

[2] The term *neossoptiles* has been applied to the down covering the newly
hatched young of many birds, in contradistinction to *teleoptiles*, the feathers
(down or contour) of the adult bird.

'powder down feathers,'[1] is found in many birds belonging to quite different groups; they are usually aggregated into special patches. These are simply down feathers of which the tops continually break down into a dusty matter. These powder down patches have been asserted to be luminous in the heron, and to aid it in attracting its prey; but the assertion seems to be void of truth.

The feathers of birds are, with a few exceptions, coloured either by the deposition of pigments alone, or by optical tints derived from the actual structure of the feathers shown up against a basis of dark pigment. The colours of birds' feathers have been chiefly investigated by CHURCH, KRUKEN-BERG, and GADOW,[2] to whose papers the reader is referred.

The arrangement of the feathers upon the wing requires a special description. They have been carefully studied by the late Mr. WRAY,[3] from whose paper both the information and some of the explanatory illustrations have been drawn. In the wing of the wild duck there is, as in all birds, a fringe of stout quills known as the *remiges*. These are attached to the fore-arm and to the hand. The border of the ulna, to which they are fixed, constantly bears impressions of the quills. Here the feathers stand out at right angles to the bone; in the hand they become more and more inclined forwards until the last of the series lies parallel with the bone (phalanx 2 of digit II.) which bears it. Of these remiges it is usual to term those which are inserted upon the ulna the *secondaries*, and those upon the hand proper *primaries*. But the term *cubitals* is gaining ground as an expression for the secondaries of many writers. The first of the remiges is much smaller than the others, and has been called the *remicle*; it nevertheless belongs to the series of remiges. The rest of the feathers of the wing are known as the *coverts* or *tectrices*. There are four series of

---

[1] L. STIEDA, 'Über den Bau der Puderdunen der Rohrdrommel,' *Arch. f. Anat. u. Phys.* 1870, p. 104.

[2] GADOW, in Bronn's *Thierreich* ('Aves'), treats of the matter in considerable detail.

[3] 'On some Points in the Morphology of the Wing of Birds,' *P. Z. S.* 1887 p. 343.

these, which successively overlap each other and the remiges.
The first series on the upper aspect of the wing are the
tectrices majores, which have a perfectly definite relation to
the remiges, there being one for each remex. To this statement there is in the duck a single exception ; this exception
is the fifth cubital (reckoning, as it is customary to do, from
the carpus) : this remex appears, by reason of the fact that
there is a gap and that the tectrix is present, to be absent.
On the under surface of the wing there is a corresponding
row of lower tectrices majores. It will be noticed that the
reference of the remicle to the series remiges is justified by
its having its proper complement of tectrices majores.

Next to the tectrices majores comes a row of feathers,
the *tectrices mediae*. These are also present on the under
surface ; the set of both, however, is not complete, that of
the second metacarpal being wanting on the upper surface,
and the distal four or five of the manus on the lower surface.
The next row on the upper face of the wing is quintuple,
and the feathers composing the five tiers are known as
the *tectrices minores*. They are scantily represented on
the manus, where in fact there is not room for them, they
being developed on the skin covering the muscles and on
the patagium of the wing. This row of feathers passes on
to the humerus and becomes there partly specialised into
two rows ; the lower of these (sometimes called *parapteron*)
are long feathers suggestive of remiges, while the row immediately above bears the same relation to the pseudo-
remiges as the tectrices majores do to the true remiges. On
the ventral surface of the wing are similar tectrices minores
with a similar specialisation of an *hypopteron* (representing
the parapteron above, and sometimes called *axillaries*), with
its row of special coverts. The patagium is mainly filled
up with several rows of feathers, which are collectively
termed the *marginals*; anteriorly, upon the pollex, they
form together with the anterior feathers of the minores the
so-called *ala spuria*. The ala spuria is specialised into four
small quills with coverts, the specialisation being quite like
that of the humerals at the other extremity of the wing.

So much then for the arrangement of the feathers in the typical bird selected ; we must now consider the divergencies from this constituted normal.   The fifth cubital, absent in

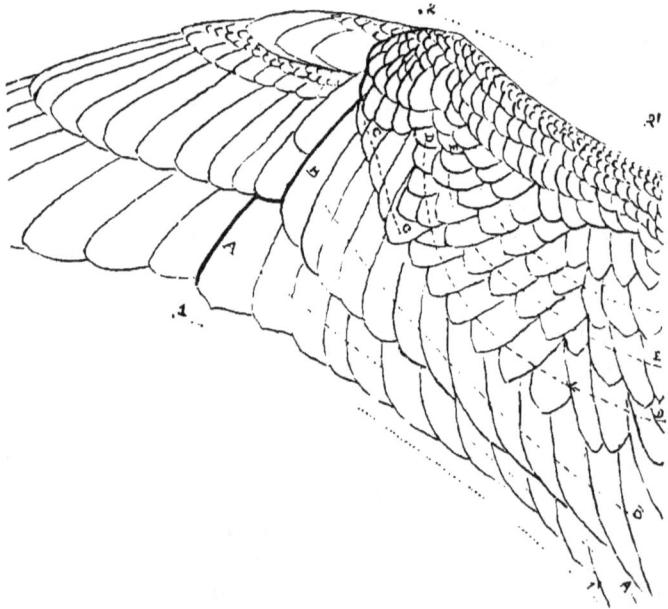

FIG. 1.—WING OF GOLDEN PLOVER (AFTER GOODCHILD).
1-1', posterior border ; 2-2', anterior border ; A, remiges ; B, greater wing coverts ; C, D, median coverts ; E, F, remaining coverts.

the wild duck, is often present in birds.   The terms 'quincubital' and 'aquincubital' have been devised to express these facts.   The following groups are quincubital : [1]

Crypturi, Galli, Rhinochetidæ, Cuculi, many Picarians. On the other hand aquincubital birds are—

Colymbidæ, Tubinares, Steganopodes, Herodiones, Accipitres, Anseres, &c.

The majority of birds, in fact, have not the fifth cubital remex.   The most remarkable fact about this missing remex is that it is either absent or present ; in no case are there any intermediate conditions, such as a small remex.

[1] See SCLATER, 'Remarks on the Fifth Cubital Remex,' &c.. *Ibis* (6), ii. 1890, p. 77.

The only explanation, so far as I am aware, of this remarkable state of affairs is contained in a suggestive paper by DEGEN.[1] DEGEN commences with the assumption that in the hands of the primitive bird all three fingers—then freely movable—were furnished with remiges. In modern birds remiges are only attached to the thumb (ala spuria) and to digit II. DEGEN also postulates a fourth finger (of which rudiments have been discovered in modern birds; see below) with its remiges.

When the metacarpal bones became fused the feathers of the third and fourth digits were, he supposed, forced back

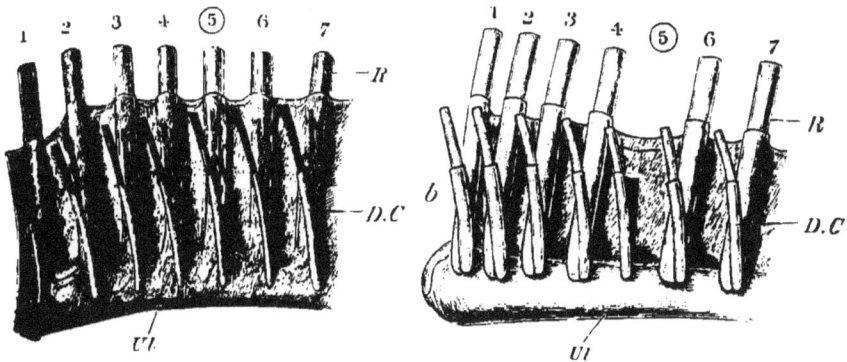

FIG. 2.—a. CUBITAL REMIGES OF PHEASANT. b. CUBITAL REMIGES OF GOLDEN EAGLE (AFTER WRAY).
R 1–7, remiges; D.C, dorsal teetrix major; Ul, ulna.

upon the ulna, as there was no longer any room for their coexistence with those of the second digit upon the compressed hand. Among these the carpal remex was also pushed back. As this remex was attached to an unstable bone or cartilage, its position was not secured, and the variability remained when the feather altered its position; hence the presence or absence of the fifth remex, which is this feather.

The carpal remex is another variable feather. It is present

[1] 'On some of the Main Features in the Evolution of the Bird's Wing,' Bull. Brit. Orn. Club, July 1894 (published in Ibis). See also for quincubitalism GERBE, 'Sur les Plumes de Vol et leur Mue,' Bull. Soc. Zool. Fr. ii. 1877, p. 289.

and fully-sized in *Nothura*. It is occasionally present but small, and sometimes even with its covert altogether absent. The remiges themselves vary in number apart from the presence or absence of the fifth cubital, but not within very wide limits. *Struthio* and the penguin alone are exceptional, and will be treated of separately and later. GADOW has published a useful table showing the number of the primaries in a very large assortment of birds belonging to all orders. The number of primaries varies only between ten and twelve. The number of metacarpals has also a small range of variation, the smallest number presenting six and the largest eight. *Casuarius* having an abbreviated hand is still further reduced, the primaries being only two and the secondaries five. The largest number of metacarpals, eight, is possessed, however, by *Apteryx*, with an abbreviated hand, and by *Struthio*. Seven metacarpals are found in the grebes, flamingoes, and several, but not all, the genera of storks. All other birds have six.

The two prominent exceptions to the foregoing statements are, as has been already mentioned, *Struthio* and the Spheniscidæ. In the ostrich (see fig. 3) there are sixteen primaries, each with its tectrix major upon the upper surface of the wing. The other rows are perfectly recognisable, as is shown in the figure. The wing of the penguin is, however, not reconcilable with the ordinary plan of structure. It has thirty-six bordering feathers, which may be termed primaries; FÜRBRINGER has suggested that these may be really ten primaries with their coverts, but in any case the wing is covered with about thirty rows of scale-like feathers.

As to the general wing feathering, GOODCHILD [1] has surveyed a large series of birds, and noted their peculiarities. Some valuable classificatory results appear to be the outcome of these investigations. Thus the plan characteristic of the humming birds resembles that of the swifts, and both are to be distinguished from the passerines. The picarian type gradually approximates to the psittacine; *Melopsittacus*

[1] · Observations on the Disposition of the Cubital Coverts in Birds,' *P. Z. S.* 1886, p. 181.

might be well referred to the picarians when judged from
the present standpoint. On the other side the birds of prey,
both diurnal and nocturnal, are parrot-like in the arrange-

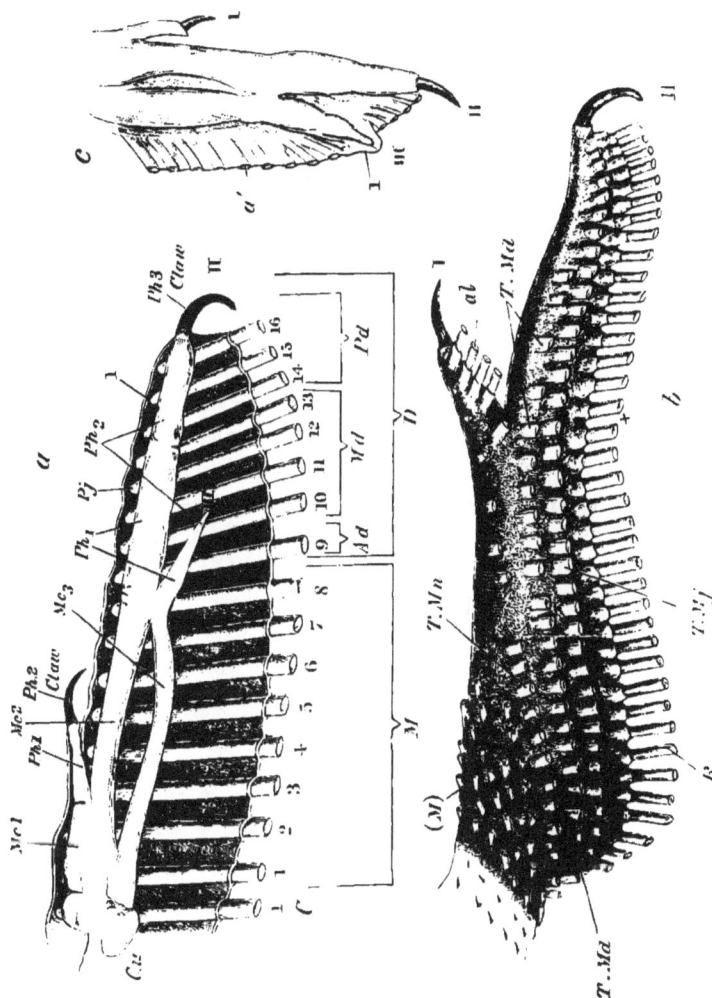

FIG. 3.—*a.* MANUS OF OSTRICH, SHOWING ATTACHMENT OF REMIGES. *b.* MANUS OF OSTRICH, UPPER SURFACE, SHOWING REMIGES AND COVERTS. *c.* DIGITS OF EMBRYO, SHOWING REMIGES AND COVERTS. (AFTER WRAY.)

Mc 3. metacarpals ; *Ph,* phalanges ; I-III. digits ; *C,* carpal remex ; *M,* metacarpal remiges ; *D,* digital remiges ; *Ad,* addigitals ; *Md,* mididigitals ; *Pd,* prodigitals ; *T.Mj,* tectrices majores ; *T.Md,* tectrices mediae ; *T.Mn,* tectrices minores ; (*M*), marginals ; *al,* ala spuria ; *d,* tectrices majores.

ment of their wing feathers. But, curiously enough, *Pernis*,
Pandion, *Gypogeranus*, and the Cathartidæ differ from their
allies. This is not the only case where the disposition
of the feathers runs counter to the affinities to be

derived from an examination of other structures ; for
while *Phalacrocorax* is quite accipitrine the other **Steganopodes** are quite different.   Herons agree with the
Accipitres, while the ciconiine pattern leads towards that
of the **Tubinares**, and is identical in some cases with that of
the American vultures.   The cuckoos should be, when
judged by the feathering of their wings, placed in the
immediate neighbourhood of the **Columbæ**, from which group
Goura ought to be separated.   *Chauna* is practically a
pigeon in these characters, while the **Limicolæ** are not far
off.   The Crypturi are gallinaceous.[1]

Pterylosis

As a general rule the feathers of birds are not distributed
uniformly over the surface of the body, but are set in the
skin in definite tracts, between which are spaces that are
entirely bare or covered only with down plumage.   The
feathered tracts are termed pterylæ, the interspaces apteria.
A few birds, such as the struthious, the penguins, and the
screamers, have an uninterrupted plumage ; but this state
of affairs, though corresponding with what one supposes
to be the original condition, is not necessarily so in the
birds under consideration.   Thus, although the ostrich has
an uninterrupted plumage in the adult state, the young
embryo, as first figured by Miss Lindsay, has definite
pterylæ, thus proving that the continuous feathering is here
purely secondary.   There is a very great variety in the
arrangement of the pterylæ among birds, and for the details

---

[1] De Meijere has devoted some pains to the arrangement of the feathers
with reference to each other, a subject which, as he says, has been hitherto
treated of only in a stepmotherly fashion.   It appears from his investigations
that the feathers are arranged in groups, as are the hairs of mammals.   For
example, upon the naked region of the head of *Numida* the feathers are
grouped in fours, a stronger feather with two hair-like feathers, one on one side
and one on the other.   This is what is generally found, a central stronger
feather with hair feathers surrounding it.   There is here a remarkable analogy
with the grouping of mammalian hairs, where a stronger hair is often surrounded by three or four more slender hairs.   Moseley also in the case of the
dodo (*q.v.*) has found the feathers to be grouped in threes.

the reader is referred to the systematic part of this work. In all, however, there are the following tracts present : —

(1) The *spinal tract* runs from the head to the oil gland. This tract is sometimes continuous at the sides of the neck with the ventral tract, to be described next. It is sometimes a single solid tract throughout, but more usually there is a space developed in it, an apterion in the back, which is of greater or less extent. Commonly there is a break, more or less distinct, between the anterior and the posterior portion of the tract, which may be complete or may consist in an abrupt transition between anterior stiffer feathers and posterior slighter feathers.

(2) The *ventral tract* is always a double tract, but the median apterium may be very narrow. The anterior part of the tract may be single, but sometimes it is double from its very origin. Very commonly on the pectoral region each half of the ventral tract gives off a lateral branch.

(3) The *humeral tract* is a band of stiff feathers running across the humerus ; it is always present and shows no particular modifications.

(4) The *femoral tract* is a corresponding band crossing the thigh. There is sometimes, as in the barbets, a small tract lying between the femoral and the spinal ; and, besides the main tracts, the patagium and the lower leg are more or less covered with contour feathers. The study of Pterylography was first taken up in a systematic manner by NITZSCH ; since his day the number of facts has largely increased, and careful figures of the pterylosis of many birds, not figured by NITZSCH, have been published by a host of observers, especially GARROD, FORBES, SHUFELDT, PYECRAFT, GADOW,[1] and others. For references to these see the descriptions of the different orders of birds.[2]

The general facts of pterylosis must be used cautiously.

[1] See also W. MARSHALL, 'Pterologische Mittheilungen,' *Zool. Gart.* xiv. xv. xvi.

[2] For the musculature of the feathers see HELM, 'Ueber die Hautmuskeln der Vögel,' &c., *J. f. O.* 1884, p. 321. These skin muscles are either limited to the skin, running from feather to feather, or are parts of skeletal muscles, such as the cutaneous branch of the *latissimus dorsi*, &c.

On theoretical grounds it might fairly be assumed that a continuous covering, without any distinctions between pterylæ and apteria, was a primitive condition. But there is evidence to show that where a continuous covering of feathers exists it is not invariably a mark of ancient stock. Thus the ostrich, as already remarked, which when adult has practically no separation into pterylæ and apteria, has, when young, very distinct pterylæ and apteria. In this case, therefore, it is clear that the uninterrupted feathering is a secondary character and not a primitive one. With the penguins, on the other hand, it is conceivable that the absence of apteria is a primitive character. As to the value of the various arrangements of apteria in the pterylosis, GADOW lays stress upon the continuous feathering of the neck, or the presence there of lateral spaces, but admits that the Indian painted snipe (*Rhynchæa*) is an exception which somewhat destroys the value derivable from the consideration of the facts. FÜRBRINGER, on the other hand, uses in his tables of characters the dorsal tract and its modifications. But the variations which occur here in a single and surely well-marked family (*e.g.* Picidæ) tend to shake our faith in the value of the exact way in which the hypothetically continuous feathering has lost its continuity. No doubt GADOW is right in saying that it is of taxonomic importance ' more in the investigation of small than of large groups.'

NITZSCH, for instance, lays some stress upon the ' furcate division and degradation of the portion of the spinal tract situated between the shoulder blades ' in the Accipitrinæ; this division includes the owls.

But on turning over his plates one is struck by the fact that the peculiarity in question is by no means confined to that group, occurring as it does in such widely removed forms as Caprimulgidæ, Charadriidæ, and *Psophia*. Nor is an undivided dorsal tract a distinctive mark of affinity, since it is to be found in such a diversified assemblage as that including *Pavo, Alcedo, Certhia, Todus,* and various passerines. The ventral tracts divide each of them upon the breast into an outer and an inner division in *Pernis apivora, Coracias garrulus, Rhamphastos, Musophaga, Gallus,* various cuckoos, *Charadrius,* &c. They do not divide in *Pandion, Cypselus, Cuculus, Opisthocomus, Buceros, Columba, Fulica,* and *Ciconia.*

GARROD made originally the apparently reasonable suggestion that the down feathers upon the apteria of many birds may be the remains of contour feathers, from which the inference is necessary that those birds with downs upon the apteria are nearer to the continuously feathered and ancestral bird than are those whose apteria are nude. But the whole matter is rather more complicated than this. There are birds with only contour feathers and nude apteria ; there are birds with contour feathers only upon the pterylæ, and down upon the apteria ; there are birds with downs everywhere ; and finally there are birds with downs only upon the pterylæ, mixed with the contour feathers. The facts, therefore, when stated thus fully are not so easy of interpretation.

The evidence derivable from *Archæopteryx*—less negative, perhaps, than ' negative ' evidence often is—may afford us a clue. So many feathers of that bird are well preserved that it seems possible that where feathers have not been preserved they were either really absent or soft down feathers. The latter suggestion seems to be the more probable, on account of the plain fact that *Archæopteryx* was a flying bird. Now the fact that the contour feathers are frequently preceded by downs points in the same direction, viz. that the primitive feathering of birds was in the form of downs. The persistence of downs, therefore, on this hypothesis is so far a primitive character, and the greater the persistence the more primitive the bird. Thus those birds which have downs everywhere will be the more archaic. This is so far promising that that group contains such apparently old types as *Palamedea, Opisthocomus, Rhinochetus,* &c. On this view the most modern of birds will be those which I elsewhere try to show are an ancient race, *i.e.* the bulk of the Pico-Passeres. But, as might be expected with an ancient race, there is every variety shown, and members of this great

C

group are found in all the divisions of birds founded upon the distribution of the downs.   This view throws a side light upon the **Struthiones**.   The feathers of those birds have been called intermediate between contour feathers and downs.   It may be that they are primitive, and that the struthious birds have arisen from some ancient type in which the modern bird's feather had hardly been evolved. Among nearly related families the details of pterylosis do at least sometimes afford indications of resemblance.   Thus, for instance, there are certain small likenesses between the barbets, toucans, and woodpeckers (see below), which help in establishing the near kinship between the three families.

The size or the presence or absence of the *aftershaft* appears to be of little use for systematic purposes.   Among the ducks, for example, some have it and some have it not.   It is as large as the main feather in the emus and totally absent in *Rhea*.   Facts like these, which might be multiplied, throw doubts upon the value of this structure in classification.   So too with the *oil gland* [1] and its feathering or absence of a tuft.   *Cancroma*, which in other points of its structure conforms to the heron type, is alone in that group in having a nude oil gland.   The gland is absent in some parrots, present in others.   GARROD at one time thought that he could correlate among the Pico-Passeres a nude oil gland with small cæca, and a tufted oil gland with the absence of cæca; to the vast majority of picarian birds there is no doubt that the correlation does

Fig. 4. —FEATHER SHOWING AFTERSHAFT (AFTER SCLATER).

---

[1] A. PILLIET, ' Sur la Glande Sébacée des Oiseaux,' &c., *Bull. Soc. Zool. Fr.* xiv. 1889, p. 115; R. KOSSMANN, ' Ueber Talgdrüsen der Vögel,' *Zeitschr. f. wiss. Zool.* 1871, p. 568.

apply. But the todies were found to be birds with a tufted oil gland and with large cæca.

It has been pointed out that when the oil gland has a tuft of feathers upon its apex the rest of the gland is unfeathered, and that, on the contrary, when the tip is nude the general surface of the gland is feathered. The oil gland is, so far as we know, a structure special to birds : it is, indeed, the only purely external glandular apparatus that exists in them. It is therefore possible, if not probable, that the organ first arose in the class—that it is not an inheritance from any ancestor. On this view it is quite possible that the absence of the oil gland may not be always due to its disappearance; birds without oil glands may or may not have lost them. It seems very likely, for example, that the usual absence of this structure among the struthious birds is rather a primitive than a secondary character. If this view of the matter is justifiable, the presence of a tuft may also be, in some cases at least, secondary ; for it is certainly a specialisation that may have appeared after the oil gland was fully developed.

Alimentary Canal

The **tongue** [1] of birds is one of the most variable organs as to size and texture. In *Plotus*, for example, it is practically altogether absent (see fig. 5). When present it is larger or smaller, more or less fleshy. The long, thin, horny tongue of the toucans characterises those birds ; the parrots (see fig. 6) have a thick and fleshy tongue ; naturally the organ is of more use to the latter than to the former. A very remarkable modification of the tongue, seen in birds quite remote in the scale, is the pulling out of the free end into a tuft of fine fibres ; this is associated with the capture of honey or minute insects from the corollas of flowers ; it is

---

[1] C. S. Minot, 'Studies on the Tongue of Reptiles and Birds,' *Ann. Mem. Boston Soc. Nat. Hist.* 1880; Nitzsch-Giebel, 'Die Zunge der Vögel,' &c., *Zeitschr. f. d. ges. Naturw.* xi. 1858, p. 19; Ludwig, Prinz v. Bayern, *Zur Anatomie der Zunge* (München, 1884).

seen in the Trichoglossinæ (see fig. 6), named so on account
of the very structure, and in the Nectariniidæ, &c. Very
frequently the tongue is more or less spiny upon its surface,
particularly towards the attached end
of the organ. A very singular modification is the extraordinarily long
tongue of the woodpeckers, which
is, of course, associated with the ex-
traction of grubs from the crevices in
trees. A detailed description of the
numerous forms of this organ would
occupy more space than can be allowed;
but the principal varieties will be
found described under the different
families. The modifications of the
tongue are not of great assistance to

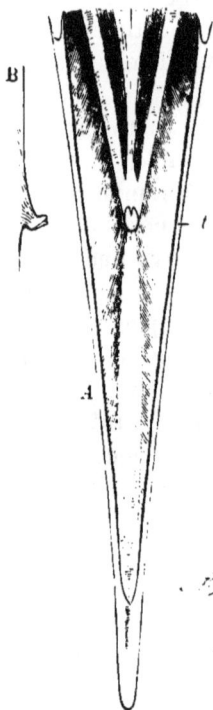

FIG. 5.—*A*, LOWER MANDI-
BLE OF INDIAN DARTER.
t, RUDIMENTARY TONGUE.
B, TONGUE IN PROFILE.

FIG. 6.—HEAD OF *Lorius*, SHOWING EXTENDED TONGUE
WITH BRUSH TIP (AFTER GARROD).

the taxonomist, except as regards smaller groups. Thus
the Plataleidæ have been distinguished from other **Herodiones**
as ' Lipoglossæ.'

Teeth are not met with in living birds. RÖSE, however,
has discovered what he believes to be a rudimentary tooth
band (' Zahnleiste ') in *Sterna*,[1] a discovery which may have

[1] This has been recently confirmed by Miss CARLSSON (' Ueber die Schmelz-
leiste bei *Sterna hirundo*,' *Anat. Anz* xii. 72), who found it to characterise both
jaws. For other embryonic traces of teeth see P. FRAISSE, ' Über Zähne u. Zahn-

some added significance in view of the possible relationship
to modern birds of the cretaceous *Ichthyornis* (cf. below).
The need for teeth seems to have disappeared with the
development of a horny bill, the replacement of the one
structure by the other being, perhaps, comparable to the
replacement of functional teeth by horny plates in the
Ornithorhynchus. Functional teeth, however, existed in
the Jurassic *Archæopteryx* and *Laopteryx* (?), and in the
toothed birds of the cretaceous epoch, *Hesperornis* and
Ichthyornis. In the latter the teeth are in sockets, in the
former in grooves. In both the teeth are numerous, but
not, perhaps, extending on to the premaxillaries; the teeth
show no specialisation in different regions; they are of
dentine coated with enamel, and in *Hesperornis* the basal
portion of the roots consists of osteodentine.

The Eocene bird (from the London clay) *Odontopteryx
toliapicus* has a strongly serrated upper jaw, a state of
affairs which is paralleled in the South American passerine
Phytotoma rara. In this latter bird, as Parker has pointed
out,[1] there is a 'row of clearly defined denticles, both along
the dentary and palatine ridges of the premaxillary.' He
suggests that in these birds and in the merganser, where
similar 'denticles' occur, the bone of the jaw has grown
into arrested dental papillæ.

The **œsophagus** dilates in a few birds into a *crop*, which
is more highly specialised in *Opisthocomus* (q.v.) than in any
other form. When the crop is well marked it consists of a
spherical to oval dilatation of the œsophagus, which in
pigeons is divisible into a right and left half and an intermediate unpaired portion. The gallinaceous birds, the
parrots, and among the **Limicolæ** the American genera
Thinocorys and *Attagis*, are provided with a crop. In other
birds a slight dilatation of the œsophagus, either permanent

papillen bei Vögeln,' *J.B. nat. Ges. Leipzig*, 1882, p. 16; A. F. J. C. MAYER,
'Zähne im Oberschnabel bei Vögeln,' &c., *Froriep's Notiz.* xx. 1841, p. 69;
BLANCHARD, 'Observations sur le Système Dentaire chez les Oiseaux,' *Comptes
Rend.* l. 1860, p. 540; M. BRAUN. 'Die Entwicklung des Wellenpapageis,' *Arb.
Zool. Zoot. Inst. Würzb.* v. 1879.

[1] In his memoir upon ægithognathous birds.

or temporary, foreshadows the fully developed crop of the birds mentioned.

The **stomach** [1] consists of two compartments following each other, the glandular *proventriculus* and the more muscular *gizzard*. The proportions of these two segments of the stomach vary, and both are much reduced in the hoatzin, whose crop appears to take on the function of a gizzard. The proventriculus is usually, but not always, separated by a marked constriction from the gizzard, and has a patch of large glands which generally forms a band lining the upper part of the sac and continuous right round it; to a proventriculus in which the glandular patch is disposed in this fashion the term ' zonary ' is applied. More rarely the patch of glands is a single oval or round patch not continuous round the proventriculus, or there may be two such patches. In *Plotus anhinga* the two patches of proventricular glands are contained in a special diverticulum of the proventriculus. In *Tantalus ibis* MITCHELL [2] has described a remarkable divergence from the usual structure of the proventriculus. In this bird the glandular areas are two, as in other storks. Above these is a row of crypts, which are partly glandular and partly lymphatic, and are believed to be organs for the absorption of water. Among the **Steganopodes** and in other birds the proventriculus is much larger than the gizzard, which follows. In certain tanagers this state of affairs culminates in the apparent absence of the gizzard as a distinct structure (see below). LUND and FORBES have mentioned a number of tanagers in which this occurs. The gizzard is more muscular in grain-eating and in some other birds than it is in flesh- and fish-eating birds. It is strong and hard and lenticular in form in the **Galli**, **Ralli**, &c., bag-like and soft-walled in the heron, &c. The lining of the gizzard undergoes a remarkable modification in certain pigeons (*q.v.*), where it may be even ossified.

[1] ' On the Proventricular Crypts of *Pseudotantalus ibis*,' *P. Z. S.* 1895, p. 271.

[2] A comprehensive work upon this organ is that of CAZIN, *Ann. Sci. Nat.* (7), iv. 1887, p. 177. See also the same, ' Structure et Mécanisme du Gésier des Oiseaux,' *Bull. Soc. Philom.* 1888, p. 19.

The **intestine** of birds varies much in proportional as well
as (naturally) in actual length. In the systematic part of
this work a number of actual measurements will be found ;
from these it is obvious that on the whole purely frugivorous
birds have a short gut, while fish- and grain-eating birds have
a long gut. To compare, for example, two birds of roughly
the same size but of different feeding habits, the touraco
and the common pigeon, we find in the former a gut of
42 c.m., and in the latter of 108–132 c.m. As the gut is
always longer than the abdominal cavity in which it lies, it
has to be thrown into folds in order to find room.

In the embryo chick the gut is straight and is supported
by a continuous dorsal mesentery of equal vertical diameter
throughout. The coiling is both lateral, which results in
lateral foldings of the mesentery, and vertical, which results
in unequal growths of the mesentery. It only affects the
middle part of the alimentary tract, the œsophagus and
stomach on the one hand, and the rectum on the other, or
at least a part of it, retaining the original straight condition.
The lateral foldings give rise to secondary connections
between different regions of the mesentery, and tend to
obscure the course of the gut ; but it is easy, by carefully
removing the entire intestine to distinguish these secondary
mesenteries from the primary sheet binding the gut to the
dorsal body wall.

When the body walls of a series of birds are removed, and
the disposition of the intestines thus shown examined, they
have been found to present great differences. These have
been studied and described by GADOW in two memoirs,[1] and
the main results extracted for the account of the digestive
system in Newton's ' Dictionary of Birds.' It is mainly
from the latter work that the abstract here given is drawn.

In a goose, for example, the main disposition of the
intestinal folds is in a longitudinal direction ; they run
parallel with each other in a direction roughly coinciding

[1] 'Versuch einer vergleichenden Anatomie des Verdauungssystemes der
Vögel,' *Jen. Zeitschr.* xiii. 1879, pp. 92, 339; 'On the Taxonomic Value of the
Intestinal Convolutions in Birds,' *P. Z. S.* 1889, p. 303.

with the long axis of the body. On the other hand the intestines of a gull, seen when the body wall is cut through and without any other disturbance, have a watchspring-like arrangement. In *Platalea* the coils are parallel, but mainly at right angles to the long axis of the body.

These and other variations have been mapped out by GADOW into seven principal schemes, which are represented

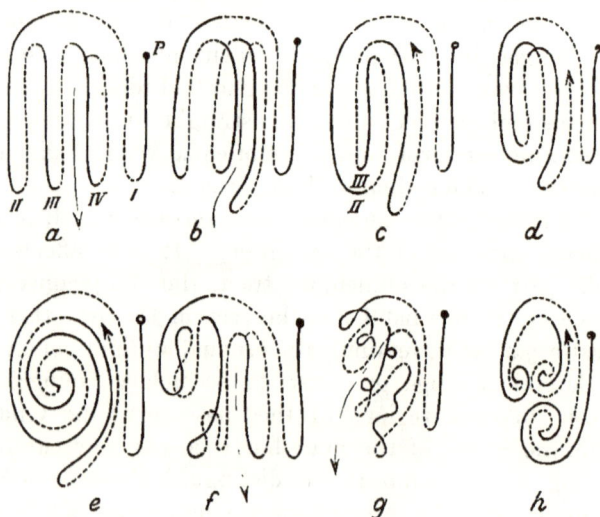

FIG. 7.—INTESTINAL LOOPS (AFTER GADOW).
a, isocœlous; *b*, anticœlous; *c*, antipericœlous; *d*, isopericœlous; *e*, cyclocœlous; *f*, *g*, plagiocœlous; *h*, telogyrous; *P*, pylorus.
N.B.- The ascending branches are dotted.

in the annexed cuts. These are, of course, not accurate pictures of the actual course of the gut, but diagrams ' of the principal relative positions ' of the intestinal loops ; and it must be further explained, in relation to another mode of mapping the intestine that we shall refer to immediately, that the diagrams, mainly if not entirely, concern the lateral foldings of the mesentery that have been already mentioned ; they are representations, in fact, of the relations of the folding of the gut to the body cavity and not to the medial line of attachment of the mesentery ; nor is any attention paid to fixed points in the intestine, such as the cœca or the vitelline duct. They express, however, an interesting series of facts. The general term of orthocœlous is applied to those cases where the folds are as a rule parallel to each other and in the long axis of the body. When they form spirals the general term of cyclocœlous is applied to them by GADOW.

The prevailing number of loops is four, of which the first, the duodenal (which contains the pancreas), is a loop which rarely undergoes additional twisting. The orthocœlous condition may be regarded as the starting point. The cyclocœlous arrangement, as will be seen by the figure, is derived from the orthocœlous by the conversion into one spiral of the second and third loops. In all the **Passeres** the cyclocœlous arrangement is arrived at by a spiral twisting of the middle or second loop only (there being but three loops); this kind of gut has been termed by GADOW mesogyrous. The **Limicolæ**, which have a spiral formed by the second and third loops, are also, of course, mesogyrous; but it is clear that a similar state of affairs has been arrived at independently. Finally, there is the telogyrous condition, in which merely the end of a given loop or loops becomes twisted into a spiral, the rest remaining straight. This is shown in the last of the series of figures on p. 24. In such cases, as in the one figured, the duodenal loop rarely, but still occasionally, undergoes a twisting. The plagiocœlous condition is an irregular twisting of the ends or of parts of the loops of an orthocœlous gut. These varied arrangements of the gut may be recognised in most birds; but there are a few exceptions of which note must be taken. In certain fruit-eating birds, such as *Carpophaga*, *Rhamphastos*, the gut is so short and wide that the number of loops is reduced, and the arrangement quite undecipherable. On the other hand the extremely lengthened gut of the fish-eating *Pandion* produces an equal confusion.

Since GADOW's description of the coils of the intestinal canal in birds the subject has been studied from another point of view by CHALMERS MITCHELL.[1] GADOW considers only

[1] 'On the Intestinal Tract of Birds,' *P. Z. S.* 1896, p. 136.

the way in which the folds of the gut are packed away in the body cavity. MITCHELL describes the actual coiling of the gut itself. In its simplest condition the gut of any animal, as is shown in their embryos, is a straight tube, passing from the stomach to the cloaca, supported by a continuous dorsal mesentery, the ventral mesentery being in nearly all vertebrates defective so far as the intestinal region is concerned. This simple condition is, however, not retained in any existing bird; in all the length of the tube is to some extent, generally to a large extent, longer than the body. The alligator (fig. 8) offers the ideally simplest condition of a coiled intestine, where additional length is achieved without any complications of the gut, merely by its being thrown into a series of folds, of which all are more or less alike. So simple a condition as this does not occur in any known bird. But there is more than one type in which this arrangement is retained with but little modification. It is a significant fact that the most primitive arrangement of the folds of the intestine, judged from the crocodilian standpoint, than which we have none other more nearly approximating to the probable reptilian ancestor of birds, is found in birds which other considerations lead us to assign a low position in the avian series. In the accompanying drawing (fig. 9) of the screamer, for example, we have a gut which is but slightly advanced from that of the crocodile. The greater part of the small intestine shows the same series of undifferentiated folds, only the duodenal loop (not missing as a specialised fold in any bird) being separated off from the general coiling. The large intestine, however, differs from the short and

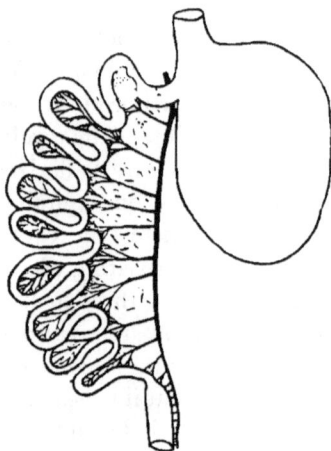

FIG. 8.—*Alligator Mississipiensis*; ALIMENTARY TRACT (AFTER CHALMERS MITCHELL).

straight large intestine of the crocodile by its convoluted course. So too with the gallinaceous bird (fig. 10), where

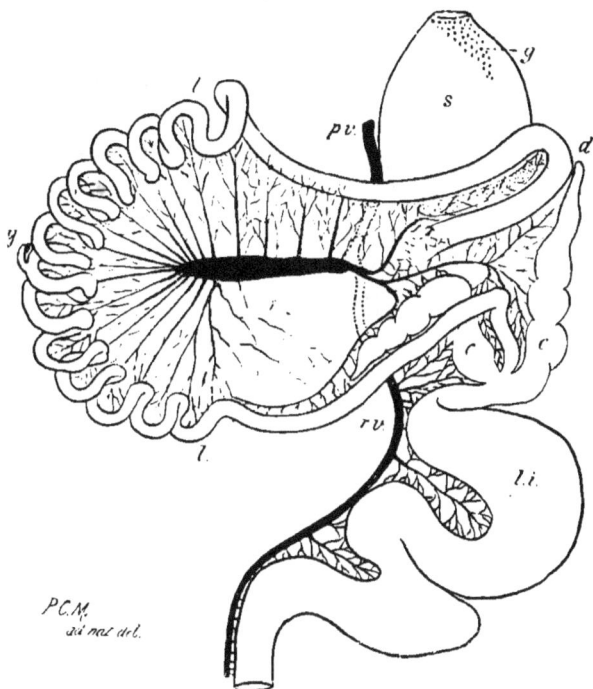

FIG. 9. —*Chauna chavaria* ; ALIMENTARY TRACT.

s, proventriculus ; g, glandular tract ; d, duodenal loop ; l·l, large loop of small intestine ; p, vitelline duct ; c, cæca ; l.i, large intestine ; p.v, portal vein ; r.v, rectal vein. (After CHALMERS MITCHELL.)

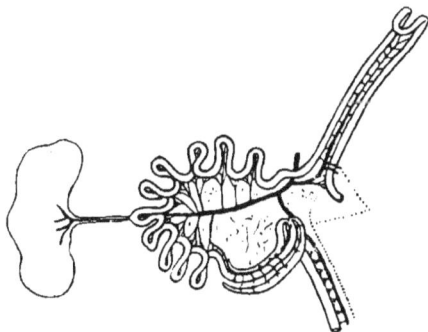

FIG. 10.— *Argus giganteus*, CHICK ; INTESTINAL LOOPS (AFTER CHALMERS MITCHELL).

the duodenal loop is again the only specialised region of the intestine.

The ostrich and the cassowary present but little modification of the same primitive type of gut. In the former we have a rather more complicated duodenal loop, which is furnished with a small subsidiary fold. Then, too, the large intestine has the same remarkable folded condition seen, but to a less extent, in *Chauna*. *Casuarius* is simpler than *Struthio*. In all these figures (taken from Mr. MITCHELL'S paper) the course of the principal blood vessels is shown ; only the veins, as the arteries were found invariably to accompany the veins. It will be observed that the principal feeder of the portal vein runs directly across the main mesenteric fold, and ends near to the vitelline duct, the rudiment of the yolk sac : from the junction of this with the main stem of the mesenteric vein

FIG. 11. *Casuarius* ; INTESTINAL TRACT (AFTER CHALMERS MITCHELL).

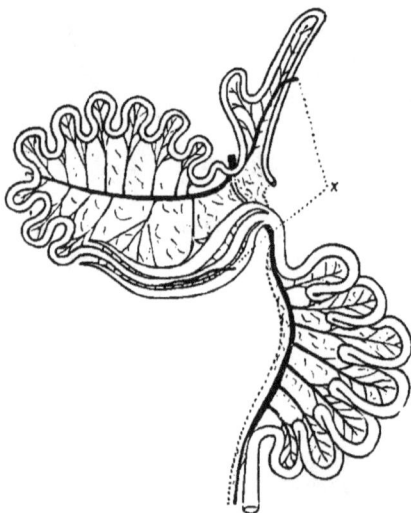

FIG. 12.—*Struthio camelus* ; INTESTINAL TRACT.

x, short-circuiting vessel cut across. (After CHALMERS MITCHELL.)

arise two other vessels ; one of these supplies the duodenal loop, the other the large intestine ; there is sometimes a 'short-circuiting' connection, as MITCHELL terms it, between the duodenal and the main mesenteric stem.

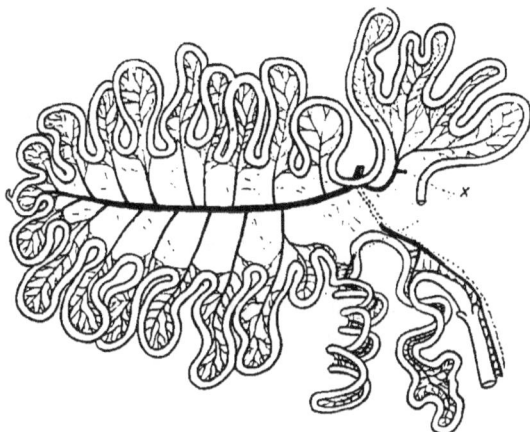

FIG. 13.—*Haliaetus albicilla*; INTSETINAL TRACT.
x as in fig. 12. (After CHALMERS MITCHELL.)

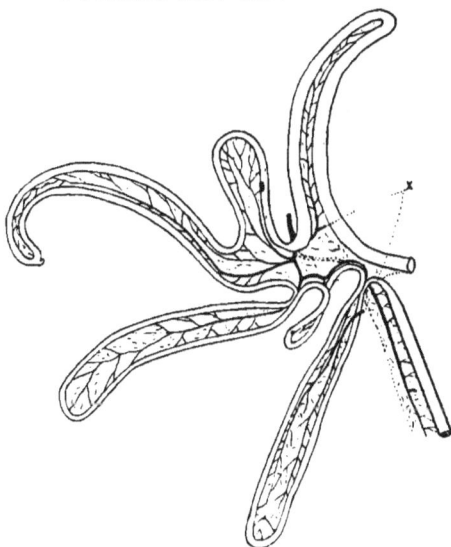

FIG. 14.—*Ara ararauna*; INTESTINAL TRACT.
x as in fig. 12. (After CHALMERS MITCHELL.)

In other groups of birds this simple state of affairs is further evolved, but in some there are remains of the primitive

folded intestine for a greater or less extent; thus in the
penguin a good deal of the intestine retains the early
system of folding. But this bird is remarkable for the
extraordinary complications of the duodenal loop, which has
become enormously lengthened, and, in *Eudyptes*, thrown
into a series of secondary loops, which in *Aptenodytes*
are arranged in a spiral—a convergent resemblance, thinks
MITCHELL, to *Haliaetus* (see above, fig. 13). *Platalea*,
which, possessing as it does the complete muscle formula of
the leg, must be regarded as a primitive type among the
storks, has the greater part of the small intestine disposed
in the primitive fashion; the duodenal loop is, of course,
distinct, and there is also a well-marked long loop just
before the large intestine. *Haliaetus albicilla*, to which
reference has just been made, is less divergent from the
primitive form than many other birds, and *Larus marinus*
has preserved considerable traces of the same. Among the
most specialised birds from the present point of view are
the parrots (see fig. 14).

In *Ara*, at any rate, there is nothing left of the original
primitive folds of the intestine; the whole of its course is
disposed into six specialised loops. The owls and the Caprimulgidæ appear, according to MITCHELL, to have preserved
much of the primitive short, convoluted loops of the lower
birds, and so, though to a less extent, has *Corythaix*. The
very short intestine of the **Passeres**, represented here by *Parus
major*, is not so easy to understand, but its general appearance
is that of a primitive gut. Some further details will be found
in the succeeding chapters, under the different families to
which they refer.

The **cæca** lie at the commencement of the large intestine, which indeed can be, as a rule, only differentiated from
the small intestine by the break thus caused. This is, however, not invariably the case, for in *Struthio* (see fig. 12, p. 28)
there is a distinct break between the two sections of the
gut, quite independent of the cæca, which can be readily
seen from an inspection of the figure cited. The cæca are
among the most variable organs of birds. Not only are they

sometimes completely absent, as in the majority of that
large assemblage of birds the picarians, but when present
they show every degree of relative size. In the **Passeres**, and
in some other birds, the two cæca are the merest nipples,
which cannot be believed to serve any function. They are
in the same way reduced in size
in the hawks and storks. On the
other hand, in the gallinaceous
birds, in the **Limicolæ**, and in some
others the cæca are large tubular
diverticula of the gut, the length
being sometimes to be measured
by inches. Among the owls the
cæca are large, and have the
additional peculiarity of being
swollen at the blind extremity.

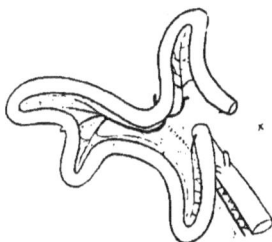

Fig. 15.—*Parus major*.
x, as in fig. 12. (After
Chalmers Mitchell.)

The cæca are most complicated in the ostrich, the screamers,
and the tinamou *Calodromas*, under the descriptions of which
birds will be found an account of this organ. The herons
are remarkable for the fact that one of the two cæca has
disappeared, the remaining one being but small.

The **liver** is invariably composed of two lobes, of which
the left often shows a more or less distinctly marked
secondary division. The size of the lobes varies greatly, as
does their relative size. Thus in some birds the liver lobes
are quite hidden by the sternum; in others again they
descend some way down below the shelter of that bone and
are apparent when the muscular walls of the abdomen are
cut through or removed. The two lobes are occasionally
equal or subequal in size; more generally there is a dis-
crepancy, the right or left, as the case may be, being the
larger, sometimes very much the larger. The two lobes of
the liver are commonly firmly attached to each other by a
bridge of hepatic tissue. In *Chauna* they are nearly separate,
being only united by a very narrow isthmus of liver substance.
The liver sometimes (*e.g.* in *Rhynchotus rufescens*) has two or
three small vessels, belonging to the portal system, entering
its substance at the free edge, a state of affairs which has a

very lizard-like appearance. The relative sizes of the
liver lobes appear to be of no importance systematically.

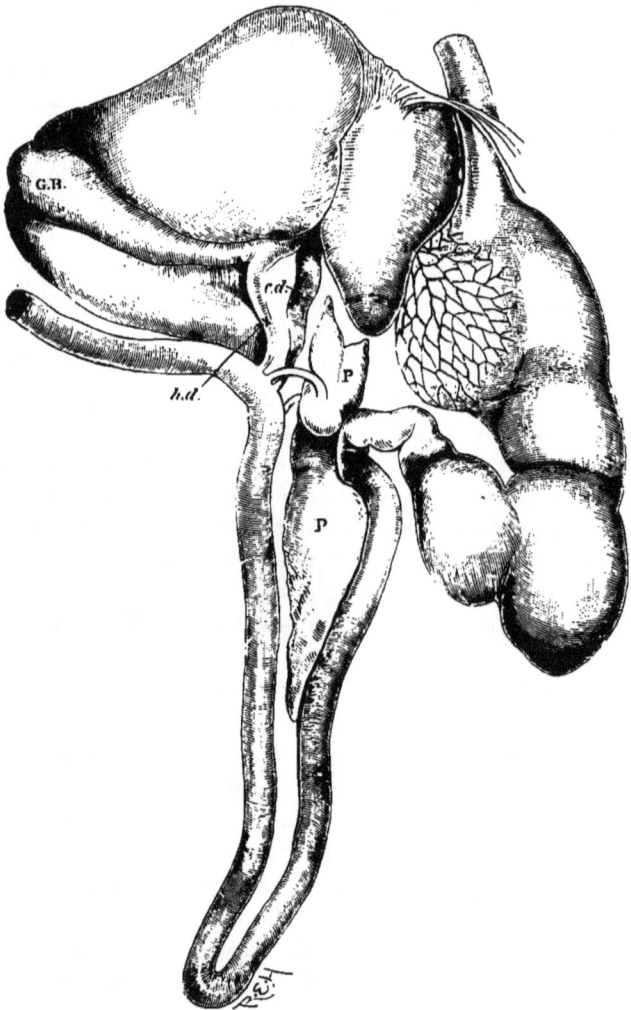

FIG. 16.—ALIMENTARY VISCERA OF INDIAN DARTER.
G.B., gall bladder ; *h.d.*, *c.d.*, bile ducts ; *P*, pancreas.

Numerous livers are figured by GADOW in Bronn's 'Thierreich.'

The **gall bladder** is an organ which is not invariably present in birds. It is even sometimes present and sometimes absent in the same family (*e.g.* parrots). As a rule this vessel is of a rounded or oval contour and is embedded on the surface of the right lobe of the liver. The Picidæ, Capitonidæ, and Rhamphastidæ are remarkable for the extraordinarily elongated gall bladder, which reaches a long way down the abdominal cavity; this is described more fully below. The penguin has an almost equally elongated

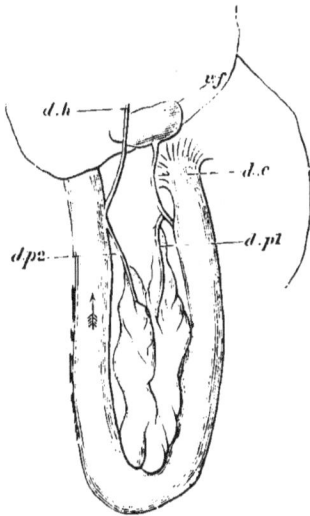

Fig. 17.—Duodenum of *Syrrhaptes*.
v.f, gall bladder; d.c, cystic duct; d.h, hepatic duct; d.p1, d.p2, pancreatic ducts. (After Brandt.)

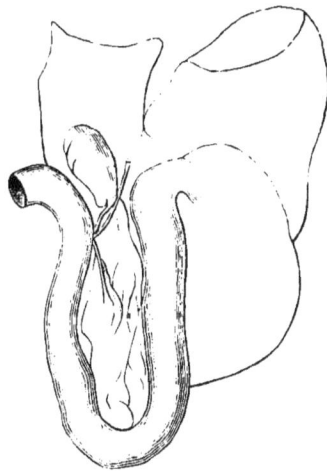

Fig. 18.—Duodenum, Bile Ducts, and Pancreatic Ducts of another *Syrrhaptes* (after Brandt).

gall bladder. The position of the apertures of the cystic and hepatic ducts upon the small intestine varies. The ostrich is remarkable for the fact that the single duct opens practically into the stomach.

The **pancreas** lies in the fold of mesentery that unites the two arms of the duodenal loop. It is commonly more or less distinctly composed of two parts, and in relation to this there are two pancreatic ducts which pour its contents into the duodenum. Apparently, however, no value can be

D

attached to either the form of the gland or the number and
position of the orifices of its ducts. In *Syrrhaptes para-
doxus*, for instance, both of the arrangements figured in the
accompanying cuts have been found by BRANDT, who investigated the structure of the bird. In one of them both
ducts open close to each other and to the cystic duct on the
ascending part of the duodenal loop; in the other the cystic
and hepatic ducts were on opposite sides of the duodenal
loop, and in common with each opened a single pancreatic

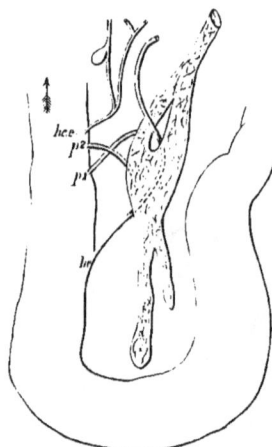

FIG. 19.- DUODENAL LOOPS OF
Rhea americana.
hc.c, *hc*, bile ducts; *p*1, *p*2, pancreatic
ducts. (After GADOW.)

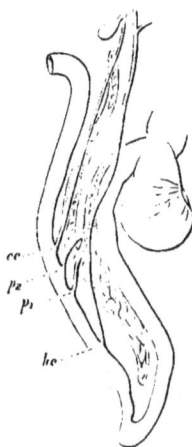

FIG. 20. – DUODENAL
LOOP OF *Rh. Darwini.*
cc, *hc*, bile ducts; *p*1, *p*2, pancreatic ducts. (After GADOW.)

duct. This latter arrangement was found by GADOW in
Pterocles. In two species of *Rhea* the relative positions of
the pancreatic and bile ducts were as is shown in the figures.
In the owl *Photodilus badius* I found that the cystic duct
opened near to the summit of the ascending arm of the
duodenal loop; below this opened the hepatic duct, and
some way below this again, and near together, the two pancreatic ducts. A good many details upon this subject will
be found in GADOW's paper on the digestive organs of birds.

The **cloaca** of birds is the terminal chamber of the
alimentary canal, which also receives the urinary and genital
ducts, and is provided with an appendix of unknown function,

the so-called *bursa Fabricii*. GADOW, in a recent work [1] upon this region of the alimentary canal, recognises three chambers in the cloaca. Above, and separated by a constriction from it, is the coprodæum, into which the rectum opens; this is divided by a constriction from the middle chamber, or urodæum, which receives the genital and urinary ducts;

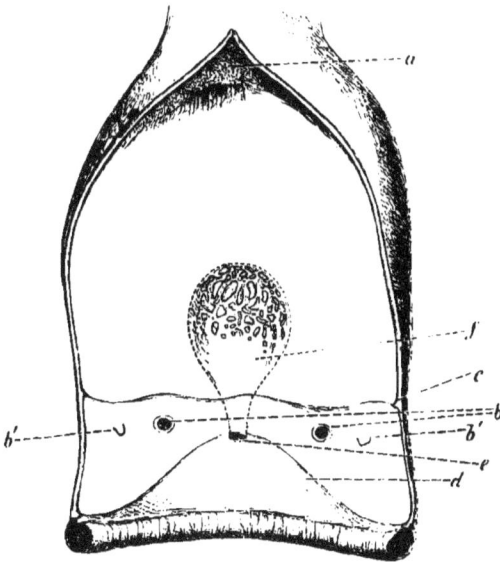

FIG. 21.— CLOACA OF *Chauna derbiana* LAID OPEN FROM IN FRONT.

a, rectum; *b*, orifices of ureters; *b'*, genital papillæ; *c*, fold separating coprodæum from urodæum; *d*, fold separating urodæum from proctodæum; *e*, opening of *f*, bursa Fabricii. (After FORBES.)

then follows the proctodæum, of which the *bursa Fabricii* is a diverticulum.

The **bursa Fabricii** has been chiefly investigated by FORBES [2] and WENCKEBACH.[3] It is a dorsal diverticulum of the proctodæum, and therefore has nothing to do with the

[1] 'Remarks on the Cloaca and on the Copulatory Organs of the Amniota,' *Phil. Trans.* vol. clxxviii. p. 5.

[2] 'On the Bursa Fabricii in Birds,' *P. Z. S.* 1877. p. 304.

[3] 'De Ontwikkeling en de Bouw der Bursa Fabricii,' *Inaug. Diss.* Leyden 1888. See also E. RETTERER, 'Contribution à l'Etude du Cloaque,' &c., *J. de l'Anat.* xxi. 1885. p. 369.

ventral bladder of other vertebrates. It is largest in young birds, and often becomes obliterated in older birds. The general relations of the bursa to the cloaca are shown in the two accompanying figures. The organ contains a quantity of lymphatic follicles, and presents us with two types. In most birds it is a diverticulum opening by a narrow neck into the proctodæum ; but in the struthious birds (in the young at any rate) it is not constricted at its orifice into the proctodæum, and the boundaries of the two are therefore indistinct. The structure and arrangement of

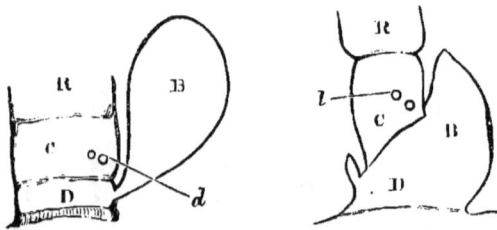

FIG. 22.—TWO TYPES OF BURSA.

R, coprodæum ; *C*, urodæum ; *D*, proctodæum ; *B*, bursa ; *d*, ureters. (After FORBES.)

the follicles and of the bursa generally have led WENCKEMANN to certain classificatory conclusions.

Reproductive and Renal Organs

The **kidneys** are so unimportant from the point of view of the present book that they can be dismissed in a few words. Each kidney is a lobulated organ lying in the pelvic region, so closely in contact with the adjacent bones that they are marked by grooves upon the dorsal surface. In some hornbills each kidney is divided into an anterior and a posterior piece, which are perfectly separated. A ureter runs from each kidney to the urodæum.

The reproductive organs consist of a pair of testes in the male, and of one, rarely two, ovaries in the female. Corresponding to the single ovary (the left) is a single oviduct, the right one remaining rudimentary. BALLOWITZ [1] has

[1] ‘ Untersuchungen über die Struktur der Spermatozoen,’ &c., *Arch. Mikr. Anat.* xxxii. 402.

figured the spermatozoa of many birds, whence it appears that their form is often characteristic, and may be of systematic use. A penis is not present in all birds; it exists in the **Struthiones**, the **Anseres**, **Tinami**, **Herodiones**, **Galli**, and *Œdicnemus*. It is a paired organ—that is to say, it is composed of two incompletely joined halves with a longitudinal groove.

The Cœlom

Birds differ both from reptiles and mammals in the complication of the subdivisions of their body cavity. The subject is one that is far from being thoroughly worked out, but enough information has been collected to allow of a certain amount of definite statement and of comparison with other animals.

When a bird is dissected in the usual way from the ventral surface, the abdominal cavity, or at least the cavity containing what are generally termed the abdominal viscera— *i.e.* liver, intestines, &c.—is seen to be divided into two by a toughish septum, which varies in extent according to the bird dissected. This membrane, to which attention has been directed by WELDON[1] under the name of '*pseudepiploon*,' has been investigated in a number of birds by me.[2] Its relations in the common fowl have been described with the aid of diagrammatic representations of sections by BUTLER.[3]

This membrane, believed by MALL[1] to be the actual homologue of the omentum of the mammal, is more or less horizontal in direction, so that it may be conveniently termed, without prejudice to its homologies, the '*horizontal septum*.' This *horizontal septum* is attached to the ventral body wall, to the oblique septa (of which see description later), and to the gizzard, which viscus appears to lie *within*

---

[1] In his memoir upon the anatomy of the storks and flamingo in *P. Z. S.* 1883, p. 638.

[2] 'Notes on the Visceral Anatomy of Birds,' *P. Z. S.* 1885, p. 836.

[3] 'On the Subdivision of the Body Cavity in Lizards, Crocodiles, and Birds,' *P. Z. S.* 1889, p. 452.

[1] *Journ. Morph.* 1891.

the thick septum, in a cavity formed by the splitting of its layers. Anteriorly the horizontal septum, passing forwards, lies beneath the liver, coming into relations on each side with a cavity which will be referred to later as the '*pulmo-hepatic recess.*'

If, therefore, the abdominal walls of the bird have been cut through anteriorly to the attachment of the horizontal septum to the abdominal walls, the only abdominal viscera exposed will be the gizzard and the liver lobes. These latter are separated from each other by the median vertical '*falciform ligament,*' which is continued backwards to divide the cavity into right and left halves.

If, on the other hand, the abdominal walls of the bird have been cut through posteriorly to the attachment of the horizontal septum to the abdominal walls, the viscera exposed will be the intestines and kidneys and *not* the liver.

In some birds—for instance, in the duck, and in many charadriiform birds—the horizontal septum is so short behind the gizzard that the latter is closely attached to the abdominal parietes by what looks at first sight almost like a pathological adhesion, due to peritonitis. On the other hand in many storks, in *Chauna*, *Cariama*, struthious and other birds, the horizontal septum is very extensive, reaching back to the immediate neighbourhood of the cloaca. Various intermediate stages are offered by other birds.

The Oblique Septa.—Reference has been already made to these structures, which are present in all birds, and concerning whose homologies there is some divergence of opinion. Their structure and relations are as follows : On either side of the body is a tough fibrous sheet of membrane, which runs an oblique course (hence HUXLEY'S name [1] of *oblique septum* '), entirely enclosing and shutting off from the abdominal cavities (dorsal and ventral) the lungs and air sacs, with an exception to be noted immediately. These oblique septa have, as HUXLEY pointed out, a tent-like arrangement, coming into contact with the median septum in front of the heart,

---

[1] ' On the Respiratory Organs of *Apteryx*,' *P. Z. S.* 1882, p. 560.

thence diverging to be attached ventrally to the sternum
along two lines, one on each side, set obliquely to the median

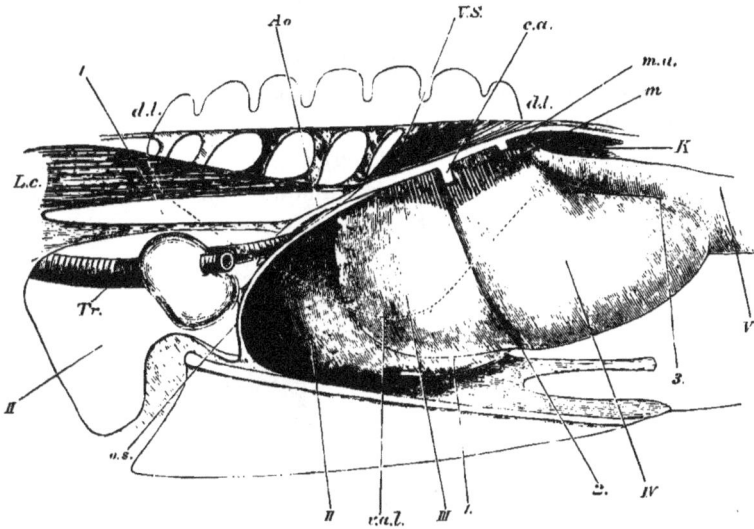

Fig. 23.—Respiratory Organs of Duck.

I V, air sacs ; *Tr,* trachea ; *d.l,* dorsal margin ; *v.a.l,* ventral margin of lungs ; 1–3, dissepiments
between air sacs ; *Ao,* aorta ; *c.a,* coeliac artery ; *m.a,* mesenteric ; *m,* muscular fibres on *o.s,*
oblique septum ; *V.S,* vertical median septum ; *L.c,* longus colli ; *K,* kidney. (After Huxley.)

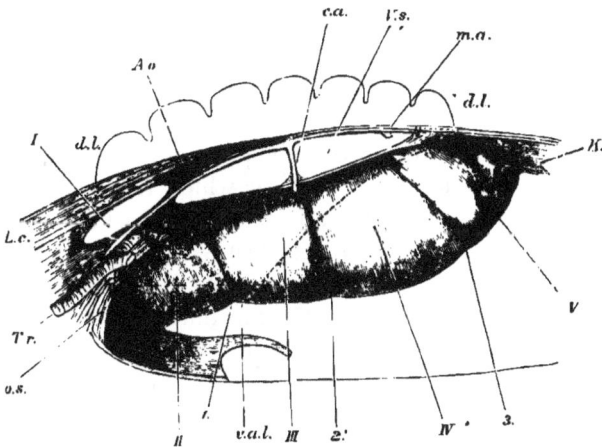

Fig. 24. Respiratory Organs of Apteryx. Lettering as in Fig. 23
(after Huxley

attachment of the falciform ligament. Dorsally they are attached to the parietes. As a general rule the abdominal air sac is the only one of the posterior air sacs (see fig. 23) which is not enclosed within the oblique septa; in all birds, so far as is known, except the *Apteryx*, the wall of this air sac 'has been apparently driven out, like a hernial sac, between the peritoneum and the parietes, and projects into the abdominal cavity.' In *Apteryx* the air sac in question is completely enclosed by the oblique septum. Another exception to the statement made above as to the completely

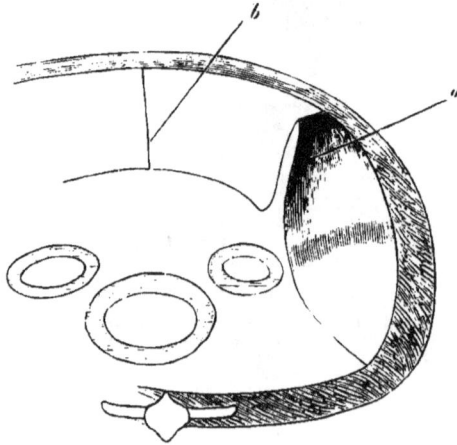

FIG. 25.—DIAGRAMMATIC TRANSVERSE SECTION OF EMU, TO SHOW THE PROJECTION OF OBLIQUE SEPTUM.
a, as a free fold ; *b*, falciform ligament.

dissepimental nature of the oblique septa occurs in a few birds (*e.g.* emu and *Cariama*), in which the posterior end of the oblique septum, though firmly attached to the dorsolateral parietes, is not so attached ventrally, but projects into the abdominal cavity as a free fold. In these cases the free fold is double (see fig. 25), the inner half being continuous with the horizontal septum. To the possible significance of this fact we propose to return later.

The oblique septa are, as has already been stated, membranous, but they are occasionally and partially invaded or covered by muscular tissue. HUXLEY speaks of ' unstriped muscular fibres' in the oblique septum of the duck ; and in the puffin [1] (*Fratercula arctica*) and the penguin (*Eudyptes, Eudyptula minor* and *Spheniscus demersus*) the posterior

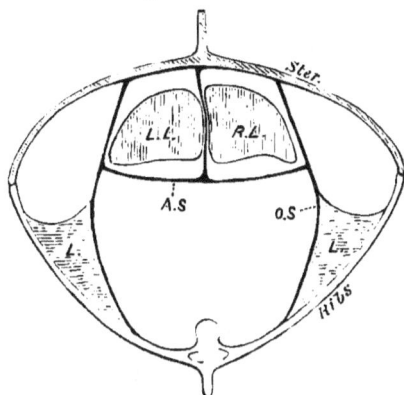

FIG. 26.—DIAGRAM OF A TRANSVERSE SECTION THROUGH THORAX OF DUCK.

L.L, R.L, lobes of liver ; *L,* lungs ; *A.S,* horizontal septum ; *O.S,* oblique septum.

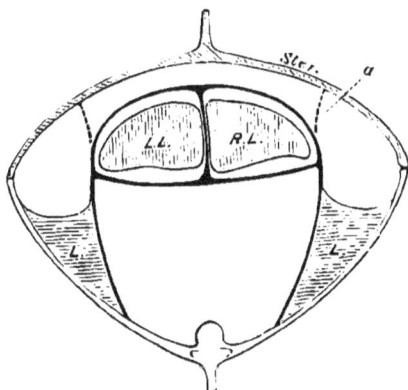

FIG. 27. SIMILAR DIAGRAM OF CROW (*Corvus capellanus*).

a, rudiments of sternal attachment of oblique septum. Other letters as in fig. 26.

part of the oblique septum is largely covered with a thickish layer of muscular fibres, which FILHOL [2] (their describer in

[1] BEDDARD, 'Notes on the Visceral Anatomy of Birds,' II. 'On the Respiratory Organs in certain Diving Birds,' *P. Z. S.* 1888, p. 252.

[2] 'Sur la Constitution du Diaphragme des *Eudyptes*,' *Bull. Soc. Philom.* (7). vi. p. 235.

Eudyptes) has termed ' muscle diaphragmatique transverse.'
These muscles are, however, striated ; but the duck is not
the only bird with unstriated fibres in the oblique septum,
for these also occur in the toucan.

In all birds, with the exception of certain passerines—
possibly of the entire group of passerines—the oblique septa
have the structure and relations that have been thus briefly
described.   In passerines [1] they have undergone what appears
to be a modification.   The oblique septa of each side, instead

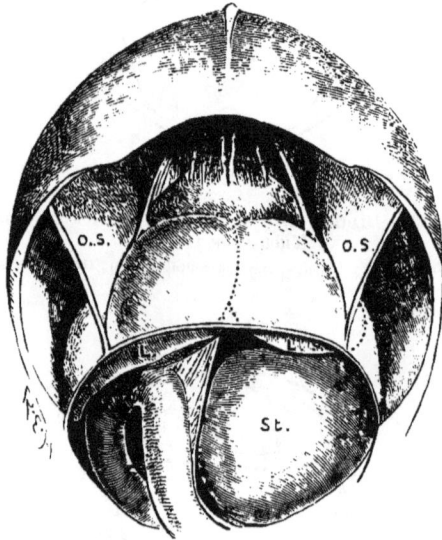

FIG. 28.-- VISCERA OF ROOK DISPLAYED BY REMOVAL OF ABDOMINAL WALLS.
St, gizzard ; *L,* liver ; *O.S,* oblique septum. The liver is covered by a membrane
continuous with the oblique septa.

of being attached independently to the sternum, become
fused with the falciform ligament in the middle line, and
form a horizontal sheet of membrane covering over the two
lobes of the liver.   The original (?) attachments of the
oblique septa are not, however, in these birds entirely lost ;
a much fenestrated membrane—sometimes, indeed, reduced
to a thread or two—remains to remind the anatomist of the

[1] BEDDARD, ' On the Oblique Septa in the Passerines, and in some other
Birds,' *P. Z. S.* 1896, p. 225.

more prevalent conditions. In the rook, however (fig. 28),
they are completely preserved. But the attachment of the
falciform ligament to the sternum in the median line is lost.

The cutting off of two lateral sections of the body cavity
by the oblique septa, and the division of the remainder by
the horizontal septum, do not, however, exhaust the subdivisions of that space. The liver lobes are attached, as is
so common among reptiles, to the oblique septa, in the
neighbourhood of the lungs, by what may be termed the
pulmo-hepatic ligaments. In making the comparison with
reptiles we assume for a moment the correctness of BUTLER's
contention that the oblique septa are in reality a portion of
the pulmonary aponeurosis, a view which will require a
careful re-examination. This ligament assists in closing
the pulmo-hepatic recess, which is really an extension forward
of the abdominal cavity, as shown by MALL's instructive
figures of casts of the perivisceral cavity of birds. They are
narrow cavities, one on each side of the body, walled and
floored by the ligament mentioned, and by the oblique and
horizontal septa. The aperture of entrance has been compared to the foramen of Winslow of the mammal, and so
named. In some birds there are no further complications of
the thoraco-abdominal cœlom, but of others there are still
a few facts to relate before dealing with the homologies of
the various spaces and membranes. In the Australian passerine *Struthidea cinerea* the liver lobes are each partitioned
into two by a transverse septum, which runs from the falciform ligament in the middle line to the oblique septum on
either side. This septum, which is clear and transparent,
does not actually divide the liver lobes; it arches over each
with a free crescentic margin. In some other birds similar
septa are present, with very nearly the same relations. In
hornbills, cuckoos, and owls—at any rate in some species
of each family—the two liver lobes are each completely
shut off from the ventral section of the abdominal cavity
(the 'submental space,' as it has been termed) by delicate
partitions, of which one only, the left, is present in some
other birds, e.g. *Chrysotis Guildingi*. The same *appearance*

is occasionally presented by birds in which the horizontal septum is very short, and therefore almost vertical in direction. But in the birds mentioned the septa are perfectly

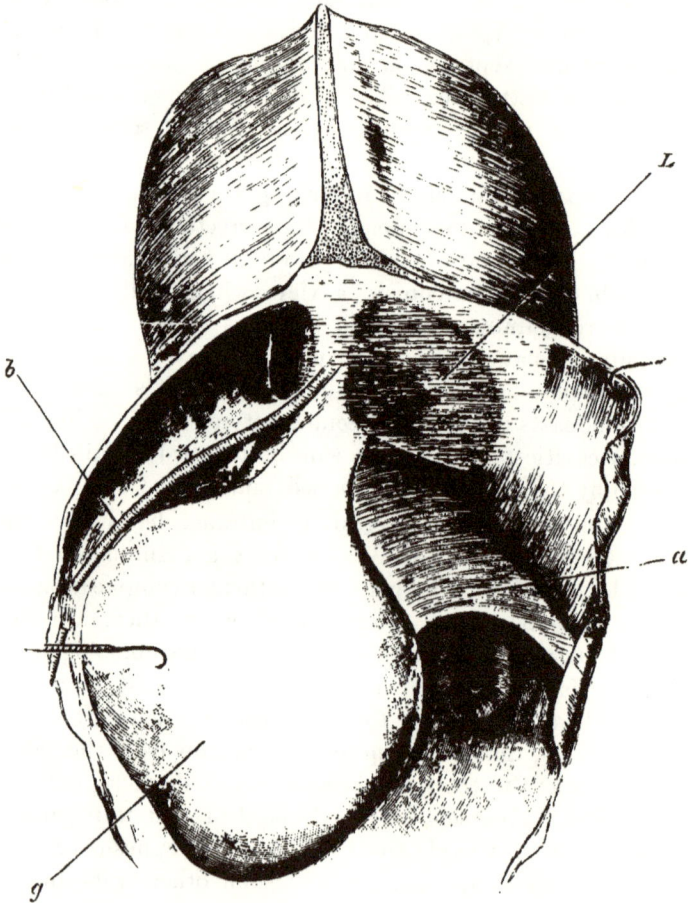

Fig. 29.—Abdominal Cavity of *Bucorvus*.

g, gizzard ; *a*, band of muscular fibre connecting it with oblique septum ; *b*, umbilical vein ; *L*, liver seen through fibrous partition mentioned in the text.

distinct from, and present in addition to, the horizontal
septum.

Homologies of Oblique and Horizontal Septa.—Sir RICHARD
OWEN in the year 1838 described the oblique septa of the
Apteryx[1] as 'a well-developed diaphragm.' His figure,
indeed, is highly suggestive of the mammalian diaphragm,
more so than are (in our opinion) the actual structures
observable in the dissection of the bird. OWEN *assumed*
that the structures corresponded, contenting himself with
indicating the principal differences between the avian diaphragm and the mammalian, and pointing out how *Apteryx*
was nearer to mammals than any other bird. This assumption
was undoubtedly based upon current opinions of the day, for
in the 'Leçons d'Anatomie Comparée' of CUVIER (ed. 2,
vol. vii. p. 21) such a comparison appears to be drawn. As
OWEN justly observed, the imperforateness of the oblique
septum of the *Apteryx* is more mammalian than in other
birds, even struthious, where the abdominal air sacs project
beyond it; the obliquity of its direction, too, is paralleled in
the dugong and manatee, and it is furthermore less oblique
than it is in other birds.

SAPPEY and MILNE-EDWARDS also use variants of the
word 'diaphragm' to describe what we term in the present
work, following HUXLEY, the oblique septa. But for the two
first-named observers the costo-pulmonary (see under description of lungs, below) muscles also form part of the diaphragm.
The older observers, impressed with certain resemblances
(such as warm-bloodedness) between birds and mammals,
regarded them as more nearly akin than is at present the
belief. HUXLEY devoted much work to the demonstration
of the nearer relationship between birds and reptiles—a
relationship which is now generally held. In redescribing
the respiratory organs of *Apteryx* HUXLEY pointed out
their thoroughly ornithic character, and remarked that 'in
this, as in all other cases, the meaning of ornithic peculiarities of structure is to be sought not in mammals, but in
reptiles.' It is on *a priori* grounds likely enough that there

[1] 'On the Anatomy of the Southern Apteryx,' *Trans. Zool. Soc.* ii. 276.

are structures in birds comparable to the diaphragm of mammals; but in that case the likeness would be due to the derivation of both from a common form, perhaps low down in the reptilian series. At any rate certain anatomical facts forbid a precise comparison of the entire oblique septa of the bird to the mammalian diaphragm, though, as will be shown in the sequel, there appear to be a few genuine resemblances which can hardly be got over. HUXLEY emphasised the fact that in the bird the heart lies behind the so-called diaphragm, which is moreover not supplied by a phrenic nerve. If comparable to any structure in the mammal, it is with the mediastinum that they should be homologised. As to the phrenic nerve, several considerations present themselves; it is, it is true, a specialised nerve, but it is spinal in origin. Now in birds it is also branches of spinal nerves which supply the oblique septa. In the mammal it may be that the pulling out of the phrenic nerve may be due to a cause similar to that which has produced the looping of the recurrent laryngeal nerve. When nerves are drawn out by a change in position, or by an elongation of the structures which they supply, there is at least a tendency for their roots of origin, of many, to fuse into a single nerve; witness, for instance, the limb nerves arising from the anterior and posterior plexuses. The fact that the spinal nerves which form the limb plexuses are not always exactly the same has not led to any very serious belief in the serial homology only of the fore limbs in any two vertebrates which show these differences. The lungs, and consequently the diaphragm (assuming for a moment its correspondence with the mammalian diaphragm), are further back in birds; hence their different nerve supply. If we look upon the posterior portion of the oblique septa, which is alone, be it observed, muscular, as the homologue of the lateral parts of the mammalian diaphragm, the rest being absent, no great violence to the mutual relations of the different structures concerned will have been done. In any case it may well be that both the mammalian diaphragm and the avian have been derived from some such reptilian structure as is to be seen in the crocodile.

Professor HUXLEY [1] has made such a direct comparison.
'As in birds, the liver [of crocodiles] lies between the
stomach and the pericardium, and has a peculiar peritoneal
investment, shut off from the great·sac of the abdomen; and,
as in the ostrich, the whole circumference of the stomach is
united by fibrous tissue with the parietes. A fibrous expansion extends from the vertebral column over the anterior face
of the stomach, the liver, and the dorsal and front aspect of
the pericardium to the sternum and the parietes of the
thorax, separating the thoraco-abdominal space into a respiratory and a cardio-abdominal cavity, and representing the
oblique septum of the bird.' Further on we read : 'A broad
thin muscle arises, on each side, from the anterior margin of
the pubes; and its fibres pass forwards, diverging as they go,
to be inserted into the ventral face of the posterior part of
the pericardium, and into the ventral and lateral parts of the
fibrous capsule of the stomach, passing between that organ
and the adherent posterior face of the liver, and being
inserted into the fibrous aponeurosis which covers the anterior
surface of the stomach and represents the oblique septum.'

Professor HUXLEY seems, according to BUTLER, to have
included the ' omentum ' with the oblique septa in his comparison with the fibrous expansion and the accompanying
muscle of the crocodile. I have already pointed out that
' the entire fibrous expansion which arises from the vertebral
column, and extends over the anterior face of the stomach,
liver, &c., in the crocodile represents both the oblique septa
and the omentum in the bird.' A justification for this
opinion is to be seen in the dissection of an emu and one or
two other birds. We have occasionally observed that where
the posterior part of the oblique septa is free from the
abdominal walls, ending, in fact, in a free edge within the
abdominal cavity, this edge is really continuous with the
horizontal septum, as shown in the cut (fig. 25). The
oblique septum is thus merely a fold of the horizontal septum;
they form one continuous structure. As to the muscles of
the crocodile mentioned in the quotation just made from

[1] 'On *Apteryx*,' *loc. cit.* p. 568.

Professor HUXLEY, we have already referred to the existence in birds of what may be considered their homologue, and we have only to add that in a hornbill the left-hand portion of the horizontal septum was muscular—that, at any rate, a strong band of muscle bound the gizzard to the left oblique septum.

Circulatory System

The bird's **heart** is very uniform in structure; there are very few and but slight differences in any part of the heart between the most and the least specialised forms. It is, however, in certain particulars equally distinctive in structure, and differs in a number of well-marked points from the heart of either reptile or mammal. As might be expected, the reptile which shows the nearest approximation in the anatomy of its heart to the bird is the crocodile, while the Monotremata are the mammals which on the other side occupy a corresponding position.

As with the mammalia the heart is completely separated into four chambers; in the bird the heart has perhaps more of an elongated form than in the mammal, the apex (which, as in the mammal, is formed by the left ventricle alone) being rather more pointed than in the heart of any mammal. In a transverse section through the ventricular walls a notable difference in the relative dimensions of the right and left ventricles for the two types is apparent. It will be noted that in the bird the cavity of the right ventricle is, as it were, partially wrapped round that of the left, and is in consequence of a decidedly crescentic form. The cavity of the right ventricle of the mammal's heart is more oval in form, and is not wrapped round that of the left. In this particular the Monotremata stand midway between the bird and the higher mammals.

The interest of the structure of the bird's heart, however, largely, for reasons of comparative anatomy, centres in that of the valve which guards the orifice from the auricle. The interior of that ventricle has fairly smooth walls, a sculpturing so conspicuous in the mammalian ventricle being almost entirely absent. The valve itself, represented in fig. 30, really consists of two parts, which are distinguished by the insertion of a large papillary muscle, which ties the entire valve to the free wall of the ventricle. It is only rarely that any other representative of the generally numerous papillary muscles and chordæ tendineæ of the mammalian heart occur, but occasionally a few muscular threads in addition to the single papillary muscle are to be found. Their existence has been noted, for instance, in *Apteryx australis* and in a few other birds. It has been usually held that the muscular right auriculo-ventricular valve of the bird's heart represents

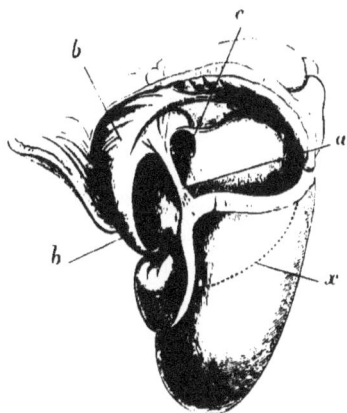

FIG. 30.—HEART OF FOWL, INTERIOR OF RIGHT VENTRICLE.

papillary muscle: *b,c,* valve. (After LANKESTER.)

FIG. 31.—HEART OF *Apteryx*, INTERIOR OF RIGHT VENTRICLE, WITH ATTACHMENT OF PAPILLARY MUSCLE CUT THROUGH.

x, flap of ventricular wall removed with muscle. (After LANKESTER.)

only one half of the complete valve of the mammalian and crocodilian hearts. In these last-mentioned animals the entire circumference of the ostium, which leads from the auricle into the ventricle, is surrounded by the valve, which thus forms a complete collar. There is, however, an exception in the case of the Monotremata, where the septal flap of the valve (*i.e.* that lying on the side of the ostium which abuts upon the interventricular septum) is partially or entirely absent. On a careful comparison, however, between

E

the bird's heart and that of the crocodile it appears that this is not the case. If the hearts of the two animals be laid side by side in a corresponding position, it will be seen that the crocodile's heart valve is furnished with a muscle which seems comparable to that lettered *a* in the bird's heart. And, furthermore, on the septal side of this muscle the fibres which in the bird constitute that half of the valve have a direction which is quite different from that of the fibres in the other and larger half. Finally, while the larger half of the valve is never, so far as is known, fibrous in character, the lesser half occasionally appears to be so wholly or partially. Thus there are some grounds for thinking that the bird's right auriculo-ventricular valve is composed of a complete outer half and of a smaller septal half, presenting, therefore, less difference in this one particular from the monotrematous than from the crocodilian heart.[1]

We can, therefore, derive the bird's heart as regards this valve from a heart like that of the crocodile, in which the septal flap has for the most part disappeared. But in one bird at any rate there appear to be traces of a still further retention of the septal half of the valve. GEGENBAUR, who some years since wrote an exhaustive paper upon the vertebrate heart,[2] made the following remarks about the heart of the condor, which in translation run as follows :—

'Only in the heart of *Sarcorhamphus* do I find a peculiarity which has interest in this connection. From the anterior origin of the muscular valve on the septum ventriculorum a fold runs backwards, which is formed by a thickening of the endocardium. The fold runs obliquely backwards and downwards, and crosses in its direction the margin of the muscular valve. The course of this fold corresponds to the line of origin of the membranous valvular flap of the crocodile ; I think it reasonable, therefore, to regard it as a

---

[1] See for a fuller account BEDDARD and MITCHELL, 'On the Alligator's Heart,' *P. Z. S.* 1895, p. 342.

[2] 'Zur vergleichenden Anatomie des Herzens,' *Jen. Zeitschr.* 1866 ; 'Notes on the Anatomy of the Condor,' *P. Z. S.* 1890. p. 142.

remnant of the structure which is further developed in the crocodile.'

I have, since that sentence was written, examined the heart of a condor, in which was found along a line corresponding to where the flap would be, were it present, 'a series of tiny yellowish spots and vesicles . . . probably pathological,' but perhaps, like other pathological structures, associated with a rudimentary structure. With this exception no trace has ever been found of a septal flap other than the small flap already described.

The left ventricle of the bird's heart has an auriculo-ventricular valve, which is completely membranous, and is tied to the parietes of the ventricle by tendinous threads attached to papillary muscles.

There is one more structure occasionally present in the right ventricle of the bird to which we must direct attention before leaving the matter. The late Professor ROLLESTON, in his Hunterian lecture, described and figured in the heart of the cassowary a muscular pillar uniting the free and fixed walls of that ventricle, to which he gave the name of moderator band. This structure occurs in a few other birds —for example, in *Chunga Burmeisteri*, where it has been figured. In the latter bird, however, there are two muscular bridges, which run in the same direction. One of them is also connected with the muscle tying the auriculo-ventricular valve to the free wall of the ventricle. This may conceivably be a rudiment of the septal half of the valve lying to the right side of the heart. In any case these moderator bands, which are also found in deer and in other running animals, seem to be, according to ROLLESTON's suggestion, a mechanism for increasing the power of the ventricle to contract, and thus ensuring a more rapid and regular flow of blood into the lungs. It is characteristic, where it occurs, of running animals.[1]

[1] For other facts about the avian heart see A. SABATIER, 'Etudes sur le Cœur,' &c., *Ann. Sci. Nat.* (5), xviii. 1873, art. No. 4 ; F. R. GASCH, 'Beiträge z. vergleichenden Anatomie des Herzens,' *Arch. f. Naturg.* liv. 1888, p. 119 ; C. RÖSE, 'Beiträge zur vergleichenden Anatomie des Herzens der Wirbelthiere,' *Morph. J.B.* xvi. 1890, p. 27.

The **arterial system** of birds [1] is chiefly remarkable for the
large number of the different arrangements of the carotids.

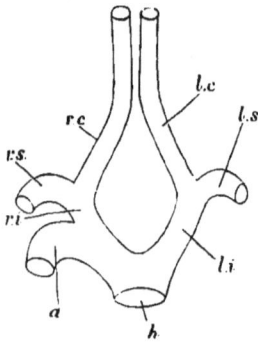

FIG. 32.—NORMAL AVIAN CAROTIDS.

r.c,l.c, carotids ; *r.s,l.s,* subclavians ; *r.i, l.i,*
right and left innominate ; *a,* aorta ; *h,*
its origin. (This and five following figs.
after GARROD.)

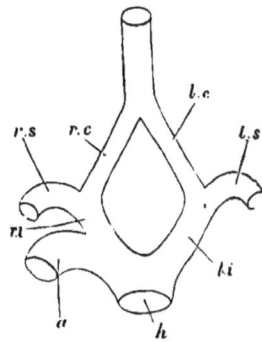

FIG. 33.—CAROTIDS OF BITTERN.
LETTERING AS IN FIG. 32.

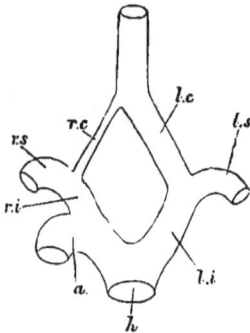

FIG. 34.—CAROTIDS OF FLAMINGO.
LETTERING AS IN FIG. 32.

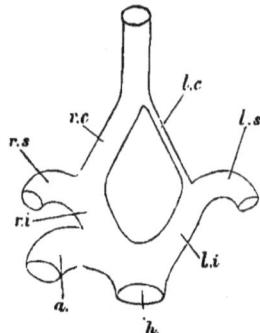

FIG. 35.—CAROTIDS OF *Cacatua.*
LETTERING AS IN FIG. 32.

Many writers, especially NITZSCH, among the earlier anatomists, have drawn attention to some of these variations.

[1] L. A. NEUGEBAUER, ' Systema Venosum Avium,' *Nov. Act. Acad. Nat. Cur.*
xxi. 1845, p. 517 ; RATHKE, ' Über die Carotiden . . . der Vögel,' *Arch. f. Anat. u.
Phys.* 1850, p. 184, and ' Bemerk. über die Entstehung, &c., der gemeinsch.
Carotis,' *ibid.* 1858, p. 315 ; GARROD, ' On the Carotid Arteries of Birds,' *P. Z. S.*
1873, p. 457 ; C. H. WADE, ' Notes on the Venous System of Birds,' *J. Linn
Soc.* xii. 1876, p. 531 ; F. HOCHSTETTER, ' Beiträge zur Entwicklungsgeschichte
des Venensystems,' &c., *Morph. J.B.* xiii. 1888, p. 575, und ' Über den
Ursprung der Subclavia d. Vögel,' *ibid.* xvi. 1890, p. 484.

But the whole matter was described at considerable length
by Garrod, who had more abundant material to work upon,
but who, nevertheless, left for his successor Forbes one out
of the eight known types to describe. The most prevalent
type is that illustrated in fig. 32. It characterises a large
number of birds. The two carotids are of equal size, and
run up the neck for the latter part of their course in the
hypapophysial canal. A modification of this (fig. 33) is seen
in the common bittern and other birds, where the carotids
are of equal size, but fuse into one trunk early in their
course. In *Phœnicopterus* the right (fig. 34) and in *Cacatua*

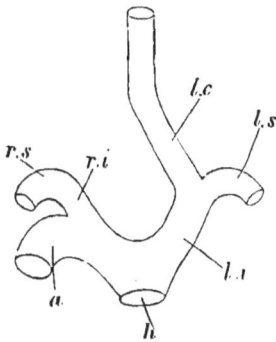

FIG. 36.—CAROTIDS OF PASSERINE.
LETTERING AS IN FIG. 32.

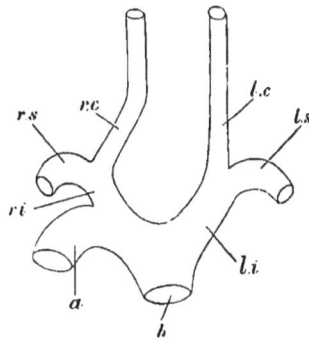

FIG. 37.—ABNORMAL ARRANGEMENT
OF CAROTIDS, WHERE THE LEFT IS
SUPERFICIAL IN POSITION.

sulphurea (fig. 35) the left of the two trunks, which are later
fused together, are very much the smaller. This state of
affairs leads to the condition shown in **Passeres**, where the
left carotid alone is present (fig. 36). Quite exceptionally,
and only seen in two species of the bustard, genus *Eupodotis*,
the right carotid alone is present.

A very curious modification of the carotids is seen (exceptionally, according to Fürbringer) in the hornbill,
Bucorvus. Here the two carotids are entirely superficial,
running up the neck in company with the vagus nerves. In
this case, as Ottley discovered, the true carotids are reduced
to the condition of white imperforate cords, the developed

carotids being the equivalents of the comes vagi nervi of
other birds.   Forbes thought that the same might be the
case with *Leptosoma*, where the carotids are unusually
small, and apparently bound together, not absolutely fused,
as also in *Opisthocomus*.   In certain parrots the left carotid
artery is superficial, while the right runs in the ordinary
way within the vertebral canal.   This is illustrated in the
accompanying figure (fig. 37).   A final variation has been
observed by Forbes in the passerine *Orthonyx*, where the
left carotid, as in all passerines, is alone present; but it runs
superficially, and there is no deep right carotid, as in the
parrots, just referred to.   These facts, striking though they
are, are unfortunately of but little value in classification, or
at least their value is not understood.   We may, however,
accept Forbes's statement that ' no passerine bird has ever
yet been found with more than a left carotid, and no pigeon,
duck, or bird of prey without two normally placed ones.'

In all birds, as is well known, the right aortic arch [1] has
alone persisted.   It is, however, a commonplace of the textbooks to mention the fact that the Raptores have often a
ligamentous rudiment of the left.   There are occasionally
(perhaps individual) remains of the left arch more conspicuous.   Thus I have found a considerable tract of the left arch
capable of being injected, and measuring quite an inch in
length, in *Spizætus* and in the hornbill *Accros*.   *Cariama*,
I may mention, has (at least sometimes) a ligament representing the otherwise aborted left aortic arch.

A variation in the thigh arteries has been noted by
Garrod, who found that in some birds the ischiadic, in
others the femoral, was the most important.   In **Passeres** the
neotropical Clamatores were termed by him Heteromeri,
since the femoral was the principal artery ; other **Passeres**
(including, however, the neotropical and clamatorial genus
Rupicola) Homœomeri, from their ischiadic artery.   The fact

---

[1] J. Y. Mackay, ' The Development of the Branchial Arterial Arches in
Birds,' &c., *Phil. Trans.* clxxix. 1889, p. 111 ; J. F. van Bemmelen, ' Die Visceral-
taschen u. Aortenbogen bei Reptilien u. Vögeln,' *Zool. Anz.* ix. 1886, pp. 526,
543.

that in different species of the same genus (*Centropus*) the same variation occurs tends to throw considerable doubt upon the value of the character, an observation that is frequently and unfortunately necessary to make in describing the anatomy of birds.

The descending aorta gives off three branches, which supply the alimentary canal. These are, in order of origin, the cœliac, the superior, and the inferior mesenteric. These arteries do not, however, supply certain definite regions of the gut. Thus in *Bernicla rubidiceps* the left cæcum receives its blood both from the cœliac and from the superior mesenteric. The right cæcum, however, appears to be supplied by the superior mesenteric alone. The cœliac is mainly concerned with the blood supply of the stomach and the liver. In *Porphyrio* the two cæca are provided with blood from both cœliac and superior mesenteric. A dissection of *Platalea leucorhodia* showed that roughly the cœliac was concerned with the blood supply of the duodenal loop and of the posterior part of the intestine ; the mesenteric, on the other hand, supplied the anterior part of the intestine, excepting the duodenal loop.

In a few birds (*e.g.* in *Buceros* and *Haliaetus*, GADOW) the cœliac and superior mesenteric arise from a common stem.

The posterior mesenteric supplies the end of the alimentary tract. The aorta also gives off two spermatic arteries, and on either side a crural and an ischiadic.

We shall deal briefly with the **venous system**, since it has not been up to the present largely used for systematic purposes. This is doubtless due in a great measure to our imperfect knowledge of the variations that occur. As in the lower mammalia, there are two venæ cavæ superiores. These are formed by the union on each side of a jugular, a subclavian, and a vertebral. The two jugulars are often of very unequal size ; the right is usually stronger than the left, but the two veins are connected in the neighbourhood of the head by a transverse branch. Sometimes the left jugular may absolutely disappear. Connected with the subclavian

are several veins which extend on to the pectoral and even
on to the abdominal region of the body. Most important
on physiological grounds of these is the abdomino-pectoral,
which on each side of the body collects blood from the
pectoral and abdominal regions, and forms in the female
during the breeding season a network of vessels with the
corresponding arteries.

The vena cava inferior is composed of the two hepatic
trunks and of an unpaired median portion, the main stem of
the vena cava. The latter traverses the right lobe of the
liver, and in its transit receives several smaller twigs from
the liver. At the commencement of the kidney the vena
cava divides into, or rather is composed of, the two venæ
iliacæ. It has been pointed out that in diving birds the
part of the vena cava which traverses the liver is wider than
in other birds, a state of affairs which is paralleled in certain
aquatic mammals.

The two common iliacs divide each of them into two veins,
of which the first to be given off is termed the vena iliaca
externa. This divides at once in the pigeon (according to
JOURDAN) into the femoral and into a trunk which runs
along the kidney, and after giving off the sciatic and
numerous branches to the kidney substance receives the
hypogastric from the pelvic region, and then joins its fellow
in the middle line : at the point of junction of the two
iliacæ externæ a median coccygeal is received, and a median
mesenteric from in front. The other branch of the iliaca
communis is purely renal. It results from what has been
said that blood entering the kidney from any of the
branches of the common iliacs may traverse the kidney
substance before reaching the heart *via* the vena cava
posterior. The suprarenal bodies also have their portal
system. The body of each side receives a branch from the
rib region and from the branch of the iliac which runs
embedded in the substance of the kidney.

The existence of a renal portal system in birds is
therefore possible, but not certain, on the anatomical facts
available ; but the liver portal system, as in all other verte-

brates, is quite certain. All the blood from the alimentary canal, however, need not reach the heart *via* the liver.

In *Aceros nipalensis* the vena cava inferior receives three veins from the liver, the abdominal vein (see below), and a smaller twig which is compounded of branches from the œsophagus. Moreover the blood from the posterior part of the intestine, at any rate, may reach the system of the vena cava inferior *via* the mesenteric vein, which, as already stated, enters the iliacs at their point of junction in the middle line.

In addition to the main portal trunk a number of small veins (five on the left side in *Channa chavaria*) may enter the liver lobes separately, a state of affairs which is precisely that found in lizards.

The umbilical vein, which is the equivalent of a part at least of the anterior abdominal vein of the lower vertebrata, is found in very various conditions of degeneration among birds. In some it appears to be fairly well developed; in others it is practically absent altogether. In a specimen of the hornbill *Aceros nipalensis* it was as well developed as I have ever found it in birds. The vein arose near to the posterior end of the abdominal cavity as a double vessel; further forward the two halves joined to form a single vessel. The vein is supported by the falciform ligament, and the upper of the two component vessels receives, not very far from the junction of the two, a recurrent vessel from the inside of the sternum. It may be that the recurrent nature of this vessel is one among many hints of the shortening of the sternum among birds. The anterior abdominal trunk does not enter the liver, but joins the hepatic vessels, and its blood is conveyed straight to the heart.

In *Platalea leucorhodia* I could find no trace of an anterior abdominal vein in the falciform ligament. In *Grus monachus* it was very small; in a large number of other birds of different families the vein was present, and large, e.g. *Crax globicera, Spizaetus coronatus, Serpentarius, Bucorvus, Chauna chavaria, Botaurus stellaris, Bernicla brenta*, &c.

Respiratory System

Trachea.—As a general rule the trachea is a straight tube passing into the thorax, where it bifurcates into the two bronchi. It is composed throughout of rings, which are cartilaginous, or may be wholly or partially ossified. The rings are generally simple rings, which are like each other and quite complete, excepting just at the bifurcation, where it is common for them to be modified in connection with the formation of the syrinx.

This is especially marked in the tracheophone Passeres and in the Ciconiidæ, to the accounts of which families (as well as to below, 'Syrinx') reference must be made for the facts. Other modifications of some of the last tracheal rings are to be seen in the cassowary, where the last few are incomplete behind, as in the mammals and in the bird of Paradise, *Seleucides* (see fig. 38), where the membranous interspaces between the rings become largely increased, and the rings ossified at both sides, but not in the middle, in a peculiar fashion.

A peculiarity of the trachea, seen in representatives of some most diverse groups of birds, is its looping. This is, of course, suggestive of the similar looping of the trachea in

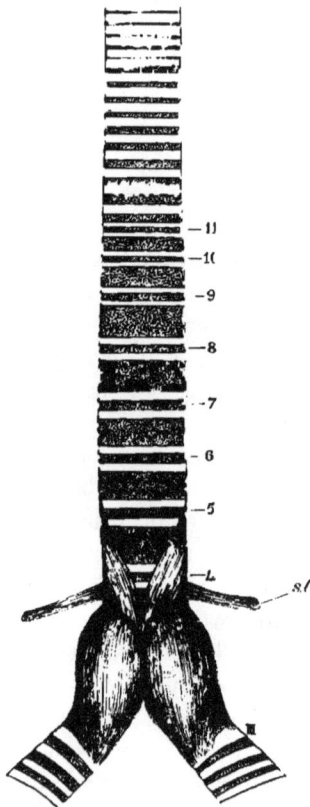

Fig. 38.—Windpipe of *Seleucides nigra.*
4-11, tracheal rings; III, third bronchial; *s.t,* sterno-tracheal muscles. (After Forbes.)

crocodiles and chelonia. The occurrence of this state of affairs is not a character of any one group, but it is found

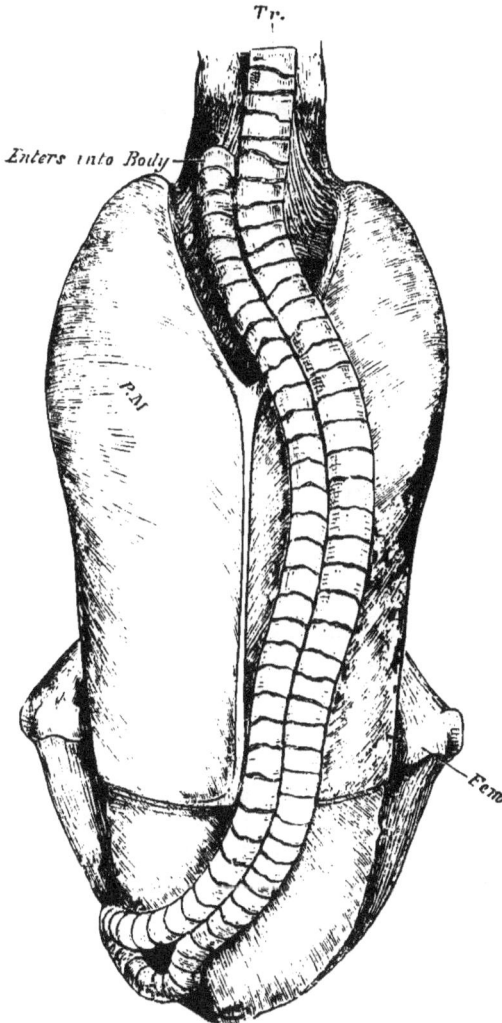

FIG. 39.—BREAST REGION OF *Manucodia*.
Tr, trachea ; *P.M*, pectoral muscle ; *Fem*, femur.

sporadically, as it were, in members of quite different groups. The facts have been collected into a general account of the

matter by FORBES.[1]   There are various grades of this
lengthening of the trachea. In certain species of *Phony-
gama* and *Manucodia* among the Oscines the loops, which
vary in complication, lie, as is shown in the figure of
Manucodia (fig. 39), beneath the skin. Many of the curassows,
a few Scolopacidæ (*Rhynchæa australis* and *R. capensis*), the
duck *Anseranas melanoleuca*, have a convoluted trachea of
the same kind. In the male of *Tetrao urogallus* the loop is
present, but is in the cervical region, not in the thoracic
and abdominal, as in the types just referred to.

The Syrinx.—The voice organ of birds, usually termed
the syrinx, is, as is well known, situated at the bifurcation
of the two bronchi. Its complexity varies greatly, though
it cannot be said that a complex voice organ necessarily
implies an elaboration of sound-producing power. Some of
the singing birds and the parrots, whose voices are capable
of emitting a great variety of tones, have, it is true, a much
specialised syrinx. But, on the other hand, there are other
passerines which have just as complicated a syrinx, but can,
like the raven, by no means vie with some of the starling
tribe, for example, in range of sound. Then, too, some of
the singing passerines have syringes which are much simpler
than those of others which sing as well and no better.
It is, however, true that in the least differentiated forms of
syrinx the bird has but one or two notes. The ostrich, for
example, which has one of the simplest syringes, can roar,
but possesses no variety of sound. The *Apteryx*, whose
syrinx is about on the same level of organisation, appears
to be absolutely mute. The sounds of the emu are due
not to its simple syrinx, but, chiefly at any rate, to a
throat pouch, to which due reference will be made later.

The syrinx of birds, as has been said, varies considerably
in structure. Many of the variations will be treated of in
the systematic part of this book, since they are more of
systematic than of morphological interest. In this place,
however, the leading modifications of the organ will be

---

[1] ' On the Convoluted Trachea of two Species of Manucode, with Remarks on
similar Structures in other Birds,' *P. Z. S.* 1882, p. 347.

shortly described. The syrinx is an organ special to birds; there are no hints of it in any reptile. In reptiles there is no modification at the bifurcation of the bronchi; the tube simply branches, and there are two sets of cartilaginous rings where there was but one. In birds the case is different, and it may be convenient to commence with what may be regarded as the typical avian syrinx, which has been termed the '*tracheo-bronchial*,' since the end of the trachea and the beginning of the bronchi take a share in its formation. It is fair to term this the typical syrinx, since it is found in the majority of the groups of birds; it occurs, for instance, in such diverse families as Passerines, Ardeidæ, Rallidæ, **Struthiones,** Picariæ, &c. The accompanying cuts illustrate this form of syrinx in a number of birds, and from an inspection of them the principal features in the organisation of this form of syrinx may be gathered. At the end of the trachea there is usually a certain amount of modification of the tracheal rings, which may be more or

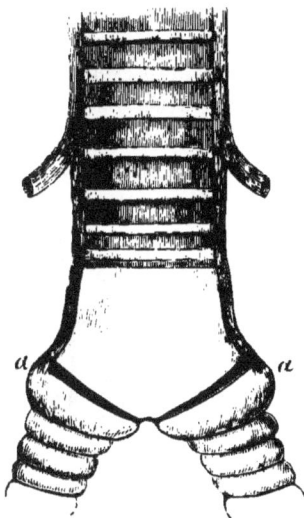

Fig. 40.—Syrinx of *Indicato*, Enlarged.
a, first bronchial semi-rings. (After Garrod.)

less marked, and may be in different directions. It is not necessary to particularise here, and we can select fig. 40 to illustrate one example of this modification, which consists in a complete fusion of the last few rings of the trachea. The bronchi are formed at first of the short semi-rings, the wide interspace being occupied by membrane, the tympaniform membrane, which closes them internally; the extent of this membrane varies, and below it the bronchial semirings become more closely applied—sometimes, indeed, becoming complete rings. The tympaniform membrane of each bronchus is separated from its fellow by a cartilaginous or bony bar, which runs across the base of the trachea, arising from the last tracheal ring anteriorly, and attached to the penultimate or antepenultimate ring posteriorly. This bar is called the pessulus. It is shown in fig. 41, which represents a syrinx seen from below. When the syrinx is cut open it may be seen that this bony bar bears a tough semilunar membrane, directed upwards between the bronchi; the voice is due to the vibrations of this membrana semilunaris.

Another external feature of the typical tracheo-bronchial syrinx is the presence of a pair of muscles which arise some way up the trachea and are inserted on to an early semiring of the bronchial series, one on each side, or even to the last or nearly the last of the tracheal series. From this starting point we can follow the modification of

Fig. 41.—Syrinx of *Cymbirhynchus*, × 5, from behind (after Forbes).

Fig. 42. —Syrinx of *Balæniceps*. *m*, ligamentous rudiment of intrinsic muscle.

the syrinx in a number of directions—in the way of complication, in the way of simplification, and in a direction of alteration which can hardly be termed either complication or simplification.

The first sign of simplification is the disappearance of the intrinsic syringeal muscles, which in many forms have completely disappeared. The disappearance is not necessarily associated with any other changes in the general structure of the organ. Occasionally, as in *Balæniceps* (see

fig. 42), the former presence of the muscle is testified to by
a thin ligament, which occupies the position that the muscle
would occupy were it present.

The syrinx of the hoatzin shows an intermediate stage;
the muscle is absent for the lower part of its course, but
present above; it is represented below by a fibrous band. It
seems from what we know of the relation of muscle to
tendon generally that this change is in the direction indicated and not in the converse direction. The fibrous band of
the syrinx of *Balæniceps* has, so to speak, been muscle; it is
not on its way to become muscle. The ostrich affords an
example of a further degeneration of the syrinx, or a retention of a very primitive and unspecialised syrinx, according
as we view the facts. The syrinx of this bird has been carefully described and figured by FORBES,[1] whose words we will
quote. 'The trachea inferior to the insertion of the sterno-
tracheales slightly narrows, having above the antepenultimate
ring a diameter of about one inch. The tracheal rings are here,
as elsewhere, entire simple rings of an average depth of about
·15 inch, and are separated only by very slight interannular
intervals. The trachea is slightly compressed and posteriorly
carinated for about the last seven rings. The last ring but
four is somewhat produced downwards in the middle line
both anteriorly and posteriorly; it is in consequence narrower laterally than elsewhere. The antepenultimate ring
presents the same features more strongly developed. In
two of the four specimens examined it sent down a small
pessuliform process of cartilage in the middle line behind,
filling the chink left between the posterior extremities of
the last two (incomplete) rings. The penultimate ring is
narrower and more cylindrical than its predecessors; it is
also wider transversely, and incomplete behind in the middle
line, its extremities, however, being closely approximated to
each other. The last tracheal ring is still wider transversely,
and more cylindrical: and it, too, is incomplete posteriorly
to a greater extent than its predecessor; viewed from the

[1] 'On the Conformation of the Thoracic End of the Trachea in the "Ratite"
Birds,' *P. Z. S.* 1881, p. 778.

side it is convex upwards, as are its immediate predecessors
in a less degree. The interannular intervals between all
these rings are, when undisturbed, mere chinks filled up

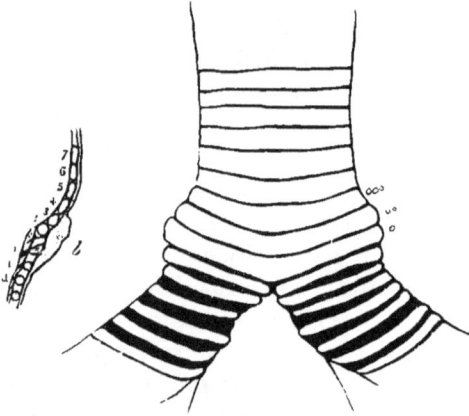

FIG. 43.—SYRINX OF *Struthio*.

a, last, *aa*, penultimate, *aaa* antepenultimate tracheal rings : *b*, section of wall of windpipe to show vocal chord ; 1-7, tracheal, I, II, bronchial rings.   (After FORBES.)

by dense fibrous and elastic tissue. There is no trace of a
pessulus, though the last tracheal ring is slightly produced

FIG. 44.— THE SAME SYRINX FROM BEHIND.

downwards in front. The first bronchial semi-ring on each
side is narrow and cylindrical, strongest anteriorly, and
somewhat attenuated posteriorly. It is separated only by a

narrow interval from the last tracheal ring. The second and third rings are similar, but are more slender and lengthy; they are convex downwards, but very slightly so; hence the interannular intervals are slight here also. Their anterior ends are very slight, inturned, impinging but to a small extent on the membrana tympaniformis, which completes the bronchial tubes internally, and, in consequence of the absence of any three-way piece, passes continuously from one bronchus to another, so closing the tracheal tube inferiorly. The fourth, fifth, and succeeding bronchial rings are similar in character, but their ends, which tend to be dilated posteriorly, are successively more and more incurved to about the tenth. Nowhere are the bronchial rings complete. There is at most only a trace of a membrana semilunaris, in the form of a feeble, scarcely raised antero-posteriorly directed fold of mucous membrane.'

This syrinx, therefore, differs from the more typical tracheo-bronchial syrinx in, at any rate, three essentials— (1) the absence of tracheo-bronchial muscles; (2) in the slight amount of specialisation of the last rings of the trachea ; and (3) in the absence of a pessulus. The only distinguishing feature of the syrinx which is present is the membrana tympaniformis. But the presence of this, and of the rudimentary membrana semilunaris, fully justified FORBES in contradicting the assertion, prevalent at the time when he wrote, that the **Struthiones** had no ' lower larynx,' an assertion, indeed, which could not possibly be made with the syrinx of *Rhea*, a quite typically tracheo-bronchial one, in existence. Still, it is undoubted that the syrinx of the ostrich is in a very simple condition, and hardly deserves the name. In the stork tribe we have a series of stages in the degeneration of the syrinx. In *Abdimia sphenorhyncha*, as in other storks, there are no intrinsic syringeal muscles, but the membrana tympaniformis is well developed and of considerable extent. In *Xenorhynchus* the membrana tympaniformis is almost, but not quite, obliterated, and, finally, in *Ciconia* the bronchial rings are *rings*, and not *semi-rings* ; there is in them no trace of a membrana tympaniformis ; but in all the

F

pessulus is present. The syrinx of the stork, indeed, and of the American vultures very nearly approaches what we should on *a priori* grounds regard as the original form of the syrinx.

In the other direction the syrinx may be further increased in complication; this is brought about by an hypertrophy of the intrinsic muscles. The simplest case is that of the plover, *Vanellus cayennensis*, figured and described by GARROD,[1] whose figure is here reproduced. It will not need

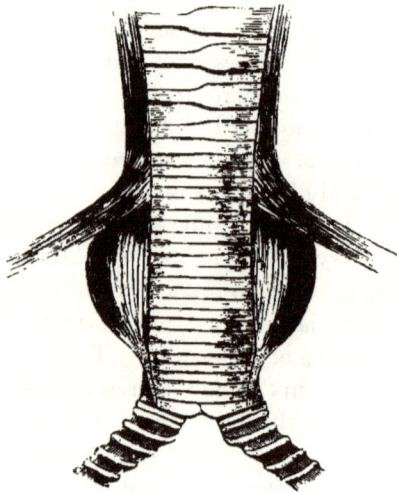

FIG. 45. —SYRINX OF *Vanella cayennensis*, FROM IN FRONT (AFTER GARROD).

much description; the principal change is in the enormously thickened pair of muscles.

This modification of the syrinx, however, is seen at its extreme in certain passerines and in the parrots. Here we meet with a multiplication of the intrinsic muscles, which may exist to the number of three or four pairs. A syrinx of this kind, when found in the **Passeres**, is frequently termed oscinine, the group of Passeres exhibiting the character being the Oscines, a term, however, which is not now used in the classification of the group.

[1] 'On the Trachea of the *Tantalus loculator* and of *Vanellus cayennensis*,' P. Z. S. 1878, p. 625.

In *Menura superba*, which has been described by GARROD, there are three pairs of muscles, which are attached to different bronchial semi-rings ; the posterior pair of muscles are attached to a ring below that which bears the insertion of the anterior pair, while the remaining middle pair are inserted higher up again.

We now come to the consideration of those modifications of the syrinx which will be spoken of neither as degenerations nor complications. They are parallel modifications, which

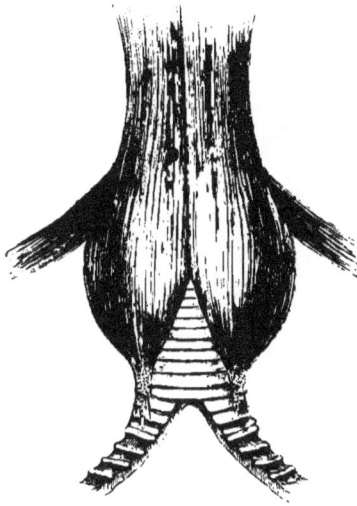

FIG. 46. THE SAME FROM BEHIND (AFTER GARROD).

can in all cases be traced to the typical tracheo-bronchial syrinx, though whether they have originated from it is, of course, a matter of question. To the two varieties of syrinx which we briefly refer to here the names *tracheal* and *bronchial syrinx* have been given, implying the fact that the modification of the windpipe has taken place in the one case mainly or entirely in the trachea, and in the other mainly or entirely in the bronchi.

The *tracheal syrinx* is distinctive of a group of **Passeres** which have been on that account called the Tracheophonæ ;

F 2

but a syrinx that presents some of the same characteristic modifications distinguishes the stork tribe. The principal modification of this type of syrinx is that a large number of tracheal rings are altered in character, in a way which will be pointed out in detail for one or two forms. The tracheal syrinx of the **Passeres** was investigated by JOHANNES MÜLLER,[1] but GARROD has added many new facts of importance to our knowledge of this kind of syrinx.

In *Hylactes megapodius* there are nine tracheal rings which are very much thinner than their predecessors; on the anterior side, however, there are twenty-three which are thus altered. Two or three of the anterior bronchial semi-rings are modified and ossified; from the second of these on each side is a ridge of cartilage, the processus vocalis, which extends up to the twelfth tracheal ring (from the bottom). As a rule these tracheophone **Passeres** possess intrinsic muscles to their syringes, but the rule is not without exceptions, of which *Conopophaga* is one,

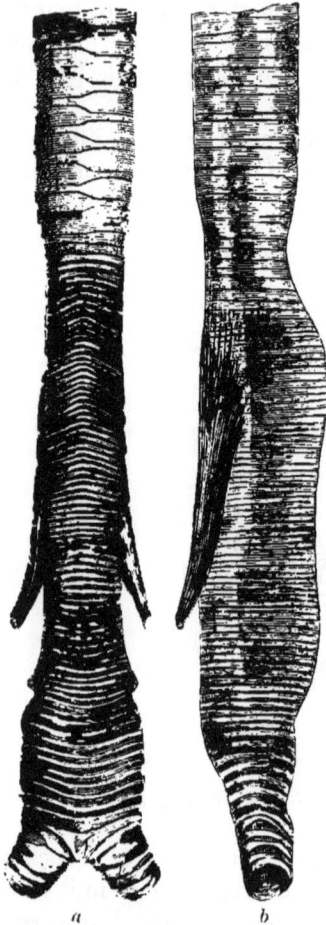

FIG. 47.   TRACHEA OF *Tantalus loculator.   a.* FROM FRONT. *b.* FROM SIDE.   (AFTER GARROD.)

as well as *Hylactes*, already described; others, such as *Furnarius*, have the intrinsic muscles. The same form of syrinx seems to exist in the storks. In *Tantalus loculator*,

[1] In *Abh. Berlin. Akad.* 1845, p. 367.

for instance (see fig. 47), the lowest seventy-eight rings of the trachea are modified through being thinner than those elsewhere, and this portion of the tube is of a greater calibre than that above. In *Ciconia alba* the lowest twenty-nine rings are thus changed in structure, and 'there is a small prolongation upwards of the lateral portions of the three lowermost tracheal rings, which forms a consolidated triangular process on each side, overlapping the next few rings and looking extremely like the rudiment of the similarly situated processus vocales of the passerine tracheophone syrinx, which resemblance is increased by the thinness of the neighbouring rings and by their being flattened from before backwards.'

The *bronchial syrinx* is seen in its most extreme development in *Steatornis* and in *Crotophaga*, where it was originally described by MÜLLER; but other cuckoos and goatsuckers, as has been shown by me,[1] possess also a syrinx which may be called bronchial: furthermore, as WUNDERLICH has shown,[2] the owl tribe resemble the goatsuckers in this respect, while there are indications of the bronchial syrinx in certain petrels.

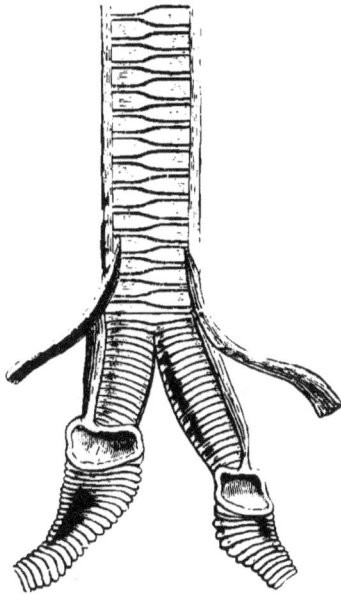

FIG. 48.—SYRINX OF *Steatornis*. FRONT VIEW. (AFTER GARROD).

The fullest description of the syrinx of *Steatornis*, which we take as a type of the perfectly formed bronchial syrinx,

[1] 'On the Syrinx and other Points in the Anatomy of the Caprimulgidae,' *P. Z. S.* 1886, p. 147; 'On the Structural Characters and Classification of the Cuckoos,' *P. Z. S.* 1888, p. 168.

[2] 'Beiträge zur vergleichenden Anatomie und Entwickelungsgeschichte des unteren Kehlkopfs der Vögel,' *Nov. Act. Acad. Leop. Cæs.* 1884.

is contained in a paper upon the general anatomy of this
bird by GARROD.[1]   From that paper we borrow the description as well as the illustration.   It will be seen from that
drawing (fig. 48) that the trachea of the bird bifurcates, as
does the trachea of a mammal, without any modification of
the rings, either tracheal or bronchial.   The latter are at first
complete rings ; it is not until the thirteenth or fourteenth
—the exact position appears to vary—that the syrinx appears ;
here the rings cease to be complete rings, and are semi-rings,
their inner ends being completed by membrane, the membrana tympaniformis.   To the first of these modified semi-
rings is attached in the case of either bronchus the intrinsic
muscle of the syrinx.

The transition between this purely bronchial syrinx and
the more usual tracheo-bronchial syrinx is afforded by
various genera of cuckoos and goatsuckers (of which a particular description will be found later), in which the membrana tympaniformis is placed, as in *Steatornis*, far down
the bronchus, but which have also a sheet of membrane
forming a continuation of the membrana upwards to the
trachea, which is due to the non-closure internally of the
earlier bronchial semi-rings ; this latter gets more and more
limited in various genera until we have the purely tracheo-
bronchial syrinx, in which the wide membrana tympaniformis commences at once on the bifurcation of the
bronchi.

The syrinx has undoubtedly some value as a test of
affinity.   As to the **Passeres**, it is, as FÜRBRINGER has
remarked, a ' classical ' object for the determination of
relationships.   In other families too it is of importance.   From
a more general standpoint, however, apparently but little
reliance can be placed on the modifications of this so variable
organ.   An approximation to the reptilian condition—in the
absence of any special modification at the bifurcation of the
bronchi—is seen in some of the struthious birds and in the
American vultures.   It is not clear, however, that this
simplicity is not a case of the reduction rather than of the

[1] ' On some Points in the Anatomy of *Steatornis*,' *P. Z. S.* 1873, p. 526.

retention of a primitive character. In special cases the form of the syrinx seems to be of not little value as a mark of affinity. The peculiar syrinx of the storks, for example, distinguishes them from their near allies the herons. The stork-like syrinx of *Tantalus* is one of the many reasons for placing it with that family. The peculiar form of syrinx termed the ' bronchial syrinx ' may seem to some to militate against the value of this organ as a test of affinity; but, on consideration, the fact that both owls and goatsuckers possess it will not seem extraordinary in view of their other resemblances, while the cuckoos are perhaps not so widely remote from those two families as they have been placed. We may have here a clue to the relationship of these three groups of birds. The complicated (as regards musculature) syrinx of the parrots is so far an indication of affinity with certain **Passeres**. The systematic position of the parrots is by no means clearly defined, and therefore this indication of a possible affinity must not be ignored.

The Lungs and Air Sacs.—The lungs of birds never depend freely in the coelom, as is the case with most reptiles. They are closely fixed to the parietes, and covered with a thin and transparent aponeurosis, which is the peritoneum. So closely are they adpressed to the body walls that when they are carefully removed by dissection an impress of the ribs is to be seen upon their dorsal and lateral surfaces. An approach to this peculiar position of the lungs in Aves is to be seen in the crocodilia and to a less extent in the Chelonia and Monitor lizards. In these animals the lungs are bound down to the parietes, and do not hang freely, as in the Lacertilia generally and in the mammalia. The lungs in birds occupy the space between the first rib in front and the anterior end of the kidney behind. They nearly meet in the middle line. Seen from below, when left undisturbed in the body, only the adjacent structures being cleared away, the lungs present two facets, an anterior and a posterior; the latter is divided from the former by a ridge which does not divide the lung into two equal halves. The anterior is considerably the smaller.

When the lung is thus bared it is seen to be provided
with a number of conspicuous orifices where the covering
aponeurosis is deficient; these are termed the ostia, and
they lead into the air sacs. Their number is variable in
correspondence with the variability in the number of air
sacs. The free surface of the lung is supplied with bands
of muscle, which have been termed 'diaphragm,' but which
are called by HUXLEY[1] costo-pulmonary muscles. These
muscles arise from the ribs and spread out over the aponeurosis covering the lung; they are, as a rule, extensive,
extending but a short distance from their origin. The
number of these costo-pulmonary muscles varies much
among birds; but little attention has been hitherto paid to
them. For instance, among the storks the muscles in
question are reduced to a minimum; there are only two
pairs at the anterior end of the lung, which arise not from
ribs, but from the end of the windpipe. In herons, on the
other hand, and in the emu, all the ribs bordering upon
the lungs give off fasciculi of fibres, which in the emu are
of considerable thickness. Each bronchus enters the lung
at a little distance from its anterior end, and sometimes, as
in the condor, the cartilaginous rings cease some little way
before it enters. The bronchus dilates somewhat when it
has entered the lung, and from the posterior end of this
dilatation a tube is continued backwards, which opens into
the posterior or abdominal air sac. This trunk is termed
by HUXLEY the mesobronchium. Further on in its course
the mesobronchium gives off another branch, which opens
into the posterior intermediate air sacs (a description of the
air sacs will be found a little further on). From the
dilatation of the mesobronchium, the vestibule, arise four
other tubes, which are called the entobronchia by HUXLEY
(whose nomenclature is adopted throughout in the present
section). The first curves forward and gives off several
branches, one of which opens into the præbronchial air sac,
while the main trunk is continued into the subbronchial
air sac. The second entobronchium passes dorsally and

---

[1] 'On the Respiratory Organs of *Apteryx*,' *P. Z. S.* 1882.

ramifies, a wide branch descending to the subbronchial
ostium. The third entobronchium runs backwards and
gives off a number of branches. Close to its origin from
the bronchus it opens into the anterior intermediate air sac
by the anterior intermediate ostium. The fourth entobronchium runs parallel with this ; it gives off branches
from its ventral wall, but ends cæcally. In addition to the
entobronchia there are the ectobronchia. These are six or
seven branches given off laterally and dorsally from the
mesobronchium. These various bronchia are in communication with each other, so that the substance of the lung is
a meshwork.

As a rule there are ten air sacs in birds, which are
arranged in five pairs, five, in fact, arising from each lung.
In front of the windpipe are the præbronchial air sacs ;
below the trachea are the subbronchial air sacs ; the oblique
septa, which have been described elsewhere (p. 38), enclose,
in the duck and all other birds except the *Apteryx*, two air
sacs, the anterior and posterior intermediate sacs. The
abdominal air sacs lie among the intestines, and are fed by
an ostium which is at the extreme posterior end of the lung ;
they have been, as HUXLEY has expressed it, pushed out
from the space enclosed by the oblique septa like a hernia.
In *Apteryx*, quite exceptionally, these air sacs are not so
pushed out, but lie within the area enclosed by the oblique
septa. The only differences that have been noticed in birds,
apart from those that have been already mentioned, appear
to consist in the number of the intermediate air sacs and in
the condition of the præbronchial. WELDON [1] has described
the breaking up of the præbronchial in the storks into a
number of sacs, at least five in number, and the complete
fusion of the subbronchial sacs into a single one. The
breaking up of the præbronchial sacs is carried to a more
complete extent in *Chauna*. In some birds there are three
instead of two intermediate air sacs. I have observed this
in *Podargus*. In many **Accipitres** the abdominal air sacs are

[1] 'On some Points in the Anatomy of *Phœnicopterus* and its Allies,' *P. Z. S.*
1883, p. 610.

peculiar in that one of the walls of the sacs has got to be firmly adherent to the ventral parietes, and the walls of the two sacs enclose between them the intestines, which have thus the appearance of being enclosed in a special compartment of the cœlom. In one or two **Accipitres** there is the same subdivision of the intermediate air sacs that I have referred to in *Podargus*.

These air sacs communicate with subsidiary spaces lying among the viscera, between the muscles, in the skin (particularly in *Chauna* and several Steganopodes), and with the bones. The skull, however, is aerated by a set of spaces which are not connected with the trachea and lungs, but with the Eustachian tubes and the nasal chambers.

The literature relating to the lungs and air sacs is large. In addition to the memoirs already quoted the following bear upon the matter :—

F. BIGNON, 'Sur les Cellules Aériennes du Crâne,' &c.. *C. R. Soc. Biol.* 1887, p. 36.

Idem, 'Recherches sur les Cavités Aériennes Cervicocéphaliques chez les Psitacidés,' *Bull. Soc. Zool. France,* xiii. 1888, p. 180.

Idem. 'Note sur les Réservoirs Aériens de l'Urubu (*Cathartes atra*),' *C. R. Soc. Biol.* (9), i. 1889, p. 39.

Idem, 'Contribution à l'Etude de la Pneumaticité,' &c., *Mém. Soc. Zool. France*, ii. 1889, p. 260.

MILNE-EDWARDS, 'Observations sur l'Appareil Respiratoire de quelques Oiseaux,' *Ann. Sci. Nat.* (5), iii. 1865. p. 137, and *ibid.* 1867, p. 12.

Idem. 'Sur les Sacs Respiratoires du *Calao rhinoceros*,' *C. R.* 1885, p. 833.

E. FICALBI, 'Alcune ricerche sulla struttura istologica delle sacche aerifere,' &c., *Atti Soc. Tosc. Sci. Nat.* (vi. 1885, p. 249).

H. FILHOL, 'Sur la Constitution du Diaphragme des *Eudyptes*.' *Bull. Soc. Philom.* (7), vi. 1882, p. 235.

N. GUILLOT, 'Mémoire sur l'Appareil de la Respiration dans les Oiseaux,' *Ann. Sci. Nat.* (3), v. 1846, p. 25.

E. SELENKA, 'Beiträge zur Entwicklungsgeshcichte d.

Luftsäcke des Huhns,' *Zeitsch. wiss. Zool.* xvi. 1866, p. 178.

H. STRASSER, ' Die Luftsäcke der Vögel,' *Morph. J.B.* iii. 1877, p. 179.

G. ROCHÉ, ' Prolongements Intra-abdominaux des Réservoirs Cervicaux chez l'Autruche,' *Bull. Soc. Philom.* (8), i. 1889, p. 111.

Idem, ' Sur l'Appareil Aérifère des Oiseaux,' *ibid.* (8), ii. 1890, p. 5.

Idem, ' Contribution à l'Etude de l'Anatomie Comparée des Réservoirs Aériens d'Origine Pulmonaire chez les Oiseaux,' *Ann. Sci. Nat.* (7), xi. 1891.

H. BOULART, ' Note sur un Système particulier de Sacs Aériens observés chez quelques Oiseaux,' *Journ. de l'Anat. et Phys.* xviii. 1882, p. 467.

Idem, ' Note sur les Sacs Aériens Cervicaux du Tantale,' *Bull. Soc. Zool. Fr.* 1885, p. 348.

Other references are contained in the paper of HUXLEY upon the lungs and air sacs of *Apteryx,* already quoted.

Muscular Anatomy [1]

The most general feature of the muscles of birds is the great length of their tendons of insertion ; the tendency of

[1] For memoirs dealing with the muscular anatomy of several types see CARUS, ' Erläuterungstafeln zur vergl. Anat.,' Leipzig, 1826 (*Astur, Falco, Cypselus*) ; P. HARTING, ' Observations sur l'Etendue relative des Ailes et le Poids des Muscles Pectoraux,' &c., *Arch. Néerl. Sci. Exact. et Nat.* iv. 1869, p. 33 ; S. HAUGHTON, ' On the Comparative Myology of Certain Birds.' *P. R. Irish Ac.* 1867. p. 524 (*Falco, Grus, Anas*) ; G. JAEGER, ' Das Os humero-scapulare der Vögel,' *J.B. K. Akad. Wiss.* xxiii 1857, p. 387 ; LEGAL and REICHEL ' Über die Beziehungen der Grösse der Flugmusk.,' &c., *Ber. schles. Ges.* 1879 ; H. J. MAGNUS, ' De musculis costarum sternique avium,' Diss. Inaug. Vratislaviæ, 1867, and in *Arch. f. Anat.,* 1869 ; H. PFEIFFER, ' Zur vergleichenden Anatomie des Schultergerüstes,' &c., Diss. Inaug. Giessen, 1854 ; J. J. PRECHTL, ' Untersuchungen über den Flug der Vögel,' Vienna, 1846 ; QUENNERSTEDT, ' Studier i Foglarnas Anatomi,' *Lund's Univ. Arsskr.* ix. 1872. p. 4 ; N. RÜDINGER, ' Die Muskeln der vorderen Extremitäten,' &c., *Nat. Verh. Holland. Maatsch. Wet.,* 1868 ; C. J. SUNDEVALL, in *K. Vet. Akad. Förh.* 1843, p. 303, and *Förh. Skandin. Naturf.* 1851 ; G. ALIX, ' Sur les Muscles Fléchisseurs des Orteils,' &c., *Bull. Soc. Philom.* xi. 1874, p. 28, and *Essai sur l'Appareil Locomoteur des Oiseaux,* Paris, 1874.

this is to mass the fleshy and heavy parts of the muscle about the centre of gravity of the body, a desideratum for a flying animal. This peculiarity of the muscular system is especially well seen in the muscles of the leg. The muscular system of birds is remarkably constant for the species, the number of variations being apparently, comparatively speaking, but slight. It is true that in but few cases has a large number of individuals been carefully dissected; but of a good many species, on the other hand, have three, four, or even more individuals been dissected from the point of view of the relations or presence of a particular muscle or muscles. The muscular system too is apt to be very constant for a given genus or even a larger division. A glance at the systematic part of this work will show how trifling are the variations even between families in some cases. All these facts lead to the inference that the muscular system in birds is of very considerable value for classificatory purposes. GARROD, FORBES, and FÜRBRINGER are the three anatomists who have laid greatest weight upon the muscular system as an index of affinity. It is, thinks Professor FÜRBRINGER, the muscles of the anterior extremity which have the greatest value of any part of the muscular system. The wing is an organ which is used in much the same way by all birds in which it is properly developed. On the other hand the uses of the muscles of the leg are manifold; we have hopping birds, climbing birds, perching birds, swimming birds, &c. &c. Nevertheless GADOW is inclined to think (with GARROD) that they are the most important. The existing knowledge of the muscles of birds is mainly confined to the muscles of the leg and of the fore limb, a knowledge which we owe almost entirely to GARROD and FÜRBRINGER, many other anatomists having, of course, filled up many details. Less is known about the muscles of the head, neck, trunk, and hyoid region.

It is curious, indeed, how very few birds have been at all thoroughly dissected. Apart from the detailed account of *Apteryx* by Sir RICHARD OWEN, and of less comprehensive memoirs by COUES on the diver, by MORRISON WATSON on the penguin, we have only two recent memoirs which contain anything like a complete account of the muscular structure of a given type. These are the book upon *Corvus corax* by SHUFELDT and a paper by CHALMERS MITCHELL and myself upon *Palamedea*. The most comprehensive general account of bird muscles is unquestionably that of GADOW in Bronn's 'Thierreich.' I shall base the following account of avian musculature largely upon the last-mentioned work, adding to it only such details as were inaccessible to GADOW at the time of its publication. The muscles known to vary will naturally be treated at greater length than those of whose comparative structure but little is registered. GADOW allows altogether 112 separate muscles and sets of muscles like those of the ribs, arranged in a serially homologous row. Some of these are, however, divided again. Of these, so far as we know at present, the following are of the greatest systematic importance, as presenting really considerable variations even to disappearance :—

Glutæus maximus, gl. anterior.

Obturator internus.

Femorocaudal and accessory femorocaudal.

Ambiens.

Semitendinosus and accessory semitendinosus.

Biceps femoris.

Semimembranosus.

Flexores profundi hallucis et digitorum.

Peronei.

Tibialis anticus.

Pectoralis primus, p. secundus.

Deltoid.

Patagialis.

Biceps.

Anconæus.

Expansor secundariorum.

Cucullaris propatagialis.

The value of muscles in classification has been highly rated by many ornithologists, especially, of course, by GARROD, FORBES, and FÜRBRINGER. It is, however, only a comparatively small number of the total series of muscles in the body that can be trusted much as evidence of affinity. The *ambiens* is unquestionably of value as it is found or not found, as the case may be, through whole groups whose mode of progression when walking or climbing is as different as can be. Its total absence from all picarian and passerine birds is a fact upon which I comment elsewhere. There are very few groups in which the ambiens may be present or absent, and in those cases it is often reasonable to separate as distinct families the genera which have it from those which have it not. This cannot, perhaps, be done in every case. Some storks, such as *Abdimia*, have no ambiens, while the majority have it. There are auks with and auks without this muscle. The same may be said of petrels, parrots, and pigeons. *Rhynchops*, the only larine bird without an ambiens, may be, perhaps, rightly elevated to the distinction of a separate family. These examples, however, are so few that they may be compared to such singular exceptions as the absence of the odontophore in the nudibranch *Doriopsis*, which does not in the opinion of any one invalidate the great importance of that structure in arranging the mollusca. In estimating the value of the ambiens the facts of its total or apparently partial suppression, referred to below, must be borne in mind. The entire absence of all trace of the muscle in the owls shows that they are not necessarily to be placed in the neighbourhood of the parrots, in which the muscle, when absent, has left traces behind.

Muscles of the Fore Limb

Pectoralis Primus.—This muscle consists of two parts, the thoracic part, arising from the sternum, and an abdominal portion, arising from the pelvis. The latter portion, well developed in lower vertebrates, is slight in birds, and is often completely absent. The *pectoralis thoracicus* arises from the sternum, the clavicles, and intermediate membranes; it is inserted on to the humerus. In ratite birds there is no origin from clavicles, but, on the other hand, an origin from coracoids not present in carinates. There is frequently an intimate connection between the pectoral near its insertion and the tendon of origin of *biceps*. The *pectoralis* is frequently divided into two portions, the mode of division being twofold. In *Apteryx* the coracoidal portion is separate from the sternal, a state of affairs which recalls some of the lower vertebrates;

in others the pectoral is divided into a superficial and deeper layer. In many 'Ciconiiformes' this is the case. The pectoral, in all birds except the ratites, gives off one or two branches to the patagium. The branch has been termed the *pectoralis propatagialis*. There are either two separate muscles split off from the surface of the pectoral (as in *Nisus*), the tendon of one going to the tendon of the tensor patagii longus, that of the other to the tensor patagii brevis; or (*Podargus*) there is but one muscle which divides into two tendons; or the origins of the two tendons are separate, one of them commencing with a special muscle, the other arising as a tendon from the surface of the pectoralis; or both may be tendinous in origin. Finally, there is in tinamous and gallinaceous birds (some) a special '*propatagialis posticus*,' joining with its tendon that of the other muscle. The *pectoralis abdominalis*, totally absent in nearly all ratites, in storks, various hawks, &c., is divisible into two parts, of which one or other is sometimes wanting. The *pars posterior* springs from the pelvis and adjacent fasciæ; it ends in front freely or comes into more or less close relations with the pars anterior. In **Anseres**, for example, the two form one continuous band of muscle, their boundaries being simply marked by a slight tendinous inscription. The pars anterior arises from the skin close to the termination of the last, or is, as already mentioned, continuous with it; it generally ends upon the humerus, near or in common with the insertion of the main part of the pectoral. In a few birds (quite remote from each other in the system, *Pelecanus, Chauna, Cathartes*) the terminal tendon is lost in the axillary region; a more remarkable modification, possibly of classificatory importance, is described later in *Crypturus*. In a variety of birds there is a slip from this muscle to the metapatagium.

Latissimus Dorsi.—This muscle is divided by FÜRBRIN-GER into three sections—

(1) *L. d. anterior.*

(2) *L. d. posterior.*

(3) *L. d. metapatagialis and dorso-cutaneus.*

The first-named muscle is totally wanting in *Apteryx* and

Alcedo bengalensis.[1] It arises in other birds from the spinal processes of a varying number of cervical and dorsal vertebræ. The narrowest area of origin is seen in *Alcedo*, Macrochires, and various passerines, where it embraces at most a single vertebra. In other birds it may arise from as many as four and a half vertebræ. The broad fleshy or tendinous, or partly fleshy and partly tendinous, insertion varies in width.

The second division is entirely wanting in *Otis, Pterocles*, many passerines, &c. It arises from the spinous processes of posterior dorsal vertebræ, ilium, and even adjacent ribs. Its origin is usually widely separated from that of *anterior* ; but there are various intermediate conditions which culminate in the cuckoos, Macrochires, and some other birds, where the two muscles form one, with, however (save in *Cypselus*), separate insertions. The insertion of this muscle is by a slender tendon in front of that of the *latissimus dorsi anterior*.

The *latissimi dorsi metapatagialis* and *dorso-cutaneus* are two slips running to the metapatagium and the neck region of the skin respectively. They are not often both present, but are in *Apteryx*, Charadridæ, Alcidæ, and some gallinaceous birds. They are both absent in ratites (excl. *Apteryx*), Macrochires, *Colii*, Bucerotidæ, &c. The *dorso-cutaneus* is the rarest, and apart from the instances mentioned is found only in the Cracidæ, piciform birds, and passerines.

Cucullaris.—This is an extensive muscle occupying the greater part of the neck. The only muscle superficial to it is the *sphincter colli*. It has two main divisions. The *pars cranialis* arises from the region of the occipital and the squamosal ; in many birds (of the most diverse orders) a branch is given off from this (the *dorso-cutaneus*), which ends on the back below the spinal pterylon, whose feathers it raises. The main part of the muscle ends upon the clavicle, or sternum, or ligaments in the neighbourhood. In some birds a part of the fibres end upon the fascia covering the *pectoralis primus*, and in those with a crop some of the deeper fibres come into relation with that

[1] Not in *A. ispida.*

organ, forming a *levator ingluviei*. A portion of the *cucullaris*
also directs itself towards the patagium, and in most Passeres
and in parrots, Pici, and *Upupa* forms a special *cucullaris
propatagialis*, joining the tensor longus tendon. The *pars
cervicalis* of the *cucullaris* arises from the dorsal edge of the
neck, and is inserted near or in common with the other part.
In many birds (*e.g.* Anseres) a slip is given off from this
which supplies the humeral pterylon. It is termed the *cucullaris dorso-cutaneus*.

Rhomboideus Externus.—This muscle arises tendinously,
the width of the tendinous part being about the same as
that of the muscular part, from the last cervical and from
the dorsal vertebræ ; it is inserted fleshily along the whole
length of the scapula. The muscle varies but slightly ; the
origin is more or less extensive, and the vertebræ from
which it arises are not always the same. Its insertion is
not always limited to the scapula, but sometimes extends on
to the furcula. In *Casuarius* and *Apteryx* the muscle arises
from the ribs.

Rhomboideus Profundus.—This muscle also arises tendinously from the neural spines of the dorsal vertebræ, or
from both dorsal and cervical. Its origin sometimes extends
as far back as to the ilium. It is inserted into the
scapula below the last. In *Casuarius* and *Apteryx* this
muscle arises from the ribs. The *rhomboideus profundus* is
occasionally divided into two distinct parts ; in *Megalæma*
there are three distinct divisions.

Serratus Superficialis.—This muscle is divided by FÜR-
BRINGER into three parts, of which two are always present,
while the third is sometimes absent. This is the *pars
metapatagialis*. The *pars anterior* arises from one or more
ribs on the boundary line between the cervical and dorsal series.
It is attached to the scapula along the ventral border, but, quite
exceptionally, in *Rhea* on to the dorsal border. The *pars
posterior* springs from a varying number of dorsal ribs, and
in several birds (e.g. *Rhamphastos*) it, with the *pars anterior*,
which can hardly be separated as a distinct muscle, springs
from a considerable number of ribs—five in the case referred to.

It has generally a broadish insertion on to the scapula, but
in *Meiglyptes*, many Passeres, &c., it is attached merely to the
extremity of that bone.   The *pars metapatagialis* is absent
in ratites (except *Apteryx*), humming birds, and a few others.
It springs from 1–4 ribs, and is inserted on to the metapatagium.

Serratus Profundus.—This muscle is highly developed
in *Struthio* and *Casuarius*, less so in other struthious birds
and in the Carinates.   It arises from a variable number of
cervical and dorsal ribs, and it passes backwards (in the
contrary direction, therefore, to the serratus superficialis) to
be inserted on to the scapula.

Patagialis.[1]—This muscle, concerned with the folding of
the patagial membrane, is present in all birds except the
struthious.   It arises from the clavicle and from the tip of
the scapula, is sometimes divided into two muscles from the
start, and sometimes arises as a single muscle, which
immediately divides into two tendons, the tensor patagii
longus and the tensor patagii brevis.   Exceptionally the
former may be absent.   In a specimen of *Crex pratensis*
the representative of tensor patagii longus was found by
GARROD to be simply the biceps slip, a muscle that will be
treated of presently.   The size and importance of this
muscle vary considerably; it is largest in the parrots,
where, indeed, it is uncertain whether a part of the deltoid
has not been converted to a similar function.   This matter,
however, is dealt with under the description of that family.
The tensor patagii longus always ends in a single tendon
which runs along the anterior margin of the patagium and is
inserted on to the metacarpal.   It is usual for the middle
part of this tendon to be of a more fibroid character and of
a yellowish colour, contrasting with the steely and typically
tendinous aspect of the other tendons arising from the
tensores.   Very commonly the entire tensor muscle is
reinforced by a tendinous or muscular slip from the pectoralis, and sometimes there is a separate slip to each of the

[1] G. Bucner, 'Première Note sur l'Appareil Tenseur,' &c., *C. R. Soc. Biol.*
1888, p. 328.

two tensores. It is also very general for one of the two
tendons, or for both before their division, to be attached by a
tendinous slip to the deltoid crest of the humerus. The
tensores patagii are of considerable use in classification.
But it must be admitted that they are apt to vary greatly
from genus to genus. The variations chiefly concern the
more or less complicated condition of the tendons of the
brevis. The simplest condition is seen, e.g., in *Rhamphastos
Cuvieri*, where the tendon is single and is attached below to
the tendon of origin of the extensor metacarpi radialis.

A further degree of complication is seen in, e.g., a cuckoo,
where the single tendon gives off, near to the fore arm, a slip
running obliquely wristwards, which is attached to the tendon
of the extensor metacarpi radialis. In the limicolous birds the
main tendon is usually divided from the first into two, of
which the anterior has the wristward slip, already referred
to ; in those birds and many others there is the further
complication of a band of tendinous fibres which arise at the
junction of the wristward slip with the fore arm, and pass
obliquely forwards and upwards to be inserted on to the
tendon of the longus. This slip is termed, in the following
pages, the 'patagial fan ; ' it is frequently of a fanlike form.
The tendon of the tensor patagii brevis has not always the
regular form that it has in the types that have been already
selected for illustration. In the tinamou, *Rhynchotus*, for
instance, the tendon is a diffuse fascia spreading out over
the greater part of the patagial membrane ; in other birds,
e.g. storks, it is a broad, rather diffuse band, as a rule with
a thicker edge or edges. A peculiar condition of the
tendons of the brevis characterises the auks, some gulls, and
at least one limicolous bird. In them (see below) one or
two delicate tendons arise from the longus tendon near to
the insertion thereon of the patagial fan, and run obliquely
backwards and downwards to be attached on to the dorsal
surface of the fore arm—the reverse side, that is to say, to
that to which the other tendons which together make up
the tensor patagii brevis are attached.

In some birds, e.g. certain passerines, the tendon of the

tensor longus is reinforced by a muscular slip ending in a
tendon which is derived from the cucullaris muscle. Another
muscle which is also related to the patagialis in an analogous
way has been termed the ' *biceps slip* ' (*q.v.*)

Anconæus Longus.—This muscle arises from the neck of
the scapula alone, by a head which is entirely tendinous or
partly muscular, or in addition from the edge of the scapula, a
little further away from its junction with the coracoid, by a
tendinous head. It is inserted by a broad tendon on to the

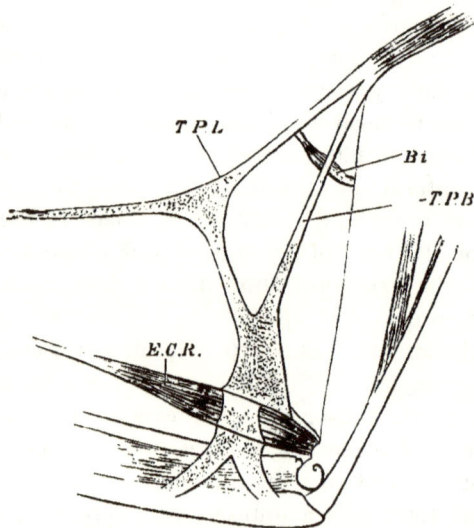

Fig. 49.—Tensores Patagii of *Phœnicopterus.*
T.P.L, tensor patagii longus: *T.P.B*, tensor patagii brevis : *Bi*, biceps
slip: *E.C.R*, extensor carpi radialis.   (After Wellaon)

elbow joint. Sometimes there is an accessory head from
the humerus, which in this case arises, as a rule, in common
with the tendon of insertion of the posterior latissimus
dorsi. In *Palamedea* this head is double, the two halves
being united by cross tendinous threads. On the whole the
humeral head is characteristic of Garrod's Homalogonatæ
and not of the Anomalogonatæ ; but there are exceptions on
both sides. The breadth of the humeral head varies greatly ;
it is sometimes reduced to a thin thread.

Triceps.—This muscle springs from the greater part of
the humerus fleshily by two heads, of which one—that from
the tuberculum minus—is often tendinous. The name
triceps, be it observed, has been given to the muscle on the
understanding that the last-described muscle is a part of it.
Their tendons of insertion join.

Expansor Secundariorum.[1]—This extraordinary muscle
appears to be partly a skin and partly a skeletal muscle.
A bundle of non-striated fibres arises near the secondary
feathers of the arm and ends in a tendon. This is occasionally reinforced by a band of striated fibres arising
from beginning of ulna. The long tendon is inserted in
various ways. The typical condition (termed by GARROD
' ciconiine ') is for it to be inserted into the middle of a ligament running from the scapulo-coracoid to the sterno-
coracoid articulation. Other modifications occur among the
gallinaceous birds (*q.v.*), &c. The muscle is totally absent
in **Struthiones, Sphenisci, Alcæ, Psittaci,** the majority of **Pico-
Passeres,** and in a few species of groups where it is usually
present.

Sterno-coracoideus.—This muscle, wanting only in the
Macrochires, runs from the anterior lateral border of the
sternum to the adjoining region of the coracoid. The muscle
shows every stage between a single muscle and a completely
double one. It is double, for example, in *Casuarius.* In
Struthio, Chauna, and some other birds where there is but
one sterno-coracoid, it is the homologue (according to FÜR-
BRINGER) of the deeper section of the double muscle.

Scapulo-humeralis Anterior.—Runs from the beginning
of the post-glenoidal region of the scapula to the beginning
of the dorsal surface of the humerus. It is a muscle which is
frequently absent. FÜRBRINGER failed to find it in **Struthiones,
Sphenisci,** *Fregata, Chauna,* **Columbæ,** *Pterocles, Chunga,
Bucorvus,* &c.

Scapulo-humeralis Posterior.—Contrary to the last this
is a large muscle and is never absent. It arises from the

[1] GARROD, 'On the Anatomy of *Chauna derbiana,*' &c., *P. Z. S.* 1876
p. 193, &c.

hinder part of the scapula and runs to the humerus, where it is inserted on to the tuberculum mediale.

Coraco-brachialis Externus.—This muscle, which is relatively larger in the struthious birds, runs from the coracoid to the beginning of the 'planum bicipitale' of the humerus, where it is generally covered by the tendon of the biceps. It is noteworthy that among carinate birds this muscle is largest in the tinamous, which thus approach the ostrich tribe. It is most reduced in the **Passeres**, in some of which, indeed, it has actually disappeared.

Coraco-brachialis Internus.—Springs from the coracoid and often from neighbouring parts of sternum. It is inserted on to the median tubercle of the humerus.

Pectoralis II.—This muscle arises from the ventral surface of the sternum, from the coracoid, and from the coraco-clavicular membrane. It is inserted by a long tendon of attachment to the lateral tubercle of the humerus. The muscle is small in ratites, large in carinates.

Deltoides Major.—Arises from the acromion and the dorsal part of the clavicle, and is inserted on to the deltoid crest of the humerus. The muscle and the length of its attachment vary much in size. It is large in **Accipitres**, **Passeres**, &c., small in **Alcæ**, **Psittaci**, &c. It appears, indeed, to be absent in *Psittacula*.

Deltoides Minor.—This is a small muscle passing from the neighbourhood of the foramen triosseum. It is absent in *Phaëthornis* and (occasionally) in *Cypselus*. It is also absent in **Struthiones**.

Biceps.—This muscle consists typically of two heads, as its name denotes. The longer of these arises from the coracoid by a long tendon. The second head arises also tendinously from the head of the humerus. The insertion of the muscle is double, on to the radius and the ulna. The division commences at a varying distance from the actual insertion.

In the penguins this muscle is totally absent. In *Colymbus*, *Pelecanoides*, *Thalassiarche*, and other petrels, in some Alcidæ, the coracoid head alone is present, the humeral

head being in some of these birds entirely diverted to form
the biceps slip to the patagialis (*q.v.*) The coracoidal head,
when it exists alone, may be divided into two quite separate
muscles, uniting only at their very origin. This is the case
with certain Alcidæ. This division of the coracoidal half of
the muscle also exists in the Laridæ and in certain **Limicolæ**,
where there is a humeral head present also. In the **Steganopodes** both heads are present, but the humeral head after
its attachment to humerus is continued on to the coracoid.
A trace of this arrangement is apparently left in some birds
(e.g. *Porphyrio*), where, though the humeral head arises from
the humerus only, a ligament passes on from that part to
the coracoid.

Brachialis Inferior.—A flat fleshy muscle arising from
the distal part of the humerus, and inserted upon the ulna.
In the penguins (where the biceps is absent) this muscle is
particularly large, and is inserted on to the radius.

Pronator Sublimis.—This muscle springs tendinously from
the inner condyle of the humerus and is inserted fleshily,
and for a varying distance in various birds, upon the second
third to second eighth of the radius.

Pronator Profundus.—This muscle lies deep of the last,
but has a similar origin and insertion. In the Ratitæ this
muscle and the last form a single muscle.

Entepicondylo-ulnaris.—This muscle, found apparently
only in **Galli** and **Tinami**, arises in common with the pronator profundus, and is inserted on to the ulna.

Ectepicondylo-radialis.—This muscle arises tendinously
from the outer condyle of the humerus and is inserted
fleshily on to the radius. It appears to be wanting in the
penguins, and to be largest in the **Galli**.

Ectepicondylo-ulnaris.—This muscle arises from the
outer condyle of the humerus, and is inserted similarly to
the last upon the ulna. Its insertion is fleshy, and in
Palamedea it is larger than the last.

Flexor Carpi Ulnaris.—This arises from the inner condyle of the humerus by a strong tendon, in which there is
a well-marked sesamoid; it runs down the inner side of the

ulna to be inserted on to the great tuberosity of the ulnar carpal bone. A thin muscle arising from it passes into a tendon which is connected with the secondary feathers. The above refers to *Palamedea*. In penguins the entire muscle is represented by a tendon only.

Extensor Digitorum Communis.—Arises from the external condyle of humerus. It splits on the hand into two tendons, of which one is inserted on to the basis of the first phalanx of digit I., the other on to the corresponding phalanx of digit II. In *Struthio* the first of the two tendons is wanting. In the penguins the muscle is represented only by a tendon which is inserted on to the outer side of metacarpale II. and on to the basis of the first phalanx of that digit.

Extensor Longus Pollicis.—This muscle arises from the proximal region of both radius and ulna. The common tendon is inserted on to the origin of metacarpale I.

Extensor Indicis Longus.—This muscle is two-headed. The longer head arises from the radius from its middle two-thirds, but sometimes also receives a few fibres from the ulna; the second much shorter head springs either from the distal end of the radius or from the os carpi radiale, or finally from the basis of metacarpale II.; the united tendons are inserted on to the head of the first and the basis of the second phalanx of digit II. The second head is absent in *Fulica* and in some other birds.

Interosseus Dorsalis.—This muscle arises fleshily from the opposed surfaces of metacarpals II. and III. The common tendon is inserted on to the base of the second phalanx of the second digit.

Interosseus Palmaris.—Has an origin from the same metacarpals as the last and is inserted on to the first phalanx of digit II.

Ulni-metacarpalis Ventralis.—It arises fleshily from the radial face of the last quarter of the ulna, and is inserted on to the head of the first metacarpal.

Ulni-metacarpalis Dorsalis.—This springs by a tendon from the distal region of the ulna, and has an insertion upon the third metacarpal.

Extensor Metacarpi Ulnaris.—Springs from the external condyle of the humerus. There is generally a second head, which, instead of being tendinous, is fleshy and rises from the humerus a little below the first head. The tendons in which the two end do not join until a little before their insertion on to the base of the metacarpal of digit I. The degree of separation of the two heads differs considerably. In the penguins only one head is present.

Flexor Digitorum Sublimis.—From the internal condyle of humerus to os carpi ulnare is a strong aponeurotic fascia, from the distal end of which springs the muscle in question, to be inserted on to phalanx I. of digit II. In *Palamedea* Mr. MITCHELL and I traced the tendon to the base of the second phalanx of the same digit. In *Psittacus* and *Columba* the tendon has the same extension. In *Struthio* the muscle is entirely absent.

Flexor Digitorum Profundus.—This springs from the middle and proximal third of the ulna, and is inserted on to the basis of the second phalanx of the second digit. In *Corvus*, &c., the muscle is two-headed, the two heads being separated by the insertion of the brachialis internus. In other birds the extent of the origin varies.

Abductor Indicis.—From metacarpal II. to basis of phalanx I. of digit II. In *Palamedea* it also arises from the flexor pollicis.

Flexor Pollicis.—From metacarpal I. to thumb phalanx. In *Palamedea* its fleshy belly gives rise to a slip which passes to the abductor indicis.

Adductor Pollicis.—In *Palamedea* it arises from the metacarpal just beyond the articulation of the thumb; it ends in the ala spuria and not on the thumb bone. In some birds it has also a connection with the thumb bone.

Extensor Pollicis Brevis.—This muscle arises fleshily from the second metacarpal. It is inserted on to the phalanx of the thumb.

Abductor Pollicis.—This muscle arises from metacarpal I. and passes to phalanx of same digit.

Flexor Digiti III.—Arises from metacarpal III. and is inserted on to basis of phalanx I. of same digit.

Flexor Metacarpi Radialis.—This muscle arises from the outer condyle and is inserted on to the ulnar border of metacarpal II. or on to the beginning of metacarpal III.

Muscles of the Hind Limb

Sartorius.[1]—This is a broad strap-shaped muscle arising from the ilium and from the fascia covering the glutæus maximus; it is inserted on to the ligament containing the patella, and on to the crest of the tibia. The muscle has an origin which sometimes extends further forwards, and is then overlapped by the latissimus dorsi. Sometimes, on the other hand, its insertion moves further back. In *Phœnicopterus* the muscle is divided into three distinct portions.

Glutæus Maximus.[2]—This often large muscle was used by GARROD[3] in his muscular classificatory scheme, and at first termed the tensor fasciæ. It has an origin which is sometimes entirely in front of the acetabulum, and sometimes extends behind it. It arises tendinously from the fascia covering the glutæus medius, and from the ridge of the ilium; its insertion is tendinous on to fascia covering thigh.

Glutæus Anterior.[4]—This muscle arises from the ridge of the ilium below the last, by which it is entirely covered; it is inserted by a tendon on to the outer face of the thigh. The most remarkable modification which this muscle undergoes is its entire conversion into tendon in *Bucorvus*, &c., in which birds it comes to be merely a thigh ligament.

Glutæus Medius.[5]—This arises fleshily from the ilium, and is inserted by a short strong tendon on to head of femur.

[1] 'Ilio-tibialis internus' (GADOW).   [2] 'Ilio-tibialis externus' (GADOW).
[3] 'On certain Muscles of the Thigh of Birds.' &c., *P. Z. S.* 1873, p. 626, and 1874, p. 111.
[4] 'Ilio-femoralis externus' (GADOW).
[5] 'Ilio-trochantericus medius et posterior' (GADOW).

Glutæus Quartus.[1]—This is a small muscle lying at its insertion between that just described and that about to be described.

Glutæus Minimus.[2]—This muscle is also small, and

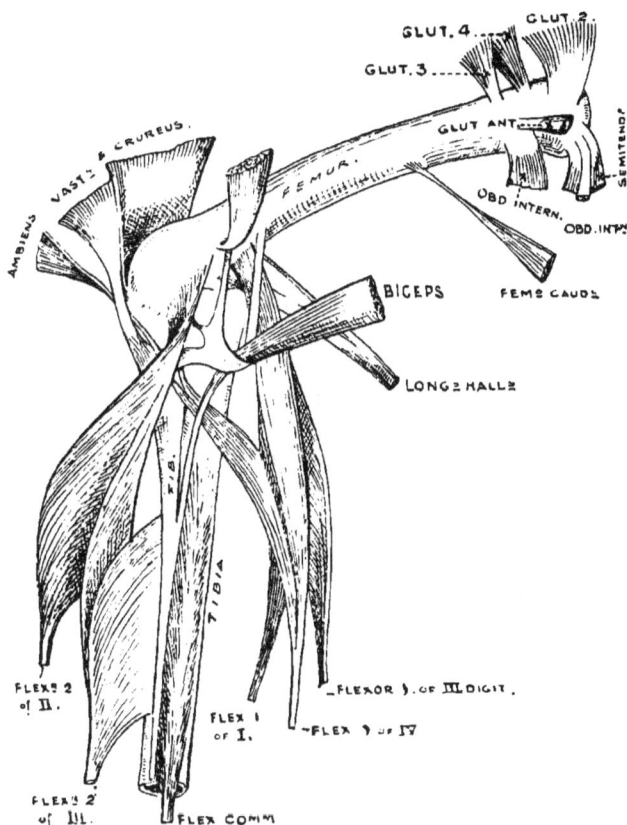

Fig. 50.— Muscles of Leg of *Palamedea*, Outer View (after Beddard and Mitchell).

arising from the ilium is inserted next to the last on to the neck of the femur.

Pectineus.[3]—This is a smallish muscle arising below the

[1] ' Ilio-trochantericus medius et posterior ' (Gadow).
[2] ' Ilio-trochantericus anterior ' (Gadow).
[3] ' Ilio-femoralis internus ' (Gadow).

origin of the glutæus quartus and inserted on to the inner face of the femur below the head.

Vastus Externus.[1]—This arises fleshily along the greater part of the outer side of the shaft of the femur; it is fused with the crureus in front, and ends with it upon the patellar ligament.

Crureus.[1]—This is tendinous on the outer surface at its origin from the neck of the femur; it also arises from a considerable part of the shaft of the femur and is inserted as has already been described.

Vastus Internus.[1]—This arises from the inner surface of the shaft of the femur and is inserted on to the tibia along with the tibial insertion of the sartorius.

Obturator Externus.[2]—This is a small deep-lying muscle, which arises from the ischium or the ilium, or even partly from the pubis; it is inserted on to the trochanter.

Obturator Internus.[3]—This muscle arises from the membrana obturatoria and from the ischium. It is inserted on to head of femur. GARROD[4] laid some classificatory stress upon the shape of this muscle, oval or triangular.

Gemellus.[5]—This is a small fleshy muscle, double or single, surrounding the tendon of insertion of the last.

Semitendinosus.[6]—In *Palamedea* this muscle arises fleshily from the ilium behind the biceps; it is half an inch broad, and, after being joined by the small accessory which springs from the femur near its distal end, sends a flat tendinous slip to the semimembranosus. The rest of the tendon of the muscle joins the middle head of the gastrocnemius. The muscle shows considerable variations in its attachments and size. It is completely absent in the owls, hawks, and swifts. The accessory head is absent in kingfishers, many **Steganopodes**, *Colymbus*, &c.

---

[1] ' Femori-tibialis ' (GADOW).   [2] ' Ischio-femoralis ' (GADOW).
[3] ' Obturator ' (GADOW).
[4] ' On the Anatomy of *Chauna derbiana*,' &c., P. Z. S. 1876. p. 195.
[5] ' Accessorii musculi obturatoris ' (GADOW).
[6] ' Caud-ilio-flexorius ' (GADOW).

Femorocaudal.[1]—This is another of the variable muscles of the thigh. Typically the muscle is two-headed, one head arising from the transverse processes of the caudal vertebræ, the other (termed by GARROD accessory femorocaudal) from the ilium. The two are inserted together upon the flexor side of the femur, as a rule by a longish tendon.

The variations culminate in the entire absence of the muscle, which occurs in *Chunga Burmeisteri* and *Leptoptilus.* In some species—for example, in most **Passeres** and picarian birds—the caudal portion is alone present. In others—*e.g.* in *Serpentarius, Otis, Phœnicopterus,* the iliac portion is alone present.

Biceps Femoris.[2]—This muscle is covered externally by the glutæus maximus where this is present ; otherwise it is the most superficial of the flexors of the leg. It arises from the postacetabular region of the ilium, and ends in a strong, generally round, tendon, which passes through a sling of tendon which is derived from the femur independently, and from the same bone in common with one of the heads of the gastrocnemius, to be inserted on to the fibula. The principal variations of the muscle concern its more or less extensive origin. It has never been known to be absent. In *Corythaix* GADOW states that it is double. In the ostrich and in the ducks and swans there is the usual sling, but before entering the sling the biceps gives off a branch, which joins one of the heads of the gastrocnemius. In *Fregata* and in some swifts quite exceptionally the sling is totally absent, but the muscle has the usual insertion. In certain auks the muscle gives off, before entering the biceps sling, a branch to the thigh superficially. In *Podica senegalensis* the muscle divides into three branches. The first of these has a considerable superficial attachment to the outside of the leg ; the second is attached to the fibula below the attachment of the third insertion, which is the normal one, through a sling. In *Heliornis* only the first and third of these are present.

[1] ' Caud-ilio-femoralis ' (GADOW).   [2] ' Ilio-fibularis ' (GADOW).

Semimembranosus.[1]—Arises from the ischium, sometimes
trenching a little upon the pubis. It is inserted upon the
tibia. The variations which it shows are mainly of size.
In *Phœnicopterus* it is two-headed, and in certain Falconidæ

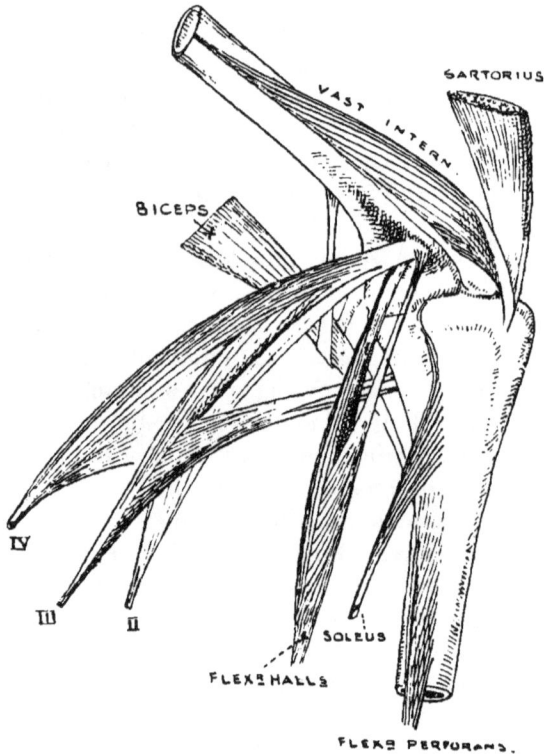

FIG. 51.—LEG MUSCLES OF *Palamedea*, INNER VIEW (AFTER BEDDARD
AND MITCHELL), ILLUSTRATING BICEPS AND ITS SLING.

completely double. It is often inserted in common with
the semitendinosus.

Adductors.[2]—These are, as a rule, two broad, flat, fleshy
bands, which arise from the pubis, and from the ischium,
and are inserted upon the inner edge of the femur. The
separation between the two parts is less in some birds than

[1] ' Ischio-flexorius ' (GADOW).   [2] ' Pub-ischio-femoralis ' (GADOW).

in others. There is also occasionally (e.g. hornbills) an
additional attachment to the gastrocnemius.

Ambiens.—This muscle, as is well known, is not present
in all birds. Though the late Mr. GARROD used it largely
in his scheme of classification, its mere presence or absence
is not an absolute guide to the systematic rank of the bird.
Broadly speaking, it is present in all the birds which
GARROD called homalogonatous, or normal-kneed, and

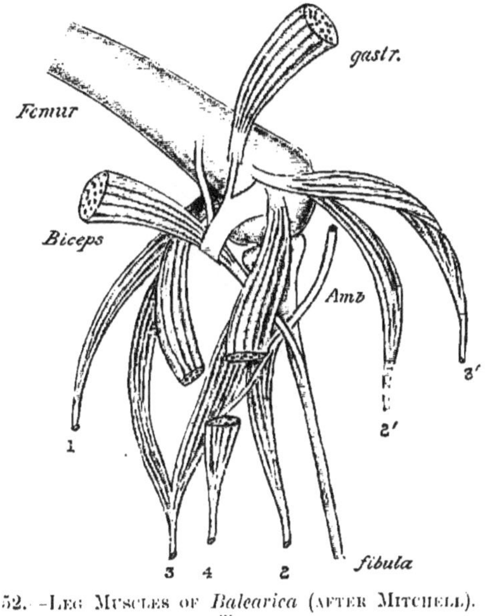

FIG. 52. —LEG MUSCLES OF *Balearica* (AFTER MITCHELL). THE
AMBIENS TENDON IS CUT.
1, flexor longus hallucis ; 2-4, flexores perforati ; 2', 3' flexores perforati et perforantes.

it is absent in all the birds which were termed by him
anomalogonatous, or abnormal-kneed. But there are exceptions, at any rate on one side. Thus while the muscle is
present in the storks generally it is absent from the nearly
related herons, and, indeed, is absent in three storks, *Xenorhynchus*, *Abdimia*, and *Dissura*. When the muscle is
present it has as a rule the relations described above; but
in a few birds it does not reach beyond the knee, thus
showing, perhaps, an incipient disappearance. The import-

ance of this muscle in classification has been much increased
by MITCHELL's interesting paper upon its exact relations
to the flexors of the leg in a series of birds.[1]   He has shown
that in *Balearica chrysopelargus* the mass of muscle which
forms the flexor perforatus arises from three distinct heads ;
one of these is, in common with the flexor longus hallucis,
from the intercondylar notch of the femur ; the second is

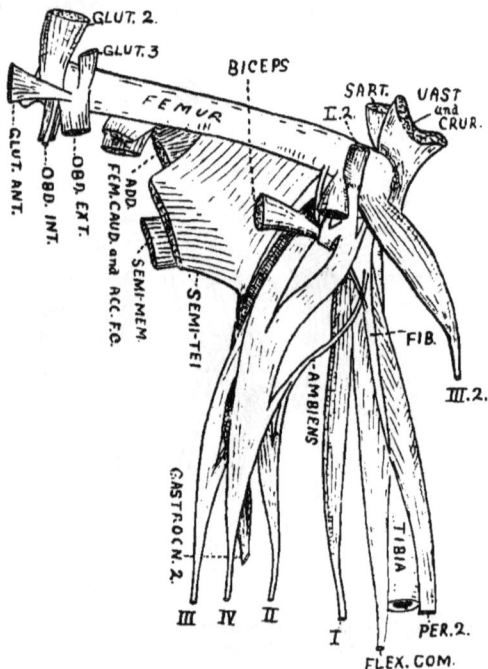

FIG. 53.   LEG MUSCLES OF *Opisthocomus* (AFTER MITCHELL).
II IV, flexores perforati.

from the outer condyle of the same bone ; the third is from
the tendon of the ambiens.   This tendon divides into three,
one for each of the three divisions of the flexor perforatus.
The arrangement will be obvious from the accompanying
cut.   Apart from slight differences in detail the same
arrangement was found to hold good for a few other birds
provided with an ambiens.

[1] 'On the Perforated Flexor Muscles in some Birds.' *P. Z. S.* 1894, p. 495.

In *Nycticorax Gardeni*, which has no ambiens, there is a difference in the origin of the flexors in question which is of great interest. The two origins from the femur are as in the crane. But there being no ambiens there can be no origin from that muscle. Nevertheless the third head of the flexors is present in the shape (see fig. 53) of a broad tendinous band arising from the fibula, which soon divides into the three tendons to the three muscles, precisely as does the tendon of the ambiens. This is highly suggestive of the rudiment of an ambiens, a suggestion which is confirmed on referring to a specimen of *Opisthocomus* without a fully developed ambiens (fig. 53). And as the herons are birds which are presumed to be really homalogonatous birds, though they have lost the ambiens, the fact is of additional interest; particularly is this so when we compare the conditions obtaining in *Nycticorax* with those which characterise *Corvus*, a clearly anomalogonatous bird, none of whose near relatives possess an ambiens. In *Corvus capellanus* it was discovered that the flexors usually connected with the ambiens, or with its rudiment, had no origin from the fibula at all, and arose by only a single head from the femur. The same was practically the case with *Bubo maximus*, only that both femoral heads were present. Now the owls, formerly relegated to the **Accipitres**, are more generally looked upon as related to the picarian birds, forming, in fact, a section of the anomalogonatæ of GARROD. The state of their ambiens is entirely confirmatory of this placing. We have some evidence, therefore, that there are degrees in the disappearance of the ambiens, which, so far as the few types that have been examined enable us to say, distinctly support the division of birds into the two great divisions of GARROD.

Peroneus Superficialis.—Confined as a rule to the tibia in its origin, this muscle sometimes springs also from the fibula. The tendon of insertion, after giving off a branch to the tarsus, becomes attached to the tendon of the flexor perforatus digiti III. This muscle is occasionally completely absent; this is the case with various Picopasseres, owls, &c.

In *Podiceps*, according to GADOW, there is no branch to the flexor tendon, the tendon of insertion ending at the ankle.

Peroneus Profundus.—This muscle arises from the lower part of the tibia, and is inserted on to the outer side of the ankle. It is completely absent in such birds as *Ciconia*, *Otis*, &c.

Gastrocnemius.—This great muscle occupies the greater part of the back of the leg. It has three heads, of which the outer arises from the outer condyle of the femur, and from the ligament which supports the insertion of the biceps; the inner head springs from the inner side of the head and neck of the tibia; the middle head is confused at its origin with the insertion of the accessory femoro-caudal. This middle head appears to be wanting in *Cypselus*. It is inserted by a strong tendon to the tarso-metatarsus, and also, dividing, to the phalanges of the toes.

Popliteus.—This muscle passes between the tibia and the fibula in most birds, but has been found to be sometimes absent (e.g. *Picus*, &c.)

Tibialis Anticus.—This muscle arises by two distinct heads. The first is entirely tendinous, and is from the external condyle of the femur. It forms a long and strong ligament, which runs over the knee; the second head is fleshy, and is from the front part of the head of the tibia. Its long tendon of insertion is attached to the metatarsal. In certain birds (*Chrysotis*, *Podargus*, and owls) the tendon and even a portion of the muscle are double.

Extensor Digitorum Communis.—This muscle arises generally from the front part of the tibia only, but sometimes its fibres of origin stray on to the patella and on to the fibula. The divisions of the tendon of the muscle are usually inserted on to several phalanges of the toes, which they supply. It is remarkable that the parrots are the only birds in which this muscle supplies the hallux as well as the other toes; it is, therefore, in them, as GADOW remarks, truly an extensor *communis*. In other birds digits I., III., IV. (when present) are the only toes supplied. The common tendon divides in various ways; in *Grus virgo* the tendon divides into two, and each of these again divides ; there are thus four tendons, of which the two middle ones supply the third toe. In *Ptilonorhynchus violaceus* the tendon first gives off a branch to the second toe, and then divides for the third and fourth. In *Rhamphastos carinatus* all three branches are given off at the same level. In *Pharomaerus moeinno* the extensor supplies only the two middle digits. In *Scopus* there is a slight variation of what is found in *Grus*, *Nothura*, &c. The tendon divides into three, and the middle one again divides into two, both of which latter supply the middle digit.

The superficial flexors of the foot consist of—

Flexor Perforatus et Perforans Indicis (fig. 51).—Arising from the outer condyle of the femur, and from the septum between itself and adjacent muscles ; its tendon is inserted on to base of second phalanx of its digit.

Flexor Perforatus et Perforans Medii.—Has two heads of origin, one as in last, the other from fibula. Its tendon perforates that of *flexor perforatus* of the same digit, and is perforated by *flexor profundus* ; it is inserted on to base of second phalanx (in *Palamedea*, third in *Ciconia*) of its digits.

Flexor Perforatus.—This muscle (see figs. 52, 53) arises from two heads, an inner head from the intercondylar notch and an outer head from the outer condyle of the femur. As will be seen from the annexed cuts, those portions of the muscle which supply digits III. and IV. have slips from both heads, but not that which supplies digit II. These muscles also arise either from the ambiens or from the fibula, as has been explained above under the description of the ambiens. The tendon to digit II. is inserted at the base of the first phalanx, that to digit III. is usually joined by a vinculum (absent in *Opisthocomus*, *Asio otus*, and *Rhytidiceros plicatus*) to tendon of *flexor perforatus et perforans medii*, and is inserted on to base of second phalanx of digit III. The third tendon has four slips of insertion, on to four proximal phalanges of digit IV. The descriptions of the insertion of these tendons applies to *Ciconia nigra*. There are variations.

Flexor Profundus.—Arises from nearly whole hinder surface of tibia and fibula, and sometimes also by a head from the outer condyle of femur.

Flexor Hallucis.--Arises by a single head, or by two heads, from the outer condyle of femur and from inter-condylar region.

The tendons of the two last-described muscles are con-

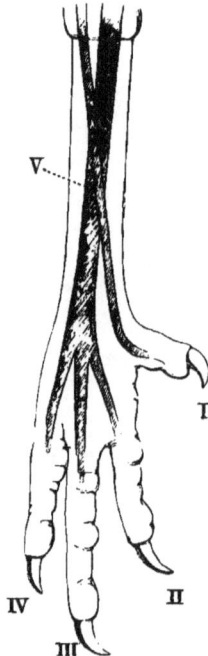

FIG. 54.- *Gallus bankiva* (AFTER   FIG. 55.—*Apteryx Mantelli* (AFTER
GARROD).   GARROD).

nected with each other in various ways, which have been described and illustrated by Mr. GARROD.[1] In using his figures I accept the seven types admitted by Dr. GADOW.[2]

(1) In gallinaceous birds, pigeons, parrots, storks, &c., the tendons cross, and are united by a simple vinculum (fig. 54).

[1] 'On the Disposition of the Deep Plantar Tendons in different Birds,' *P. Z. S.* 1875, p. 339.

[2] In Newton's *Dictionary of Birds*, p. 617.

(2) In *Apteryx*, &c., the vinculum is very strong, and forms the direct continuation of the tendon of the flexor hallucis; the tendon to the hallux has the appearance of being a branch of this (fig. 55).

(3) In many Accipitres the flexor hallucis divides into two parts at the lower end of metatarsus; one of these goes to hallux, the other blends with the branch of the flexor communis

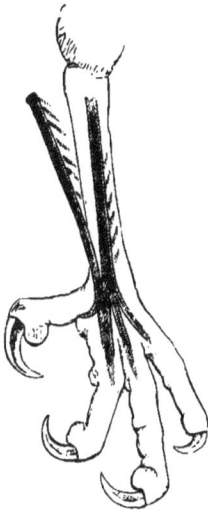

FIG. 56.—*Tinnunculus alaudarius* (AFTER GARROD).

FIG. 57.—*Buceros rhinoceros* (AFTER GARROD).

which supplies digit II.; there may be in addition a strong vinculum, uniting tendons before their splitting (fig. 56).

(4) In *Rhea*, *Phœnicopterus*, &c., where hallux is small or absent, the two tendons are present, but fuse together, branching later to supply the digits present.

(5) In *Buceros*, *Podargus*, *Sarcorhamphus*, &c., the two tendons fuse completely before supplying the digits (fig. 57).

(6) In *Megalæma*, *Rhamphastidæ*, &c., the vinculum is present, but flexor digitorum supplies only digit III.,

the others being supplied by trifurcate flexor hallucis (fig. 58).

(7) Both tendons present, but no vinculum at all in Passeres (fig. 59).

Of these seven types there are naturally modifications and intermediate conditions, which will be described in the systematic part below.

Short Flexors.—Of these there are two series ; one set

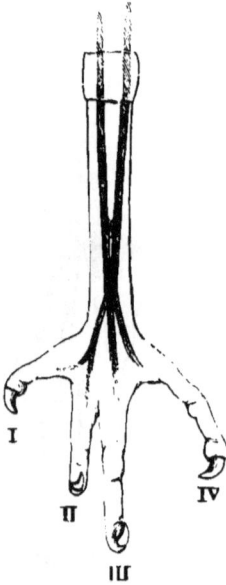

FIG. 58.— *Megalæma* (AFTER GARROD).

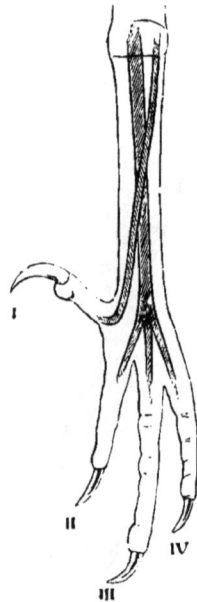

FIG. 59.—A PASSERINE FOOT (AFTER GARROD).

arises from the bones of the metatarsus, the others from the long tendons. We shall commence with the first set.

The *flexor hallucis brevis* is often (*Palamedea, Aceros*) composed of two distinct muscles with separate tendons and insertions on to the first digit.

Flexor (adductor) digiti II. is inserted on to the median side of the base of first phalanx of digit II.

Flexor (adductor) digiti IV. arises near to the last, and has a similar relation to the basal phalanx of digit IV.

Flexor secundus (adductor) digiti IV. is a muscle which appears to be generally absent (e.g. *Ciconia*), and is not, according to GADOW, mentioned by authors. It is present in *Rhea* and *Bucorvus*.

The last-named muscle may be the equivalent of the first of the two short flexors which arise from the deep long flexor tendons. Of these there are two.

(1) MITCHELL has described in *Opisthocomus* a muscular slip which leaves the longus hallucis tendon and runs to the fourth digit. A similar muscle is present in *Ardea cinerea* ;[1] I have found it in the hornbill, *Ceratogymna elata* (not in *Aceros nipalensis*).

(2) Another muscle (*flexor brevis digiti III.* of GADOW) arises from tendon of *flexor profundus* in *Opisthocomus*, *Rhea*, &c., and passes to third digit.[2]

Of short extensor muscles there are at most six.

The *extensor hallucis*, generally single, is two-headed in *Pandion*, and formed of two distinct muscles in *Palamedea*.

The *extensor proprius* and *extensor brevis digiti III.* both supply the third digit. In *Aceros* and *Bucorvus* a single but two-headed muscle appears to represent both.

The *extensor (adductor) digiti II.* is not always present.

The same may be said of the *extensor digiti IV.*

Muscles of the Neck and Trunk

These muscles are, many of them, not easy to isolate and describe. There is, in consequence of this, some divergence in the published accounts. Furthermore, insufficient data have been collected for the estimation of the use of these muscles in classification. The account which follows is an almost verbatim transcript from a paper upon *Palamedea*[3] by Mr. MITCHELL and myself.

[1] It is suggested by MITCHELL that these muscles (which require further study) may ' throw light upon the origin of the very peculiar modes of distribution of the hallucis tendon in some groups of birds, as it has been repeatedly shown that a tendon may be the homologue of a muscle.'
[2] GARROD MS. [3] Quoted on p. 108.

Biventer Cervicis.—The two muscles are perfectly separate from each other. They arise tendinous from the spinous process of the first dorsal vertebra. Then follow a tendon of an inch long, a belly of two inches, again a tendon of four inches, then another muscular belly of one and a half inch, which is inserted fleshy on to the occipital below the *complexus.* In some kingfishers (*q.v.*) the two muscles are joined by a tendinous limb.

Complexus.—This muscle arises from the transverse processes of the third and fourth cervical vertebrae, and from the fibres covering the *intertransversarii* of the same. It is inserted, separated from its fellow by a septum, on to the transverse ridge of the occipital. The muscle is entirely fleshy.

Longissimus Dorsi.—It arises by a series of fleshy fibres from the front edge of the ilium, becomes tendinous in the middle, and then is inserted by fleshy fibres on to the lateral surface of the vertebral spine next in front ; the next anterior part arises tendinously from the spinous process of the most posterior uncovered dorsal vertebra, and is inserted on to the vertebra next in front : then follow two of precisely similar relations ; the next is carried on to the dorsal surface of the longissimus dorsi, as also is the last or most anterior portion.

Ilio-costalis.—This complex muscle lies laterally to the foregoing muscle : it is fused at the edge with its fibres. It arises from the ilium and from the transverse process beside the attachment of the rib ; two similar slips in front of this arise from the transverse process and from the adjacent surface of the rib. The ends of the slips are inserted partly on to the surface of the ribs and partly pass on to the lateral musculature of the neck.

Cervicalis Ascendens.—This is the lateral muscle anterior to the ilio-costalis. It consists of five distinct slips arising from the transverse processes of vertebrae XVI.–XI., with the exception of XII. The two posterior are inserted on to the vertebrae next in front ; the next two are inserted on to the surface of the oblique muscles next in front ; the last one on to the oblique muscle next but one in front. Behind these slips, which were obvious, there were indications of additional slips both in front and behind, but these were not sufficiently differentiated from the adjacent muscles for separate description.

Longus Cervicis.—This median muscle arises from the forward continuation of the longissimus dorsi and from the median underlying part of the spinalis complex.

Spinalis Complex.—This system of muscles lies deeper than the foregoing. It is divisible into three parts. Part I. (sometimes called the *spinalis dorsi*) arises apparently only from the longissimus dorsi ; it gives off six fleshy bellies which increase in length from the posterior to the anterior ; they are inserted on to the upper posterior surface of the oblique processes of cervicals X.–XVI. In addition the superior fibres from these heads form a well-marked rounded muscular cord, which runs forward to form the *longus colli posticus.* Part II. consists of only four well-differentiated slender bellies ; these arise from the spinous processes of cervicals XIII.–XV., and they are inserted on to a continuous longitudinal band, the posterior part of which sends slips to the three posterior branches of the *spinalis dorsi*, while the anterior end is inserted on to the oblique processes of cervicals X., XI., at the roots of the anterior two *spinalis dorsi* bellies. Part III. (*longus colli posticus*) arises from the sides of the spinous processes of cervicals II.–XI., and from part I. of the spinalis complex ; it is inserted by digitations which merge with the intervertebral muscles in front of its origins. It has been specially described and figured by GARROD for *Plotus.*

Rectus Capitis Posticus.—It arises from the spinous process of atlas and axis ; its fibres spread out over the occipital under the complexus.

Intertransversales.—These muscles are obvious all the way along from the ilium to the neck, running between the transverse processes of the vertebrae.

Obliqui (Transverso-spinales).—They are clearly differentiated only from the last to the seventh cervical. They are large fleshy digitations arising from the transverse processes, and inserted on to the lateral face of the spinous processes next but one in front.

Rectus Capitis Anticus Major.—It arises all along the neck from the hypapophyses and from fascia; about the middle of the neck it grades into the *longus colli*, from a slip of which it first arises about the level of the seventh vertebra. Its broad fleshy insertion is tendinous on the outside, is fused with its fellow in the middle line, and extends for about a quarter of an inch on the anterior outer edge of the basi-occipital.

Rectus Capitis Anticus Minor.—This is a fleshy broad muscle underlying the preceding. Its origin is fleshy and continuous from first four vertebræ. It has a broad fleshy insertion to the extreme outer posterior face of the ridge behind the meatus auditorius.

Longus Colli.—It arises from the middle of the centrum of the second dorsal vertebra tendinously, and then by a series of tendons from each vertebra up to the overlap of the *rectus capitis*. It is inserted by a series of slips to the vertebræ in front of its origins.

Intertuberculares.—These are a series of short muscles forming the deepest layer of the neck musculature.

Interappendiculares Costarum.—The first arises from the end of the last free rib, and runs backwards and downwards to the lateral anterior process of the sternum; the second from the junction of the sternal and costal parts of the first complete rib; it shortly fuses with the third, which arises from the costal part of the next rib. These two are then inserted together. The fourth arises from the third, fourth, and fifth costal ribs and from the space between them, and is inserted immediately behind the others. The posterior ones are smaller.

Intercostales Externi.—These are confined to the whole of the costal part; the fibres run from above in front and downwards towards the caudal end.

Intercostales Interni.—These are confined to the lower half of the costal ribs, and are chiefly tendinous.

Costi-sternales.—Four slips arising tendinously from the sternal ribs, and inserted fleshy to the sternum.

Costo-sternalis Externus.—This peculiar muscle, apparently found only in Palamedeidæ, replaces physiologically the uncinate processes, as its broad ribbon-like belly runs diagonally across the outer surface of the ribs. It arises by a very thin flat tendon from the third, fourth, and fifth ribs, and from the interspaces between them. It is inserted to the costal edge of the sternum half an inch from the posterior end.

Caudal Muscles

The caudal muscles (in *Palamedea*) are illustrated in the accompanying figure.

Levator Coccygis.—This arises on each side from ilium, from lateral faces of spinous processes, and from transverse processes of caudal vertebræ. It is inserted on to membrane covering rectrices.

Ilio-coccygeus.—On each side there are two parts of this muscle, both entirely fleshy. They arise from the ilium and the ilio-sacral ligament, and are inserted on to outer rectrix.

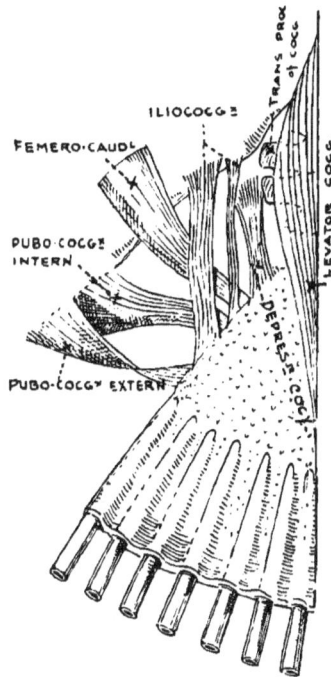

Fig. 60.—CAUDAL MUSCLES OF *Palamedea* (AFTER BEDDARD AND MITCHELL).

Pubo-coccygeus Externus.—This is the most posterior of the muscles of the tail. It arises from the pubis and is inserted on to external rectrix.

Pubo-coccygeus Internus.—Arises in front of and below the last-mentioned. The origin extends also on to the ischium. The muscle is inserted on to last one or two caudal vertebræ.

Depressor Coccygis.—This springs from the transverse process of the last sacral vertebra, and from the adjacent surface of the ilio-sacral ligament. It is inserted on to the transverse processes of the last three or four caudal vertebræ.

Abdominal Muscles

Obliquus Abdominis Externus.—The muscle arises from the ribs and from their uncinate processes. It ends by an aponeurosis upon the pubis.

Obliquus Abdominis Internus.—Lies between the last muscle and the next. It passes from the pubis, extending on to ilium to the last true rib. A separate slip of this is described as the quadratus lumborum, running from the last false rib to the crest of the ilium.

Transversus Abdominis.—This is the deepest of the abdominal muscles. It springs from the pubis and preacetabular ilium, and its aponeurosis ends in that of its fellow in the linea alba.

Rectus Abdominis.—Springs from last sternal rib and from sternum, and is attached to pubis.

Transverso-analis.—This passes across the abdomen in front of the cloacal aperture, and meets its fellow. It arises either from the pelvis or from the transverse processes of certain caudal vertebræ.

Hyoidean Muscles

Our knowledge of the muscles of the hyoidean apparatus and the neighbourhood is chiefly due to GIEBEL,[1] GADOW,[2] and MITCHELL.[3] I mainly follow the latter in his account of these muscles in *Opisthocomus* and in *Palamedea*.

[1] 'Die Zunge der Vögel,' &c., *Zeitschr. f. d. ges. Naturwiss.* xi. 1858, p. 19.
[2] BRONN'S *Thierreich*, 'Aves.'
[3] 'A Contribution to the Anatomy of the Hoatzin (*Opisthocomus cristatus*),' *P. Z. S.* 1816, p. 618; BEDDARD and MITCHELL, 'On the Anatomy of *Palamedea cornuta*,' *P. Z. S.* 1894, p. 536. See also G. L. DUVERNOY, 'Mémoire sur quelques Particularités des Organes de la Déglutition de la Classe des Oiseaux,' &c., *Mém. Soc. Hist. Nat. Strasbourg*, ii. 1835; J. KACZANDER, 'Beiträge zur Entwicklungsgeschichte der Kaumuskulatur.' *Mth. Embr. Inst. Wien*, 1883.

The *mylohyoid anterior* is a sheet of muscles passing across between the rami of the mandible anteriorly.

The *mylohyoid posterior* is composed of two layers, a deeper and a more superficial ; the latter is a broad sheet of muscle nearly reaching the mylohyoid anterior in front and

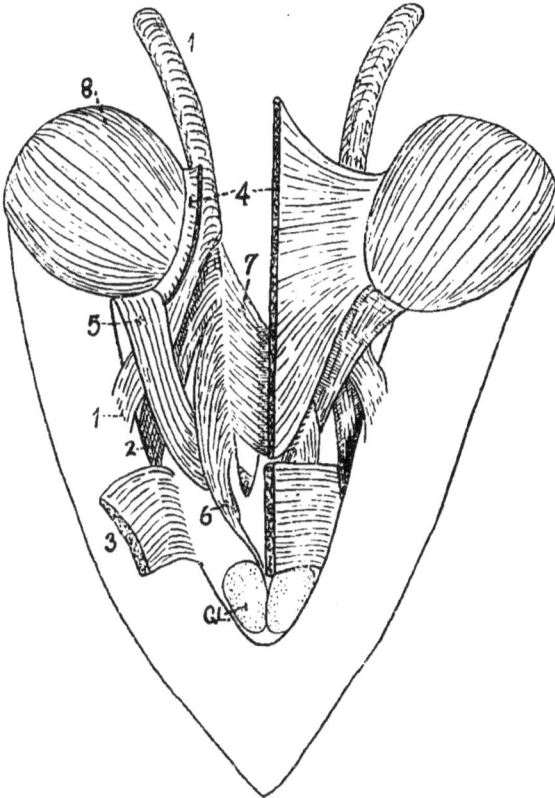

FIG. 61.—HYOIDEAN MUSCLES OF *Opisthocomus* (AFTER MITCHELL).

1, 2, geniohyoid ; 3, 4, 5, mylohyoid ; 6, ceratoglossus ; 7, ceratohyoid ; 8, depressor mandibulæ ; *Gl*, mandibular gland.

covering the space between the rami of the mandible in the region where it is developed ; the deeper layer arises in common with the superficial layer from the rami ; it is a narrower muscle and runs forward.

The *geniohyoid* is in two distinct portions; the posterior division arises on each side from a ramus of the mandible; it passes backwards and is wrapped round the ceratohyoid to its tip. The anterior portion also arises from the rami of the jaw; it is inserted upon the ceratohyal partly under and partly distally to the insertion of the posterior division. This muscle is sometimes a single muscle, as, for example, in *Palamedea*.

The *genioglossus* (entirely absent in *Palamedea*, &c.) is a slight muscle, springs from the middle line near the chin, and passes to the os *entoglossum*.

The *ceratoglossus* is a strong muscle, arises from the outer side of the ceratohyal; it ends in a tendon which is inserted along the side of the tongue almost to the tip. The muscle in *Palamedea*, &c., is divided into two parts.

The *ceratohyoid* arises from the inner side of the cerato-hyal and is inserted on to the urohyal.

The *sternohyoid* (in *Palamedea*) is a band of muscle which arises from the basihyal and entoglossus; it spreads out over the thyroid cartilage and trachea.

The *hypoglossals* are in *Palamedea* indistinguishable from the ceratoglossus.

Muscles of the Head

The *temporalis* (in *Palamedea*) is divided into two portions: the superficial part arises from the whole temporal fossa and from the external and internal surfaces of the midtemporal process; it runs to the outer upper surface of the lower jaw; the deeper part is a pyramidal muscle ending in a stout tendon attached to the lower jaw; in addition to this another portion, deeper still, runs across from the forward process of the quadrate to the inner side of the ramus, and a wide band of muscle bridges the interval between the inner edges of the forward process of the quadrate and the wall of the orbit behind the optic foramen.

The *pterygoid* is divided into several layers which connect the lower jaw with the palatines and pterygoids.

The *depressor mandibulæ* consists of two parts which have been termed digastric and biventer. It connects the under surface of the occiput with the lower jaw.

Osteology [1]

Vertebral Column.—Highly characteristic of birds is the saddle-shaped ('heterocœlous') form of the vertebral centra. The existence of this mode of articulation, though confined to birds, is not found everywhere and in all of the existing members of the order, and is not found at all in some of the extinct forms, *Archæopteryx* and *Ichthyornis*. As to existing birds, the opisthocœlous form is frequently met with, but so irregularly as not to be of much use from a classificatory point of view. The existence of such vertebræ, pointing towards reptiles, may, however, be accepted as some indication of an archaic position in the order. The matter has been lately summed up by the late Professor PARKER,[2] from his own investigations, and from those of others.

The dorsal vertebræ have been found to be opisthocœlous in penguins and auks, in Limicolæ (including Laridæ), but not in petrels; among the parrots PARKER met with this condition in several forms, where it was found to be combined with terminal epiphyses, a character which is just commencing in the lowest mammalia (*Ornithorhynchus*) and

[1] The following are a few works which deal with the general osteology of a number of forms. More special treatises will be referred to in their proper place:—
E. BLANCHARD, 'Recherches sur les Caractères Ostéologiques des Oiseaux,' &c., *Ann. Sci. Nat.* xi. 1860, p. 11; J. F. BRANDT, 'Beiträge zur Kenntniss der Naturgeschichte d. Vögel,' &c., *Mem. Acad. Sci. St. Petersb.* (6). iii. 1840; EYTON, *Osteologia Avium*, London, 1858-81; A. B. MEYER, *Abbildungen von Vogelskeleten*, Dresden, 1879-96; MILNE EDWARDS and GRANDIDIER, *Hist. Nat.*, &c., *de Madagascar*, 'Oiseaux,' Paris, 1879-85; C. L. NITZSCH, *Osteographische Beiträge*, &c., Leipsic, 1811; P. GERVAIS, 'Description Ostéologique de l'Hoazin, du Kamichi,' &c., *Zool. in Voy. de Castelnau*, Paris, 1855. Besides FÜRBRINGER, *Untersuchungen zur Morphologie v. Syst. d. Vögel*, and GADOW, 'Aves,' in Bronn's *Klassen v. Ordnungen des Thierreichs*, for brief and largely osteological definitions of birds see SEEBOHM, *Classification of Birds*, London, 1889, and SHARPE, *Osteological Catalogue of College of Surgeons Museum*, London, 1891.
[2] 'On the Vertebral Chain of Birds,' *Proc. Roy. Soc.* xliii. 1888, p. 465.

entirely exceptional among birds. *Steatornis* is another bird
with opisthocœlous vertebræ. So too are the cormorants and
darters. Another form of vertebral articulation met with
in reptiles—namely, the procœlous articulation—also exists
in birds. In all birds, of course, the atlas is procœlous,
articulating with the convex occipital condyle. In many
birds 'the last two movable joints in the caudal series
become procœlous.'

The biconcave form of vertebræ characterises the extinct
Archæopteryx,[1] and the gull-like *Ichthyornis*, called by its
name on this very account. The concavities, however,
according to FÜRBRINGER, seem rather to have been produced
by the maceration out of a plug of cartilage than to have
characterised the unaltered vertebra. It is doubtful, in fact,
whether the spaces left in the fossil vertebræ were filled
during life with copious remains of the notochord, as in
fishes.

The vertebral column of birds can be distinguished into
four series, as in the higher vertebrates generally. It is
customary to regard as cervical those vertebræ which either
have no movable ribs or, if they have, do not become connected through their intermediary with the sternum. The
rib-bearing vertebræ are the dorsal series, while those which
articulate with the pelvis are usually termed sacral. But
it seems better to reserve the term 'sacral' for the two
vertebræ which, in the chick, bear the ilium.

The number of true sacrals is not, however, always two.
J. J. PARKER[2] describes in the young *Apteryx* three vertebræ,
which abut upon the ilium, and are the only ones in which
there are separate rib-like ossifications at the ends of the
transverse processes. These vertebræ, which are quite conspicuous in the adult (fig. 62), are regarded as the true
sacrals. There are also three in the ostrich (fig. 63). In
other birds (e.g. *Larus, Chionis*) there is apparently only
one sacral vertebra.

Behind the sacrum are the caudal vertebræ. *Archæopteryx* is unique among birds for its long tail, composed of

<hr />

[1] Not certainly.   [2] 'Development of *Apteryx*,' *Phil. Trans.* 1891.

separate vertebræ. In all other birds the tail is short and does
not extend far beyond the sacrum. In the majority of carinate
birds the terminal vertebræ are fused together into the highly
characteristic ploughshare bone (urostyle or pygostyle).[1]

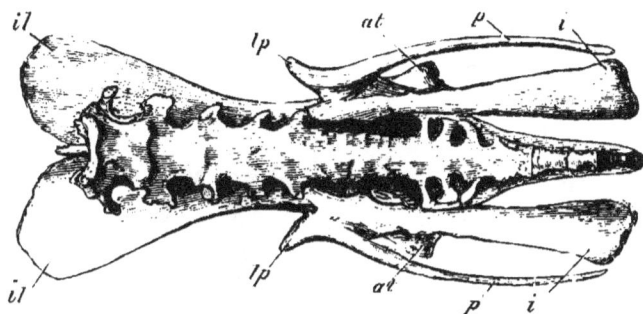

FIG. 62.—PELVIS OF APTERYX. FROM BENEATH. (AFTER MIVART.)
il, ilium ; *p*, pubis ; *i*, ischium ; *lp*, prepubic process.

There is, however, a closer correspondence between
the tail of *Archæopteryx* and that of the carinate bird

FIG. 63.—LUMBAR AND SACRAL VERTEBRÆ OF AN IMMATURE OSTRICH
(AFTER MIVART).
8, 9, 10, sacral vertebræ : *p*, parapophyses ; *d*, diapophyses.

than might be assumed from the last-mentioned differences. The first four caudal vertebræ of *Archæopteryx* have
strong transverse processes, which are weaker, but present,
on the fifth, which thus affords a transition to the remaining
sixteen, upon which there are no such processes. In the

---

[1] W. MARSHALL, 'Untersuchungen über den Vogelschwanz.' *Ned. Arch. f.
Zool.* i. 1873, p. 194.

same way the free caudals of carinate birds have transverse
processes, which are at most faintly represented upon the

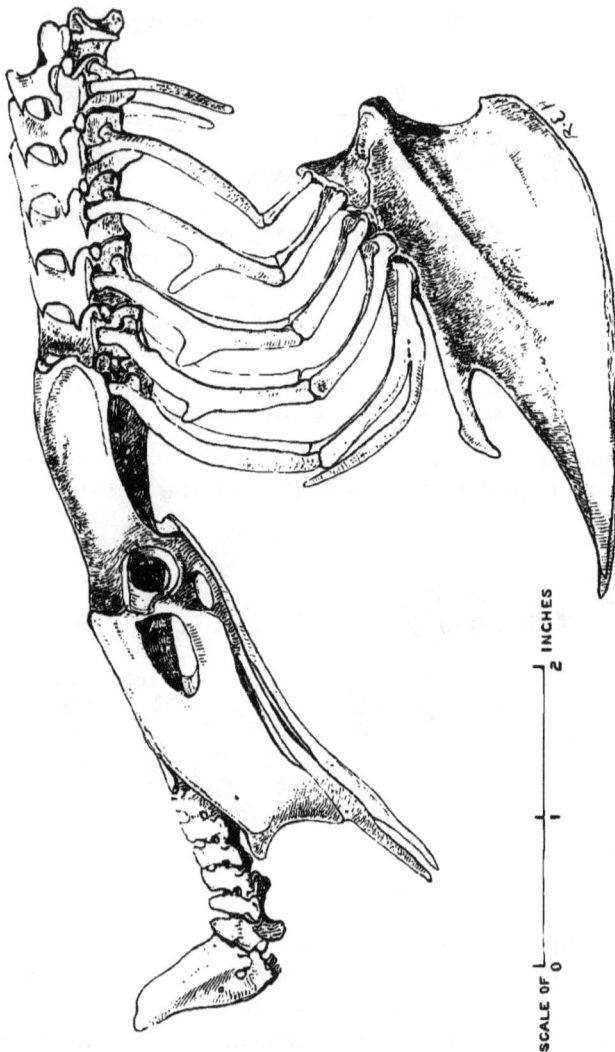

Fig. 61.—Ribs, Sternum, and Pelvis of *Chunga Burmeisteri* (after Beddard). The Upturned Pygostyle in its Natural Relations is shown.

fused posterior set of caudals which form the pygostyle.
Four of the posterior caudals of *Archæopteryx* have fine

splints of bone lying on one side, which have been compared to the ossifications in tendons found among the ptero- dactyles. They may conceivably be misplaced chevron bones.

The pygostyle varies much in the degree of its develop- ment. It is weakest in various aquatic birds, such as the auks and grebes, where it is thin and narrow; in more purely flying birds it is very thick at the base, and is turned upwards instead of, as in the auks, carrying on the line of the tail. In the grebe there is really no more definite a ploughshare bone than in the ostrich. The number of vertebræ which are fused together to form the urostyle varies. In the ostrich MARSHALL finds four, five in the grebe and hornbill, six in the duck and in *Eurylæmus*.

The total number of vertebræ [1] in the column varies greatly; the extremes are something like thirty-nine and sixty- four (reckoning the urostyle as one). The greatest number characterises the ratites, and the smallest some of the higher arboreal birds. *Archæopteryx* had only about fifty vertebræ. While, therefore, it may be generally true to put down as older types those with the largest number of vertebræ, it is evident that on this view *Archæopteryx* must be regarded as a parallel branch to the existing birds, and not as their ancestor. The number of vertebræ, though it may perhaps be considered from this general point of view, is not of the faintest use for the systematic arrangement of existing forms. The number varies so extremely that among the Gruidæ Professor PARKER found no two alike. Rather more fixed, but still subject to variation among the species of a genus, are the cervical vertebræ; and some account will be taken in the pages which follow of this fixedness. The results must, however, be tempered by the reflection that while the common swan has twenty-five the black-necked swan has twenty-four.

Between the successive centra are the 'intervertebral

[1] GIEBEL (' Die Wirbelzahlen am Vogelskelet,' *Zeitschr. f. d. ges. Nat.* xviii. 1866, p. 20) gives a long list; see also 'Der letzte Schwanzwirbel des Vogelskeletes,' *ibid.* vi. 1855.

discs,' the 'intercentra' or 'basiventral' elements. These are, as has been shown,[1] originally the portions of each vertebra with which the ribs articulate, from which they are outgrowths. But as the ribs come to articulate with the centra these structures degenerate. In the development of *Apteryx* T. J. PARKER found a postoccipital and a post-atlantal intercentrum, and two in the caudal region, which ossify so as to retain their independence in the adult skeleton.

Intercentra in the caudal region of the bird's vertebral column are by no means so rare as might be inferred from some published statements upon the matter. They are especially conspicuous among the Limicolæ and the nearly allied auks, and in most water birds. In *Numenius femoralis*, for example, there are three small osseous nodules lying between caudal vertebræ 1–5. Behind these are a series of hypapophyses, which are a continuation of the same series, but much more pronounced and ankylosed to the vertebræ.

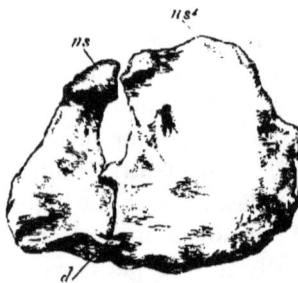

FIG. 65.—LAST TWO VERTEBRÆ OF STRUTHIO (AFTER MIVART).
ns, neural spines ; *d*, osseous bridge.

They exist also in the duck tribe. In *Biziura lobata* there are three distinct intercentra in the form of largish nodules. I have found intercentra also in **Palamedeæ, Tubinares, Steganopodes, Colymbi, Herodiones**, *Opisthocomus*.

These free intercentra are rare among the Pico-passeres, but in a few of them are present. Thus in *Tccus* there is a distinct intercentrum lying between the last free caudal vertebræ.

The hawk tribe have not these bonelets as distinct structures.

In the cuckoos, parrots, **Ralli, Otides, Columbæ**, and the

---

[1] This matter of the composition of the vertebra has been recently gone into by GADOW (on the 'Evolution of the Vertebral Column of Amphibia and Amniota,' *Phil. Trans.* 1896, p. 1), who quotes previous literature.

tinamous I have not seen in the adult skeleton any free
intercentra; nor in the **Grues**, excepting *Chunga*.

Further details on this matter will be found in the paper
cited below.[1]

Though free intercentra are by no means universal
among recent birds, hypophyses of the last caudals are
almost so. That these latter are derived from intercentra,
and are, therefore, not comparable to the hypapophyses of the
cervicals and dorsals, is clear from such cases where the gradual transition between free intercentra and fixed hypapophyses is shown. In reptiles the intercentra are in the tail region
constantly in the form of chevron bones, which are **V**-shaped,
articulating with the vertebral column by the free ends of
the **V**. This form of the hypophyses of the caudal region is
not so common as a simply bifid condition, but does occasionally occur. I have seen it, for example, in **Tubinares**,
Accipitres, and **Cuculi**.

The first vertebra of the cervical series is called the
atlas (see fig. 66) ; it is a ring on bone, of which the greater
part of the 'centrum' is formed by the projecting odontoid
process (see fig. 67), the rest being formed by a pair of
intercentra. In the hornbills the atlas is fused with the
following axis vertebra.[2] Generally the atlas has not what
the succeeding vertebræ have, a vertebrarterial canal, but
this is sometimes present (see under 'Ribs,' p. 119). The
odontoid process sometimes notches the lower part of the
atlas, and sometimes perforates it. These two conditions
are illustrated by figs. 66 and 68. It sometimes happens that
the neural arch of the atlas is incomplete, e.g. *Chunga*,
Colius, Pandion. As a rule it is perfect.

In the cervical vertebræ the chief facts which appear to
be of systematic importance are the relations to each other
of the paired processes, to which MIVART has applied the
name of *catapophyses*. These are sometimes inconspicuous processes of the transverse processes on the under
side. Very often the last one or two pairs of them closely

[1] BEDDARD, 'Note upon Intercentra,' &c., *P. Z. S.* 1897, p. 465.

[2] In a specimen of *Chunga* I have found the same fusion.

approach each other in the middle line, as in *Psophia*.
Sometimes a number of these processes unite to form a
canal ; this occurs in the **Steganopodes**, and most **Herodiones**
(but not in *Scopus umbretta*) ; but the classificatory significance of the fact is marred by the occurrence of a similar
canal similarly formed in some Picidæ, and in the case of
one vertebra in the parrot, *Eclectus polychlorus*.

It is sometimes the case that the last of the catapophyses
is consolidated into a thick process, which is bifid at the
extremity ; this process forms a transition to the following
hæmapophyses (or hypapophyses).   These latter are un-

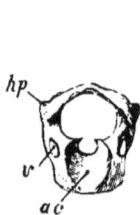

Fig. 66.—Atlas of
Emu   (after
Mivart).

ac, articular surface ;
v,   vertebrarterial
canal ; *hp*, hyper-
apophyses.

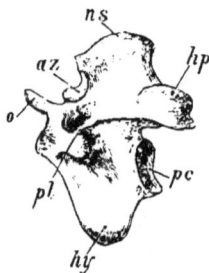

Fig. 67.—Axis of Emu (after
Mivart).

o, odontoid process ; *ns*, neural spine ;
az, anterior zygapophyses ;   *pl*,
pleural lamella ; *pc*, articular surface ; *hy*, hypapophysis ; *hp*, hyper-
apophysis.

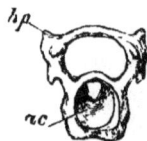

Fig. 68.—Atlas of
Cassowary (after
Mivart). Letters
as in Fig. 66.

paired median processes, which commence upon the cervical
vertebræ, and extend for a variable distance back along the
dorsal vertebræ. They are very feeble, and sometimes
limited to the cervical region, in the **Herodiones**.   They are
most highly developed in **Sphenisci, Colymbi, Alcæ**, and some
Anseres, being in these cases continued to the end of the
dorsal series, and even being found upon some of the lumbar
vertebræ.   In many cases these processes are flattened out
at the free end like an inverted **T**, or are trifid at the same
place.   This is seen to be due to the gradual shifting in
position of a posterior set of catapophyses, which at first are
at the sides of, and far from, the hæmapophyses, but get

nearer and nearer, until at length they mount upon the
hæmapophysis itself and pass to its very end. Details of the
formation of the vertebræ will be found in the systematic
part of this book.[1]

Ribs.[2]—The ribs of birds vary greatly in number. There
are as a rule three series of ribs to be distinguished. The
last cervical vertebræ, more or fewer of them, are furnished
with short ribs which do not reach the sternum. Behind
these are, again, a variable number of true ribs, which do
reach and articulate with the sternum. These true ribs
consist of the vertebral portion, which articulates with the
vertebra, and of a sternal portion, which is articulated with
the vertebral half of the rib above and with the sternum
below; it is bent at an angle with the vertebral portion.
Attached to, originally separate from, and sometimes per-
manently separate from, the vertebral half of the rib is the
uncinate process, of which there are a variable number.
These processes are absent[3] in *Archæopteryx* and in the
Palamedeæ only. Behind the true ribs, which articulate
with the sternum, are a variable number, in all degrees of

[1] The relationship of the so-called catapophyses to the unpaired hæmapophyses varies, and suggests—what has been advanced on other grounds—an
occasional excalation of vertebræ. Without wishing to commit myself to a
belief in the actual dropping out of a vertebra from the middle of the series, I
may mention some of the facts which may be regarded as pointing in this
direction. In the grebe *Æchmophorus* the catapophyses form on certain
vertebræ a complete ventral canal for the carotids. The summit of the arch
thus formed gradually acquires a median dorsal process. This increases, and
the catapophyses finally end in the obliteration of the canal which they sur
round, and a solid arch is formed; the hypapophysis of the succeeding vertebra
is single and no longer retains traces of its evolution from a ring of bone sur-
mounted by a process. In other cases the catapophyses suddenly end and the
hypapophyses begin without such intermediate stages. An intermediate stage
is seen in certain types where the catapophyses end suddenly, but the first
hypapophysis is double, either formed of two clearly fused pieces or with merely
a bifid spine. These latter cases suggest the dropping out of one or more
vertebræ, effecting the transition between the paired catapophyses and the unpaired hypapophyses.

[2] In all birds except *Archæopteryx* the ribs are two-headed with a capitulum
and tuberculum.

[3] They have been often said to be absent in *Dinornis*, but they are not.

W. BEHRENS, *Untersuchungen über den Processus uncinatus der Vögel und
Crocodile*, Inaug. Diss., Göttingen, 1880.

degeneration, of floating lumbar ribs. Morphologically equivalent with ribs are the processes firmly ankylosed to the cervical vertebræ, which form a canal for the vertebral artery ; these are, as a rule, absent from the atlas, but are present on that vertebra in the **Anseres**, *Opisthocomus*,[1] *Triponax Feddeni, Dromæus*[2] (fig. 66). *Archæopteryx* is alone in possessing the abdominal ribs of the crocodiles and other reptiles.

The Shoulder Girdle.[3]—The shoulder girdle of birds consists of at any rate three separate elements—the *scapula* above ; the *coracoid*, articulating with the sternum ; and the *clavicles*, generally united into a **U**-shaped piece. Of these the first two are preformed in cartilage, the last in membrane.

The *scapula* is a thinnish sword-shaped bone which is attached by muscles to the ribs and to the vertebræ, and lies in a direction, as a rule, nearly parallel to the long axis of the body. The scapula does not show great variability of form among birds ; the most considerable variation is to be seen in the penguins, where the bone is, comparatively speaking, of enormous width. A free supra-scapula has been noted by PARKER in *Opisthocomus*. The *coracoids* articulate on the one hand with the scapula, and on the other with the sternum, where they are received into grooves on its anterior margin. There are some variations in the way in which these grooves are arranged : in some birds the two coracoids at their insertion are not in contact at all ; in others they are in contact ; and finally they may overlap, as in reptiles.

The coracoid has in many birds a *procoracoid process*, which is believed to be the equivalent of the *procoracoid* of reptiles. This is especially prominent in the ostrich, but is present in a large number of other birds, though more reduced in extent. But its large or small size is so capricious

---

Stated by PARKER to be absent.

[2] MIVART figures a canal on one side of the atlas of the ostrich.

[3] A. SABATIER, *Comparaison des Ceintures et des Membres Antérieurs et Postérieurs dans la Série des Vertébres*, Montpellier, 1880 ; PARKER, quoted below. See also LÜDER in *J. f. O.* 1871, p. 321.

in its relations to other structural similarities and dissimilarities that the fact is not of great use in classification.

The same remark may be made about the *foramen coracoideum* perhaps, which, again, is found in many birds and absent from others. As to the morphological significance of this foramen, which transmits a nerve twig to the pectoralis secundus, it may perhaps be regarded as the boundary between the coracoid and the procoracoid.

In the course of the development of the common fowl, according to Miss LINDSAY, whose figures are here reproduced, there is a very considerable trace of the procoracoid. The three elements of the shoulder girdle are perfectly distinct from each other in the young embryo, but become fused (the scapula and the coracoid), again to get separate in the older chick. This temporary fusion may be significant of the struthionic condition to be described later. The intermediate piece is, it will be noticed, triangular in form, the elongated aspect of the adult coracoid being acquired later. Miss LINDSAY is of opinion that this change of form is to be correlated with the disappearance of the anterior section of the bone, as indicated in the accompanying diagram, the disappearing (shaded) part being the equivalent of the procoracoid.

The interrelationship of the scapula to the coracoid offers facts of some importance. In the ostrich tribe the two bones are firmly ankylosed; this is not the case with the young, but it is plainly the case with the adult. In carinate birds, on the other hand, there is not ankylosis, but a close union by means of fibro-cartilage. It appears, however, that in *Didus* (exceptionally ?) there is an actual synostosis, which of course bears out the suggestion that the synostosis of the ratite birds has something to do with their loss of the power of flight. In the ratite birds and in *Hesperornis* the scapula and the coracoid are nearly in the same straight line, the angle in *Apteryx* varying from 150 to 122°, whereas in the carinates the two bones are at right angles or at an even acute angle. That this is not a morphological distinction, but is distinctly related to the development of the shoulder muscles, has been clearly pointed out by T. J. PARKER.[1] He discovered an approximation to the struthious condition in several of the flightless rails and other birds. But the state of affairs which characterises the **Tubinares** warns us against placing too much reliance upon this apparently sound generalisation ; for in them we are informed by FORBES that ' the angle it ' (the scapula) ' forms with the coracoid varies much in different genera, being most acute in *Pelecanoides*, whilst in the Oceanitidæ it is hardly if at all less than a right

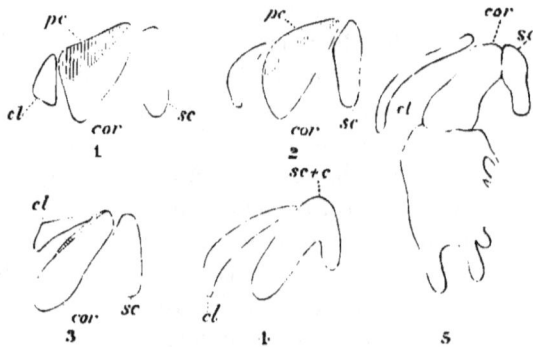

FIG. 69.- DEVELOPMENT OF SHOULDER GIRDLE OF CHICK (AFTER LINDSAY).
cl, clavicle ; *pc*, procoracoid ; *cor*, coracoid ; *sc*, scapula.   1 3, fifth day ; 4, sixth day :
5, late on sixth day.

angle.' The widest angle in a carinate bird is 106°, so there is a difference of only 16° between extremes of carinates and ratites.

The two *clavicles*[2] sometimes spoken of collectively as the *furcula* vary much in their degree of development. They are totally absent in the *Apteryx*. In the emu and in certain parrots they are distinct and smallish bones which do not come into contact with each other ; but in the majority of birds they form a single **U**- or **V**-shaped bone.

The furcula varies in the expansion or non-expansion of the base to form a circular hypocleidium. In some birds the

[1] ' On *Notornis*,' in *Tr. N. Zealand Inst.* xiv. 1882.

[2] A. WEITZEL, ' Die Furcula : ein Beitrag zur Osteologie der Vögel,' *Zeitschr. f. d. ges. Naturw.* xxv. 1865, p. 317.

furcula is connected with a ligament (e.g. *Psophia*) by articulation (*e.g.* **Herodiones**) or by direct synostosis with the carina sterni. In the gallinaceous birds the furcula does not come into near relations with the carina, and in *Opisthocomus* - quite exceptionally--the bone is ankylosed with the spina sterni on the one hand, and with the coracoids, so firmly that 'no trace of the primitive distinctness of the bones is discernible.' There are also considerable variations in the degree of the connection between the furcula and the coracoid and scapula.

In a few birds the ends of the clavicles where they articulate, or at least are connected, with the coracoid and scapula have a process, the acrocoracoid process of the clavicle. This is seen, for instance, in the **Anseres**, and the rudiment of such a process in the flamingo appears at first sight to be a duck-like character in that bird. But the same process is also developed, and to a great extent, in the Alcedinidæ, a fact which must be borne in mind before coming to any such conclusions.

Fürbringer has devoted some space to describing and illustrating the relations at their articulations between the clavicle, coracoid, and acrocoracoid. The two extremes may be seen in *Phalacrocorax* and *Psophia*; in the former the clavicle articulates with the acrocoracoid only, and does not reach the scapula ; in the latter, where the procoracoid is well developed, the clavicle comes into contact with all three.[1] Further details will be found under the description of the different groups.

The Fore Limb.—The fore limb is present in all birds except most Dinornithidæ, where up to the present no trace of one has been discovered. In *Hesperornis* only the humerus appears to exist ; in *Apteryx*, *Dromæus*, and *Casuarius* there is but one finger. With these exceptions the wing of birds consists of a humerus, radius and ulna, carpus, metacarpals, and three fingers (with sometimes a rudiment of a fourth) ; even *Archæopteryx* has not been definitely shown to possess

[1] A. Tschan, *Recherches sur l'Extrémité Antérieure des Oiseaux et des Reptiles*, Diss. Inaug., Geneva, 1889.

more than the typical three fingers (see, however, below). The relative length of the arm varies much in birds; it is longest in the flying gulls, terns, &c., whence the name applied to the former of Longipennes. In the struthious birds it is the shortest, and in many running birds the wing is reduced in length. There is too variation in the relative lengths of the humerus, fore arm, and hand. In the divers, for instance, the upper section of the arm is the longest, in the gulls the fore arm, and in the penguins the hand. The length of the hand in the **Macrochires** is so great that it equals that of the humerus and fore arm together. The exact reverse is seen in the ratites, where the length of the humerus is greater than that of the rest of the wing. PYE-CRAFT has brought out the interesting fact that during the growth of *Opisthocomus* the proportions of the different sections of the wing alter.

A study of the relative lengths of the different parts of the arm shows that a reduction of the wing, and a consequent decay of its powers as an organ of flight, do not invariably follow the same path. In the ostrich the middle segment is the shortest, in the cassowary the hand.

The length of the *humerus*, the exact form and degree of development of the deltoid ridge, and of the tubercles for the insertion of muscles, furnish systematists with reliable points for the identification of genera and species. So much of our knowledge of extinct birds depends upon fragments of this and others of the 'long' bones that the value of slight characters of this description has been thoroughly appraised. A glance, for instance, at LYDEKKER's recently published 'Catalogue of the Fossil Birds in the British Museum' will reveal the importance of the power of discriminating species by such slight indications, which furnish the student of affinities between families or genera with nothing tangible.

The *radius* and the *ulna* are always separate bones, of which the ulna is the longer; it is frequently marked on its outer surface with tubercles, to which the quill feathers are attached. The most striking modification of the radius is seen in the Parridæ (fig. 70, p. 125), where it is prolonged on the outer side into a strong flat process, the upper surface of which is slightly grooved for the reception of the tendon of the extensor metacarpi radialis muscle.

The *carpus* of recent birds consists only of two separate bones. But in the embryo there are six separate cartilages. The two bones which persist are looked upon by ZEHNTNER as ulnare + intermedium and radiale + centrale. The three distal carpals, according to PARKER, fuse with their three metacarpals. In the emu, according to PARKER, there are no carpal elements either in the young or adult, in *Casuarius galeatus* there is a small ulnare.

Though no bird has more than three more or less com-

FIG. 70.—RADIUS AND ULNA OF *Metopidius* (AFTER FORBES).

plete digits, there is commonly a trace of a fourth metacarpal, found by ZEHNTNER in the development of the swift, and by STUDER in the development of the penguin. PARKER was convinced of the existence of a prepollex and of intercalary digits, but WIEDERSHEIM looks with no favour upon this broadening of the hand.

The *metacarpals* are free only in *Archæopteryx* and *Gastornis*; in all other birds they are partially fused. The formula for the *phalanges* in *Archæopteryx* is I. 2, II. 3, III. 4. In some other birds it is I. 1, II. 2, III. 1, with the exception of the ostrich,[1] *Numenius*, and the embryo duck,[2] where digit III. has a small additional phalanx. In other birds the formula is I. 2, II. 3, III. 1, and in the apteryx digit II. shows three phalanges in the course of its develop-

---

[1] PARKER, 'On the Structure and Development of the Wing in the Common Fowl,' *Phil. Trans.* 1888, p. 385, where previous literature is quoted.

[2] BAUR, *Science*, vol. v. p. 355.

ment, of which one (apparently the middle one) is not recognisable in the adult.

LEIGHTON has contributed to this question with a study of the development of the wing of *Sterna*.[1] He finds here too a rudimentary fourth digit, which in the first stage, which he figures, is as long or nearly as long as the first digit. A rudimentary metacarpal even persists attached to the side of the last metacarpal in birds just before hatching. In the carpus there are never more than four distinct cartilages ; there are in the first place a radiale and ulnare,

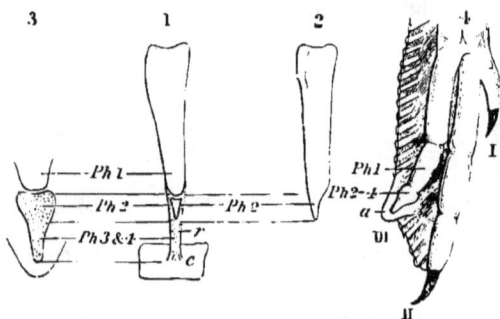

FIG. 71. — DIGITS OF OSTRICH (AFTER WRAY).

1, phalanx 1 (*Ph*1) and rudimentary phalanx 2 (*Ph*2) of digit III.; *c*, connective tissue. 2, another specimen with phalanges ankylosed. 3, distal part of digit III. of embryonic manus (4).

and distally two cartilages, of which one appears to represent the combined distalia of the two first digits, and the second that of the third digit. In birds, just before hatching, all the distalia have fused into one mass. The cartilages lettered respectively radiale and ulnare in the figures are thought, however, by the author to be really radiale + intermedium and ulnare + centrale ; and in support of this view is the partial separation between the two supposed elements of each, which is, however, never carried very far.

As to the homologies of the digits in the adult with those

---

[1] 'The Development of the Wing of *Sterna Wilsonii*,' *Tufts Coll. Studies*, 1891. Previous literature is here quoted.

of the reptilian hand, LEIGHTON leans to the view that the supposed pollex is really the index. In putting forward this opinion he rests first of all upon the fact that the radial artery is absent, thus indicating a reduction of the radial side of the hand; the second argument is derived from the fact that in animals with a reduced manus the first digit is the first to go, and then is followed by the last; thus in *Orohippus* there are four digits, the first having disappeared, while in *Protohippus* the fifth has vanished. In addition this view is moreover strengthened by a consideration of the most reduced manus that occurs in birds; in *Apteryx* and *Casuarius* the reduction has similarly occurred on both sides of the large persisting digit, which is thus to be regarded as No. III.

Sternum.—The sternum in its most complicated condition consists of the following regions (see fig. 72): Anteriorly it ends in a moderately narrow extremity which is known as the manubrium sterni or *rostrum*. On either side of this is a forwardly directed process, the costal processes or *anterior lateral processes*. In the middle of the sternum, and forming the great projecting keel, is the lophosteon or carina sterni, or *keel*. The sternum ends in a median process behind (sometimes, but wrongly, called the xiphoid process), to which are appended two processes on each side, which may be termed middle and external xiphoid processes, or these may be termed, for reasons which will appear later, the *posterior lateral process* and the *accessory process*. The nomenclature first used in the preceding brief descriptions is that of HUXLEY; the second set of terms which will be used throughout in the descriptions which follow are those used by Miss LINDSAY in her paper upon the development of the avian sternum.[1]

The sternum is subject to much modification among birds, of which the principal varieties will be now described. The birds which show perhaps the greatest difference from the gallinaceous type, selected for the above description, are the ratite birds. In them there is no keel developed, hence

[1] 'On the Avian Sternum,' *P. Z. S.* 1885, p. 684.

the name ratite (raft-like), or at most, as in *Rhea*, a slight protuberance, which, however, as will be pointed out immediately, is not really comparable to the keel of the carinate birds. There are, however, other birds, such as the extinct *Cnemiornis* and the living *Stringops*, in which the keel is absent, its absence being associated with the loss of the

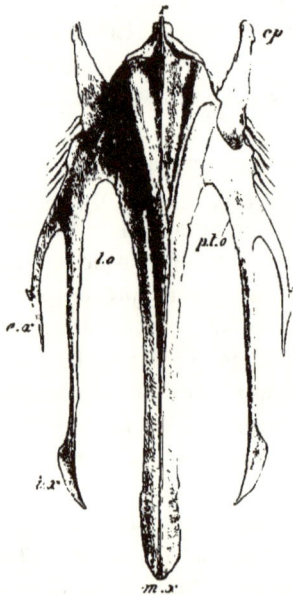

FIG. 72. STERNUM OF *Lophophorus impeyanus* (AFTER HUXLEY).

r, rostrum ; *cp*, anterior lateral process ; *p.l.o*, posterior lateral process ; *e.r, i.r*, its inner and outer divisions ; *l.o*, carina.

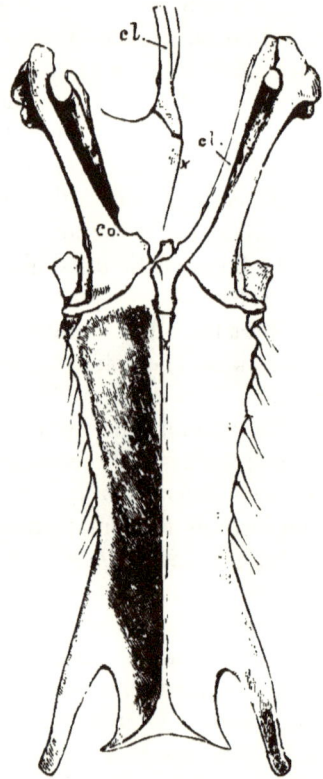

FIG. 73.—STERNUM OF *Podica senegalensis* (AFTER BEDDARD).

cl, clavicle ; *co*, coracoid ; *cl, x*, articulation of clavicle.

power of flight. In the singular *Opisthocomus* the anterior part of the keel is, as it were, cut away (the enormous crop resting here), the posterior region being retained. The four posterior lateral processes of the sternum figured above are not always present in birds. The extremest modification is as seen in the goose and the crane, where the posterior

margin of the bone is entire, without any processes at all.
In passerine birds generally, and in some others also, there
is but a single pair of these processes ; while, finally, by
excessive growth of the parts concerned the processes have
joined and converted the notches into foramina. The converse course of events has been suggested—*i.e.* that deficient
ossification leads to the fenestrated condition, whence to
the posterior notches is an easy step. Development, however, shows that the former view is the more correct. The
diversities in the form of the sternum undoubtedly must

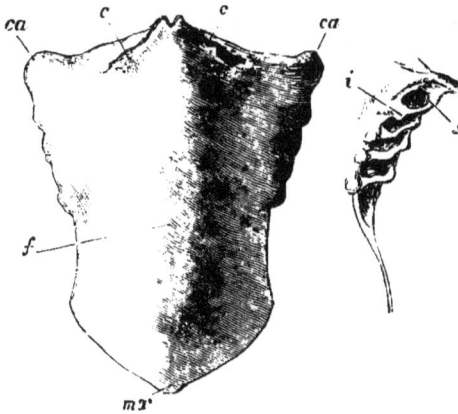

Fig. 74.--Sternum of Emu (after Mivart). ½ Natural Size.
ca, anterior lateral process ; *c*, grooves for coracoids ; *f*, elevation in centre ; *m.r*, posterior
end ; *is*, lateral view showing articulation of ribs.

have some relation to the muscles which are inserted on to
and take their rise from the margins of the bone. Thus, as
already mentioned, the flat sternum of the Ratitæ is associated
with the slight development of the pectorales muscles and
the consequent loss of capacity for flight. It has been
ingeniously suggested that the relative development of the
posterior lateral processes of the sternum has possibly an
analogous explanation. The muscles that are attached
thereto are mainly the pectorals and the abdominals. Now
the pull of these two is in an opposite direction. The
tendency of the action of the pectorals would be to

K

straighten the posterior margin of the sternum, while
that of the abdominals would be to pull it out, perhaps
irregularly.   Hopping and walking birds might therefore
be expected to have a more notched sternum than purely
flying birds ; that there is some relation of this kind seems
possible when we contrast the sternum of the running
gallinaceous bird with that of the essentially aerial eagle.
Moreover, since the pull of the abdominal muscles is in two
directions, one antero-posterior (recti), the other oblique (the
obliqui), we might expect to find what we actually do find, a
direction of the xiphoid processes which corresponds with
the resultant of these two forces, as is indicated in the
annexed diagram.   There are a few other modifications in
the shape of the sternum which have been made use of for
systematic purposes, besides the keel and the notches, or
excavations of the posterior border.   The rostrum of the
bone is sometimes very pronounced, and sometimes practically absent altogether.   According to its position, more
dorsally or more ventrally, the process has been called by
FÜRBRINGER spina externa, or spina interna sterni.   In
the gallinaceous birds the two are combined in a vertically
compressed plate of bone which arises both from the lower
and from the upper side of the sternum.   In the passerines
and in the todies, and a few of the allies of these groups of
birds, the anterior process of the sternum is more or less
distinctly bifurcate.

The sternum of birds arises, as does that of other vertebrates, in the first place between the ends of the ribs which
fuse together.   Birds invariably have a few ' floating ' ribs
at both ends of the sternum which are no longer connected
with it, this connection being often lost ontogenetically.
There is, in fact, usually a shortening of the sternum during
development.   The keel arises from the conjoined edges of
the two sets of fused ribs ; it is not preformed separately as
a median piece.   This seems to settle in the negative an
earlier view that the carina sterni was the surviving representative of the interclavicle of the reptiles, a view which
commended itself to more than one anatomist of distinction,

and appeared to be strengthened by the occasional connection by ligament and bone with the hypocleidium.

The most recent modification of this view is put forward by PARKER, who has shown in *Opisthocomus* a needle-shaped splint of bone lying upon the keel, and therefore independent of it.

The development of the sternum throws a light upon the homologies of its different parts in different birds, and in other vertebrates.

It is plain in the first place that the spina externa and the spina interna have nothing whatever to do with the manubrium sterni of the mammal; for they are (in the bird) secondary outgrowths, and not, as in the mammal, part of the primitive sternum formed by concrescence of the ribs. The same holds good of the posterior median region of the bone, which is a secondary outgrowth, and can therefore have no relations with the xiphoid process of the mammalian sternum; what does correspond to the latter are the posterior lateral processes of the avian sternum.[1]

Pelvis.[2]—The pelvis consists of three pairs of bones, the ilium, ischium, and pubis. In the young embryo these bones form a continuous sheet of cartilage, but are all separate distally; the ilium is directed in an antero-posterior

---

[1] The literature of the sternum is large, and is to a considerable extent to be found under the several groups. Memoirs of a wider scope are W. K. PARKER, ' A Monograph on the Structure and Development of the Shoulder Girdle and Sternum in the Vertebrata,' *Ray Soc. Publications*, 1868; L'HERMINIER, ' Recherches sur la Marche d'Ossifications,' &c., *Mém. Ac. Sci.* 1830. The history of the development of knowledge concerning the ossification of the sternum and the classificatory results therefrom is treated by NEWTON in *Dict. Birds*, ' Introduction.' Miss LINDSAY's paper, already quoted, contains references to the chief memoirs upon the subject. See also R. DIECK, *De Sterno Avium*, Diss. Inaug., Halæ, 1867.

[2] C. GEGENBAUR, ' Beiträge z. Kenntniss des Beckens der Vögel,' *Jen. Zeitschr.* vi. p. 157; MEHNERT, ' Untersuchungen über die Entwicklung des Os pelvis d. Vögel,' *Morph. J.B.* xiii. 1888, p. 259; A. JOHNSON, ' On the Development of the Pelvis Girdle, &c., in the Chick,' *Quart. J. Micr. Sci.* 1883, p. 399; B. HALL, *Jemförande Studier öfver Foglarnes Bäcken*, Lund, 1887, and ' Morphologisk Byggnoden af Ilium,' &c., *Act. Lund. Univ.* xxii. 1887, p. 1 G. BAUR, ' Bemerkungen über das Becken d. Vögel v. Dinosaurier,' *Morph. J.B.* x. 1885, p. 613; A. BUNGE, *Untersuchungen zur Entwicklungsgeschichte des Beckengürtels*, &c., Diss. Inaug., Dorpat, 1880.

direction; the ischium and the pubis look downwards and slightly backwards; at the end of the pubis, near to where it comes into contact with the iliac portion of the cartilage, is a forwardly directed process, the prepubic process. This primitive state of affairs has been most nearly preserved in *Apteryx* and *Dinornis*; in these birds the pubis and ischium are free from each other distally and from the ilium; their direction is, however, more backwards than in the embryo, and the prepubic process is relatively smaller. In all other birds the pubis and the ischium lie in a line more parallel

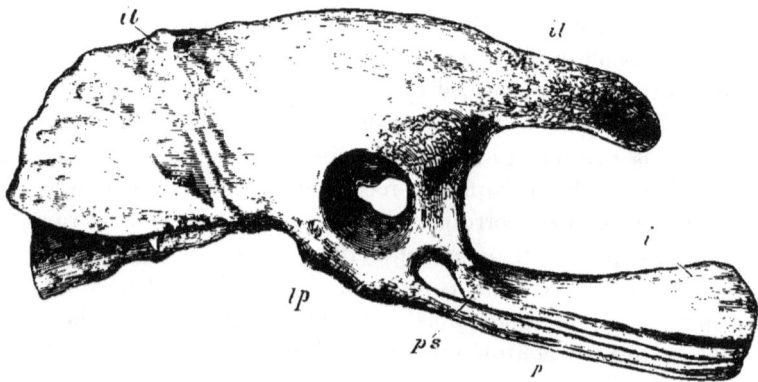

FIG. 75.--PELVIS OF DINORNIS (AFTER MIVART). ¼ NATURAL SIZE.
il, ilium; *p*, pubis; *ps*, interobturator process; *lp*, pectineal process.

with the ilium, and there is a greater or less connection between the several bones. This is seen in a less developed condition in the struthious birds and tinamous than in any others. In the tinamous, in fact, the pubis and ischium are quite free from each other distally, and from the ilium. In the ostrich the pubes unite in a ventral symphysis; in *Rhea* there is a remarkable modification induced by the meeting of the ischia. The details of the struthious pelvis will be found described under the description of that group. In carinate birds the ischium is for the greater part of its extent fused with the ilium, a foramen only—the ischiadic foramen —being left anteriorly. The pubes join to a less extent with the ischia, and are sometimes (e.g. *Colymbus*) quite free from them. The prepubic or pectineal process is large in the struthious birds, in the tinamous, and in a few carinates, such as *Geococcyx*. It has received much attention as the possible equivalent of the reptilian pubis, the bird's so-called pubis being in that case the homologue of the posterior pubic process of the reptile. The dinosaurians seemed at one time to have been the means of solving the questions involved; for in some of them there is a backwardly directed pubis, not quite so bent as in the bird, from the anterior and upper end of which a stout bone, considered to be the homologue of the pectineal process, is directed forwards. This latter, however, is clearly a part of the pubis, while the pectineal process is at least not always a product of the pubis, being sometimes purely iliac in origin, sometimes partly pubic and partly iliac. It may be that the missing prepubic process of the dinosaurian pelvis is represented by the remarkable bones ankylosed to the ostrich's pelvis anteriorly, and continued forwards by a cartilaginous tract, which were discovered by DARWIN and GARROD. This will reduce the pectineal process to the level of a mere projection of the pelvis of no particular significance save as a secure hold for the important ambiens muscle which is there attached. In any case it is safe to assert that wherever that process is present and long the ambiens is also present and well developed.

Among carinate birds the pelvis does not show a great variability of form. The pelvis is either broader or narrower, being excessively narrow in the **Colymbi**. The proportions of the preacetabular and the postacetabular regions also differ, as do their relative breadths.

It is noteworthy that the pubis is sometimes defective in the middle, appearing then to consist of a proximal and distal portion unconnected in the dry skeleton. The fact also that but few muscles arise from the pubis seems to show that it is in a condition of degeneration.

In *Archæopteryx* alone are the elements of the pelvis not ankylosed together.

The acetabulum of all birds except *Dromæus* is perforated ; but the perforation is reduced in size in *Geococcyx*, *Tinamus*, and *Hesperornis*.

Hind Limb.—The hind limb of birds consists of femur, tibia and fibula, tarsus, metatarsus, and phalanges.

In all birds the femur is shorter than the tibia, the proportions varying much. It seems impossible to place those birds in which the difference is least at the base of the series, on account, of course, of a resemblance so far to reptiles, since relative importance of the fore and hind limb appears to have something to do with the matter. From FÜRBRINGER's tables [1] it is to be gathered that *Fregata* is the bird in which the difference between those two segments of the leg is least. It is most pronounced in the divers, flamingo, and **Tubinares**. A bone of some classificatory importance is the patella,[2] a sesamoid on the upper surface of the knee. This bone is not ossified at all in *Colymbus*, but is enormous in the grebes and in *Hesperornis*, in which latter, as in *Phalacrocorax* and *Biziura lobata*, it is perforated by the tendon of the ambiens. In *Plotus* the patella is grooved only for this tendon.

In no bird (except as an occasional abnormality) is the fibula a complete bone. It fails below, and does not reach the tarsus. It is usually more or less coalescent with the tibia. The latter is a strong bone with a crest in front, which is enormous in the divers and *Hesperornis*. The distal end of the tibia is formed by a portion of the tarsus, of which the remaining portion is coalesced with the metatarsus. The *tarsus* in the embryo [3] consists of three chondrites, a tibiale, a fibulare, and a distale. The latter represents the separate distal elements of the tarsus fused. The tibiale sends up-

[1] KESSLER (' Osteologie der Vogelfüsse,' *Bull. Soc. Nat. Mosc.* 1841) has also given tables. The value of the 'long bones' of the leg for 'defining orders, families, and often genera' is plainly set forth in this paper. See also MILNE-EDWARDS, *Oiseaux Fossiles de la France*, where further information is to be found.

[2] J. KACZANDER, ' Beitrag zur Lehre über die Entwicklungsgeschichte der Patella,' *Mt. Embr. Inst. Wien*, (2) ii. 1887, p. 12.

[3] G. BAUR, 'Der Tarsus der Vögel u. Dinosaurier,' *Morph. J.B.* viii. 1883, p. 417; E. G. MORSE, 'On the Carpus and Tarsus of Birds,' *Ann. Lyc. New York*, x. 1873, p. 141, and 'On the Identity of the Ascending Process of the Astragalus,' &c., *Anniv. Mem. Bost. Soc. Nat. Hist.* 1880.

wards an 'ascending process,' found also in the dinosaurs, which is the equivalent of the intermedium, while the centrale is represented by a distinct osseous nodule in the adults of the **Struthiones** (including *Dinornis*) and tinamous. In *Apteryx* T. J. PARKER found two osseous centralia.

The number of toes and phalanges has been already described above.

The Skull.[1]—While presenting many characteristic features of its own, the skull in birds shows certain fundamental likenesses to the skull of the reptilia. As in them, and contrary to what we find in the mammalia, the skull of birds—

1. Articulates with the spine by a single occipital condyle.

2. Possesses a quadrate bone for the articulation of the mandible.

3. The mandible itself is composed of at least a dentary angular and articular portion.

4. The columella auris is very similar.

The bird's skull is, however, distinguishable by a number of characters, of which the following are the most important :—

1. The bones of the cranium are very closely united and fused, this being less marked in the penguins and ratites.

2. The brain case is large as compared with that of reptiles.

3. The bones of the skull, as are those of the skeleton in general, are light and contain air spaces.

4. The columella and the os transversum of the reptiles are absent.[2]

5. There is no distinct postfrontal.

The bones of the bird's skull, as that of other vertebrates, may be distinguished into four categories - - (1) those of the cranium ossified from its cartilage ; (2) those of the sense capsules ; (3) those of the visceral arches, and (4) membrane bones connected with the several regions enumerated.

---

[1] H. MAGNUS, ' Untersuchungen über d. Struktur d. knöchernen Vogelkopfes , *Zeitschr. wiss. Zool.* xxi. 1871.

[2] See however below, under Passerine skull.

1. The bones formed by ossification from the chondro-cranium are four occipitals, viz. basioccipital, two exoccipitals, and the supraoccipital, forming a complete ring of bone round the foramen magnum; basisphenoid, with two wings, the alisphenoids; presphenoid, with two wings, the orbito-sphenoids (occasionally atrophied, e.g. Apteryx); meseth-moid, with two lateral wings, the ectethmoids (sometimes termed prefrontals); these are occasionally absent as distinct ossifications, and may sometimes, on the other hand, be very large and even appear on the frontal surface of the skull, marking the orbit anteriorly; in those cases they take the place of the orbital part of the lacrymal and have a better claim to be called prefrontal.

2. The investment of the auditory capsule, termed collectively the periotic bone, consists of three separate elements, the prootic, opisthotic, and the epiotic, which last is absent in the Apteryx.

3. The first visceral arch, the mandibular, has but two bones[1] ossified from its cartilage, the quadrate and the articulare of the lower jaw. The second and third arches form the hyoid apparatus; the ossifications are, first, the columella auris, a bone corresponding physiologically, if not also morphologically, to the ear bones of mammals; secondly, a median piece in front, composed of two fused pieces, the basihyal, with sometimes lateral processes, the ceratohyals; thirdly, the basibranchial, with two long lateral outgrowths, of which the nearest regions are ossified to form the cerato-branchials; thirdly, a single median piece (sometimes absent), the urohyal, a remnant of the third arch.[2]

The membrane bones of the bird's skull are numerous, and may be referred to the same categories as the cartilage bones.

1. Associated with the cartilaginous cranium are pos-teriorly the parietals; in front of these the frontals; with the frontals articulate the lacrymals, of varying development,

---

[1] A mento-meckelian has been recently discovered in hawks.

[2] For modifications of hyoids see especially Gadow, in Bronn's *Thierreich*, and Giebel, *Zeitschr. f. d. ges. Naturw.*, xi. 1858.

and in the hawks bearing a second and separate bone behind. These bones appear sometimes to have a definite relation to the cartilaginous ectethmoids. I do not refer so much to the fact that they sometimes entirely fuse with them (and with the skull wall) as to the varying size and relations of the two. In the kingfishers, for example, where the ectethmoids are small, the lacrymals are large, and have below an expanded plate which supplies the place of the feeble ectethmoid. When the lacrymal does not reach the orbital margin, as in *Corvus*, the ectethmoid does, and, as it were, takes its place. In many birds belonging to quite different orders there is a small bone connecting the lower end of the lacrymal, or of the ectethmoid, with either the palatine or the jugal bar; this bone has been termed ' os crochu,' os uncinatum, os lacrymo-palatinum, and will be described in detail in those birds where it is to be found.[1] It may be that the os uncinatum should have been de-scribed as one of cartilaginous bones of the cranium. In some birds (tina-mous, *Menura*, *Psophia*, and *Arbori-cola*) there is a set of supraorbital bones margining the orbits above. The base of the brain case is protected by a large basitemporal, which has sometimes (e.g. *Apteryx*)

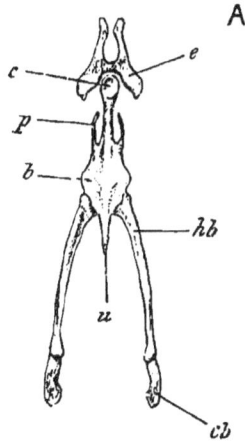

Fig. 76.—Hyoid of *Latha-mus discolor* (after Mivart).

b, basihyal : hb, hypobranchial ; cb, ceratobranchial ; u, urohyal ; e, entoglossum : p. parahyal ; c, concavity in e.

a long rostrum in front. The maxillæ are sometimes sepa-rate from each other, and at times united across the middle line by more or less extensive ossifications, of which a pro-minent one, and with the appearance of a separate bone, is the maxillo-palatine. The premaxillaries in front of these send back a long process extending as far as the nasals.

2. To this category, perhaps, belong the squamosal,

[1] See under *Cariama*, **Tubinares**, **Steganopodes**, **Musophagi**, where cross references will be found.

referable to the auditory capsule, and the nasals above and
the vomers below to the nasal capsule. The vomers are
often paired bones, and near to them are other small and
independent ossifications, such as the septomaxillaries (see
under Passeres) and the mediopalatines (see under Cuckoos).
Connected with the eye—ossified, indeed, in the sclerotic—is
the sclerotic ring.

3. As membranous ossifications connected with or in
the neighbourhood of the visceral arches are the pterygoids
and palatines of the first arch, and perhaps the quadrato-
jugal and jugal, connecting the quadrate with the maxillae.
The pterygoids are, as a rule, style-like bones, flat, however,
in the penguins, and may or may not articulate with the
basis cranii by means of the basipterygoid processes. The
palatines may be completely separate from each other, or
fused for a greater or less extent. They are broader or
narrower, as the case may be. The lower jaw has a number
of membranous ossifications ; these are the splenial, dentary,
angular, supra-angular, and coronary. One or two additional
elements may be present.

With these general characters the skull of birds shows a
considerable number of differences of minor importance in
different families and genera. The first, perhaps, of these
in degree of importance—certainly in general estimation, if
not actually so—is the series of modifications of the avian
palate which were worked out in detail by HUXLEY, and
after him by PARKER, but which had been previous to
HUXLEY's [1] well-known paper studied with some degree of
success by CORNAY.[2]

HUXLEY distinguished among the palates of birds the
following principal modifications :—

1. *Dromæognathism.*—This characterises not only the
ratites but also the tinamous, and has been indeed the

[1] In *P. Z. S.* 1867. See also paper by NEWTON, *Ibis*, 1868, p. 85, and reply,
ibid. p. 357.

[2] 'Considérations Générales sur la Classification des Oiseaux,' &c., *Rev.
Zool. Soc. Cuvierienne*, 1847 ; see also HEERWAGEN, *Beiträge z. Kenntn. d.
Kiefergaumenapparates d. Vögel*, Diss. Inaug., Nürnberg, 1889.

principal reason for the close association of these birds by subsequent writers. In these birds (see fig. 77) the vomer is broad posteriorly, and thrusts itself between the pterygoids and palatines on the one side and the basisphenoid rostrum on the other, and thus prevents their articulation. This is the typical dromæognathous state ; but there are certain modifications which will be described in detail later. The ostrich, for example, is only dromæognathous in that the pterygoids and palatines do not articulate with the basisphenoidal rostrum ; for the vomer in this bird is short and does not reach back far enough to prevent (so to speak) the union.

2. *Desmognathism.*—In a variety of birds belonging to many orders the vomer has either disappeared or is very small ; the two maxillo-palatine plates come into contact in the middle line, as, indeed, they do in the dromæognathous skull. As with all the types of skull to be enumerated, the pterygoids and palatines at the point of their union with each other articulate with the basisphenoidal rostrum. (This kind of skull is illustrated in fig. 78.)

3. *Schizognathism.*—This type is almost as prevalent as desmognathism. The vomer, well developed, terminates, as a rule, in a point anteriorly. The maxillo-palatines, variable in size and shape, do not meet across the middle line with each other or with the vomer. (See fig. 79.)

4. *Ægithognathism.*—Found typically in ' finches ' and in passerines generally ; is very like the last type. The distinguishing character (fig. 80) is that the vomer is broad and truncated anteriorly, lying between the separate maxillo-palatines. The skull is thus ' schizognathous ' etymologically. To these four divisions Professor PARKER has added—

5. *Saurognathism.*—Exemplified in the woodpeckers. The maxillo-palatines are extremely slight, hardly extending inwards from the maxillæ ; hence the skull is widely schizognathous. The vomers are delicate paired rods.

As stated in the foregoing brief epitome of the characters of the several types of skull, the facts seem to differentiate the five types fully. ELLIOT COUES remarks of desmo-

gnathism that 'it does not fadge so well as any other one of
the palatal types of structure with recognised groups of
birds based on other considerations.' This might be really
said of saurognathism also ; for the woodpeckers are not so
far removed from other picarian birds as the structure of

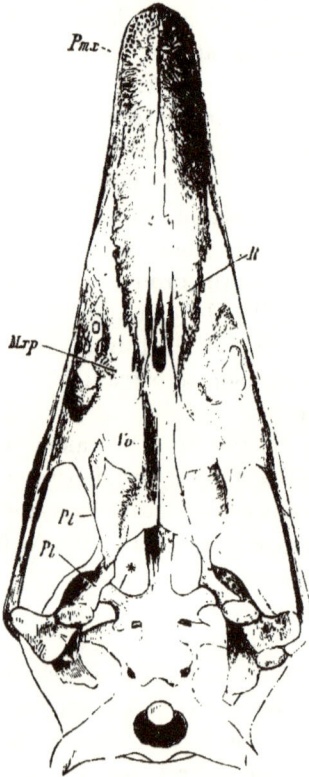

FIG. 77.—SKULL OF *Rhea*. VENTRAL   FIG. 78.—SKULL OF *Dacelo* (AFTER
VIEW. (AFTER HUXLEY).   HUXLEY).

Pmx, premaxilla ; *Mxp*, maxillo-palatine ; *R*,   *La*, lacrymal. Other letters as in previous
rostrum ; *Vo*, vomer ; *Pl*, palatine ; *Pt*,   figure.
pterygoid ; *, basipterygoid process.

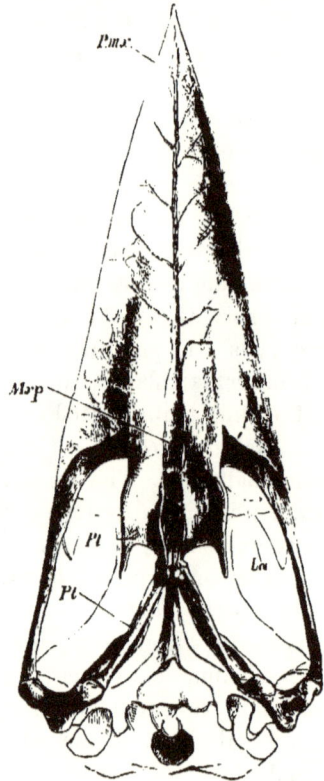

their skull would lead us to believe.   Neither are any of the
subdivisions, except that of the dromæognathæ, really satis-
factory from the classificatory point of view.   Their in-
efficiency, however, is rendered harmless by the fact that

they are in reality not such hard and fast distinctions as might be gathered from the foregoing abstract and from textbooks in general.

PARKER has distinguished four categories of desmognathism—(*a*) perfect direct, the maxillo-palatines uniting below in the middle line ; (*b*) perfect indirect, maxillo-palatines separated by a chink in the middle line ; (*c*) imperfectly direct, maxillo-palatines sutured together in the

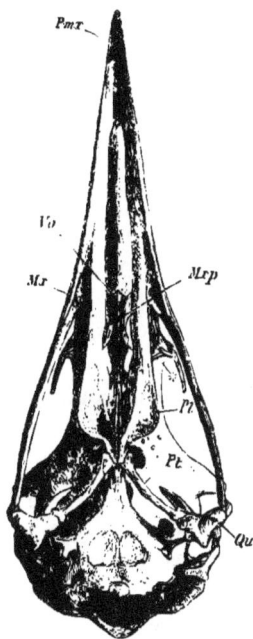

FIG. 79.—SKULL OF *Alca* (AFTER HUXLEY).
*, prefrontal (ectethmoid); *Qu*, quadrate, *Mx*, Maxilla. Other letters as in previous figures.

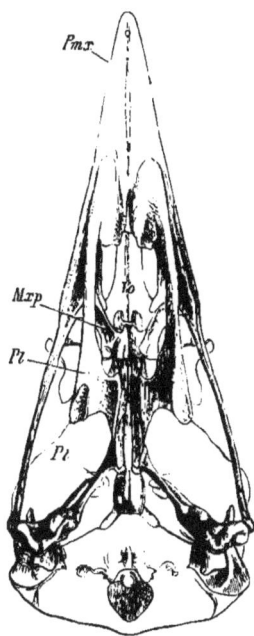

FIG. 80.—SKULL OF *Corvus* (AFTER HUXLEY). LETTERING AS BEFORE.

middle line; (*d*) imperfectly indirect, maxillo-palatines closely articulated with and separated by the median septomaxillary. There is also the exaggerated desmognathism (' double desmognathism ') of the hornbills, &c., where not only the maxillo-palatines but also the palatines are united across the middle line ; and finally the spurious desmognathism of certain ægithognathous birds (*Megalæma*), in which the vomers are ægithognathous, but the maxillo-palatines are united.

As of desmognathism so of ægithognathism, PARKER distinguishes four kinds—

Incomplete ægithognathism (Hemipodes) is distinguished by the fact that the broad double vomer has a septomaxillary at each angle, which is only 'strongly *tied*' to alinasal cartilages.

Complete Var. 1.—The vomers are distinct from the often long alinasal walls and turbinals : a small septomaxillary appears on the angle of the alinasal cartilage, but does not run into it.

Complete Var. 2.—Here the vomers are grafted upon the nasal wall.

Compound, where in an ægithognathous palate desmognathism is produced by ankylosis of the inner edge of the maxillaries with a highly ossified alinasal wall and nasal septum.

Among the higher families the septomaxillaries are often absent.

It is clear, therefore, that a very narrow boundary line separates desmognathism in some of its forms from schizognathism, and that there is a direct relationship between desmognathism and ægithognathism. The only type of skull which is really distinct is the dromæognathous.

In classificatory importance perhaps next comes the condition of the nasal bone. GARROD distinguished birds into those with holorhinal and those with schizorhinal nostrils. These terms refer to the posterior edge of the bony nostril, which in one set of birds, the holorhinal, ends behind with a clear oval outline (fig. 81), or in the schizorhinal birds runs back as a gradually narrowing chink ; this latter arrangement is shown in fig. 82. In the holorhinal bird a straight line, drawn across the face from the posterior boundary of one nostril to that of the other, passes in front of the termination of the nasal processes of the premaxilla. It is not always the case that a line drawn similarly to that of the holorhinal birds passes behind the end of the premaxillary process, but it is generally so. In the schizorhinal skull it often appears, as in the typical schizorhinal cranes and charadriiform birds, as if the outer part of the nasal bone were a distinct bone ; for it joins the

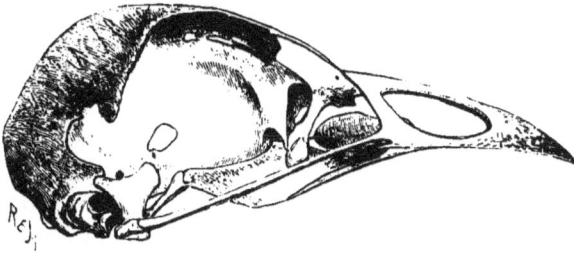

FIG. 81. SKULL OF *Psophia*. LATERAL VIEW. (AFTER BEDDARD.)

inner lamina at an angle. In the typical holorhinal skull, on the other hand, as, for example, in the Rallidæ, the two parts of the nasal come smoothly together, leaving the clear-

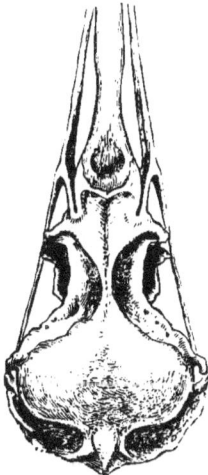

FIG. 82.—SKULL OF *Larus*. DORSAL VIEW. (AFTER GARROD.)

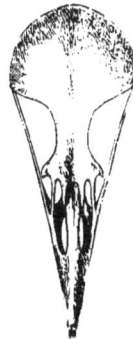

FIG. 83.—SKULL OF *Furnarius*. DORSAL VIEW. (AFTER GARROD.)

cut, ovally contoured nostril. It is not, however, always easy to distinguish so clearly as can be done between *Rallus* and *Grus*. Thus GARROD admits the schizorhiny of *Furnarius* and some other Passerines, in which the bony opening of the nostrils, although, as he figures it, rounded off at its termination, ends behind, or at least on a level with, the ends of the nasal processes of the premaxillæ. In the same way an intermediate condition is offered by *Thinocorus* and *Glareola*, in which there is much the same kind of arrangement. But one of the most striking instances which have come to my personal knowledge is that of *Chunga*. Its near ally *Cariama* is, as correctly stated by GARROD and others, holorhinal, which in view of its relationship to the cranes is unfortunate. But in *Chunga* it is clear that the holorhiny is secondary, being produced by a slight modification of schizorhiny.

A careful examination of *Chunga* shows that the two parts of the nasal bone do not join evenly above the opening of the nostril, but that the outer descending lamina of the bone is divided for some little distance by a crack from the premaxillary portion, the two running up in close contact— so close that no actual space is left between them, only a line of junction to mark their original separateness. In the skulls of *Cariama* that I have examined there is no trace of this ; but, considering the nearness to each other of the two birds, it seems probable that it is merely disguised. This fact favours FÜRBRINGER'S idea that schizorhiny is more primitive than holorhiny, and is so far adverse to GARROD's view that ' the schizorhinal disposition is most certainly one which is a secondary development upon the normal holorhinal nares.'

It is clear too that the holorhiny of such a bird as *Opisthocomus*, where the ossified alinasals produce the rounded margin of the bony nostril, cannot be accurately compared with the holorhinal nostril of a gallinaceous bird, where it is the nasals themselves that bound the orifice.

The presence or absence of *basipterygoid processes* is another matter upon which some stress is usually laid from a systematic point of view. One assumes that the existence of these processes is the original condition and that their loss is secondary. The presence or absence of basipterygoid processes is most capricious. Thus among the limicoline series they are absent from the skulls of the gulls and—more unexpectedly perhaps—from the Œdicnemidæ and Thinocoridæ. Among the birds of prey the secretary bird and the American vultures have these processes, while the Falconidæ have them not. The goatsuckers may be similarly divided into those without and those with basipterygoid processes. The TUBINARES, again, show variation in this respect, as do the trogons. It is therefore, in the first place, impossible to compare directly all birds which are without these processes, just as it is impossible to put together all birds without an ambiens. We may note, however, that it is only among groups of birds which show a considerable range of structural variation that there is this variation of the basipterygoid processes. It is not, so to speak, lightly that they have gone. The reason for the assumption that the basipterygoid processes are primitive is their existence in reptiles and in such widely separated types as chameleons, pterodactyles, and *Hatteria*, and in addition I may point out that there is a significant correspondence between a primitive arrangement of gut and the presence of these structures. It cannot be said that every bird with basipterygoid processes has the most primitive arrangement of gut, but we do find both in the ratites, *Chauna*, the gallinaceous birds, the charadriiform birds, the owls, and the goatsuckers. The falconiformes (*Haliaetus*) are, it is true, an exception; but it must be remembered that this group is one that has basipterygoid processes (*Serpentarius*, *Cathartes*), though they are absent in the true falcons.

The principal variations exhibited by the cranium of birds, apart from those that have been already considered, concern the existence of *supra-orbital bones*, the existence or non-existence of *occipital fontanelles*, the marks of the *supra-orbital glands*, and the presence or absence of a hinge line between the skull proper and the face.

The existence of *supra-orbital bones* in the form of a longer or shorter chain of ossicles was first pointed out by PARKER as a reptilian character of occasional occurrence.

L

In the tinamous there are a series of these bones, which in *Psophia* are reduced to a smaller number, and in the passerine *Menura* to three only; on the other hand in *Perdix* they are again more numerous. It is possible that in such birds as *Rhinochetus* and *Œdicnemus*, in which the edge of the bony orbit is very sharp, the thinness of the edge is due to the fusion of a set of these bonelets with the true margin of the frontal. As will be seen, the existence of these bones is so rare as to render them of not great service from a systematic point of view; but it is only among birds which may be fairly on other grounds regarded as archaic that they are to be met with.

Fig. 84.—Skull of *Chloephaga Magellanica*. Back View. (After Garrod.)

The *occipital fontanelles* (fig. 84) are mostly developed in water birds, though any connection between them and the habit of the birds is at least not obvious. They are most general among the **Limicolæ** and the duck tribe, but are found also in the Plataleidæ and exist temporarily in the gulls; they are also found in the flamingo, Gruidæ, and among the auks.

Almost the same remarks may be made of the *impressions for the supra-orbital glands*. They are very marked in the **Limicolæ**, being more usual than the occipital fontanelles; they are met with in the auks, divers, and penguins; in the cranes and Plataleidæ they are present, but not so conspicuous.

The hinge line between face and skull is seen in its most fully developed condition in parrots, where the face is actually movable on the head. But it is commonly met with elsewhere, particularly among the Anomalogonatæ; it is associated with holorhiny and with comparatively short nasal processes of the premaxillary.

Apart from their relations to the vomer the *palatines* and *pterygoids* show some variations in structure. As to the pterygoids, the most prominent difference concerns the place where the articulation with the basipterygoid facets occurs; it is usually towards the middle of the bone: but in the emu, in the **Anseres** and **Galli** it is quite at the end of the bone, the end nearest to the palatines. The palatines differ greatly in shape and in breadth. Among the **Anseres** and **Galli** the internal lamina which meet in the middle line above the basisphenoids are practically absent, being only represented by a more or less faintly marked ridge. In other birds these regions of the palatines are well marked, and may meet for a considerable space in the middle line, and, as in *Steatornis*, an anterior portion of each bone may come into contact in the middle line further forward, thus giving rise to what has been termed double desmognathism, the maxillo-palatines of course forming the other junction in the middle line in front of them again. In most birds perhaps the palatines are gradually rounded off behind, but in the Ardeidæ and in *Rhinochetus*, &c., they are as it were sharply cut across behind this edge, being at right angles to the long axis of the bone. In many Passeres and in some other birds there is a well-developed postero-lateral process of the bone, which has been termed transpalatine; this, if it came into contact with the jugal arch, as it nearly does in some Passeres, would bear the strongest likeness to the transverse bone of reptiles. The palatines are occasionally, but rarely, fenestrate, e.g. *Anous*, *Eurypyga*.

The *vomer* varies from complete absence (*Colius*) to very striking presence. When present there is every grade between a thin splint and a broad flat bone, which in the latter case is often obviously formed of two lateral halves. These, however, do not appear to remain completely separate except among the **Pici** (*q.v.*), where they are smallish splints. But not only among the ostrich tribe, but in *Hesperornis* and passerines the vomers are double in the young. A series of small bones lying between the palatines, and called interpalatines and septo-maxillaries by PARKER, may be regarded as belonging to the vomerine series. They are unpaired ossicles which continue, though with a hiatus, the vomer backwards.

The degree of development of the *lacrymals* and the *ectethmoid processes* varies greatly among birds, and is at times of use for systematic purposes. In some birds, as, for example, among the cranes, there is no junction between those bones; in the Charadriidæ, on the other hand they form a complete ring; in *Pterocles,* &c., the two are firmly blended into a square plate of bone which bounds the orbit anteriorly.

The lacrymal is occasionally joined to the palatine or to the jugal by a small independent ossicle, which PARKER has termed the uncinate bone, and thinks to be the homologue of the anterior connection of the palato-quadrate arch with the skull in the tadpole, &c. This bone is so variable in its presence (e.g. *Cariama*, **Tubinares**) that it can hardly be regarded as of much systematic importance. The last matter to which we may refer as of classificatory importance is the form of the quadrate, which Miss WALKER has shown to vary much and characteristically in different groups. It had been long known that the single-headed articulation with the skull was a character of struthious birds, excepting *Apteryx*, and of *Ichthyornis*, and to a less extent of gallinaceous birds and ducks.

The value of the base of the skull in classification has been continuously debated since the facts were first so clearly set forth by HUXLEY.

From the complications introduced into the originally simple series of modifications of the skull instituted by HUXLEY, by PARKER, and from the varied criticisms of fact and conclusion of a classificatory kind based upon fact, we may disentangle one conclusion that many ornithologists will agree with—that is, the more lacertilian character of the skull in the struthious than in other birds. In them the palatines are borne off from the basisphenoidal rostrum by the vomer (with the exception of *Struthio,* in which, however, the palatines are still remote from the rostrum), and the general disposition of these parts is, as HUXLEY wrote, 'more lacertilian than in other birds.' Furthermore in the struthious birds the double character of the vomer is more

universally retained than in other birds. As T. J. PARKER
has shown in *Mesopteryx*, *Anomalopteryx*, and (according to
OWEN) *Dinornis torosus*, and finally in the young of *Emeus
crassus*, there are distinct paired vomers. The strong anterior
and posterior bifurcation of the vomer in living Struthionidæ
is an indication of the partial fusion of the primitively
separate halves of the single vomer. A double vomer,
however, is not a feature of the struthious birds alone; in
the young *Hesperornis* the same occurs; in the woodpeckers
(hence Saurognathæ in PARKER'S nomenclature) there are two
distinct vomers, while PARKER has recently committed himself to the general statement that ' in most birds that have
a large or wide vomer it is double at first.' The inference
is that the single thread-like vomer of many birds is in a
degenerate condition. In any case a large vomer, double or
single, and a lacertilian palate generally mark all the struthious birds, and justify their generally recognised position
somewhere not far from the root of the avian series.

As for desmognathism and schizognathism, Professor
PARKER has aptly remarked : ' The use of such a taxonomic
character as desmognathism or schizognathism is very
extensive in some groups and very limited in others ; and
there is no sharp line of demarcation between the two. The
most lacertilian palate for openness is that of the woodpecker ;
the most modified by intense ossification is that of the
toucan ; yet these two types, each specialised to the utter-
most, have a postcephalic skeleton, not indeed identical, but
extremely similar.'

That the toucans and woodpeckers are exceedingly near
akin is also shown by many other features of their organisation (see below). HUXLEY claimed for his Schizognathæ
that they were a natural group, but hardly claimed so much
for the Desmognathæ. He admitted, however, that *Cariama*,
a schizognathous bird in the totality of its organisation, had
a palate approximating to the desmognathous. The Cracidæ
and *Rhinochetus* are in the same anomalous position. The
Trogonidæ, whose nearest allies are the desmognathous
birds, have a schizognathous palate (as was pointed out by

FORBES), while the goatsuckers show both desmognathism and schizognathism among the members of the group. On the other hand schizognathism is not on the taxonomic side sharply marked off from ægithognathism. HUXLEY's definition of the vomer in schizognathous birds was that it is ' pointed in front ; ' this is not the case with many Schizognathæ, e.g. *Hæmatopus, Numenius* (cf. below). And, on the other hand, PARKER showed that another peculiarity of the ægithognathous skull, the union of the vomer with ossified alinasals, was also found in the ' Turnicomorphæ,' a group which for other reasons should be placed among HUXLEY's Schizognathæ.

It appears, therefore, undesirable to lay too much stress upon the modifications of the palate, as seen in the three groups just discussed, as a basis of classification.

HUXLEY defined desmognathism as follows : ' The vomer is often either abortive or so small that it disappears from the skeleton. Where it exists it is always slender and tapers to one point anteriorly. The maxillo-palatines are united across the middle line either directly or by the intermediation of the ossifications on to the nasal septum.'

This definition applies perfectly well to the **Anseres** and **Palamedeæ**, but not to all other groups of birds.

But there is the form of desmognathism found in the **Steganopodes** and in the owls and **Accipitres**. In the former group there is no desmognathism (in the sense of HUXLEY) except in *Pelecanus*. The maxillo-palatines of *Phaethon* and *Fregata* (see below) are perfectly free ; in front of them the palate is complete, but that completeness is formed by a union of lateral extensions of the maxillæ which are distinct from the rounded maxillo-palatines. In the cormorant, which is admittedly an ally of *Phaethon*, there appears at first sight to be a true desmognathism, a fusion, that is to say, of the maxillo-palatines. The plates of bone in question are beneath the ends of the palatines, but, instead of running horizontally in the same plane as the palatines, they run obliquely upwards (when the skull is regarded from below). If it were not for *Fregata*, these bones might be

looked upon as the homologues of the maxillo-palatines of
Phaethon. But in *Fregata* both structures are present.
Coexistence clearly disproves homology ; we must, therefore,
place the 'desmognathism' of *Phalacrocorax* upon an
entirely different footing from that of the **Anseres**, where
the maxillo-palatines unite in the typical fashion across the
middle line.

In addition to these forms of desmognathism it appears
to me that we should distinguish that of the American vultures, where there is no union at all of the maxillo-palatines,
but only of the alinasals.

. In fact there appear to be three ways of bridging over the
palate which may be termed desmognathism, though the
word becomes merely descriptive of a condition and not
necessarily indicative of affinities ; these are -

(1) Union of maxillo-palatines (more or less complete).

(2) Union of alinasals.

(3) Union of maxillæ in front of the maxillo-palatines.

These three are usually combined in various ways ; *e.g.*

In hornbills there are all three methods of union ; in
some **Steganopodes** only numbers 2 and 3.

Brain and Nervous System.

The brain of recent birds chiefly differs from that of
reptiles by the large size of the hemispheres, the large size
of the cerebellum, and by the fact that, owing to the great
development of these two regions, the optic lobes are pushed
to one side. T. J. PARKER found in studying the development of *Apteryx* that the latter feature is acquired during
development, and that the optic lobes have originally the
reptilian position. The extinct cretaceous birds had brains
with smaller cerebral hemispheres and with larger optic lobes
than existing birds, and were in these particulars more reptilian. On the other hand, if the cast present in the London
example of the *Archæopteryx* be really, as it was first surmised
to be by Sir JOHN EVANS, the brain of the bird, it was less
reptilian than *Hesperornis*, a fact which may possibly

suggest the inference that the brain of the cretaceous bird
furnishes an example of degeneration rather than of the
retention of an archaic character.

The hemispheres, although smooth in the majority of
birds, show faint indications of what may correspond to
furrows in some others.   In the duck, for example, as
figured by GADOW (after BUMM), there is a central raised area
marked off by a furrow from the surrounding parts of the cerebrum ; faint traces of the same occur in *Buteo vulgaris*. The
weight of the brain as compared with that of the whole body
has been studied by several observers, and according to their
results the Passeres and parrots take the highest place.
But GADOW justly remarks that ' the attempts to sort birds
according to the proportion of brain to body have led to no
practical results, chiefly because the variable conditions of
fat and lean subjects have not been considered.'

The brachial and lumbar plexuses, particularly the
former, have been studied in a large number of birds ;
FÜRBRINGER has published a quantity of drawings of the
former, but no classificatory results of reliability appear to
follow from the facts collected with so much diligence. The
brachial plexus varies in position and in complexity. The
former variations are largely correlated with the varying
length of the neck : thus in *Columba* the first spinal nerve
entering into the plexus is the tenth, in *Phœnicopterus* the
seventeenth. The number of nerves which together form
the plexus varies from only three in *Bucorvus* to six in
Columba.

The Eye.—The eye of birds presents many resemblances
to that of reptiles. The minute structure of the retina
presents many points of similarity, as also the ring of ossifications in the cornea, and the pecten. The latter is a folded
process of pigmented tissue which projects into the vitreous
humour through the choroidal fissure, which is in the embryo
the gap left between the edge of the optic cup and the lens
on one side. The pecten offers differences in various birds
upon which perhaps some little stress can be laid. In
Apteryx it is entirely absent. There are very few folds in

the owls (4 to 7) and in the goatsuckers ; .this might lead, in connection with the total absence of the pecten in *Apteryx*, to the conclusion that the less or greater development of the organ had some relation to nocturnal or diurnal habits. But the existence of only four in the emu seems to throw some doubt upon this suggestion. Among the passerine birds the largest number (raven 30) of folds in the pecten is found.

Certain small ossifications in the cornea near to the entrance of the optic nerve seem to be peculiar to the Passeres and to certain picarian birds. The eyes have both lacrymal and Harderian gland ; the eyeball is moved by four recti and two oblique muscles ; the membrana by two muscles, the *quadratus* and the *pyramidalis*, both innervated by the sixth cranial nerve.

The Ear.—Birds have no external ear (concha), but in many a flap forming a valve projects into the meatus from the outer margin, a state of affairs which recalls the conditions found in the crocodile. This outer ear is especially well developed in the owls, in which birds also the ear region of the skull is often markedly asymmetrical.

The inner ear has the three semicircular canals of all higher vertebrates, but the cochlea is not coiled. The auditory ossicles consist of a single structure, partly bony and partly cartilaginous, called the columella.

The Affinities of Birds

In considering the relationship of birds to other vertebrates it is probably safe to leave out of consideration the mammalia and the amphibia. Points of likeness have, it is true, been urged in favour of the latter view of an affinity between birds and amphibia by the late Professor PARKER ; but apart from warm-bloodedness and the resemblance of some of the more simple forms of feathers to hairs there is nothing to be said on behalf of a kinship between birds and mammals. As to the likeness with amphibians, it is possible that the divergence of birds from

the reptilian stem was at a time when the characters of the
amphibian had been incompletely thrown off, and at a time
also when the mammals diverged on their own path from a
point near to that whence the birds took their origin. The
general belief is in the origin of birds from some reptile stem,
but there is not an absolute agreement as to precisely which
group of reptiles birds are most nearly akin to. The researches
of MARSH and HUXLEY, besides those of COPE, SEELEY,
HULKE, and some others, have led to a general acceptance of
a nearer kinship with the dinosaurs than with any other
group of reptiles. In considering the question, then, which
forms the subject of the present chapter, we shall commence
with the dinosaurs. The dinosaurs, ranging in size from
vast creatures of 70 or 80 feet in length to a diminutive reptile
half the size of the domestic fowl, are first known from
the Trias, persisted though the Jurassic and finally came to
an end in the Cretaceous epoch, later than which no
remains have been found. So far as we are aware birds
came into existence in the Jurassic period; hence there is
no anachronism in considering them from the dinosaurian
aspect.

It was formerly held that birds antedated the Jurassic
period; for some of the celebrated tridactyle footprints in
the sandstone of the Triassic period were put down to birds.
It seems, however, to be now fairly certain that those footprints are of dinosaurs.   Still with so specialised a form as
Archæopteryx certainly was, and as *Laopteryx* probably was
in the Jura, it would not be surprising to meet with genuine
avian remains in the Trias.   But even then there are
undoubtedly dinosaurs belonging to that period, so that the
question of relationship would resolve itself into a common
origin, not a derivation of birds from dinosaurs.

The part of the skeleton in which most resemblance is
shown between birds and dinosaurs is the pelvis.   The
dinosaurian pelvis consists of apparently three elements, like
that of birds, but the pubis is an L-shaped bone, constructed
of two pieces, one directed forwards and the other backwards and parallel with the ischium of its side.   The latter

is generally regarded as the homologue of the pubis of birds, while the forwardly directed half of the bone is considered to be the equivalent of the pectineal process. The alternative is to look upon the prepubis of the dinosaur as the pubis of birds, and the postpubis as having disappeared altogether in them. In this case the crocodile will be an intermediate form ; for in this reptile the prepubis is the 'pubis,' while the postpubis is represented by an inconspicuous process upon the pubis. The former alternative commends itself to us. There are birds in which the pectineal process is practically absent ; in others, as *Apteryx, Geococcyx*, &c., it is large ; on the other hand there are dinosaurs, *e.g. Laosaurus consors*, in which the prepubis is generally reduced—to a length of not more than one-third of the postpubis (pubis of birds). The opposite extreme is reached by *Triceratops*, where it is the postpubis which is less than a third of the prepubis, and *Ceratosaurus*, where it is even further reduced.

It is not, however, only in the pubes that the pelvis of the dinosaurs is like that of birds. In those reptiles the ilia were extended forwards and backwards, as in birds, and in *Ceratosaurus* at any rate the three bones were all firmly ankylosed, as in all birds save *Archæopteryx*, where the separation of the bones conforms to what is found in the vast majority of the dinosaurs. *Laosaurus*, which has just been mentioned, is one of the most birdlike of dinosaurs. ' The two species of the genus first described by the writer,' remarks Professor MARSH, ' show these avian features best of all, and it would be difficult to tell many of the isolated remains from those of birds.' Of the cretaceous dinosaurs the same author observes, ' Others were diminutive in size, and so birdlike in form and structure that their remains can be distinguished with difficulty, if at all, from those of birds.' *Ornithomimus*, as its name denotes, is one of those especially annectent dinosaurs. In this genus the third metatarsal is crowded backwards behind the second and fourth, as in many birds. But the dinosaurian metatarsals which are most strikingly like those of birds are of *Ceratosaurus*, which MARSH has figured side by side with those

of a penguin, where the bones show greater traces than
in any other birds of their distinctness, and are furthermore
shorter than is the rule. In dinosaurs generally those bones
are separate, but not in *Ceratosaurus*, where the degree of
fusion is almost exactly that of *Aptenodytes*.

Furthermore in some dinosaurs, as in birds, the intermedium is prolonged upwards as the ascending process of
the astragalus.

The proportions of the long bones of the hind limb are
distinctly birdlike in some dinosaurs. In *Laosaurus*, for
example, the femur, as in birds, is shorter than the tibia, the
reverse occurring in most forms. In the same animal and
in others the fibula is commencing to degenerate ; it is
decidedly smaller than the tibia.

In their skulls the dinosaurs show no marked approximation to birds ; there are nevertheless one or two features
which may be remarked upon in this connection. The
earliest known forms from the Trias have perhaps the most
birdlike forms of skull. MARSH comments upon the light-
ness and avian appearance of the skull of *Anchisaurus*; it
has moderately developed basipterygoid processes instead of
those of such great length that are apt to characterise the
dinosaurs. The great extension backwards of the premaxillaries in some dinosaurs is an avian characteristic ; this is
seen especially well in *Diplodocus* and *Claosaurus*. In the
former animal, as in some others, the two vomers diverge
widely posteriorly, as in many birds ; and in the restoration
of the under surface of the skull in this dinosaur the vomers
have a very birdlike appearance.

Finally the height at which the transverse processes of
the vertebræ are borne seems, as HUXLEY has pointed out,
to suggest birdlike respiratory organs, while the hollowness
of many of the bones in many dinosaurs points in the same
direction. It is, however, undoubtedly in the pelvis and
hind limb that the most striking likenesses to birds are
shown by the dinosaurs. It has been attempted to put this
down merely to bipedal progression. 'It may be said,'
remarked Professor HUXLEY, ' that all birds stand upon

their hinder feet, and that, as the Ornithoscelida did the same, the resemblance of structure arises from a resemblance of function. But I doubt if the majority of the Dinosauria stood more habitually upon their hind limbs than kangaroos or jerboas do ; and, unless there was some genetic connection between the two, I see no reason why the hind limbs of Ornithoscelida should resemble those of birds more than they resemble those of kangaroos.' In addition to this it may be pointed out that *Hallopus*, which appears to have been very probably a leaping dinosaur, has not the specially ornithic form of limb ; it has large pubes and no postpubes.

A recent description by Mr. E. T. NEWTON [1] of the skull, brain cast, and cast of the auditory organ in a pterodactyle, *Scaphognathus Purdoni*, shows certain most interesting resemblances between the pterosaurians and birds. It is possible that this pterosaurian, like *Pteranodon*, possessed a horny beak and no teeth. But the presence or absence of a beak or of teeth is no more distinctive of birds (cf. *Archæopteryx*) than of reptiles. The skull shows more positive points of likeness. In the first place, the bones of the pterodactyle cranium are early ankylosed and well ankylosed, being in this particular avian and not lacertilian. The large size and backward extension of the single premaxillary bone (= two fused premaxillæ) agree with that of birds and contrasts with that of lizards. The palate shows also certain interesting resemblances, more especially to both emu and cassowary. As in the struthious birds and in lizards also the palatines are borne off from the middle line by the pterygoids ; the latter bones, moreover, as in the emu, articulate *at their posterior ends* with both quadrate and basipterygoid processes. The vomer too is birdlike in being pushed backwards, owing to the extent of the premaxillæ, and in being thin, apparently single and bifurcate posteriorly. Other general resemblances in the skeleton are the development of air cavities in the bone, the large size of the orbit —which may, however, in the pterodactyles have had some

---

[1] ‘On the Skull, Brain, and Auditory Organ of a New Species of Pterosaurian,’ *Phil. Trans.* 1888, B, p. 503.

relation to nocturnal habits—and the presence of a keel upon the sternum.

It is possible, though far from certain, that *Scaphogna-thus* had not that characteristic reptilian bone, the *os trans-versum.*

Finally, the general shape of the scapula and the angle that it makes with the coracoid are birdlike in the pterodactyles.

The brain of the pterodactyles seems also to have presented avian characters ; the optic lobes are pushed aside by the large cerebellum, which had well-developed floccular lobes. In the reptile's brain the optic lobes intervene between the cerebrum and cerebellum.

The pelvis of the pterodactyles has some likenesses to that of birds. The ilium has an extension in front of as well as behind the acetabulum ; and, if the opinions of SEELEY are to be agreed to, there is a rather backwardly directed pubis, more or less fused with the ischium, and a long and thin forwardly directed piece, the prepubis of dinosaurs and the pectineal process of birds.

The main difficulty, however, in the way of comparing pterodactyles and birds is in the fact that both can fly, and that each has acquired the power of flight by a different method. Having acquired the power of flight it seems clear that certain of the points of resemblance between them may easily be due to that mode of life and may have been independently arrived at.

THE CLASSIFICATION OF BIRDS

PROFESSOR NEWTON'S article 'Ornithology' in the 'Encyclopædia Britannica.' and the preliminary sketch of Dr. GADOW in Bronn's 'Thierreich,' contain a digest of, and criticisms upon, the main schemes of classification of this group which have as yet appeared. I shall, therefore, refer the reader to those works for the history of the subject. There can be no question, in my opinion, that birds must be primarily divided into two great divisions, viz. **Saururæ** and **Ornithuræ**, the first to contain *Archæopteryx* and possibly *Laopteryx*, the latter the rest of birds, both living and extinct. As to the Ornithuræ, while there is a very general agreement with the main subdivisions—no one probably will quarrel seriously with the divisions adopted in the present work— no one has (to my mind) satisfactorily arranged the different groups with reference to each other. More especially does it appear to me that the majority of ornithologists are in error concerning the position of the picarian and passerine birds.

In considering a scheme of classification it is clear that we must bear in mind indications of the descent of birds. Existing schemes have savoured too much of a mere sorting by combining in various ways characters which are distinctively bird characters. However unsuccessful the construction of phylogenetic trees has been, it is abundantly plain that that must be the line to take in arranging a group scientifically. It follows, therefore, that in sketching, at any rate, the main outlines of our scheme attention must be paid only, or chiefly, to those characters which birds have inherited from their reptilian ancestors.

Now this at once lands us in a difficulty, which has been too lightly regarded by many systematists. Phylogenetic schemes used to be boldly linear, and even so recently as the attempt of FÜRBRINGER the family tree savours a little too much of the linear arrangement. Now the imperfect remains of birds that have come down to us from tertiary times show that the modern types of birds were fully differentiated even then in addition to a few extinct forms, such as *Odontopteryx toliapicus* (if this be not a steganopod). But beyond that point there is the most scanty record of bird life, limited to the Cretaceous Ichthyornithidæ and Hesperornithidæ, with a few obscurer forms, and to the Jurassic *Archæopteryx*. So emphatically were all these creatures birds that the actual origin of Aves is barely hinted at in the structure of these remarkable remains. Moreover, at least in the case of *Ichthyornis*, they depart fully as widely from any bird with the required ' mixed' characters as any living group, while *Hesperornis* can with safety be relegated to the neighbourhood of the existing divers. We get, therefore, no help whatever from the Cretaceous birds, and, if any, only the scantiest assistance from *Archæopteryx*, in determining what are archaic characters in birds. There are no criteria by which we can assert with any degree of safety the relative positions of this and that existing group; nor has the study of the comparative embryology of birds as yet advanced sufficiently far to give any results, except in isolated characters; such indications are the relatively primitive character of the basipterygoid processes, at least in certain groups; for the gulls which are without them when adult, have them as young chicks, as have in the adult condition most of their near allies, the **Limicolæ**. It may, therefore, so far be inferred that the gulls are a modification of the limicoline type and not *vice versa*.

It would be perhaps held that any type in which a number of undoubtedly reptilian characters had survived would be on a lower level of organisation than other types in which fewer or no such characters could be discovered. But the few specially reptilian features in the organisation of birds have, so to speak, been distributed with such exceeding fairness through the class that no type has any great advantage over its fellows. PARKER has collected together some of the reptilian survivals, and to his series a few others may be added. The rudimentary organ of Jacobson found by T. J. PARKER in *Apteryx* is a suggestion of the reptile; but we do not know enough about the development of other birds (except the struthious, where PARKER found the same rudiments himself) to lay much weight upon the discovery as indicative of the low position of *Apteryx* in the avian system. The supra-orbital chain of bones is also on the same grounds an archaic character; but they exist in such widely different types as tinamous, *Psophia*, *Menura*, and quails. The double vomer is reptilian; this bone is double or nearly so in struthious birds, in *Hesperornis*, woodpeckers, and even in passerines and other types with a broad vomer. PARKER has compared the 'os uncinatum' with the anterior suspensory cartilage of the tadpole's jaw apparatus; this is found in so many and such various forms as *Cariama*, *Fregata*, **Tubinares**, Musophagidæ, *Steatornis*, &c. Basipterygoid processes are distinctly reptilian, found as they are in so many forms of reptiles. But among birds they occur in nearly every big group, and are therefore most undistinctive. The pectineal process, if it is invariably—as T. J. PARKER says it is in *Apteryx*—the joint product of pubis and ilium, is not exactly comparable to the supposed corresponding process of the dinosaurian pelvis; but in any case it is found in *Geococcyx* and some other birds far away from the **Struthiones** and the tinamous. A large number of vertebræ in the tail is reptilian; but not only *Archæopteryx* but also the swan has a long tail. Opisthocœlous vertebræ are found in the **Alcæ**, penguins, and gulls, not to mention the darters and parrots.

As to the viscera, HUXLEY showed the close likeness between the various membranes which divide the cœlom and the corresponding membranes in the crocodile, and I endeavoured to show that the ostrich is not in these particulars more reptilian than many other birds. The partial persistence

M

of the left aortic arch is an approach to the reptile ; but birds
belonging to the most diverse orders have this arch left in
varying degrees of perfection ; so that no stress can be laid
upon this anatomical fact as a mark of low position.   It has
been pointed out that the **Struthiones** are unlike other birds
in the absence of a syrinx ; and in the absence of this
specially bird organ they approach so far to the lower
forms. This is not, in the first place, true of all the **Struthiones**,
for *Rhea* has, as has been pointed out, a syrinx fully as typical
as that of most birds, while the American vultures and even
the storks have nothing in the way of a specialised syrinx.
In fact, without going into further detail, it seems impossible
to select any existing group of bird which is distinctly more
reptilian than any other.

Since positive characters appear to fail us in discriminating between the relative positions of the several groups of
birds, it seems to be not unreasonable to turn for light to
negative characters.   Birds as birds have many peculiarities
of organisation, which are impossible in other animals ; for
example, patagial muscles cannot exist without a patagium
to contain them.   It may therefore be permissible to draw
with caution some inferences from the absence or simplicity
of certain peculiarly ornithic structures which, it appears
obvious, must have originated within the class.   The lower
types will surely possess fewer of these essentially ornithic
organs or modified organs.

There is a general belief in the modified character of all
the birds which GARROD placed in his subclass Anomalogonatæ.
Nevertheless there is something to be said in favour of their
primitive nature.   Without absolutely urging the acceptance
of this view, it may be useful to refer briefly to certain
reasons which might be alleged in support of such a placing
of the Pico-Passeres.   Their small to moderate size is to
some extent an argument.   The most ancient mammalia
and reptiles are small as compared with some of their later
and more modified representatives.   Universal distribution
is another argument, as is possibly chiefly arboreal life.   In
anatomical structure we find that many essentially ornithic

characters have not yet put in an appearance, or have done so only to a small extent. There is no member of the group, wide though it is, in which there is an ambiens, a special bird muscle. It may be that GADOW's discovery of a small independent slip of the rectus femoris, which he interprets as a rudimentary ambiens, is really the beginning of this characteristic muscle. The remarkable peculiarities of this muscle seem to forbid the notion that it is the direct descendant of anything reptilian. And we have the undoubted fact that, apart from the possible rudiment already referred to, it is present in no pico-passerine bird. If it had disappeared in them there would be here and there a rudiment left. But nothing of the kind has been hinted by GARROD and FORBES, who between them dissected so many of these birds, and who would have been especially on the look-out for such a point. I should therefore be disposed to disagree at once with GARROD's opinion that those birds which have lost the ambiens ' may be set down as having possessed the muscle in their ancestral form.' The ambiens is so purely a bird muscle, though it may doubtless have its homologue among reptiles, that one cannot but think that it was acquired within the class ; and the facts discovered by MITCHELL (see above) entirely support this way of looking at the matter, and indeed suggested it. As to the muscular system of the wing, a highly characteristic muscle is the expansor secundariorum ; this was supposed for some time to be absent in the group under consideration, but it has been found to occur in some of them. In the majority of those in which it does occur its structure is decidedly more rudimentary than in some of the Homalogonatæ. It is true that FÜRBRINGER regards this muscle as the abortive remnant of a reptilian muscle. But this statement cannot be made about the patagial muscles, which are essentially ornithic.

Now it is noteworthy that, with the exception of the colies, not a single bird referable to the great group of Anomalogonatæ has a biceps slip, while in the majority of them the tendon of the tensor brevis is exceedingly simple, being without the numerous subdivisions so often observable in the same muscle in the (hypothetically) higher birds. Nor is there the patagial fan, the junction between the tendons of the longus and the brevis, a character, again, so frequent among the larger birds. The complications in this case seem to be much more likely an effect of specialisation than that the simple conditions observable in the picarian and other allied birds should be due to a process of degeneration.

The simplicity and relative shortness of the gut is a matter which is perhaps of importance in this connection. The average relative length of the gut is much less on the whole in the Pico-Passerines than in any other group. There are, of course, some startling exceptions, but the general statement holds. As to the cæca, it must be confessed that they are as a rule small or absent. But the Coraciidæ and the Todidæ are exceptions. The fact that both lobes of the liver are frequently in this group of birds enclosed in separate compartments, separated from the subomental space, seems to me to be a vestige of a condition such as that which is found in the crocodile. On the other hand certain of the organs of the body show great variety and specialisation ; particularly is this the case with the syrinx ; but to find most of the organs of the body in a primitive condition, while others are greatly specialised, is precisely what is so often found in what are believed to be archaic types.

There can be no doubt that the *Archæopteryx*, far though it may have diverged from the ancestral stock, has retained more of the reptile than any other form known to us. One or two of the characters are shared by that large assemblage of birds which has been termed the **Anomalogonatæ**.

In the first place the structure of the foot of the *Archæopteryx* is that found in passerine birds. That the primitive bird was arboreal seems likely, and it is not surprising to find that this mode of life has led to various specialisations in the foot, such as we find in the hornbills, &c. The

Archæopteryx has the smallest number of cervical vertebræ of any bird (ten), a fact which recalls the nine vertebræ or so of the cervical region of lizards and crocodiles. Now, among other recent birds there are none which have a smaller number than that possessed by certain passerines. Fifteen is perhaps the average number of these vertebræ among birds; but among passerines the low number of thirteen is to be met with. Nearly all the **Anomalogonatæ** are holorhinal, as was the *Archæopteryx*. It is doubtful, however, whether this particular fact advances my argument, as there are reasons (see p. 144) for considering the schizorhinal arrangement to be the older, and for looking upon the holorhinal as a derivative.

There remains for consideration the large assemblage of birds which, taken together, correspond to the Homalogonatæ of GARROD. In a preliminary way we may regard, as I have pointed out above, four characters at any rate as primitive.

These are the presence of basipterygoid processes, the possession of two carotid arteries and of the fifth cubital remex, and the simplicity of the intestinal coils. The only birds which have all of these are certain **Galli** and certain **Turnices**. Allowing for the degeneration of the wing, the struthious birds may be referred to a nearly equally low place in the system. In many groups, however, we find a near approximation to this presumably primitive condition. Thus among the anserine birds *Palamedea* is deficient only in the fifth remex. Among 'Grallæ' it is only the same character that is wanting in certain forms to complete the four requisite characters. And some of them—*e.g. Cariama, Psophia*—have this elsewhere missing feather, though those particular forms have not some of the other characters. *Opisthocomus*, the cuckoos, and Musophagidæ are not far off from the base of the series, while the trogons, if they had both carotids, would be among the (hypothetically) lowest groups. The facts in question may be thus tabulated :—

—	Carotids	Basipt. Proc.	Fifth Remex	Intestine
Colymbi .	2 or 1	—	—	Complex
Sphenisci .	2	—	+ ?	Complex
Tubinares .	2	+ or —		Complex
Steganopodes .	2 or 1	—	—	Complex
Herodiones	2	—	—	Simple (in Platalea)
Anseres .	2	+	—	Complex
Palamedeæ	2	+	—	Simple
Accipitres .	2	+ or —	—	Complex
Tinami .	2	—	+	Complex
Turnices .	2 or 1	+	+	?
Galli .	2 or 1	+	+ or —	Simple
Ralli .	2	(*Podica*)		Complex
Grues .	2	—	+ (*Psophia*, &c.)	Complex
Otides .	2 or 1	—	—	Complex
Limicolæ .	2	+ or —	—	Simple (some)
Pterocletes	2	+	—	Complex
Columbæ .	2	+ or —	or —	Complex
Alcæ .	2 or 1	—	.	?
Opisthocomi .	2		+	Complex
Cuculi .	2	—	+	?
Musophagi .	2	—	+	Complex
Struthiones .	2 or 1	+	+ (?)	Simple
Psittaci .	2 or 1	—	—	Complex
Striges .	2	+	—	Simple
Caprimulgi .	2	+ or —	—	Simple
Pico-Passeres .	2 or 1	— or +	+ or —	Simple (?)

I have italicised in the above list those groups which come near to the supposed primitive condition; and it will be observed that those groups which are italicised are enough to account for the ancestry of the rest. The main difficulty is perhaps presented by the **Cuculi**; but it may well be that that group will ultimately prove to have a simple intestinal tract. It will be observed also that the groups in question comprise representatives of all the five large divisions of existing birds admitted by GADOW, and of all the correspondingly large divisions of FÜRBRINGER. Incidentally, therefore, I find myself in reassuring agreement with those authorities. In the systematic part of this work I have to some extent discussed the mutual affinities of these different groups.

Group ORNITHURÆ

ANOMALOGONATÆ [1]

Definition.— Generally quincubital. Ambiens and accessory femoro-caudal always absent ; biceps slip rarely present. Cervical vertebræ, 13-15. Atlas generally perforated by odontoid process. Skull holorhinal.

This group of birds is equivalent to the similarly named group of GARROD, with the sole addition of the **Striges**. The total absence of the *ambiens*, even of all traces of that muscle (see p. 95), is to my mind a sufficient reason for bracketing together all these birds. I am of opinion that the ambiens is not degenerate in them, but that it has not yet appeared.[2] It must be admitted that there are not many other characters that run through the whole group. There are, nevertheless, certain peculiarities of structure that are confined or nearly confined to this group. Thus, with the sole exception of the parrots, a forked manubrium sterni is a peculiarity of the Anomalogonatæ ; so too, with the same exception, is the presence of a cucullaris propatagialis. Again, it is only here and in the parrots that the syrinx has so complicated a musculature. The very prevalent shortness of the intestine is a fact (not, it is true, without exceptions) not to be ignored in considering the claims of this group to existence. The feet are nearly always anisodactyle or zygodactyle, there being but few other birds not referable to this group in which that structure of foot is to be found ; and those groups will be treated of later as possible allies of the present. Powder-down patches are exceedingly rare in this group, and but few possess the expansor secundariorum.

I allow nineteen separate families of this group (whose main characters are given in the table), of which some may be united more closely than others.

---

[1] SEEBOHM, ' An Attempt to diagnose the Pico-Passerine Group of Birds,' &c., *Ibis*, 1890, p. 29.

[2] A case which appears to contradict this statement is dealt with on p. 163.

	After-shaft	Oil Gland	Fifth Remex	Leg Muscles	Flexor Tendons	Exp. Sec.	Cuculiaris Patagialis	Cæca	Carotids	Skull	Vomer	Basipterygoid Processes	Geographical Distribution
Passeres	+,0	Nude		A×(Y)	1. vii.	Rud.	+,r.	R	1	Æg.	+	0	Universal
Coraciidæ	+	N.		A×Y	v.	+	R.	+	2	Desm.	Narrow	0	Old World
Meropidæ	+	N.	+	A×Y	v^b.	+	R.	+	2,1	Desm.	Narrow	0	Old World
Todidæ	—	Feathered	+	A×Y	v^b.	+	R.	+	2	Desm.	Rud.	0	West Indies
Trogonidæ	0	N.	+	A×	viii.	0	R.	+	1	Sch.	+	+, rud.	America, Africa, Asia
Alcedinidæ	0	F. n.	+,0	A.	v, v^b.	+,0	R.	0	2	Desm.	0	0	Universal
Bucerotidæ	0	F.	+	A.Y	v, v^b.	0	0	0	2,1	Desm.	0	0, rud.	Africa, Asia
Coliidæ	+	F.	+	A×Y	v.	0	R.	0	1	Desm.	0	0	Africa
Upupidæ	0	F.		A.Y	vii.	0	+	0	1	Desm.	0	0	Old World
Picidæ	+	F.		A×(Y)	vi.	0	+	0	1	Saur.	Double	0	Universal, exc. Australia, Madagascar, and Polynesia
Rhamphastidæ	+,0	F.	—	A×(Y)	:.	0	+	0	1	Desm.	+	0	South America
Capitonidæ		F.	+	A×Y	vi.	0	+		1	Æg./Desm.	+	0	Central and South America
Bucconidæ	0	F. n.	—	A×Y	vi.	?	?	?	2	Desm.	Rud.	0	America, Africa, Asia
Galbulidæ	+	N.	+	A×(Y).		+	?	+	2	Desm.	Rud.	0	Central and South America
Momotidæ		F. n.	+	A×Y	v^b.		R.	0	2	Desm.	Stout	0	Central and South America
Cypselidæ	+	N.	+,0	A	v.	0	R.	0	2,1	Æg.	+	0	Universal
Trochilidæ	—	N.	+	A	v^c.	0	R.	0	1	Sch.	+	0	South America
Caprimulgidæ	+,0	N. 0	0	(A)×Y	v.	+,0	R.	+	2	Desm./Sch.	+	+,0	Universal
Strigidæ	+,0	N.	0	A	1	0	R.		2	Desm.	+		Universal

In the above table I have used twelve characters. The
Picidæ, Rhamphastidæ, and Capitonidæ agree in ten of these,
and are undoubtedly nearly allied birds. The Bucconidæ are
unfortunately not well known, but of the nine characters
which are set forth in the table they agree in eight with one
or other of the three families just mentioned.

The Coraciidæ and Meropidæ agree pretty well in all the
characters except the exact arrangement of the deep flexor
tendons, the carotids being variable. There will be but little
violence done if these groups are associated. The Cypselidæ
and Trochilidæ clearly come near together, agreeing as they
do in nine of the selected characters.

The Caprimulgidæ on the one hand, and the owls on the
other, each form a distinct group with no such near affinities
to any of the others as those which we have been considering.

The trogons are the only other group with basipterygoid
processes; they do not, however, come very near to the
Caprimulgidæ or to the owls; out of the selected twelve
characters they have at least four in which they totally
differ from the first, and five in which they differ from the
second. This group may be left as equivalent to the other
compound groups already considered.

The Todidæ are placed by GADOW close to the motmots;
by FORBES, on the other hand, they are widely separated.
They agree with them in ten out of the twelve, showing
thus, it appears to me, a considerable nearness. They agree
equally closely with the Meropidæ on the one hand and the
Galbulidæ on the other. These four groups appear to me to
be worthy of association into one larger group.

We have left the Bucerotidæ, Upupidæ, Alcedinidæ, and
Coliidæ. The kingfishers undoubtedly come near to all of
these, in only at most four of the characters differing from
any one of them; but they are as near to the motmots and
Rhamphastidæ. They should form a group apart.

The same may be said of the colies; they are very near
to the hornbills and kingfishers, but equally near to the motmots and Rhamphastidæ; we may therefore place them in a
group apart. On the other hand the Bucerotidæ come nearer

to the Upupidæ than to any other groups except the colies
and Alcedinidæ ; we may therefore unite them.

Finally as to the passerines. It is not possible to place
them very definitely nearer to one than to another of the
groups enumerated. They differ at the lowest in four
characters from any. They are, perhaps, furthest away from
the Bucerotidæ ; the two groups, in fact, are typical members
each of them of the piciform and passeriform birds of
GARROD. Their leanings are perhaps to the Cypselidæ and
to the Pici.

These conclusions may be tabulated as follows :—

Group A.—Aftershaft present ; fifth remex always present.
Muscle formula, AXY.   Expansor secundariorum present.
Desmognathous. Vomer present.

This group contains the families Coraciidæ, Meropidæ,
Todidæ, Galbulidæ, and Momotidæ.

By both GADOW and FÜRBRINGER the Galbulidæ are
placed nearer to the Bucconidæ, and by inference to the Pici
(Picidæ, Rhamphastidæ, and Capitonidæ). They differ, however, from these by the presence of the expansor secundariorum, which is so rarely present among anomalogonatous
birds that when present it seems to be of special importance.
The most salient point of agreement between these latter
birds is the form of the deep flexor tendons.   But the
Bucconidæ are so little known that they may be found to
differ more than is at present suspected from the Pici.   In
this case it may be desirable to separate both Galbulidæ and
Bucconidæ from the group with which I now associate them
and place them nearer together.   Of the group as at present
constituted the Todidæ have perhaps the most claims to be
regarded as the most primitive forms.   They have a feathered
oil gland and long cæca, which two characters do not coincide
in any other of the families now under consideration.

The next group, that of the Pici, may be thus defined :—

Group B.—Fifth remex always present ; deep flexor
tendon of type VI.   Expansor secundariorum absent.
Cucullaris propatagialis present.   No cæca.   Vomer
present.

There are other peculiarities that unite these birds, which will be found mentioned on p. 183.

The third group, containing the hornbills and hoopoos, may be thus defined :—

Group C.— Oil gland feathered ; fifth remex present. Muscle formula, AXY. No expansor secundariorum. Cæca absent. Skull desmognathous.

The colies are the only birds among the Anomalogonatæ besides the Caprimulgidæ which possess the biceps slip ; there is a rudiment of this structure in *Bucorvus.*

The remaining families are treated of separately and need not be defined here.

This arrangement, nearly coincident with that of GADOW, is widely different from that of GARROD and FORBES. The former divided the birds (excl. Striges) into the three groups of Piciformes, Passeriformes, and Cypseliformes. They were thus defined :—

Piciformes.—Oil gland tufted ; cæca absent ; external branch of pectoral tract given off at commencement of breast. Muscle formula, (A)XY. Picidæ, Capitonidæ, Upupidæ, Buccerotidæ, Coliidæ, Alcedinidæ, Momotidæ.

Passeriformes.—Oil gland nude ; cæca present ; pectoral tract simple or with external branch given off beyond middle of breast. Muscle formula, AX(Y). Passeres, Bucconidæ (?), Galbulidæ, Coraciidæ, Meropidæ, Trogonidæ.

Cypseliformes.— Oil gland nude ; cæca absent. Muscle formula, A.

To these FORBES added a fourth group Todiformes, on account of its having at once cæca and a tufted oil gland. This latter group was regarded by him as most nearly representing the ancestral anomalogonatous bird.

PASSERES

Definition. **Oil gland nude. Skull ægithognathous. Atlas perforated by odontoid process. One carotid, left. Cæca present, small.[1] Muscle formula, AXY.[2] No biceps slip or expansor secundariorum.**

This is an enormous group of birds, numbering over 6,000 species, which are spread over the entire globe. As a rule they are of small or moderate size; but some large species, such as the raven and other Corvidæ, are included in the assemblage. In spite of the numerous species and of a certain amount of differentiation in external form, the group is structurally a uniform one, the difference being in characters which have not as a rule been regarded as of primary importance.

In all **Passeres** the *foot* has this structure : the first toe is directed backwards, and none of the other toes are ever changed in position. *Cholornis* is exceptional in that the *fourth* toe is abortive.

GARROD noted, some years since, another peculiar character of the group, which may possibly be universal, with the exception of *Menura* and *Atrichia*. This concerns the arrangement of the tendon of the patagialis brevis ; and the passerine disposition may be understood from a comparison of the two descriptions. In the passerine the tendon of the muscle does not end upon the tendon of the extensor, as it does in the picarian bird, but, though attached to it firmly, retains its independence and runs back to be attached near it to the extensor condyle of the radius. This difference, though small, appears to be constant to the **Passeres**. Another character dealt with on p. 41 of the present work may be also an exclusively passerine character. In birds belonging to the present family the oblique septa, instead of having a separate attachment to the sternum, are either not attached

---

[1] In a specimen of *Gracula intermedia* the cæca were as long as half an inch, an exceptional length.

[2] Very rarely AX −.

at all, lying loosely over the liver, or have but one attachment, which they share with the falciform ligament.

It is very general—but there are a few exceptions, which
will be dealt with later—for the sternum to have a forked
manubrium in front and a single pair of notches behind

The number of *rectrices* present among passerine birds
varies. They are, indeed, completely absent in wrens of the
genus *Pnoepyga*. Twelve is the usual number, but ten only
occur in *Xenicus*, *Phrenotrix*, and *Edolius*, while *Menura
superba* has sixteen. The aftershaft is 'very weak and downy'
when present, and is sometimes (e.g. *Paradisea rubra*) absent
altogether. As to the *pterylosis*, we may take *Ampelis
cedrorum*, recently described by SHUFELDT,[1] as an example
of passerine pterylosis, mentioning afterwards such variations from this type as are met with. In the bird in question
the dorsal tract is exceedingly narrow from the origin in the
fairly continuous feathering of the head down to a point in
the pelvic region, where it is greatly dilated to form a
diamond-shaped area; this again contracts to the original
dimensions, and the tract concludes a little way in front of
the base of the oil gland. From the lateral angles of the
diamond-shaped area a tract runs to the feathering of the
legs. On either side of the oil gland arises a short tract,
which does not leave the trunk, but appears to be the hinder
part of the femoral tract of some other birds. It is perhaps
noteworthy that it is very similar to the corresponding one
of the Bucconidæ and Capitonidæ, also, however, of the
kingfishers.

The ventral tract divides early on the neck, and on the
breast is increased in breadth, the outer rows of feathers
being much stronger than the inner set; the latter (not the
former) are continued down to the cloacal orifice, which
they completely surround by a narrowish band of feathers.
The humeral tracts, which are strong, are connected with
the head feathering by a special neck band as wide as the
dorsal tract. NITZSCH does not figure this connection or
that of the diamond-shaped dorsal area of feathers with the

---

[1] In a paper dealing with Macrochires in *J. Linn. Soc.* vol. xx.

femoral feathering; but apart from this the same type of pterylosis occurs in many types, such as *Motacilla, Certhia, Oriolus*, &c. The principal variation is offered by those passerines in which the spinal widened area is not solid, but encloses a space.

We find this in *Coracina cephaloptera* and *Seleucides*. In others (e.g. *Eurylæmus*) there is the same ephippial space within the dorsal tract, but the posterior sides of the diamond-shaped space are formed by a single row of feathers, which contrast with the mass of feathers which form the antero-lateral boundaries of the space. This arrangement culmi-nates in, for example, *Hirundo* and *Diphyllodes*, where the dorsal tract forks, and there is no connection between the ends of the fork and the single posterior part of the spinal tract.[1]

The *skull* of **Passeres** has been mainly investigated by PARKER and by SHUFELDT.[2] *Corvus* may be taken as a type, and the divergences therefrom noted later. As are all passerines, it is ægithognathous; the maxillo-palatines extend obliquely outwards and backwards; they approach each other in the middle line, and expand over the vomer. The vomer is broad and bifurcate both anteriorly and posteriorly; from the anterior horns a small separate piece of bone goes

---

[1] For passerine pterylosis, see, in addition to NITZSCH, GIEBEL, 'On Ptery-losis of *Paradisea*,' *Zeitschr. f. d. ges. Naturw.* xlix. p. 143, and for *Philepitta* ibid. p. 490; SHUFELDT has described *Chamæa, Journ. Morph.* iii. 1889, p. 475; HELLMANN, 'Beitrag zur Ptilographie u. Anatomie der *Hirundo rustica*,' *J. f. O.* iv. 1856, p. 360; GADOW, 'Remarks on the Structure of Certain Hawaian Birds,' in WILSON and EVANS's *Aves Hawaienses*, ii. Sept. 1891.

[2] For osteology of **Passeres** see PARKER, 'On the Structure and Development of the Crow's Skull,' *Month. Micr. Journ.* 1872, p. 217; 'On the Development of the Skull in the Tit and Sparrow Hawk,' ibid. 1873, pp. 6, 45; 'On the Development of the Skull in the Genus *Turdus*,' ibid. 1873, p. 102; 'On Ægithognathous Birds,' *Trans. Zool. Soc.* ix. p. 289, x. p. 251; MURIE, 'On the Skeleton and Lineage of *Fregilupus*,' *P. Z. S.* 1874, p. 474. LUCAS, 'Notes on the Osteology of the Thrushes,' &c., *P. U. S. Nat. Mus.* xi. 1888, p. 173.

SHUFELDT, 'Osteology of *Eremophila*,' *Bull. U. S. Geol. Surv.* vi. p. 119; 'Osteology of *Lanius*,' ibid. p. 351; 'On the Skeleton in the Genus *Stur-nella*,' &c., *J. Anat. Phys.* xxii. p. 309; 'Osteology of *Habia Auk*,' v. p. 438; 'Osteological Notes on Puffins and Ravens,' ibid. p. 328; GIEBEL, 'Zur Osteo-logie d. Gattung *Ocypterus*,' *Zeitschr. f. d. ges. Naturw.* xxi. p. 140.

on each side to the maxillo-palatines. These are the septo-maxillaries of PARKER, and appear to remain perfectly distinct in *Corvus cornix*, but not in *Corvus frugilegus*. There are no basipterygoid processes; the pterygoids have a long foot-like attachment (as long as the free part of the bone) not only to the palatines, but to the interorbital septum also. The nares are holorhinal and pervious. In *Corvus cornix* the lacrymals reach the jugal bar; there is practically no orbital portion, the descending limb being closely attached to, but not fused with, the broad and thick ectethmoid. The mandibular rami have a large oval perforation near to the articular surface.

Among genera nearly related to the Corvidæ are various slight modifications of skull structure. In *Manucodia*, for instance, the rostrum is broadly ossified and fused with the co-ossified palatal plates of the maxillæ. The nasal septum is complete, and the conjoined ectethmoids and lacrymals are enormously swollen. In *Ptilonorhynchus violaceus* the ectethmoids and lacrymals are separate, though in contact; contrary to what is found in *Corvus*, it is the latter and not the former which border the orbit above. The palatal conditions of *Manucodia* are repeated and emphasised in *Gymnorhina* and *Strepera*. There is a firm union across the middle line in front of the vomer, with which, indeed, the anterior horns of the vomer are ossified in *Strepera*. In both birds, moreover, the pterygoids are fused with the palatines, and the nostrils are partly obliterated by bony growth.

The 'desmognathism' thus produced in the crows of 'Notogæa' is not limited to that family. In *Pheucticus* and in *Cracticus cassicus* there is the same state of affairs. Other features in which the passerine skull shows variations are the maxillo-palatines, vomer, and pterygoids; in *Gracula javanensis*, for example, the pterygoid has a very limited area of articulation with the palatine; there is no expanded foot, as in crows, &c.; the maxillo-palatines are very long and slender, actually reaching the inner plate of the palatines. The vomer is narrow in the body, though the two anterior 'cornua' are thick. *Trochalopteron* is almost desmognathous

in the sense of HUXLEY; the two maxillo-palatines, dilated
at their ends, come absolutely into contact under the vomer,
the middle part of which they completely cover. In *Guis-
calus versicolor* the same arrangement occurs, the maxillo-
palatines being extraordinarily slender before they dilate to
overlap each other and to cover the vomer.

As to the rest of the skeleton, *Corvultur albicollis* may
serve as a type. There are fourteen *cervical vertebræ*; un-
impaired hypapophyses extend from the tenth to the first
dorsal. Five ribs reach the sternum, to the keel of which
the furcula is joined.

The most salient variations from this plan are shown by
many genera in which the furcula does not join the keel of
the sternum, by *Corcorax*, in which six ribs articulate with
the sternum, and above all by *Gymnorhina*, where there is
a catapophysial canal, beginning with the seventh and
ending with the tenth vertebra. In this and other forms
the hypapophyses of the last cervical and the first dorsal
vertebræ are reinforced by strong lateral catapophyses.

In the table on the opposite page are some intestinal
measurements in inches. The most noteworthy fact in the
anatomy of the alimentary tract is the absence of the gizzard
in certain tanagers.[1]

In almost all **Passeres** [2] the flexor hallucis is absolutely
independent of the flexor communis, there being no vinculum
at all.

These characters are—adding to them those used in the
diagnosis of the family—the only ones that are universal, or
nearly so, among the **Passeres**. There are, however, a number
of anatomical features in which the passerines show differences among themselves. The most abnormal **Passeres** on

[1] Cf. FORBES, 'On the Structure of the Stomach in certain Genera of
Tanagers,' *P. Z. S.* 1880, p. 143, who quotes LUND's earlier (1829) paper on the
same matter.

[2] For myology of passerines see, in addition to GARROD, KLEMM, 'Zur
Muskulatur der Raben,' *Zeitschr. f. d. ges. Naturw.* xxiii. p. 107; SHUFELDT,
The Anatomy of the Raven, London, 1890; C. L. NITZSCH, ' Ueber die Familie
d. Passerinen,' *Zeitschr. f. d. ges. Nat.* xix. p. 389; C. B. ULRICH, ' Zur
Characteristik d. Muskulatur d. Passerinen,' *ibid.* xlv. p. 28.

—	s. I.	L. I.	Cæca
Myiophoneus Horsfieldi	29	·9	·2
Geocichla citrina .	9	·5	·1
Nesocichla eremita .	15	·8	·4
Rimator malacoptilus .	5·1	·5	—
Cracticus cassicus .	10·2	1·2	·2
Pastor roseus . .	13·75	1	·2
Seleucides nigra .	14·85	1·3	·4 , ·5 ,
Manucodia atra . . .	14·15	1	·3
Uranornis rubra . . .	16·2	1·3	·3
Entomyza cyanotis . .	12	1	·25
Garrulax albogularis . .	13	1·2	·1
Cyanocorax cyanopogon .	13	·5	·25
Cissopis leveriana . .	10	—	·25
Struthidea cinerea . .	. ...	1	·2
Hirundo rustica . . .	6·25	·75	·6
Anthornis melanura . .	6·25	·75	·12
Strepera graculina . .	20	-	·5
Barita destructor . . .	13	—	·15
Tanagra sayaca . . .	8	·5	·12
„ festiva . .	6·5	—	. .
Ptilonorhynchus holosericeus	10·25	·75	·25
„ „	10·5	1·25	·5 (F.)
Gracula javanensis .	23	1·5	·25
Gymnorhina leuconota	22	1·5	·4
Corvus coróne . .	29·75	2·5	·5
Ampelis garrulus . .	7	·5	·2

the whole are the broad-bills Eurylæmidæ and the Australian *Menura* and *Atrichia*, which form a sub-family, Menuridæ. There are differences of opinion as to which of these is most independent of the normal **Passeres**. FÜRBRINGER separates the Menuridæ, GARROD and FORBES [1] the Eurylæmidæ. We should explain that in first of all discussing this particular point we are not proceeding in historical sequence. It was the syrinx that first of all attracted the attention of JOHANNES MÜLLER, whose divisions of the **Passeres** were the earliest to be based upon anatomical structure ; and in the sequel we shall show that his divisions are in the main correct, even allowing for our greatly extended knowledge. It is, however, in our opinion, beyond cavil that the major subdivisions of

[1] GARROD's contributions to our knowledge of passerine birds are as follows : ' On some Anatomical Peculiarities which bear upon the Major Divisions of the Passerine Birds,' i. *P. Z. S.* 1876, p. 506; 'Notes on the Anatomy of Passerine Birds.' ii. *P. Z. S.* 1877, p. 447 ; iii. *ibid.* p. 523 ; iv. *ibid.* 1878, p. 143. The following papers are due to FORBES : ' Contributions to the Anatomy of Passerine Birds,' i.- vi. *P. Z. S.* 1880-2.

the group first concern one of the two views that we have referred to above. The reasons which lead us to agree with GARROD and FORBES's separation of a group Desmodactyli, as opposed to the remaining **Passeres**, which are to be so-called Eleutherodactyli, are as follows : The Menuridæ (Pseudoscines of SCLATER and FÜRBRINGER) are clearly in some respects degenerate forms. The clavicle has become rudimentary, and the muscles of the syrinx, while approaching the typical oscinine form, where these muscles are numerous and strong, have become to some degree weakened by loss.

On the other hand the Eurylæmidæ, while they have retained the typical mesomyodian syrinx—typical, because it

FIG. 85.—SYRINX OF *Eurylæmus*. FRONT VIEW. (AFTER FORBES.)

FIG. 86. SYRINX OF *Cymbi-rhynchus*. SIDE VIEW. (AFTER FORBES.)

is distinctive of the vast majority of birds—have retained the plantar vinculum,[1] which in other passerines has been lost ; they have also a simple manubrium sterni, this appendage being forked in other passerines. In the feet too the third and fourth toes are largely bound together, giving to the group the name of desmodactyli.

The family Eurylæmidæ [2] contains the genera *Eurylæmus*, *Calyptomena*, *Serilopha*, *Psarisomus*, *Corydon*, and *Cymbi-rhynchus*, all Old-World. They have no aftershaft, and the oil gland is, of course, nude. There are twelve rectrices in

[1] Occasionally absent in *Calyptomena viridis*.

[2] FORBES, ' On the Syrinx and other Points in the Anatomy of the Eurylæ-midæ,' *P. Z. S.* 1880, p. 380.

Cymbirhynchus. There is a wide ephippial space, elongated and oval in form; the narrower parts of the tract behind are two feathers wide.

The tongue (of *Cymbirhynchus*) is bifid at tip and elongated and cordate.

The following are intestinal measurements of two species :

	Cymbirhynchus macrorhynchus	Eurylæmus ochromelas
Small int.	7·75	5·75
Large int.	1·25	·75
Cæca	·1	·05

Left liver lobe is the smallest.

The following scheme gives the classification of the Passeres according to GARROD and FORBES :—

	New World	Old World
I. DESMODACTYLI		Eurylæmidæ
II. ELEUTHERODACTYLI		
A. Mesomyodi		
a. Heteromeri	Pipridæ Cotingidæ	
b. Homœomeri		
Haploophonæ	Tyrannidæ (*Rupicola*)	Pittidæ Philepittidæ Xenicidæ
Tracheophonæ	Dendrocolaptidæ Furnariidæ Pteroptochidæ	
B. Acromyodi		*Abnormales.* Atrichiidæ Menuridæ *Normales* (Oscines)

Of the remaining Passeres the **Mesomyodi** (also sometimes called Oligomyodi) are divided into two subdivisions, according as to whether the chief artery of the leg is the femoral or sciatic. In the **Heteromeri** (with the exception of *Rupicola*) it is the femoral, in the others (**Homœomeri**) the sciatic. All the Mesomyodi have but one pair of muscles upon the syrinx or none at all. But the name of Mesomyodi is derived from the fact that these intrinsic muscles are attached to the middle of the bronchial semi-ring which bears them. The **Haploophonæ** are those mesomyodians in which the syrinx is quite

N 2

of the normal fashion ; the **Trachephonæ** are those in which
the last rings of the trachea are much modified, and the
syrinx may be termed tracheal.

The Mesomyodi with a tracheobronchial syrinx comprise
representatives from both the Old and the New Worlds.   In
them the syrinx presents a varied form, coupled with the fundamental resemblance indicated.   JOHANNES MÜLLER has
figured and described a number of genera.   GARROD has
figured and described others.   In *Lipaugus cineraceus* (of
the family Cotingidæ) the intrinsic muscle is of great width,
which seems to foreshadow its division in the Oscines into a
complex of muscles ; it is attached to the third bronchial
semi-ring.

The first and second bronchial semi-rings resemble the
tracheal in their flatness, depth, and close approximation.
Those which follow are slightly ossified throughout.   In
Heteropelma and *Chiromachæris*, which are Pipridæ, the
syrinx is very similar.   In *Pipra leucocilla* the intrinsic
muscle has a tendency to split into two, a further approximation to the Oscines.   In *Hadrostomus aglaiæ*, a cotingid, the
wide and thin intrinsic muscle is attached to the first bronchial semi-ring.   This semi-ring is close to the last tracheal
ring, and is like it in structure, being deep.   The next
bronchial semi-ring is separated by a considerable interval,
and the third by a wider interval still, from the ring in
front.

In the Madagascar *Philepitta*, which FORBES [1] was the
first to refer definitely to the present group of passerines, the
structure in some respects recalls that of the Eurylæmidæ.
The manubrium sterni is but slightly bifid ; but it has in
the normal passerines no vinculum.   The syrinx, on the
other hand, differs from the Old-World Mesomyodi by the
details of its structure (see figs. 87, 88).   The different arching
of the bronchial semi-rings leaves great membranous spaces
in the wall of the syrinx.   The first two semi-rings are very
concave upwards ; the two following are not so markedly
concave ; the next is concave in the reverse direction.   The

---

[1] 'On some Points in the Structure of *Philepitta*,' &c., *P. Z. S.* 1880, p. 387,

intrinsic muscles are wide and thin, nearly in contact with each other before, and behind they are attached to the first semi-rings everywhere but at their tips. *Pitta* has a much simpler syrinx, approximating so far to that of the Eurylæmidæ ; the muscles are thin and accurately median in insertion. *Pitta* is unique among passerine birds by reason of the deep temporal fossæ of the skull, which nearly meet

FIG. 87.—SYRINX OF *Philepitta*. SIDE VIEW. (AFTER FORBES.) FIG. 88.—SYRINX OF *Philepitta*. FRONT VIEW. (AFTER FORBES.)

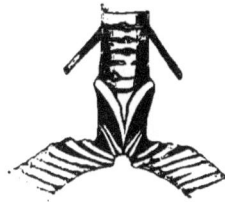

behind, in a way that is seen in some other birds not passerine.

In the New Zealand *Xenicus* and *Acanthisitta* [1] there are

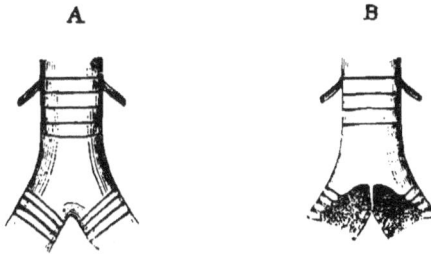

A B

FIG. 89.— SYRINX OF *Xenicus*. A. FRONT VIEW. B. BACK VIEW. (AFTER FORBES.)

only ten rectrices, twelve being the number characteristic of the majority of the **Passeres**. The syrinx of *Xenicus* as seen in the annexed figure is quite typically mesomyodian. The last few tracheal rings are consolidated into a large box, to the top of which the intrinsic muscles (small and median in insertion) are attached.

[1] FORBES, ' On *Xenicus* and *Acanthisitta*,' &c., P. Z. S. 1882, p. 569.

We now come to the Tracheophonæ, which are exclusively American in range. Their distinguishing mark is, of course, the tracheal syrinx, whose general structure has been already explained (see p. 67). Besides these general agreements the Tracheophonæ show variations in structure. *Heterocnemis*, for instance, is unique among them and the Mesomyodi in general by the existence of a bilaminate tarsus, as in the Oscines. *Conopophaga* and a few others[1] have a four-notched sternum, while *Furnarius*, *Synallaxis*, and a few others have, as has no other passerine, a schizorhinal skull. Again, the maxillo-palatine of the Dendrocolaptidæ, Furnariidæ, and Pteroptochidæ are like those of oscinine Passeres in being slender and curved backwards, instead of being comparatively wide and blunt, as in other Mesomyodi. *Phytotoma* is unique among passerine birds for the nasal gland groove on the frontal bones, as in so many water birds.[2]

The remaining group of Passeres, the **Acromyodi**, are sometimes called **Oscines** and sometimes **Polymyodi**, the latter term having reference to the numerous intrinsic muscles of the syrinx. It was discovered by KEYSERLING and BLASIUS that, with the exception of the Alaudidæ, the **Oscines** have a bilaminate tarsus, the hinder surface being covered by two closely apposed scutes. It sometimes happens (in the forms which are on that account spoken of as 'booted') that the anterior face of the tarsus is covered by a single scute. The syrinx of these birds is complex in the multitude of its muscles, of which there are four or five pairs. The only exception to the muscle formula AXY − exists in this group; in *Dicrurus* there is the reduced formula of AX −. *Ocypterus*, the only passerine with powder down, is referable here. Referable to this group, but separated from the more normal members as '**Abnormales**' by GARROD, are the two genera *Atrichia* and *Menura*, which are also regarded as the types of separate families. These two anomalous birds are by

---

[1] FORBES, 'On some Points in the Anatomy of the Genus *Conopophaga*,' &c., P. Z. S. 1881. p. 485.

[2] PARKER, *Ægithognathous Birds*, ii. p. 258.

some systematists placed in a group **Pseudoscines**, equivalent to the remaining Passeres.

The anatomy of the two genera has been mainly investigated by GARROD, who studied principally the syrinx. They are purely Australian in range. According to FÜRBRINGER this group of passerines is in many respects intermediate between the other passerines and the **Pici**. With the latter group *Menura* probably and *Atrichia* certainly agree in the following myological points: the origin of the rhomboideus profundus from the pelvis; in the tendon of insertion of the supra-coracoideus upon the shoulder joint; in the origin of the latissimus dorso-cutaneus from the ilium and its covering by the leg musculature. GARROD also called attention to the fact that the patagialis brevis was picine and not passerine.

The syrinx of *Menura*[1] has three pairs of intrinsic muscles, which are inserted respectively into the last tracheal ring and on to the second and third bronchial semi-rings. There are three modified bronchial semi-rings. So too in *Atrichia*, where, however, there are but two pairs of syringeal muscles, as shown in the figure.

The clavicles are rudimentary in *Atrichia*; there is no hypocleidium in *Menura*, another picine character. *Menura* has, furthermore, a chain of three supra-orbital bones.

PICI

Definition.—Feet zygodactyle; aftershaft small or rudimentary; oil gland tufted. Muscle formula of leg, AXY (AX); Gall bladder elongated. Skull without basipterygoid processes.

The woodpeckers, which form the first family of this assemblage, **Picidæ**,[2] are a well-marked group of birds, containing about three hundred and fifty species, as allowed by the late Mr. HARGITT.[3] They inhabit most parts of the world, excluding only Madagascar, Australia, and Polynesia.

---

[1] For passerine syrinx see JOH. MÜLLER in *Abh. Berlin. Akad.*, 1845; HERME, *Dissertatio de Avium Passerinarum Laryuge Bronchiali*, Gryphiæ, 1859; and GARROD and FORBES in papers already quoted.

[2] W. MARSHALL, *Die Spechte.* Leipzig, 1889.

[3] *Brit. Mus. Catalogue*, vol. xviii.

Twelve rectrices is the rule for this family, but the outer ones are sometimes feeble, and I found only ten in *Tiga Shorei*. The aftershaft is present. The *pterylosis* varies somewhat. But NITZSCH has pointed out that this 'peculiarity . . . which occurs almost universally among them, is the presence of a small inner humeral tract running along upon the most elevated points of the shoulder parallel to the very broad main tract.' That this second humeral tract also occurs among parrots may be a matter worthy of consideration. It appears, at any rate, to distinguish the woodpeckers from other picarian birds.

In *Picus viridis* the narrow dorsal tract passes down the middle of the neck and ends abruptly at the end of the scapulæ or a little before. Behind this are two oval wide patches, which correspond to the foot of the Y in other birds, which have a dorsal median apterion. There is a break between these patches and the posterior end of the dorsal tract, which runs to the base of the oil gland undivided. On either side of this and of the oil gland is a fainter and narrower tract. In *Tiga Shorei* there is no break between the several regions of the dorsal tract; the anterior part narrows and forks into two branches, consisting of but one row of feathers, which immediately after dilate into the wide interscapular tracts; from the lower angle of each of these a single row of feathers joins the median posterior part of tract. The diamond-shaped spinal apterion is thus completely enclosed within the dorsal tract.

Sphyrapicus nuchalis [1] has a solid spinal tract dilating between the shoulders into a rhomboidal but still solid area, as SHUFELDT says, like a passerine.

In *Centurus striatus* the arrangement is more like that of *Picus viridis*, but slight scattered feathering unites the middle and posterior portions of the dorsal tract.

The ventral tracts of *Picus* divide early in the neck and at the commencement of the pectoral region; each gives off a stronger outer branch. In *Tiga Shorei* the separation between these branches is not nearly so marked as is shown in

[1] 'Observations on the Pterylosis of Certain Picidæ.' *Auk*, 1888. p. 212.

NITZSCH'S figure of *Picus viridis*. This woodpecker is so far much more like a parrot.

In *Jynx* the interior part of the dorsal tract forms a continuous Y, of which the fork is hardly wider than the handle. This is completely *dis*continuous with the median posterior portion of the tract.

The *tensores patagii*[1] are very simple. The tendon is single. There is a conspicuous *cucullaris patagialis*, but no *biceps slip*. Each tendon, both longus and brevis, is reinforced by a tendon from the pectoralis. In *Centurus striatus* the slip to the longus is muscular in origin; it seems to be more usually tendinous.

The *deltoid* is an extensive muscle.

The *latissimus dorsi posterior* appears to be totally absent, as in *Indicator*.

As to the leg muscles, the *accessory femoro-caudal* is always absent. The *femoro-caudal* and the *semitendinosus* are always present; the accessory to the latter may or may not be present, its occurrence in different genera being shown in the following list :—

With an Accessory Semitendinosus	Without an Accessory Semitendinosus
Gecinus viridis	*Picus major*
Gecinus vittatus	*Picus minor*
Leuconerpes candidus	*Picoides tridactylus*
Melanerpes formicivorus	*Sphyrapicus rarius*
Chloronerpes yucatanensis	
Mulleripicus fulvus	
Hypoxanthus Rivolii	
Jynx torquilla	
Dryocopus martius	
Picolaptes affinis	
Tiga Shorei	
Tiga javensis	
Centurus striatus	
Melanerpes erythrocepha-	
lon	
Colaptes mexicanoides	

[1] NITZSCH-GIEBEL, 'Zur Anatomie der Spechte,' *Zeitschr. f. d. ges. Naturw.* xxvii. (1866), p. 477.

The *deep flexor tendons* are like those of toucans and barbets (cf. p. 101).

The *tongue* [1] is elongated, and so are the ceratohyals in relation to it, overlapping and grooving the skull. As a rule among the woodpeckers the right lobe of the *liver* is larger than the left; they are equal in *Xypoxanthus Rivolii*. The *gall bladder* appears to be absent in *Leuconerpes candidus* and in *Xypoxanthus*; it is long and intestiniform (like that of a toucan) in *Gecinus viridis*, *Dryocopus martius*, and *Jynx*.

The *intestines* are without cæca; [2] the following are some measurements :—

Chloronerpes yucatanensis	8½ inches
Picus minor . .	. 12 ,,
Xypoxanthus Rivolii	. 12 ,,
Dryocopus martius	. 20 ,,

The *syrinx* has no remarkable characters. It is quite typically tracheo-bronchial with a pair of extrinsic and a pair of intrinsic muscles.

As to the *skull*, the woodpeckers have an unusual palatal structure, which led PARKER [3] to invent the term Saurognathous in order to express this peculiarity. The palate of the woodpecker has, however, been variously interpreted. Professor HUXLEY, in his paper on the bird's skull, directed attention to an apparent vacuolation of the palatines, which is illustrated in the accompanying figure (fig. 90); a slender bar of bone passes backwards and comes into near relations with an equally slender projection forwards of the palatine bone, thus enclosing a space which in the fresh skull is filled with membrane. The anterior process in the dried skull is sometimes continuous (fig. 90), but sometimes not continuous with the ascending lamina of the palatine. Where it is not continuous Professor HUXLEY found that its appa-

[1] J. LINDAHL, ' Some New Points in the Construction of the Tongue of Woodpeckers,' *Am. Nat.* 1879, p. 43.

[2] Exceptionally present in *Gecinus viridis*.

[3] 'On the Morphology of the Skull in the Woodpeckers,' &c., *Tr. Linn. Soc.* (2), i. 1875, p. 1. See also KESSLER, ' Zur Naturgeschichte der Spechte,' *Bull. Soc. Nat. Mosc.* 1844, p. 285.

rent continuation was a separate bone, which he regarded as the vomer—the vomers thus being paired.

A few years later a different complexion was given to the subject by a paper written by Professor GARROD.[1] He confirmed HUXLEY's description of the supposed vomers, but regarded them merely as the perfectly ossified edge of the imperfectly ossified palatines. This opinion was chiefly based upon the discovery of a small median bone lying between the posterior ends of the palatines (*x*. fig. 90), not observed by HUXLEY. This bone, identified by GARROD with the vomer of other birds, occupies, as he admits, a somewhat posterior position, which is parallel, however, as he also points out, at any rate in *Megalæma*.[2]

PARKER's paper upon these birds is a long and elaborate one, but contains no reference to that of GARROD. The bone discovered by GARROD is figured and described as the medio-palatine, and is invariably figured as distinct from the palatines, between which it lies. PARKER adopts HUXLEY's identification of the vomers, but finds that they are often divisible each into several splints of bone : the connection with the palatines in front and a median series of bonelets (collectively termed by HUXLEY the ossified internasal septum) are spoken of as septo-maxillaries. It may be sometimes noticed —I have observed it in *Leuconerpes candidus*—that the anterior ends of the pterygoids, which in the Pici run for a considerable way over the outside of the palatines, come into actual contact with the commencement of the vomers.[3]

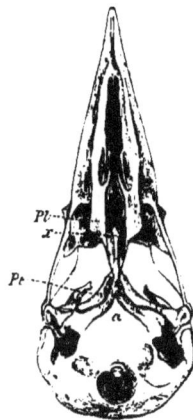

Another peculiarity of the woodpecker's skull is the fre-

FIG. 90.—SKULL OF WOODPECKER (*Gecinus viridis*). VENTRAL VIEW. (AFTER GARROD.)

Pl, palatines ; *Pt*, pterygoids ; *x*, vomer.

---

[1] 'Note on some of the Cranial Peculiarities of the Woodpeckers,' *Ibis*, 1872, p. 357.

[2] See under 'Capitonidæ,' p. 196.

[3] SHUFELDT also ('On the Question of Saurognathism of the Pici,' &c., *P. Z. S.* 1891, p. 122) argues against HUXLEY's view.

quently Y-shaped form of the pterygoids. An outer forwardly directed process is given off near to their articulation with the quadrate. The extraordinary hyoid connected with the long and exsertile tongue is, of course, a marked feature of the group. The long distal piece of the branchials curves over and furrows the cranium of those species in which it exists.

In the wryneck (*Jynx*) the vomers are small, but thoroughly 'picine.' The pterygoids have, however, no forwardly directed process, and do not extend quite so far over the palatines.

There are fourteen *cervical vertebræ*. Hæmapophyses do not extend behind the first dorsal. There is a complete canal for the carotids formed in the cervical region by the hæmapophyses. Four vertebræ enter into the formation of this in *Thriponax Feddeini*.[1] The *sternum* has two notches on either side, and has the spina exterior only, which is bifid. Four ribs reach it. The *clavicles* form a U-shaped furcula, but are expanded above, as in the toucans and barbets.

The **Bucconidæ** is a South American family of zygodactyle birds whose anatomy[2] is at present but little known, and whose affinities are therefore doubtful. It is only provisionally that I place them in the present position.

Bucco maculatus has a nude *oil gland* and twelve *rectrices*. The inferior *feather tract* starts from the symphysis, leaving a bare space on either side; it divides at the angle of the jaw, and thence the two halves remain separate. Each of them gives off a stout outer pectoral branch. The dorsal tract is in two quite separate parts, separated by a good space; the posterior part is forked for half its length, tapering anteriorly to a point.

In *Monasa panamensis* each part of the posterior fork

---

[1] This character is nearly unique among anomalogonatous birds. I describe it, however, above (p. 176) in *Gymnorhina*. The nearest possible ally in which it occurs is a parrot.

[2] GIEBEL, ' Zur Osteologie der Gattung *Monasa*,' *Zeitschr. f. d. ges. Naturw.* xviii. 1861, p. 121.

is connected with the anterior part of the dorsal tract by a single row of very small downy feathers.

In *Malacoptila fusca* ' the *oil gland* has a few fine hairs at the apex' (NITZSCH). The *aftershaft* is absent.

The *tongue* of *Bucco maculatus* is long, flat, and thin; the tip is not lacerated; the base is spiny.

The *muscle formula* of the leg is given as AXY--. Two *carotids* are stated to be present.

The *deep flexor tendons* are like those of the Capitonidae. The flexor hallucis supplies digits I., II., IV. The flexor profundus supplies only digit III. The two where they cross are connected by a small vinculum.

Monasa nigrifrons has fourteen *cervical vertebrae*. The *atlas* is perforated for the odontoid process. The eighth vertebra has two catapophyses, which come into contact at their bases in the middle line; on the ninth is the first median hypapophysis; these continue to the first dorsal; on the thirteenth cervical are large catapophyses, which are present, but much smaller, on the hypapophysis of the next vertebra. There are only three free dorsals, all of which bear ribs reaching the deeply two-notched sternum, which has a long spina externa.

The *clavicles* are expanded into a wide, thin, roughly triangular plate at their attachment to coracoids; there is a thin but broadish hypocleidium where they meet below.

The *skull* is desmognathous, and in front of the conjoined maxillo-palatines is a gap of about half an inch in length in the bony palate. The process of the squamosal very nearly reaches the jugal bar. The lacrymal completely reaches the jugal bar in front, and the descending process is in contact, but not fused, with the strong and swollen ectethmoids. The interorbital septum is quite complete, as is the intranarial.

The **Rhamphastidæ** (toucans), with their large bills, serrate slightly at the edge, bear a certain superficial resemblance to the hornbills, with which, however, they have no specially intimate connection. They are, on the other hand, undoubtedly nearly related to the Capitonidæ, and show slighter but still recognisable points of affinity with the Passeres. The toucans, which are purely tropical American, consist, perhaps, of only one well-marked genus, *Rhamphastos*, which, however, has been subdivided into *Pteroglossus*, *Aulacorhamphus*, *Selenidera*, and some others. The toucans have a *tufted oil gland*, an *aftershaft*, and *ten rectrices*.

The *pterylosis* (cf. NITZSCH) is characterised by the wide lateral neck spaces.

The dorsal tract does not divide on the neck, nor is it as a rule [1] continuous throughout. There is a break between the straight anterior portion and the Y-shaped posterior portion. The ventral tract divides in the middle of the neck, and there is a strongly marked outer branch to the pectoral tracts. The femoral tract does not, according to my own experience, arise so early from the spinal tract as NITZSCH figures.

The *intestinal tract* of the toucans is short but voluminous, eighteen inches in *Rhamphastos dicolorus*, nineteen inches in *R. carinatus*. There are no *cæca*.

The *proventriculus* is zonary; the stomach is a weak muscular bag.

In the *liver* the right lobe is larger than the left, three times larger in *R. carinatus*.

The most characteristic feature in the anatomy of the family concerns the *gall bladder*; this has been principally investigated by FORBES.[2] It will be seen that the gall bladder is tubular, and of very great length (4·15 inches in a specimen of *Rhamphastos dicolorus*), as also in the Capitonidæ and a few Picidæ (*qq.v.*) I found in *Rh. tocard* an exceptional state of affairs. As in all (?) toucans, the gall bladder is long, but from the upper extremity two separate cystic ducts arise, which soon fuse to separate again;[3] they open, however, in

---

[1] NITZSCH mentions as an exception an unidentified species. And in *Selenidera* and *Aulacorhamphus* I could find no break.

[2] 'Note on the Gall Bladder, &c., of the Toucans and Barbets,' *P. Z. S.* 1882, p. 94.

[3] This may perhaps be regarded as a faint indication of the *rete* found in some reptiles.

common with each other and with the hepatic duct into the intestine. There are two pancreatic ducts in this species, and one of them is peculiar in having a distinct, though short, diverticulum near to its intestinal orifice.

The *tongue* is long, horny, and feathered along the margins.

The toucans as a rule possess only one *carotid*, the left. But in an example of *Rh. dicolorus* FORBES found both carotids present. The second is much smaller; it is quite pervious, however, and blends with the left at the entrance to the neck.

The *syrinx* is in some respects peculiar. As in some other picarian birds, the last three or four tracheal rings are fused to form a bony box, which is marked by a deep furrow mesially and behind. This region of the syrinx is much compressed antero-posteriorly. The intrinsic muscles are attached to the upper part of the box.

The first bronchial ring is ossified in front; posteriorly each half is L-shaped and fibrocartilaginous. The descending limbs of each L are in the same straight line as the trachea; they are closely applied to each other, but really separated by a membrane of very short extent, the bronchidesmus. They are in close contact with the inner end of the ossified half of the ring. At the lower end of these cartilaginous pieces is a separate rounded bit of cartilage, which is similarly connected with the second ring.

The *tensores patagii* are very simple. The tendon of the brevis is quite undivided. Each has a tendinous slip from the pectoralis. There is a cucullaris patagialis.

The *deltoid* is particularly long in its insertion.

The *anconæus* has no humeral attachment. It arises by fleshy fibres (not a Y-shaped tendon) from the scapula. The *glutæus maximus* extends below the acetabulum; there is no *glutæus externus*.

The muscle formula of the leg is AXY−. The semitendinosus is attached to the leg bone separately from the semimembranosus; it has also a tendinous insertion on to gastrocnemius. The *deep plantar tendons* are as in *Megalæma*.

There are fourteen *cervical vertebræ*. The atlas is perforated by the odontoid process ; strongish median hypapophyses exist on vertebræ C11–D2. In *Rhamphastos ariel* there is in addition a paired hæmapophysis on C10, not present in *Pteroglossus Wiedi*. Five *ribs* reach the *sternum*, which is four-notched and has a spina externa. The *clavicles* are short and do not unite in the middle line ; they are expanded above, as in the barbets. There is a small accessory scapular nodule, from which deltoid in part arises. The *skull* is desmognathous, holorhinal, without basipterygoid processes. The ethmoids are large and quite fused with the descending process of the lacrymal, itself fused with the skull wall. The vomer is truncated at its anterior extremity.

Fam. Capitonidæ.—This family of birds, with which I include, following GADOW, *Indicator*, is of the tropics, both Old World and New.

All the members of the family that have been examined have *ten rectrices*, save *Indicator*, which has twelve. The *oil gland* is feathered, and there is an *aftershaft*.

The *pterylosis*[1] shows some differences in different genera ; in all, however, the ventral tract is single as far as the posterior end of the neck, where it divides into the two pectoral tracts ; each of these, again, gives off, as in the toucans, Bucconidæ, Picidæ, &c., an outer branch. This soon terminates, but the inner branches are continued as far as the cloaca. The apteria being entirely devoid of down feathers, the tracts are easily defined, but in spite of this NITZSCH has, in my opinion, fallen into some errors.

The typical arrangement of the dorsal pterylosis[2] (shown in fig. 91) may be considered to be that of *Megalæma*, and was found to characterise the species *M. asiatica*, *M. virens*, *M. Hodgsoni*, *M. jarensis*, *M. Franklini*. The spinal tract is single and narrow upon the neck, thus leaving con-

---

[1] GIEBEL, ' Pterylose von *Tetragonops*,' *Zeitschr. f. d. ges. Naturw.* li. 1878, p. 377.

[2] ' On the Pterylosis of Certain Barbets and Toucans,' *P. Z. S.*, 1896, p. 555.

spicuous lateral neck spaces ; behind the scapulæ it forks, and the posterior part of the spinal tract is not in continuity with the interior. The former is at first a single tract, but it divides some way in front of the oil gland, which it surrounds, as in the Picidæ and Rhamphastidæ ; but in the

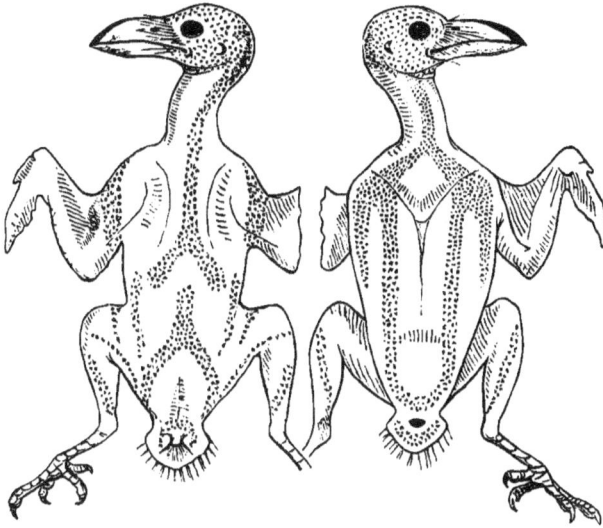

FIG. 91. FEATHER TRACTS OF *Megalæma asiatica*. THE RIGHT-HAND FIGURE SHOWS THE VENTRAL SURFACE, THE LEFT THE DORSAL.

former family there is also a median continuation of the spinal tract, which stops at the base of the oil gland. On each side of the spinal tract is a very narrow lateral tract, which is figured by NITZSCH as existing also in the Rhamphastidæ and the Picidæ. *Xantholæma rosea* shows differences from *Megalæma* ; the spinal tract divides in the usual way, but the tracts rejoin, so as to enclose a diamond-shaped space ; they then again diverge immediately, and end at the sides of the oil gland in the usual way.

Xantholæma has a faint lateral tract on either side. In spite of NITZSCH's figures I am disposed to think that this tract, so universal in the barbets, is at most feeble in the toucans. In the continuity of the anterior and posterior parts of the spinal tract *Xantholæma* is more toucan-like than any barbet.

In *Capito* and *Trachyphonus* NITZSCH has figured an anterior bifurcation of the posterior part of the spinal tract, much more marked in the former genus, where, indeed, I am inclined to suspect a more detailed resemblance to *Xantho-læma* than is suggested by NITZSCH's figure. If this is not the case, the New-World genus will be distinguished from the Old-World genera by a double posterior spinal tract divided by a break from the forked anterior tract.

The *tensores patagii* consist of a slighter tensor longus and a wider brevis ; the tendons of both are simple, and each is reinforced by a tendinous slip from the pectoralis. The brevis tendon is simple, and terminates, as in the last two families, upon the tendon of the extensor metacarpi radialis.

In *Capito*, at any rate, there is a *cucullaris propatagialis*.

FÜRBRINGER had indicated a point of difference among the barbets which concerns the *rhomboideus profundus*. This muscle in *Meiglyptes* (Picidæ) and *Rhamphastos* is a large fan-shaped muscle, arising from the tip of the ilium, as well as from the spinous processes of certain of the dorsal vertebræ, and is inserted from the tip to about halfway down the scapula. In *Capito* this muscle is plainly divided into two—an anterior and a posterior—of which the latter arises from a few vertebræ and from tip of ilium. In *Megalæma* the anterior section of this muscle is itself again divided into two quite distinct parts. This family, like the wood-peckers, toucans, and Passeres, has a *latissimus dorso-cutaneus*, but no *metapatagialis*.

The *deltoid* extends some way down the arm, as in other allied birds and in Passeres ; it has no special scapular slip.

The barbets, like the woodpeckers and toucans, possess a sesamoid ossicle, called the 'scapula accessoria,' which is developed in the scapulo-humeral ligament ; from it arise some of the fibres of the deltoid. This bone, again, is also found in Passeres and owls, and in a host of other birds.[1] The

---

[1] See FÜRBRINGER, p. 229. FORBES laid too much stress upon its classifica-tory value in the present case.

glutæus maximus is both pre- and post-acetabular in origin.
There is no *gl. externus.* In the leg the formula is the
typical picarian one of AXY—. Both *peroneals* are present,
as in the Rhamphastidæ, and with the normal attachments.
The *deep plantar tendons* have been already described and
displayed (fig. 58, p. 102).

The *tongue* (in *M. virens*) is sagittate, with horny
margins, and slightly lacerated apically. In a specimen
of *Megalæma asiatica* the horny apex was bifid, quite
regularly so.

The right lobe of the *liver* in *M. virens* is a little larger
than the left. The greatest peculiarity of the liver, however,
concerns the *gall bladder.* As in the toucans and some
Picidæ it is of great length and intestiniform, *i.e.* of narrow
and regular calibre : in a specimen of *M. virens* it was two
and a half inches long. The presence of a similar gall
bladder has been noted in *M. Franklini* and in *Xantholæma rosea.*

The *intestines* (devoid of cæca) are voluminous but short
—seventeen inches in *M. virens*, twelve in *M. asiatica.*

The *syrinx* of *Megalæma (asiatica)* is of a very simple
tracheo-bronchial form ; the last rings do not fuse at all, but
remain perfectly distinct ; there are no intrinsic muscles.
The extrinsic muscles are attached to the tips of the costal
process.

The *skull* is 'ægithognathous, with a desmognathous
tendency,' holorhinal, and without basipterygoid processes.
The desmognathous tendency is shown by the fact that (in
Megalæma asiatica) the maxillo-palatines may or may not
blend with the nasal septum. In others, e.g. *Pogonorhynchus
bidentatus*, the two bones (maxillo-palatines) blend completely across the middle line. These forms are, therefore,
genuinely desmognathous, except as concerns the vomer.
This bone is truncated, as in the ægithognathous skull, and
its two forward limbs join the maxillo-palatines, as in *Indicator.*

It is single (except, of course, for the anterior bifurcation)
in *Megalæma* ; broader and double in *Gymnobucco calvus.*

As Garrod has pointed out, the vomer of the Capitonidæ differs from that of the Passeres in being truncated *behind* the posterior line of the palatines; in the **Passeres** the truncation is in front of the line. The skull of *Megalæma asiatica* has been described and figured by Parker.[1] Outside the Y-shaped vomer there is on each side a small ' septo-maxillary ' splint, and in front a median unpaired septo-maxillary. The former are, perhaps, the equivalents of the vomers of Picidæ (*q.v.*); the vomer in that case may have something to do with the medio-palatine of *Picus*. The nostrils are impervious and very much reduced by long growths. In *Megalæma virens* and *Gymnobucco* the lacry-mals and prefrontals entirely fuse with each other and with the skull walls to form a solid and imperforate plate of bone, as in *Pterocles*, pigeons, toucans, &c.

There are fourteen *cervical vertebræ*. The *atlas* is per-forated by the odontoid. The axis and the three following vertebræ have median hypapophyses; the next vertebra (the sixth) has a bifurcate one. All these processes are more conspicuous than in the toucans, where, however, they are present. Vertebræ C11–D6 (last) and L1 have strong hypapophyses, that of D2 being bifurcate. In *Gymnobucco* only three dorsals have hæmapophyses. Five *ribs* reach the *sternum*, which is four-notched and has a spina externa; the clavicles do not meet below. The head of the clavicle expands into a wide flat triangular plate.

Indicator is considered to be the type of a distinct sub-family of the Capitonidæ, which also contains the genus *Protodiscus*. *Indicator* has been chiefly investigated by Garrod.[2] Nitzsch, however, previously and Fürbringer subsequently added to our knowledge of the bird. It has *twelve rectrices*; the *pterylosis* has a marked gap between the anterior and the posterior parts of the spinal tract; and the latter appears to encircle the oil gland, as in the Capitonidæ.

[1] 'On the Structure and Development of the Bird's Skull,' *Trans. Linn. Soc.* (2), i. p. 122.

[2] 'Notes on the Anatomy of *Indicator major*,' *P. Z. S.* 1879, p. 930.

The *syrinx* is shown in the figure on p. 61 (fig. 40). The trachea at its lower end is formed into a bony box by a fusion of several rings; the first bronchial semi-ring, to which the single pair of intrinsic muscles are attached, is larger than those which follow. The *muscular anatomy*, excepting for minute details, is the same; one of these small details is that the pectoral slip to the tensor patagii is muscular and not tendinous.

The *skull* is ægithognathous; the vomer is forked in front and joins the separated maxillo-palatines, as in *Megalæma asiatica*; but GARROD has remarked upon the fact that in this character (the widely separate maxillo-palatines) the palate of *Indicator* is more like that of the Picidæ than are the Capitonidæ.

ALCEDINES

Definition.—**Aftershaft absent. Muscle formula of leg, AX. Cæca absent. Both carotids present. Skull desmognathous.**

The kingfishers form a natural assemblage of birds— Alcedinidæ—which, however, show some variations in their structure.

Thus, while they generally agree with the bulk of their relations in having a characteristically tufted oil gland, *the genus Tanysiptera* (three species at any rate) *has a nude oil gland.* In this genus too the rectrices are ten, twelve being the more usual number.

As to pterylosis, the kingfishers are characterised by the possession of down feathers not only on the pterylæ but also on the apteria. There is a difference between *Alcedo* and *Dacelo* in that the latter has a weak dorsal tract in the middle region of the back, it being stronger in front and again near the oil gland. In *Alcedo* the tract is of uniform strength throughout. The kingfishers, like the swifts, sometimes possess and sometimes do not possess the fifth cubital remex.

In *Dacelo gigantea* there are fifteen cervical vertebræ and only three ribs reach the sternum. In *Dacelo Gaudi-*

chaudi there are fourteen cervical vertebræ, and four ribs
reach the sternum—a difference which appears to me to
justify the separation of the latter into a distinct genus,
Sauromarptis. *Halcyon smyrnensis* agrees in these points
with *D. gigantea*.

The last cervicals (from 10 [1]) and the dorsal vertebræ
(including the first lumbar) have well-developed, often
bifurcate, sometimes trifurcate hæmapophyses. The four-
notched sternum has a well-developed spina externa.
As Dr. CUNNINGHAM has pointed out, the lacrymals of
Dacelo are larger than those of *Ceryle*,[2] and, I may add, of
Halcyon and *Todirhamphus* (= *Sauropatis*). The skulls of
the genera above mentioned are remarkable for the fact that
the temporal fossæ as nearly as possible meet behind : there
is only a slight bridge dividing them. The lacrymal,
moreover, dilates into a wide plate, with a notch, on the inner
side of which is received the small flat-pointed prefrontal
process of the ethmoid. I have observed these skull characters in the Coraciidæ. The clavicle reaches the scapula
and gives off a longish acromial process. The nares are
impervious.

In *Sauropatis albicilla, Dacelo gigantea, Cittura cyanotis*,
and *Pelargopsis* (FÜRBRINGER) the tensores patagii have
the somewhat complicated arrangement shown in the figure
(fig. 94). There are two tensor brevis tendons, of which
the anterior, after giving off a wristward slip, is continued
over the arm to the lower side, fanning out as it goes.
There is in all a passeriniform tendinous slip. *Dacelo* has
a muscular pectoralis propatagialis. *Sauromarptis Gaudi-
chaudi* and *Sauropatis sanctus* are quite similar. There is,
in *Pelargopsis* at least, a tendinous slip from pectoralis I.
to both longus and brevis tendons.

In *Callalcyon rufa* (fig. 93) there is a simplification,
only the anterior of the two tendons being present ; the
passeriniform slip is barely marked. In *Alcedo* there is a
still further 'degeneration ;' not only is the passeriniform

---

[1] On C10 there is a paired hæmapophysis in *Dacelo*.
[2] SHUFELDT, 'On the Osteology of *Ceryle*,' *J. Anat. Phys.* xviii. p. 279.

slip absent, but the wristward slip is hardly shown. There is a fleshy pectoralis propatagialis joining the longus tendon.

In *Ceryle alcyon* the two tensor brevis tendons form a broad diffused tendon, to which is joined, before it gives off the wristward slip, a peculiar long (hornbill-like) pectoralis propatagialis tendon.

In *Alcyon Lessoni* we have the most simple form of tensor brevis without any branch.

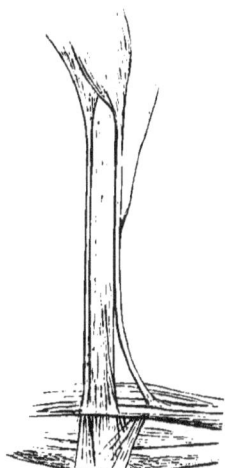

Fig. 92.—Tensores Patagii of *Ceryle alcyon*.

Fig. 93.—Tensores Patagii of *Callalcyon rufa*.

Fig. 94.—Tensores Patagii of *Sauropatis albicilla*.

In *Syma* the tensor brevis consists of two tendons, but the anterior has no wristward slip at all.

There is never a *biceps slip*. The deltoid has a scapular slip.

The *expansor secundariorum* appears sometimes to be absent. But it is present in *Dacelo, Tanysiptera, Syma*, and *Cittura*. *Dacelo*, at any rate, has no humeral slip to the *anconæus*.

The leg muscle formula is, without exception, AX· .

In *Dacelo* there is but one peroneal, which is the brevis.

The *deep flexor tendons* vary somewhat. In *Dacelo gigantea* the arrangement of these tendons is, as Professor

GARROD pointed out, precisely that of *Momotus*. But in *Halcyon ragans* the two tendons blend completely before the branches to the toes, all arising approximately at the same level, are given off. The structure of the conjoined tendon, however, seems to suggest that the flexor hallucis is concerned with the supply of digits III., IV.

In some kingfishers there is a myological peculiarity not found in any other group of birds. Dr. R. O. CUNNINGHAM discovered in *Ceryle stellata* a strong transverse tendon uniting the two biventri cervicis close to the upper belly of the muscles. He failed to find this junction in *Alcedo ispida* and *Dacelo gigantea*. *Tanysiptera* and *Cittura* [1] have this link, but not *Syma*, *Halcyon*, or *Sauropatis*.[2]

The following table gives the intestinal lengths of two species :—

Ceryle amazona	. 24 inches
Halcyon sp. .	. 14½ ..

The right lobe of the liver seems to be always larger than the left.

I have examined the syrinx in *Dacelo cervina*. The last tracheal rings are completely fused in front to form a bony box, which shows no traces of the number of rings of which it is composed. These rings appear to be five or six in number, and, with the exception of the last, are fused together in the middle line behind. In front of this box the tracheal rings interlock in the usual fashion. The first bronchial semi-ring, which is ossified, is firmly united to, but not fused with, the tracheal box; the succeeding rings are cartilaginous. The syrinx has two pairs of intrinsic muscles; the most anterior is the more slender; the wider muscle arises from the trachea, just where the extrinsic muscles are given off, and is attached to the first and apparently also to the second bronchial semi-ring.

In *Ceryle alcyon* there are no great differences, but the

---

[1] In one of two specimens of *S. ragans* it was present.
[2] 'Notes on some Points in the Anatomy of the Kingfisher,' *P. Z. S.* 1870, p. 280. See also BEDDARD, *P. Z. S.* 1896, p. 603.

rings forming the box are not so completely fused, and the larger intrinsic muscle arises much lower down—in fact, at the commencement of the box.

COLII

Definition.—**Aftershaft present ; oil gland tufted. Muscle formula, AXY; biceps slip present. Cœca absent. Skull desmognathous.**

Of the family **Coliidæ** there is only a single genus, African in range and including something like nine species.

The toes are remarkable for the fact that the zygodactylous condition can be assumed ; the first toe can be directed forwards,[1] the fourth backwards. There are ten *rectrices* ; a *tufted oil gland* and an *aftershaft* are present.

The *pterylosis* described by NITZSCH is remarkable for the width of the pterylæ. The ventral tract almost completely covers the ventral side of the body; towards the outside the feathers are stronger, but there is no outer branch. The spinal tract is narrow and strongly feathered upon the neck ; on the occiput is a bare space, reminding one of that in a similar position in the Trochilidæ. There is no median spinal apterion.

In their *myology* the colies are remarkable for possessing a fleshy *biceps slip*. The *tensor patagii brevis* muscle is very extensive, and reaches nearly as far as the fore arm. Its very short and single tendon sends back a 'passerine' slip, oblique in direction, and is also continued over the arm.

The *pectoralis* slips to both longus and brevis are present.

The *deltoid*, as in so many allied birds, is very extensive.

The *leg muscular formula* is AXY−. There is only one *peroneal*.

The *semimembranosus* is inserted below and independently of the *semitendinosus*. The latter gives off a tendinous slip to the *gastrocnemius*.

The *deep flexor tendons* blend before giving off branches to the toes.

---

[1] Hence the term 'pamprodactylous,' sometimes used for this family.

Besides the hornbills and **Macrochires** the colies are the only flying birds in which the *latissimus dorsi metapatagialis* is absent.

The colies have only the left *carotid*.

The *stomach* is not very muscular. The *liver* is small and has a *gall bladder*. There are no *cæca*. The *intestines* are short, but capacious, measuring nine inches.

The *syrinx* has been figured by JOHANNES MÜLLER.[1] It is quite typically tracheo-bronchial.

The skeleton and the affinities of *Colius* have been elaborately treated of by MURIE.[2]

There are thirteen cervical vertebræ. Four ribs reach the sternum.

The *skull* is holorhinal, without basipterygoid processes, and desmognathous. After a careful macceration GARROD[3] was unable to find a vomer, the presence of which had been previously[4] asserted (see fig. 95, p. 203).

TROGONES

Definition.—**Feet zygodactyle by reversion of second toe. Skull schizognathous with basipterygoid processes. Oil gland nude. Left carotid alone present. Cæca short. Ambiens absent. Of deep plantar tendons Fl. hall. supplies I. and II., Fl. dig. III. and IV. Vinculum joins them before bifurcation of each.**

This family is chiefly American, but also African and Asiatic.[5]

The feathers of the trogons have very well developed *aftershafts*. The *pterylosis* is remarkable for the non-bifurcation of the spinal tract, which is continuous as a single tract to the base of the naked oil gland. It is dilated to form a rhomboidal area behind the scapula.

There are twelve *rectrices*.

[1] 'Ueber die bisher unbekannten typischen Verschiedenheiten der Stimmorgane der Passerinen,' *Abh. k. Akad. Wiss.* 1845.

[2] 'On the Genus Colius, its Structure and Systematic Place,' *Ibis*, 1872, p. 263.

[4] 'Notes on the Anatomy of the Colies (*Colius*),' *P. Z. S.* 1876, p. 416.

[3] By MURIE.

[5] *Trogon gallicus* is an extinct species from the Miocene of France.

Of the muscles of the thigh which Professor GARROD regarded as of importance there are present the *femorocaudal* and the *semitendinosus*, the *accessories* of both being absent. The femorocaudal is proportionately larger than in almost any bird. There is no *glutæus primus*. The *obturator internus* is small and oval. The singular arrangement of the *deep plantar tendons* is used in the definition of the family. The two tendons concerned each supply two digits, this arrangement being unique. In the fore limb there is no biceps slip to the patagium. The patagial muscles and tendons are much complicated; they have been figured by GARROD for *Trogon puella*. The very powerful tensor brevis muscle runs as a muscle nearly to the extensors of the fore arm; it has a short broad tendinous insertion on to the fascia of the outer surface of the arm, and this is specially developed, a line running back to the humerus, as in the **Passeres** (see p. 172). Deeper than this are two parallel

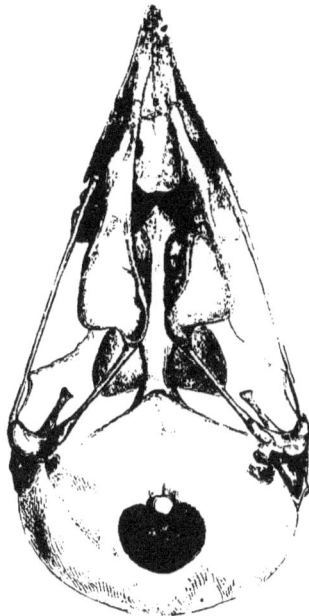

FIG. 95.—SKULL OF *Colius castanonotus*. VENTRAL ASPECT. (AFTER GARROD.)

tendons: of these the one nearer the humerus terminates exactly like the single one of the **Passeres**; the other tendon ends as in the **Pici**, elsewhere described. There is no expansor secundariorum.

The *tongue* of the trogons is short and three-sided. It is pointed in front. The left lobe of the *liver* is a little the smaller. Among GARROD's notes are the following measurements of the intestines and the cæca in three species of trogons, which we reproduce: -

—	Trogon mexicanus	Tr. puella	Pharomacrus mocinno
Intestine .	10·5 inches	8·5 inches	16 inches
Cæca .	1·25 and 1 inch	1·1 inch	1·75 inch

There is no *crop*; the *gizzard* is thin-walled and large; the *proventriculus* is zonary.

The most remarkable matter concerning the *osteology* of the trogons is the curious mistake which was originally made as to the nature of the palate. HUXLEY, in his paper upon the classification of birds, came to the conclusion, from a single incomplete skull of *Trogon Reinwardti*, that the skull was, like its presumed allies, desmognathous. Later FORBES [1] was able to show in five species that the maxillo-palatines were not united across the middle line, but that they terminated in a spongy expansion some way from each other. The end of the vomer is thin and filiform. The lacrymal is somewhat styliform, and reaches the jugal bar; there appear to be no ossified ectethmoids. The palatines of the trogons are peculiar. Instead of being flat plates, as in *Coracias*, for example, the outer portions of the bones are bent upwards, and cling closely to the basis cranii. The two palatines are, moreover, fused posteriorly, and the pterygoids where they articulate with them are expanded. They are holorhinal with impervious nares. The trogons have fifteen *cervical vertebræ*. The *atlas* is perforated by the odontoid process; four or five ribs reach the sternum. The sternum has two incisions behind, and the bifid spina externa.

CORACIÆ

Definition.— **Aftershaft present   Muscle formula, AXY ; expansor secundariorum present. Cæca generally present. Desmognathous.**

The **Coraciidæ** are entirely Old-World birds, chiefly massed in the Ethiopian region, but extending as far as the

---

[1] 'Note on the Structure of the Palate in the Trogons,' *P. Z. S.* 1881, p. 836.

Australian. The genera allowed by DRESSER in his recent
monograph of the family are *Coracias, Eurystomus, Brachy-
pteracias, Atelornis,* and *Leptosomus.* They are distributed
by him in three subfamilies : the first two genera constitute
the first, the next two the second, while *Leptosomus* is placed
in a subfamily by itself. The rollers have an anisodactyle
foot ; the feathers have an *aftershaft* ; but the oil gland is
nude. The *pterylosis* has been studied by NITZSCH, FORBES,
and by myself.[1] In *Eurystomus orientalis* the ventral tracts
commence as two from the very first : at the angle of the
mandible they are double. Though NITZSCH has figured
the pterylosis of the throat of *Coracias garrulus* as if it were
continuous, I do not find any difference from *Eurystomus* in
this particular. On the breast the two divisions of the
ventral tract remain single : there is hardly a trace of the
outer branch. The tracts are here rather wide. The dorsal
pterylosis narrows gradually until between the shoulders,
where the feathering is very strong, and where it divides
into two branches ; these unite again just at the articulation
of the femora, and finally terminate a little way in front of
the oil gland.

Leptosomus has a slightly different pterylosis ; the
ventral tract is single to about an inch behind the junction
of the mandibular rami ; for a considerable distance the
ventral band is continuous with the dorsal, so that the
lateral neck spaces do not commence until about three-
quarters of an inch above the shoulder. About the middle
of the sternum the pectoral tract of either side gives off an
outer branch some four feathers wide and slightly stronger
than the main tract. The two forks of the dorsal tract
run in between each other, the narrower posterior portion
between the limbs of the wider anterior portion, as is the
case with so many birds. FORBES has noted that *Atelornis*
has a pterylosis which agrees with that of *Eurystomus,*
already described.

Leptosomus differs, however, from the remaining Coraciidæ
in the possession of *powder-down patches,* which were first

[1] See anatomical preface to DRESSER'S monograph of the group.

described by SCLATER.[1] There are two patches in the lumbar region, lying between the dorsal and the femoral tracts.

The *tongue* of the rollers wants the spiny fringe at the base which is so common a feature of this organ in other birds; it is horny in front and entire at the tip. The *liver* has a larger right lobe and a *gall bladder*.

The following are intestinal measurements:—

	Small Intestine	Large Intestine	Cæca
Leptosomus .	9¾ inches	2¼ inches	2¼, 2¾ inches
Eurystomus	10 ,,	1 inch	1½ inch
Coracias .	11 ,,	?	2·2 inches

The *tensores patagii* of *Leptosomus* are figured in the accompanying cut.[2] The brevis tendon gives off a wrist-

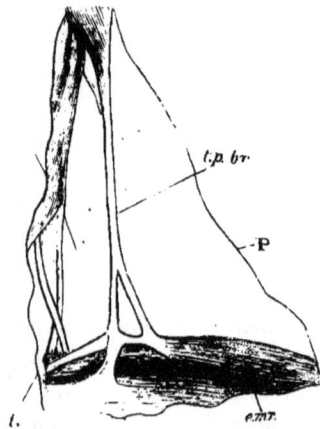

FIG. 96. TENSORES PATAGII OF *Leptosomus* (AFTER FORBES).

P, patagium ; *t.p. br*, tensor patagii brevis ; *e.mr*, extensor metacarpi ; *t*, its origin.

ward slip just before its attachment to the tendon of extensor metacarpi radialis longus, over which it passes to be inserted below. There is no biceps slip. *Coracias garrulus* is the same save for the fact that there are two separate brevis tendons, from the first of which the anterior gives off the wristward slip; the inner thinner tendon of these does not cross the wrist. *Eurystomus* does not differ. In these birds there is a well-developed *expansor secundariorum* of the fully developed type, which Professor GARROD has called 'ciconiiform.'

The muscle formula of the leg is AXY−. The *glutæus maximus* does not reach below the acetabulum; the *gl.*

[1] 'On the Structure of *Leptosoma discolor*,' P. Z. S. 1865, p. 682.
[2] From FORBES, 'On the Anatomy of *Leptosoma discolor*,' P. Z. S. 1880, p. 464.

externus is present in *Eurystomus*, represented by a ligament in *Coracias*. The deep flexor tendons are of type V., where the two tendons blend before the four branches are given off to the toes. Both *peroneals* are present. The *carotids* of the Coraciidæ are two. In *Leptosomus* FORBES found that the two arteries run up close together, but are not fused in the hypapophysial canal. He thinks that they may be, like those of *Bucorvus*,[1] no longer functional as blood vessels.

The *syrinx* of the Coraciidæ is quite typically tracheo-bronchial. In *C. garrulus* the intrinsic muscles are attached to the first bronchial semi-ring. These semi-rings are ossified; the rest of the bronchial semi-rings are more slender and not

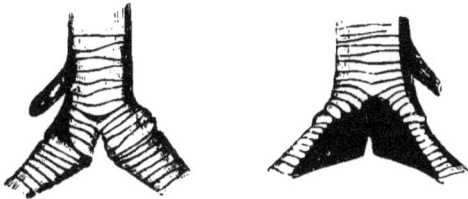

FIG. 97.—SYRINX OF *Leptosomus* (AFTER FORBES). THE LEFT-HAND FIGURE FROM IN FRONT, THE RIGHT FROM BEHIND.

ossified. In *Eurystomus* the only difference is in the fact that the three semi-rings following the first are closely attached to it and to each other, and appear to be ossified; after these are the broader soft cartilaginous semi-rings. The syrinx of *Leptosomus* (fig. 97) is rather different; it appears to be an extreme development of the type found in *Eurystomus*. The first three bronchial semi-rings, like the last tracheal rings, are ossified; the first of them appears to be nearly, if not quite, a complete ring. The fourth and the succeeding semi-rings are cartilaginous; to the former are attached the intrinsic muscles. In the commencing formation of a 'bronchial syrinx' *Leptosomus* evidently gives a hint of cuckoo affinities, to which group, however, its structure in general does not incline.

There are fourteen *cervical vertebræ* in *Leptosomus*,

[1] Cf. p. 215.

thirteen or fourteen in other Coraciidæ. The *atlas* (*Coracias*) is notched for the odontoid process; C2–4, C10–D2 have hæmapophyses. On C13 and 14 there are also a pair of downward processes (catapophyses), one on each side of the hæmapophysis, which, in the case of C14, arise from a common base with it and on D1 from its tip.

In *Eurystomus* the atlas is perforated. Four (*Leptosomus*[1]) or five (some other forms) ribs reach the *sternum*, which is singly or doubly notched on either side, and has a spina externa but no spina interna. The *skull* is desmognathous, holorhinal, without basipterygoid processes. The rollers have the same peculiar lacrymal that has been referred to above in the kingfishers. The bone expands enormously below and comes into near relations, but does not fuse, with a slight ectethmoid : the lacrymal reaches the jugal. Another peculiarity of the coraciid skull is the very large postfrontal process, which descends in a straight line and actually reaches the jugal. These remarks apply not only to *Coracias*, but to *Eurystomus* and *Atelornis*, in which latter, however, the postfrontal process is not quite so long.

The family **Meropidæ** consists of the genus *Merops*, and of a few others which have been carefully monographed by DRESSER. Like the rollers the bee-eaters are an exclusively Old-World family, ranging through the Palæarctic, Ethiopian, Oriental, and Australian regions, but again, like the rollers predominating in the Ethiopian.

As to external characters, the *oil gland* is nude: the *rectrices* are twelve; the feathers have an *aftershaft*. The *pterylosis* (described by NITZSCH and by myself[2]) is as follows :—

The spinal tract is wide, and is at first connected round the neck with the ventral tract. About halfway down the

---

[1] The osteology (and some of the viscera) of *Leptosomus* is described and figured by MILNE-EDWARDS in the *Histoire Naturelle de Madagascar*. See also for the family NITZSCH and GIEBEL, 'Zur Anatomie der Blauracke,' *Zeitschr. f. d. ges. Naturw.* x. p. 318.

[2] In anatomical preface to DRESSER's monograph.

neck it becomes separate, and terminates in a truncate or
sometimes bifurcate extremity between the shoulder blades ;
at this point there is a break and the rest of the spinal
tract is double, enclosing a space bounded by two distinctly
conical tracts, which gradually get narrower to their point
of fusion, a little way in front of the oil gland. The ventral
tract is double from close to its point of origin ; the two
tracts get wide upon the pectoral region, whence they
gradually dwindle to a single feather wide close to their
termination ; the pectoral tracts have no outer branch.

Nyctiornis has two *carotids*; most of the others have
only one, the left. The bee-eaters have long *cæca*, like the
Coraciidæ. In a specimen of *M. ornatus* with the intestines
only 5½ inches they measured ½ inch; in a larger species
1 inch.

The proventriculus is zonary : the right lobe of the *liver*
the larger, and with a *gall bladder*.

The *tensor patagii brevis* tendon gives off a wristward
slip, and a ' passeriform ' slip to the humeral at its insertion
to the fore arm, which it does not cross. There is no biceps
slip, but there is a fleshy slip to longus from pectoralis ; that
to brevis is entirely tendinous, there being in both an
agreement with *Coracias*.

The *deltoid* extends a long way down the humerus ; it
receives a tendinous slip from the scapula, which passes under
the *latissimus dorsi* and over the *anconæus longus*. The
latter muscle has a humeral head, but not in *Nyctiornis*.

The *expansor secundariorum* is present and ' ciconiine.'

The leg formula is AXY − . The *deep flexor tendons* are
as in fig. 55, p. 100 ; the flexor hallucis gives off a slip to the
hallux before it fuses with the flexor communis.

The Meropidæ have fourteen *cervical vertebræ*. The
atlas is perforated by the odontoid process. There are
hæmapophyses on C2-4, C10-D1 : that of C14 is trifurcate.
Four ribs reach the *sternum*, which has two lateral notches,
of which the outer is the deeper, and has both external and
internal spina, the latter being bifid, as in Passeres and
some other birds. The clavicles have an acromial process,

P

as in the kingfishers. The *skull* is desmognathous, holorhinal, without basipterygoid processes. The descending limb of the lacrymal nearly unites—is connected by cartilage—with the slender ectethmoid, thus forming a ring. The nostrils in the dried skull are pervious.

The desmognathism of *Merops* is different from that of its allies. The maxillo-palatines are long, slender, recurved plates, like those of passerines. They are fused in the middle line to a broad plate of bone, but the free ends of the maxillo-palatines extend backwards for some distance independently of this. The palate in front of the maxillo-palatines is to some extent vacuolate. The vomer is a single rodlike bone.

The **Momotidæ** [1] are South and Central American, comprising the genera (perhaps subgenera) *Momotus*, *Hylomanes*, *Baryphthengus*, and some others. They are placed by GADOW in close association with the todies, but there are various points in which they differ from that family, upon which stress has been laid by FORBES. It is mainly to the last-mentioned observer [2] and to GARROD [3] that the existing knowledge of the family is due. The external characters of the family show some variation; in *Momotus* the *oil gland* is quite nude; in *Hylomanes* and *Eumomota* the apex is furnished with a few small plumes. *Momotus* has twelve *rectrices*; *Hylomanes*, *Prionorhynchus*, *Baryphthengus*, ten. A remarkable characteristic of the motmots are the two central racket-shaped rectrices, which matter was investigated twenty years ago by SALVIN. [4] It appears that the original account given by WATERTON of the birds nibbling off the vanes is perfectly correct, for it was observed by BARTLETT at the Zoological Society's Gardens. As a rule the bird only nibbles at the two long central rectrices, but SALVIN reports a case where a bird had sought fresh fields and had attacked others of its feathers. It is a very remark-

[1] J. MURIE, 'On the Motmots and their Affinities,' *Ibis* (3), ii. 1872, p. 383.

[2] Collected papers, *passim*.

[3] Collected papers, *passim*, 'On the Systematic Position of *Momotidæ*,' *P. Z. S.* 1878, p. 100.

[4] 'On the Tail Feathers of *Momotus*,' *P. Z. S.* 1873, p. 429.

able fact that when the rectrices in question first appear they are narrower at the points where the nibbling occurs, and where they will be ultimately denuded, than they are elsewhere. But an inheritance of this particular acquired character can hardly be asserted.

An *aftershaft* is present, but small.

The *tensores patagii* are simple, and there is no *biceps slip*. There is a fleshy *pectoralis propatagialis*. The tensor brevis consists of two parallel tendons, the anterior of which does *not* give off a wristward slip. The fan to the ulna arises in *M. brasiliensis* and *M. æquatorialis* as a continuation of the hinder of the two tendons, in *M. Lessoni* between them. *Hylomanes gularis* agrees with the first. The *anconæus* has a humeral slip. The somewhat rudimentary *expansor secundariorum* only reaches the margin of the teres.

The *deltoid* is large ; its attachment to humerus is longer in *Hylomanes* than in *Momotus* ($\frac{3}{4}-\frac{1}{2}$). There is a separate tendinous scapular slip.

The muscular formula of leg is AXY−. Both *peroneals* are present. The *deep flexor tendons* of the motmots are rather peculiar in their structure. It will be observed that the slip to the hallux is given off before the flexor hallucis joins the flexor longus.

The first gluteal (*gl. maximus*) is only present in front of the acetabulum. The *glutæus externus* is absent as a distinct muscle, but the insertion of *glutæus II.* extends so far round the head of the femur that it may represent also the otherwise missing muscle.

The gizzard is stronger in *Hylomanes* than in *Momotus*, and is almost 'ptilopine' in section.

The *tongue* is long, bifid at the apex, and worn into filaments. In the *alimentary canal* the *cæca* are absent ; the *intestines* measure fourteen inches in *M. Lessoni*, eighteen inches in *M. brasiliensis*. The right lobe of the *liver* is the larger, and there is a *gall bladder*.

There are two *carotids*. The femoral vein is abnormal.

The *syrinx* has been described by GARROD and is figured by him. It does not apparently differ widely from genus to genus. The last few tracheal rings are fused, but there is not a complete pessulus.

The motmots have fifteen *cervical vertebræ*. The *atlas* is perforated by the odontoid process. C1–5, C11–D3 have median hypapophyses; on C14–D2 are a pair of inferolateral processes, which gradually approach the median line until they spring from a common base in D1, and are just visible as rudiments towards the tip of the hæmapophysis of D2. In *Hylomanes* there is also a double hypapophysis on C10. Three or four ribs reach the *sternum*, which has (*Momotus*) four foramina, or (*Hylomanes*) two notches and two foramina, and a bifid spina externa. The *skull* is desmognathous and holorhinal. The lacrymals are rudimentary; the ectethmoids are very small and do not nearly reach them. Nares impervious, pervious in *Hylomanes*.

The West Indian todies (**Todidæ**) form a very distinct family; their structure has been chiefly investigated by MURIE [1] and FORBES.[2] They are small birds with feet in which the syndactylism is more marked than in motmots and some others. The annexed cut shows that the digits IV. and V. are united together as far as the end of the third phalanx of the one and the second of the other. The *oil gland* is tufted.

The *skull* is very imperfectly desmognathous. The two maxillo-palatines are not united together; they are completely free from each other and from a median ossified nasal septum. There seems to be no vomer. The descending portion of the lacrymals is large and broad; the ectethmoids, on the other hand, are small. The interorbital septum is widely fenestrate. There are fifteen cervical vertebræ. The *intestines* are, according to Mr. FORBES, remarkably short, not measuring altogether more than $3\frac{1}{2}$ inches; on the other hand the *cæca* are as remarkably long (considering the sys-

---

[1] 'On the Skeleton of *Todus*,' &c., *P. Z. S.* 1872, p. 664.

[2] 'On some Points in the Anatomy of the Todies (Todidæ), and on the Affinities of that Group,' *ibid.* 1882, p. 442. See also REICHENOW, ' Ueber das Genus *Todus*,' &c., *Journ. f. Ornith.* xxxi. 1883, p. 430.

tematic position of the bird) ; they measure about one-third of an inch. The cæca are narrowed at their origin from the gut, and, as in the owls, &c., dilated apically. The *deep plantar tendons* vary from the arrangement common to the group in that the slip to the hallux is given off before the blending of the two. The arrangement, in fact, is as in the motmots. There is an *expansor secundariorum* ceasing at the axillary margin of the teres, in the gallinaceous fashion. The *syrinx* has at the middle a bony box, which is formed of three or four bronchial rings united with about two tracheal rings. It is

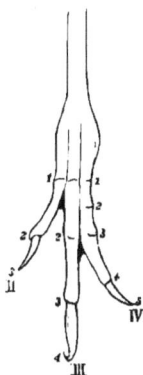

Fig. 98.—Foot of *Todus* (after Forbes). The Digits and the Phalanges are numbered.

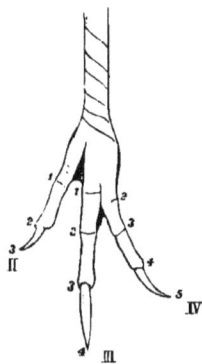

Fig. 99.—Foot of *Momotus* (after Forbes). Letters as in Fig. 98.

only ventrally that the fusion is complete. The intrinsic muscles cease at the last tracheal ring but one.

The **Galbulidæ** are a family of neotropical birds, comprising the genera *Galbula, Urogalba, Jacamerops*, &c., and known as puff birds.

They are *zygodactyle*, with a nude *oil gland*, twelve *rectrices*, and a small *aftershaft*.

The *pterylosis* of *Galbula rufociridis* is as follows :—

The inferior tract is double from the angle of the jaw ; just before leaving the neck each tract gives off a short branch, about six rows of two feathers, which runs on to the margin of pectoralis.   The main tract itself is also only two feathers broad.   It sends off, about halfway down sternal keel, a short curved outer branch, which runs outwards and then forwards towards axilla, nearly meeting a second outer branch which is given off by the tract on opposite side to the inner branch, already spoken of.   The dorsal tract has a slight break ; the very short interscapular fork is of strong feathers and connected with posterior part only by a very few feathers which are weak and arranged uniserially.

The *tongue* is long and thin, tapering to a filament ante-riorly ; a *gall bladder* seems to be absent.   In a specimen of *G. rufoviridis* the *intestinal measurements* were as follows : s. i., 4·15 inches : l. i., ·75 inch ; cæca, ·7 inch.

The Galbulidæ have an *expansor secundariorum*, but no *biceps slip*.   The *tensor patagii brevis* tendon of *Galbula* has no wristward slip.   It is merely a single tendon ; in *Urogalba* there is a wristward slip.

In the leg the formula of *Galbula* is AXY, of *Urogalba* AX, both birds, of course, lacking the ambiens.   The *glutæus* I. and V. are absent, at any rate in *Galbula*.   The *plantar tendons* are picine.

Both *carotids* are present.

The *skull* of the Galbulidæ is very like that of the Bucco-nidæ ; but there are nevertheless points of difference.

In *Urogalba paradisea* there is a long gap in the bony palate in front of the conjoined maxillo-palatines, as in Bucco-nidæ ; but the palatines are more sloped off posteriorly, and their posterior halves are more closely in contact.   The de-scending process of the lacrymal is broader, and it is perforated by a large foramen.   In *Jacamerops* and *Galbula*, on the other hand, the descending process of the lacrymal is very slender.

The ectethmoids are large and the interorbital and intra-narial septa complete.

There are fourteen *cervical vertebræ* : the *sternum* has two pairs of incisions.

BUCEROTES

Definition.—Oil gland tufted. Muscle formula, AXY. Cæca absent. Skull desmognathous.

The well-marked family **Bucerotidæ** contains at least two distinct genera, *Buceros* and *Bucorvus*. The latter (the ground hornbill) is entirely African; the former, which has been much subdivided, is both African and Asiatic.

The great *casque*,[1] not always equally developed, and the long bill, frequently serrated along its margins, and the largely black and white or black plumage distinguish these birds. But the small *Toccus* is a less typical form. The syndactyle foot, in which the second and fourth toes are united to the third—the latter for several joints, the former for only one—is highly distinctive, and is repeated in the ground-living *Bucorvus*.

The *oil gland* is tufted. The feathers have no *aftershaft*. There are ten *rectrices*.

The *pterylosis* of *Bucorvus abyssinicus* has been described and figured by NITZSCH.

The neck is completely feathered, except at its lower end, both dorsally and ventrally. The former is the commencement of the very narrow dorsal space of a long oval form, but not extensive. The pectoral tracks diverge at end of neck, but are subsequently undivided.

The *carotids* are double in *Bucorvus*; the left only is present in others. The remarkable obliteration of the carotids in the former genus, and their replacement by a pair superficial in position, have been described by GARROD[2] and OTTLEY.[3]

The *tensores patagii* are in some ways characteristic of the Bucerotidæ. In *Buceros convexus* (cf. FÜRBRINGER)

[1] OWEN, 'On the Anatomy of the Concave Hornbill,' *P. Z. S.* 1833, p. 102.
[2] 'On a Peculiarity in the Carotid Arteries . . . of the Ground Hornbill, *P. Z. S.* 1876, p. 60.
[3] 'A Description of the Vessels of the Head and Neck in the Ground Hornbill,' *ibid.* 1879, p. 461.

there is no patagialis longus. The *brevis* receives rather low down a very strong slip from the pectoral ; near to its insertion it gives off a wristward slip, which is attached to a special tendon arising from the lower end of the humerus. The main tendon passes over this, not attached to it, to the ulnar side. The absent *longus* is represented only by a thinnish tendon arising from the pectoralis.

The same structures are found in *B. malabaricus, B. coronatus, B. bicornis,* and in *Toccus.* In *B. atratus* there is in addition an excessively small *patagialis longus* muscle, arising with brevis and consisting indeed of but very few fibres. In *Bucorvus,* on the other hand (fig. 100), the tensor patagii longus is well developed. Each tendon has a slip from pectoralis (*a* and *a'* in fig. 100), but that which joins brevis receives a tendinous slip from the biceps. This, however, as it is not figured by FÜRBRINGER, is possibly individual.

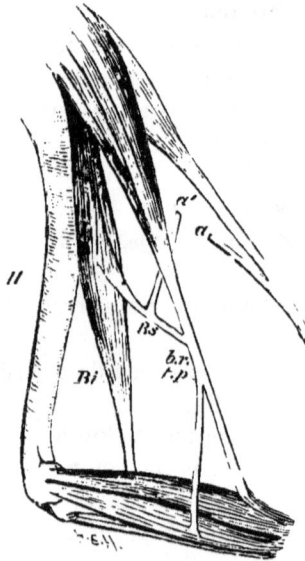

FIG. 100.—PATAGIAL MUSCLES OF *Bucorvus* (AFTER BEDDARD).

H, humerus; *Bi,* biceps ; *Bs,* biceps slips (?); *b.r, t.p,* tensor patagii brevis tendon ; *a. a'.* slips from pectoralis.

Quite exceptionally—among anomalogonatous birds—many hornbills have a broad humeral attachment of the *anconæus.* The muscle itself arises from the scapula by a Y-shaped (*Bucorvus, Buceros*) or flat, non-divided (*Acceros*) head. The humeral ' ankerung ' is found in *B. subcylindricus, B. bicornis,* not in *B. elatus, B. malabaricus, B. atratus, Bucorvus abyssinicus, Acceros,* or *Toccus.* The deltoid has no scapular slip.

The leg formula of *all* hornbills is AXY−.

The *gluteus maximus* is quite absent; the *gluteus externus* is only present as a ligament.

The *biceps* is occasionally (e.g. *B. elatus*) double at its origin, the tendons being separated by quite a quarter of an inch.

The arrangement of the semitendinosus and adductor in *Aceros nipalensis*, which is somewhat complex, will be understood from the accompanying drawing (fig. 101).

The *semitendinosus* (*St*) is inserted on to the tibia by a long, thin, flat tendon; another tendon, joining the first just where it passes into the muscle, is attached to the *gastrocnemius*.

The *accessory semitendinosus* is in two parts : the larger half (*Ast*) is attached to the semitendinosus just behind the origin of the tendon of insertion of the latter ; the second half appears to arise from the tendon which connects the semitendinosus with the gastrocnemius, it passes up towards the thigh, and just in front of its (tendinous) insertion on to the femur it receives a tendon from the adductor. This latter muscle (the *adductor longus*) is inserted by three tendons—(1) to the femur; (2) a small tendon which has already been described as joining the second half of the accessory tendinosus; and (3) near to the origin of one of the internal heads of the gastrocnemius ; to this tendon is also attached the inner head of the gastrocnemius.

The corresponding muscles [1] of *Bucorvus abyssinicus* are rather simpler than in *Aceros nipalensis*. The *adductor longus* is only inserted at two places : first, by a fleshy insertion along a considerable length of the lower border of the femur ; second, by a tendon in common with the innermost head of the gastrocnemius. The *semitendinosus* is attached by a thin tendon to the tibia, as in *Aceros*, and by a short tendon, also as in that species, to the gastrocnemius. The *accessory semitendinosus* arises chiefly from this latter tendon, but there is no division between this part of the muscle and that which takes its origin from the fleshy part of the semitendinosus.

In *Buceros atratus* there is, again, some little difference

<hr>

[1] Gadow figures most of these muscles in Bronn's *Thierreich*, 'Aves,' Bd. vi. Abth. iv. Taf. xxiii. *b*, fig. 1.

from both the types already described, although the resemblances are on the whole closer to *Aceros*.

The *adductor longus* is attached by two tendinous heads ; the upper one of these, as in *Aceros*, is attached to the lower border of the femur ; this corresponds to the fleshy insertion of the muscle in *Bucorvus* ; the lower tendon is fused on its way with the inner head of the gastrocnemius, which is continued upwards and reaches the femur, and then bifurcates into two tendons of insertion.   The relations of the

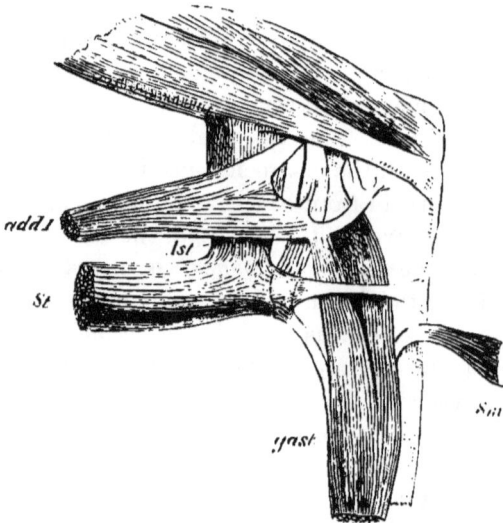

FIG. 101.—LEG MUSCLES OF *Aceros* (AFTER BEDDARD).
add, adductor ; *St*, semitendinosus ; *Ast*, its accessory ; *Sm*, semimembranosus : *gast*, gastrocnemius.

semitendinosus and of the accessory semitendinosus are as in *Aceros nipalensis*.

In *Toccus* these muscles are much the same as in *Buceros*.

In *Ceratogymna elata* I find a closer resemblance to *Aceros* than to any other of the genera mentioned in this paper, but there is an agreement with *Bucorvus* in the *fleshy* insertion of the *adductor longus* on to the lower border of the femur.   The *accessory semitendinosus* is distinctly double, as in *Aceros*, and is attached by a short tendon to the adductor, though the direction of this tendon is somewhat different from what is found in *Aceros*.

The *skull* is doubly desmognathous. There are distinct basipterygoid processes, large in *B. rhinoceros*, almost vanished in *A. nipalensis*, with which, however, the pterygoids do not articulate. The interorbital septum is widely fenestrate. The fused lacrymals and ectethmoids together make a large plate of bone; the postfrontal processes are large. The *atlas* is fused with the axis. The hæmapophyses are slight. There are fourteen or fifteen *cervical vertebræ*; they are median in C12–D2. C14 and D1 have a median and two lateral processes.[1] The *sternum*, which is faintly two- or four-notched, has both spina externa and interna.

The *liver* lobes present some differences in different hornbills.

Commencing with *Bucorvus abyssinicus*, in which the right lobe is larger than the left, the series terminates with *Buceros coronatus*, in which the left lobe is larger than the right. The following table shows the relations of the liver lobes in such hornbills as have been examined:

Bucorvus abyssinicus.	R > L.
Aceros nipalensis.	R > L.
Buceros bicornis.	R > L.
Sphagolobus atratus.	R > L.
Hydranistes subcylindricus.	R > L.
Buceros plicatus.	R = L.
Buceros rhinoceros.	R = L.
Buceros coronatus.	R < L.

I have noticed a peculiarity in several species of hornbills which is not found in all other birds. In all birds the two lobes of the liver are completely separated from each other by the umbilical ligament, which bears the umbilical vein (this appeared to be particularly large and well developed in all hornbills which have been dissected by me); and in addition one liver lobe – the right – is commonly separated from the abdomen by a thin membranous septum. In hornbills both lobes of the liver are thus shut off; I have figured

---

[1] In *Dichoceros bicornis* there is a tendency towards the formation of a hypapophysial canal.

this condition in *Bucorvus abyssinicus* ;[1] it is exactly the same in one or two other species which I have subsequently studied.

The Syrinx, Aceros nipalensis.—The last rings of the trachea are fused together to form a solid box, at the sides of which, however, the individual rings are recognisable. In front the last three rings are thus fused, but behind two additional rings fuse with the others to form a wide and deep bony plate. The tracheal rings lying in front of these five show the dovetailing arrangement which is so often found in the tracheal rings. The pessulus is well developed and bony, but, owing to the complete fusion of the tracheal rings both posteriorly and anteriorly, it is impossible to say from which rings it is developed.

The intrinsic muscles of the syrinx are attached near to the boundary line between the last and the penultimate tracheal rings.

The bronchial semi-rings are cartilaginous, and there is a considerable interval between the first of these and the last tracheal ring.

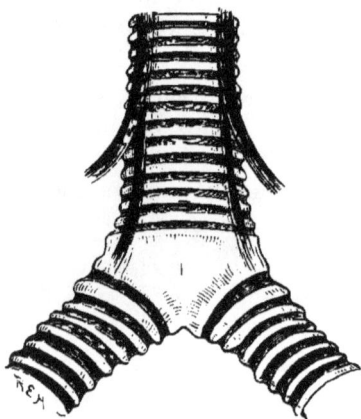

FIG. 102.—SYRINX OF *Aceros nipalensis*. FRONT VIEW. (AFTER BEDDARD.)

Bucorvus abyssinicus.—The syrinx of this hornbill (fig. 103) differs in many particulars from the last. The tracheal rings are not ossified, and there is no box formed by their fusion. Only posteriorly are the penultimate ring and the two in front of this fused just at the origin of the pessulus ; anteriorly the pessulus is fused with the ante-penultimate tracheal ring, which forms with it a three-way piece ; the last two tracheal rings do not meet in front. The

---

[1] 'Notes on the Visceral Anatomy of Birds. I. On the so-called Omentum,' P. Z. S. 1885, p. 842 ; and above, p. 44, fig. 29.

slender syringeal muscles are attached to the anterior margin of the last tracheal ring.

The peculiar-shaped tracheal rings are hardly recognisable until about the fourteenth from the end.

Buceros rhinoceros has a syrinx which is not very different from that of *Aceros*. The same rings are fused to form an ossified box; but the fusion between the several rings is hardly so extensive as in *Aceros*; furthermore the syringeal muscles are attached to the posterior border of the last tracheal ring.

In *Sphagolobus atratus* there is very little fusion between any of the last tracheal rings;

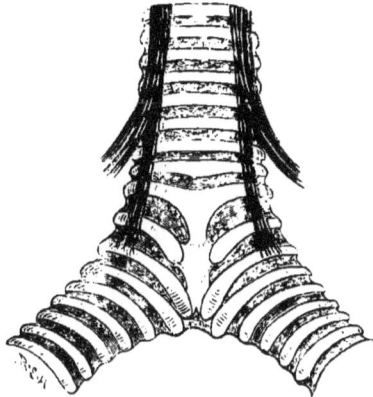

FIG. 103.—SYRINX OF *Bucorvus abyssinicus*. FRONT VIEW. (AFTER BEDDARD.)

the last three rings, which alone show any signs of ossification, are fused for a very short space anteriorly; posteriorly there is no fusion at all, and the pessulus can be plainly seen to be connected with the antepenultimate ring. Although the last tracheal rings are not fused, they are very closely applied together, and no membranous interspaces are left.

Ceratogymna elata, which is, like the last, a comparatively small species, has a very similar syrinx; indeed, I can find no differences sufficiently tangible to be described.

Buceros lunatus and *B. bicornis*, which are both large species, hardly present any differences from *B. rhinoceros*.

Bycanistes subcylindricus has a syrinx which, although of about the same size as that of *Ceratogymna elata*, shows certain differences which are worth putting on record. In the first place, the syrinx is much compressed from side to side at the level of the last tracheal ring; in the second place, the last tracheal ring is very much more arched than usual; it forms, indeed, almost a complete semicircle. The intrinsic muscle of the syrinx in this, as in the other smaller hornbills, is very much larger relatively than in the larger species.

Anthraceros malayanus, again, is a little different from all the types hitherto described. The last tracheal rings are but little fused posteriorly; only the penultimate and ante-penultimate rings are so fused, so that it is impossible to be certain as to the origin of the pessulus. The intrinsic muscles are slender.

Torcus presents certain peculiarities which I have not yet observed in any other hornbills; the trachea has *two pairs* of extrinsic muscles, given off about half an inch apart. This condition seems to me to be so remarkable that I have preserved the specimen which shows it, though unfortunately the insertions of the anterior pair of muscles are lost, and I have no recollection of where the point of insertion was. The intrinsic muscles are relatively small. There appears to be no fusion between any of the tracheal rings.

Cryptornis of the upper Eocene of France is held by MILNE-EDWARDS to be a hornbill.

The family **Upupidæ** [1] contains only the well-known hoopoe (*Upupa*) and the but little known *Irrisor* and *Rhinopomastus*.

There is a large feathered *oil gland*, but the *aftershaft* is absent or rudimentary. There are ten *rectrices*.

The *feather tracts* are narrow. The ventral tract divides very early on the neck, and gives off on each side in the pectoral region an outer branch. At the base of the neck a triserial tract is given off to the humeral tract, and just below it a uniserial tract to the patagium. Between the outer and inner branches of the ventral tract is a single row of feathers.

The dorsal tract encloses a spindle-shaped space, the pterylæ enclosing which are somewhat dilated in the middle.

[1] STRICKLAND, 'On the Structure and Affinities of Upupa and Irrisor,' *Ann. Mag. Nat. Hist.* xii. (1843), p. 238; MURIE, 'On the Upupidæ and their Relationships,' *Ibis*, (3) iii. 1873, p. 181.

The tendon of the *tensor patagii longus*[1] gives off a wristward slip; the main tendon crosses the fore arm. There is a *cucullaris patagialis*, besides slips from the pectoralis, but no *biceps slip*.

The *anconæus* has an attachment to the humerus.

In the hind limb the formula of the muscles is the typical picarian AXY−. The passerine character shown by the existence of a well-marked cucullaris propatagialis is paralleled in the hind limb by the absence of any vinculum between the *deep flexor tendons*.

The *tongue* is short and the *intestines* are without *cæca*. The left *carotid* alone is present.

There are fourteen *cervical vertebræ*. The *sternum* has a single pair of notches or fenestræ and both spinæ. The *skull* is pseudo-holorhinal, desmognathous, without vomer and basipterygoid processes.

The conjoined maxillo-palatines are rather delicate fenestrated bones, and the bony palate for a little way in front is somewhat vacuolate. The palatines have long postero-external angles, which reach back to a point corresponding to rather beyond the middle of the pterygoids.

The lacrymals are small and ankylosed to the skull. The ectethmoids are very large plates, and the distal end is segmented off, and is apparently the equivalent of the os uncinatum of many other birds; it reaches the jugal. I describe the nostrils as pseudo-holorhinal, because, though rounded at their extremities, they are unusually long, and reach, or very nearly reach, the ends of the nasal processes. They are obliterated in the middle by bony alinasals. There is a largish median foramen[2] a little way above the foramen magnum, and a minute one just above the latter.

---

[1] Nitzsch and Giebel, ' Zur Anatomie des Wiedehopfs,' *Zeitschr. f. d. ges. Naturw.* x. p. 236.

[2] This was present in only one of three specimens, in which also alone the os uncinatum was present. It has a shorter bill and may be a different species.

MACROCHIRES

Definition.—Rectrices, ten ; oil gland nude ; aftershaft present. Muscle formula of leg, A —. Expansor secundariorum, sterno-coracoideus, and biceps slip absent. Cæca absent. Manus very long. Sternum unnotched.

This group of birds contains two well-marked types— the humming birds and the swifts,[1] the former confined to America, the latter world-wide in distribution.

In external characters the generally minute size, the frequently brilliant metallic plumage, and the long slender bill distinguish the Colibris from the swifts. But Dr. SHUFELDT has found[2] in a nestling humming bird a bill hardly longer than that of a swift.

The *rectrices* are ten, and in all these birds the *oil gland* is nude. There is an *aftershaft*. In the swifts there are down feathers upon the apteria : in the humming birds there are not.

The *pterylosis* of the group has been chiefly studied by NITZSCH, to whose account Dr. SHUFELDT has added details of value.

The throat is completely feathered in the swifts, the two ventral tracts, however, becoming distinct at the beginning of the neck. The ventral tracts widen out in the pectoral region, but there is no outer branch or trace of one. The narrow dorsal tract bifurcates between the shoulders and reunites again to enclose a narrowish spinal space. There are well-marked femoral tracts.

In the humming birds the ventral tract is double up to the symphysis of the mandibles, or nearly so ; the dorsal tracts are very much wider and form a diamond-shaped patch, within which is a very slight dorsal apterion : there appear to be no femoral tracts, and there is a naked space in the nape of the neck, dividing the dorsal tract.

---

[1] GIEBEL, ' Ueber einige Eigenthümlichkeiten in der Organisation der Kolibris,' *Zeitschr. f. d. ges. Naturw.* l. 1877, p. 322; W. K. PARKER, ' On the Systematic Position of the Swifts,' *Zool.* (3), xiii. 1889, p. 91.

[2] 'Studies of the Macrochires,' &c., *J. Linn. Soc.* 1888.

The swifts are among the very few birds which are partly quintocubital and partly aquincubital.

Among the more obvious external characters are the magnificent metallic colours which are so usual a feature of humming birds. It is on account of the latter character mainly that they have been placed in the neighbourhood of —or more properly confused with—the Nectariniidæ. It is, however, practically the universal opinion that these two families have no near relationship, and the feathers of both have been lately submitted to a careful examination by Miss NEWBIGGIN.[1] The metallic colours of humming birds occur in both sexes, though more brilliant in the male; they are mainly to be found on the throat and on the head as a crest. It has been pointed out that the rapidly vibrating wings would destroy all advantage (in sexual selection) of the development of these tints upon the wings. The colours are of every shade, and gold and red are often present, two colours which are not found among the Nectariniidæ. The striking difference between the two families, however, consists in the fact that while the Nectariniidæ have the ends of the barbs affected by the metallic colour it is the basal part of the barbs which is so coloured in the humming birds. Hence the barbs have cilia in the latter case and not in the former; for this reason the rectrices in the humming birds can show metallic colours and yet not have their efficiency as flight feathers destroyed. The interlocking apparatus is there in the form of the cilia. The barbs which are thus metallically coloured are in both groups of birds (and in other birds which show the same kind of coloration) composed of a series of roof-shaped laminæ, in the cavities of which the dark brown pigment essential for the due production of the metallic colour is located. Further details may be found in the interesting memoir cited.

The *tensores patagii* show a striking resemblance among

---

[1] 'Observations on the Metallic Colours of the Trochilidæ and the Nectariniidæ,' P. Z. S. 1896, p. 283.

the swifts and humming birds. In *Chætura, Cypselus*,[1] and *Phaëthornis* and other humming birds the tensor brevis is fleshy for almost its whole extent. In *Dendrochelidon* the tensor brevis has still a larger muscular portion than is usual, but the tendon is more evident and has a passerine slip to the humerus. In the Trochili, however, the muscle is inserted on to a special tendon upon the fore arm,[2] and not on to the extensor metacarpi. In the leg muscles the birds of this group agree in only possessing the *femoro-caudal* of those used by GARROD in his classification ; the formula, therefore, is A—. GARROD, however, has left a note to the effect that in *Chætura caudacuta* the femoro-caudal passes through a muscle arising from both pubis and ischium, which is thus possibly a combined semitendinosus and semimembranosus.

Though the *semitendinosus* is as a rule absent, there seem to be traces of its accessory in a few swifts. Thus in *Cypselus alpinus* and *Chætura Vauxi* the gastrocnemius has an origin between the biceps loop and the main body of the sciatic nerve from the femur.

Another peculiarity in the leg of certain swifts (cf. also *Phaëthon*) is the absence of a biceps loop ; but the value of this character may be gauged from the following table :—

Without Biceps Sling	With Biceps Sling
Chætura caudacuta	*Chætura spinicauda*
Panyptila melanoleuca	*Chætura zonaris*
Dendrochelidon coronata	*Cypseloides fumigatus*
Macropteryx mystacea	*Cypselus alpinus*

The *biceps femoris* of humming birds—at any rate of *Patagona gigas*—is peculiar in the fact of its being two-headed.

The *deep flexor tendons* in the swifts vary. In the majority of forms the two tendons completely blend ; in

---

[1] For *Cypselus* muscles see NITZSCH-GIEBEL, 'Zur Anatomie d. Mauer-schwalbe,' *Zeitschr. f. d. ges. Nat.* x. 1857, p. 327.

[2] This tendon looks like a degenerate representative of the abductor pollicis.

Macropteryx, however, Lucas[1] has described the flexor hallucis as going to the first digit only after giving off a vinculum to the tendon of digit IV. As to the humming birds, there has been some confusion. It appears, however, that the flexor hallucis before going to digit I. gives off a slip to flexor communis of digit II., and (according to Gadow[2]) III. and IV. also.

Both swifts and humming birds have, as a rule, only one *carotid*, the left. The following swifts have two carotids : *Chætura rutila* (right carotid larger), *Cypseloides fumigatus*. In *Micropus* (? *Panyptila*) *melanoleuca* Shufeldt has described the left carotid (the only one present) as crossing over to the right and being until the middle of the neck free of the hypapophysial canal.

The chief peculiarity of the vascular system concerns the femoral vein. In *Panyptila melanoleuca* and in *Chætura zonaris* the femoral vein, instead of running deep of the femoro-caudal muscle, comes to join the sciatic artery and nerve immediately it has passed the obturator externus superficial to the femoro-caudal tendon ; *Cypseloides fumigatus* is the only other swift which has been shown to be characterised by this structural abnormality.

The large size of the heart of the humming birds as compared with that of the swifts is commented upon by Shufeldt.

The *syrinx* of the swifts is not in any way remarkable ; it is tracheo-bronchial, with the usual pair of intrinsic and extrinsic muscles. The former are attached (at any rate in *Chætura caudacuta*) to the first bronchial semi-ring. In *Cypseloides fumigatus*, however, a swift which is in other ways abnormal, there appear to be no intrinsic muscles.

The humming birds, on the contrary, have an unusual form of syrinx, which is remarkable in two ways.

In the first place the trachea bifurcates very high up in the neck, recalling the characteristics of *Platalea rosea* (see below). Each bronchus in *Trochilus colubris* (according

[1] *Ibis*, 1895. p. 298.   [2] *Ibid*. p. 299.

to MacGillivray [1]) has as many as thirty-four rings, which are complete and not semi-rings. There seem to be two pairs of extrinsic muscles, which form a very prominent muscular mass, as in Passeres. Dr. Shufeldt was unable to find any sterno-trachealis.

The *tongue* in the swifts is short and sagittate, with a spiny base. It is constantly bifid at the tip.

In the humming birds, as is well known, the long tongue is tubular, and for its support the hyoids are bent over the top of the skull, as in the woodpeckers. The tongue itself ' is double right down to the unpaired part of the os entoglossum, whilst each of the two distal prolongations of the entoglossal bone or cartilage is surrounded by a horny sheath, which is curled upwards and inwards, in a similar fashion to what we have seen in the Nectariniidæ. In many species the outer and inner edges of these tubes, however, are entire and not laciniated. Thus the Trochilidæ have developed the highest form of tubular tongue ' (Gadow [2]).

The *gizzard* of the humming birds is remarkably small; that of the cypselids presents no remarkable characters, and Shufeldt has remarked upon the large size of the *liver* in the humming birds as compared with the swifts; in both the right lobe is larger than the left, and there is a *gall bladder* in the swifts.

Cæca are entirely absent in the **Macrochires**.[3] The following are intestinal measurements of the swifts :—

Cypselus apus . .	6·25 inches.
Dendrochelidon coronata	4·30 ,,
Chætura caudacuta	10 ,,
Cypselus alpinus	10 ,,
Chætura Vauxi . .	3·25 ,,

A careful account of the trochiline and cypseline skeleton will be found in a memoir by Shufeldt.[4] Though this

[1] In Audubon's *Birds of N. America*.

[2] ' On the Suctorial Apparatus of the Tenuirostres,' *P. Z. S.* 1883.

[3] Crisp, ' On some Points relating to the Anatomy of the Humming Bird (*Trochilus colubris*),' *P. Z. S.* 1862, p. 208, observed a ' rudimentary appendix.'

[4] ' Contribution to the Comparative Osteology of the Trochilidæ, Caprimul-

observer is disinclined to allow a very near affinity between the birds, it is undeniable that there are resemblances.

The *skull* is schizognathous in the humming bird, ægithognathous in the swifts. But the ægithognathism in the latter is a little abnormal. GARROD has pointed out in describing [1] the osteology of *Indicator* that that bird, in common with the Capitonidæ, has a truncated vomer, in which the truncation occurs behind the line joining the maxillo-palatines, while in the true Passeres the truncation is in front of this line. The swifts are intermediate, the truncation being, as is shown in the accompanying figure (fig. 104), about on a level with the line joining the maxillo-palatines. It is true that the lateral processes so characteristic of the ægithognathous skull are better developed in the swifts than in the swallows : but, on the other hand, it must be borne in mind that the undoubtedly ægithognathous *Indicator* is without these processes. In both swifts and humming birds the skull is holorhinal and without basipterygoid processes. As to the vomer, HUXLEY described it as truncated ; but SHUFELDT finds it to end in an excessively fine point. In swifts the vomer is, as already stated, truncated. But as to this difference and its value as a means of separating the birds cf. the manifold vomer of **Limicolæ**.

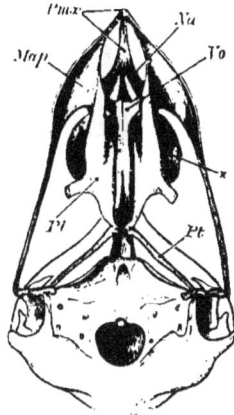

FIG. 104.—SKULL OF *Micro-pus melanoleucus*. UNDER VIEW. (AFTER SHUFELDT.)
Pmx, premaxilla ; *Mxp*, maxillo-palatines ; *Vo*, vomer ; *Na*, nasal ; *Pl*, palatine ; *Pt*, ptery-goid.

The humming birds have fourteen or fifteen (*Trochilus Alexandri*) *cervical vertebræ*. The Cypselidæ have thirteen or fourteen. Four ribs [2] join the *sternum* on each side

gidæ, and Cypselidæ,' *P. Z. S.* 1885, p. 886, and 1886, p. 501. See also ZEHNTNER, ' Beiträge z. Entwicklung von *Cypselus melba*,' *Arch. f. Naturg.* lvi. 1890, p. 189 (transl. in *Ibis*, 1890, p. 196).

[1] *Loc. cit.* (on p. 196.)   [2] FÜRBRINGER says five or six.

in both groups of birds. The *sternum* in both is unnotched and broader behind than in front.

In the fore limb the length of the hand distinguishes

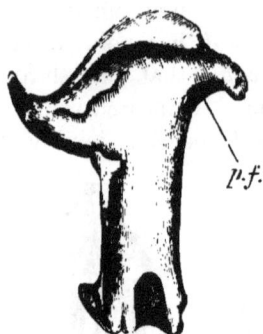

FIG. 105.—ANCONAL ASPECT OF LEFT
HUMERUS OF *Micropus melanoleucus* (AFTER SHUFELDT).
p.f. pneumatic fossa.

FIG. 106.—PALMAR ASPECT OF SAME
BONE (AFTER SHUFELDT).

both the families of the **Macrochires**, whence, of course, the name. The nearest approach in length of hand is shown in the swallows, petrels, and, oddly enough, in the penguins.

FIG. 107.—ANCONAL ASPECT OF LEFT
HUMERUS OF *Trochilus Alexandri*
(AFTER SHUFELDT).
p.f. pneumatic fossa.

FIG. 108.—PALMAR ASPECT OF SAME
BONE (AFTER SHUFELDT).

The humerus in both families is extremely short; the radial crest is well developed in both into a long process which curves over the shaft in the Trochilidæ, but over the head in the swifts.

CAPRIMULGI

Definition.—**Anisodactyle. Oil gland nude.[1] Rectrices, ten. Aquinto-cubital. Aftershaft present. Skull holorhinal. Both carotids present. Cæca[2] large. Ambiens and accessory femorocaudal absent. Deep flexor tendons of type V.**

This group of birds shows a considerable amount of structural variation, which allows of the separation of the genera into at least two families; they are, however, all united by the characters in the above definition. The

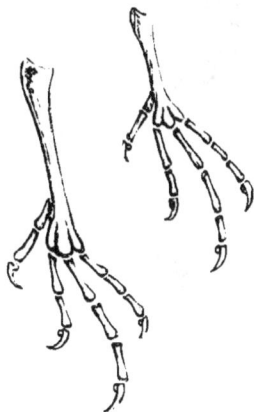

Fig. 109.-- Left Feet of *Antrostomus vociferus* (Right-hand Fig.) and *Nyctidromus albicollis* (after Sclater).

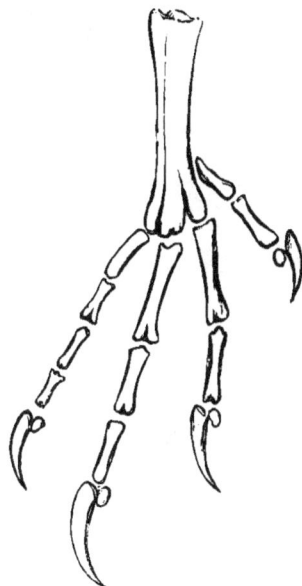

Fig. 110.--Right Foot of *Podargus Cuvieri* (after Sclater).

external aspect too of these birds, with the widely gaping mouth and their generally softly tinted grey and brown plumage, enables them to be readily distinguished from other groups.

[1] Sometimes said to be absent in Podargidæ, but Fürbringer found it in *Batrachostomus.*

[2] Absent in *Ægotheles.*

In the typical Caprimulgidæ (fig. 109) the claw of the
middle toe is serrated and the fourth toe has but four phalanges. There is no serration and five phalanges in others.[1]
The *aftershaft* is present; in the aberrant *Steatornis* it is

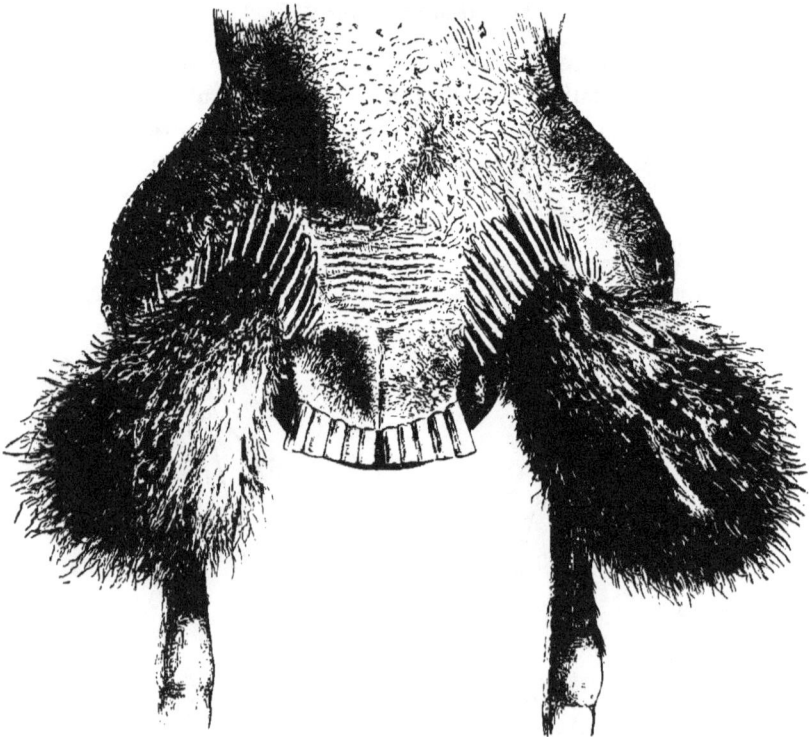

FIG. 111. -POWDER-DOWN PATCHES OF *Podargus* (AFTER SCLATER).

not absent (as GARROD asserted). *Podargus* is remarkable
for the possession of *powder-down patches*, of which there are
two, one on either side of the rump (see fig. 111), first discovered by Mr. SCLATER.[2] The patches of this bird are well
defined and very compact, and have not the diffuse character
that is seen in, for example, *Rhinochetus*. Powder downs
are also found in *Batrachostomus* and *Nyctibius*. The

---

[1] SCLATER. 'Notes upon the American Caprimulgidæ,' *P. Z. S.* 1866, p. 120.
[2] ' Additional Notes on the Caprimulgidæ,' *ibid.* p. 581.

pterylosis has been elaborately described by GARROD for
Steatornis;[1] so we shall select that bird, though it is, as

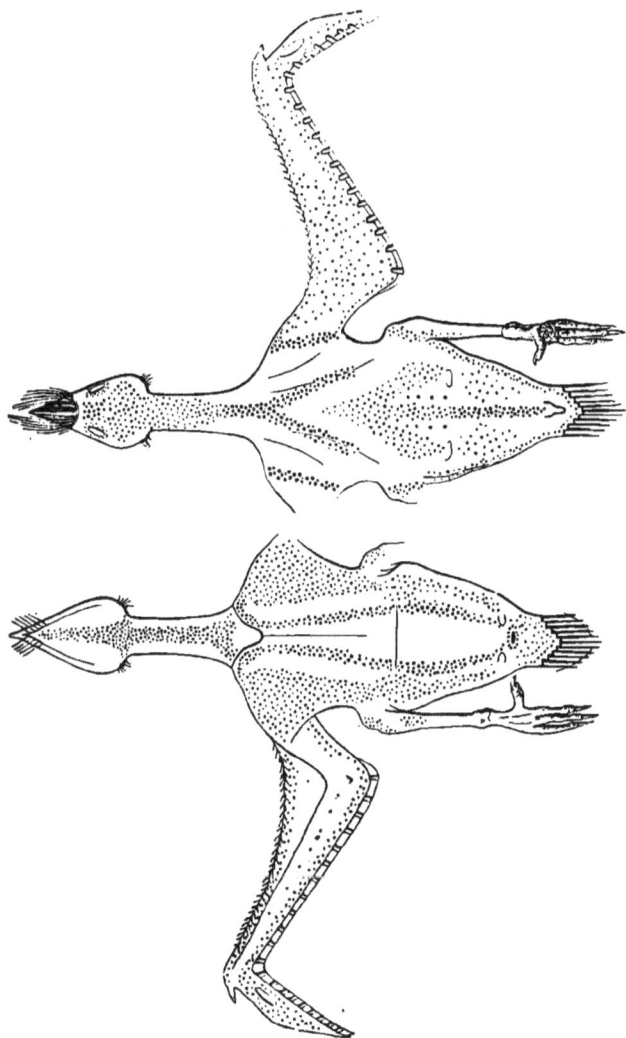

FIG. 112.—PTERYLOSIS OF *Steatornis* (AFTER GARROD). THE RIGHT-HAND FIGURE REPRESENTS THE DORSAL VIEW; THE LEFT, VENTRAL.

[1] For notes on *Steatornis* see, in addition to papers quoted, N. FUNCK,
'Notice sur le *Steatornis caripensis*,' *Bull. Ac. Belg.* 1844, p. 371; F. STOLZ-
MANN, 'Observations sur le *Steatornis* Péruvien,' *Bull. Soc. Zool. France*,
v. 1880, p. 198; HUMBOLDT, 'Sur le *Steatornis*,' *Bull. Soc. Philom.* 1817, p. 51;

already hinted, in many ways an aberrant form as a type. The dorsal tract (see fig. 112) gradually narrows as it passes down the neck, but the feathers get stronger; it bifurcates between the scapulæ to form a well-defined fork, which ends ultimately, having become weaker. Between this fork, and not connected with the rest of the dorsal tract, appears a spear-headed patch of feathers. The 'shaft' of the 'spear' becomes stronger as it descends to end abruptly at the base of the oil gland. The ventral tract is narrow between the mandibles; it is undivided upon the neck. At the beginning of the breast it divides into a wide outer and a narrow inner portion, the latter being more strongly feathered. The two converge on each side towards the cloacal aperture, but do not reunite.

In *Caprimulgus*, on the other hand, the ventral tract bifurcates in the neck, and the two ventral tracts are single, and there is no such abrupt break between the two parts of the dorsal tract as has been described above in *Steatornis*. Nor is there in *Antrostomus*, where there is no narrowing of the posterior part of the dorsal tract. *Nyctidromus* is much the same.[1]

The *tongue* in the goatsuckers is more or less abortive; in *Podargus* it is a curious tough but transparent membranous organ.

As mentioned in the definition of the group the *cæca* are large; but as a unique exception *Ægotheles* seems to be entirely without them. In all the genera the left lobe of the liver is rather the smaller; and all, save *Chordeiles*, have a gall bladder. The intestinal measurements in inches are given on the following page.

The *intestine* (according to MITCHELL) is primitive and owl-like, while the *cæca* are dilated apically, as in owls.

J. MUIE, 'Fragmentary Notes on the Guacharo, or Oil Bird,' *Ibis* (3), iii. p. 81; L'HERMINIER, 'Mémoire sur le Guacharo,' *Nouv. Ann. Mus.* iii. 1834, p. 321, and note 'Sur la Classification Méthodique du Guacharo,' &c., *Rev. Mag. Zool.* (2), i. p. 321; J. MÜLLER, 'Ueber die Anatomie des *Steatornis caripensis*,' *M.B. k. Akad. Wiss. Berlin*, 1841, p. 172, and 'Anatomische Bemerkungen über den Guacharo,' *Arch. f. Anat. u. Phys.* 1842, p. 1.

[1] For further details of feathering see CLARK, 'The Pterylography of certain American Goatsuckers and Owls,' *Proc. U. S. Nat. Mus.* xvii. 551.

	S.I.	L.I.	Caeca
Nyctidromus albicollis .	7·5	1·4	1·3, 1·4
Caprimulgus europaeus	10·5		1·6
" "	9·5		1·75
Chordeiles texensis	7		1·1, ·9
Podargus Cuvieri .	. 16		2
Steatornis caripensis .	. 18·5	·2	1·25
" "	18		1·75

The *syrinx* is highly characteristic in the Caprimulgi. Like the nearly related (?) cuckoos, we have both the tracheo-bronchial and the purely bronchial syrinx. Indeed, the stages are almost identical in the two groups. *Cuculus* and *Caprimulgus* correspond with a tracheo-bronchial syrinx ; then we have *Centropus* and *Podargus*, and finally the culmination in *Crotophaga* and *Steatornis* of a syrinx furnished with a membrana tympaniformis, which does not commence until many rings below the bifurcation of the tube, the intrinsic muscles being attached to the first ring which borders upon it. It will be necessary to describe the various syringes in some detail ; they have been studied and figured by myself.[1] In *Nyctidromus albicollis*, which will serve as a type of the tracheo-bronchial syrinx which exists in the Caprimulgidæ (s.s.), the last four tracheal rings are closely applied in contradistinction to the preceding, which are separated by copious membranous intervals. The last two tracheal and the first five bronchial semi-rings are ossified. To the first of the latter are attached the intrinsic muscles. In *Batrachostomus* we have the intermediate type of syrinx, which may, however, be called bronchial. The first six bronchial semi-rings and the last three tracheal are ossified,

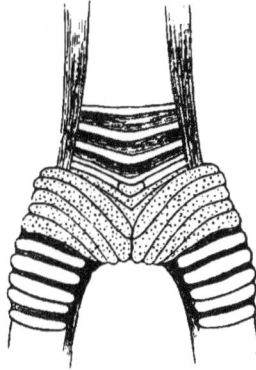

Fig. 113.—Syrinx of *Nyctidromus albicollis* (after Beddard).

[1] 'On the Syrinx and other Points in the Anatomy of the Caprimulgidæ,' P. Z. S. 1886, p. 147.

and bear much resemblance to each other, which will be apparent from the illustration (fig. 115).

The intrinsic muscles are attached to the seventh bronchial ring, which is soft and cartilaginous ; where the bronchial rings change their character is a constriction of the membrana tympaniformis ; it is, however, of equal breadth before and after the change.   In *Podargus Cuvieri* there is a further approach to the purely bronchial syrinx of *Steatornis*.   The

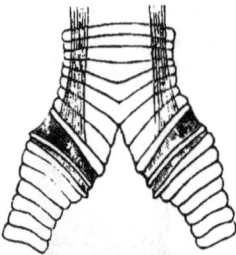

FIG. 114.—SYRINX OF *Ægotheles*
(AFTER BEDDARD).

FIG. 115.   SYRINX OF *Batrachostomus*
(AFTER BEDDARD).

first two bronchial rings are complete.   These and the fourteen following are closely applied to each other and ossified. The intrinsic muscles are attached to the last of this series. *Ægotheles* really belongs to this section of the Caprimulgi, though the intrinsic muscles are attached very high up upon the bronchi ; but the two rings immediately preceding the attachment are complete rings.   The final development of the bronchial syrinx is seen in *Steatornis* (see fig. 48, p. 69), where all the rings in front of the attachment of the intrinsic muscles low down upon the bronchi are closed and complete rings, as in *Crotophaga*.

The *tensor patagii* shows certain differences among the goatsuckers.   In the genera *Caprimulgus*, *Nyctidromus*, and *Chordeiles* there is a *biceps slip*, absent in the rest.   Of these three genera the arrangement of the tendon is shown

in the annexed cut (fig. 116). *Steatornis*, as will be seen
(fig. 117), hardly differs, and *Podargus* agrees with it. In
Ægotheles there is a slight difference in that there is hardly
any trace of the wristward branch of the tendon. *Steatornis*
has an *expansor secundariorum*, apparently absent among
the other genera. The muscle is attached to the teres by its
long tendon. The insertion of the *deltoid* is extensive, and
it receives a tendon from the scapula. In many Caprimulgidæ the *biceps* is split for some distance before its insertion,
the bifidity even invading the muscle itself and not being

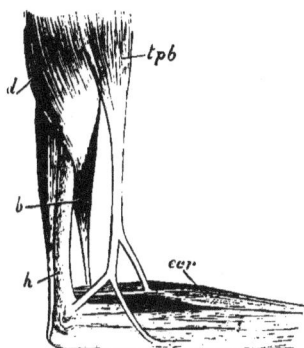

FIG. 116. PATAGIAL MUSCLES OF
Caprimulgus (AFTER GARROD).
d, deltoid ; *b*, biceps ; *h*, humerus ; *tpb*, tensor patagii brevis ; *ecr*, extensor carpi radialis.

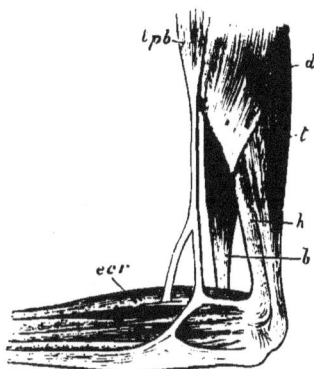

FIG. 117.—CORRESPONDING MUSCLES
OF *Steatornis*, BUT OF LEFT WING
(AFTER GARROD).
t, triceps. Other letters as in fig. 116.

limited to its tendon. In *Podargus* FÜRBRINGER describes
a special slip of *rhomboideus profundus*, arising separately
from ilium. The *anconæus* has a tendinous humeral head.
In the thigh the muscle formula is either AXY− (most Caprimulgi) or XY− (*Steatornis*). The *tibialis anticus* tendon of
Podargus is double. The glutæus I. extends over biceps in
Nyctidromus, &c., not in *Ægotheles*. No glutæus V. There
are sometimes both and sometimes only one of the two
peroneals present. In *Steatornis* and *Ægotheles* only the
brevis is to be found, in *Nyctidromus* only the longus ; in
Podargus both.

The goatsuckers have by no means a uniform skull structure.

In *Caprimulgus* [1] the skull may be termed schizognathous. The palatines are enormously expanded, and between their posterior extremities (not indicated in the figure) are a small anterior and posterior medio-palatine, a state of affairs recalling the Picidæ (*q.v.*) The vomer is a long bone, distinctly paired in the young bird, rounded in front. It articulates

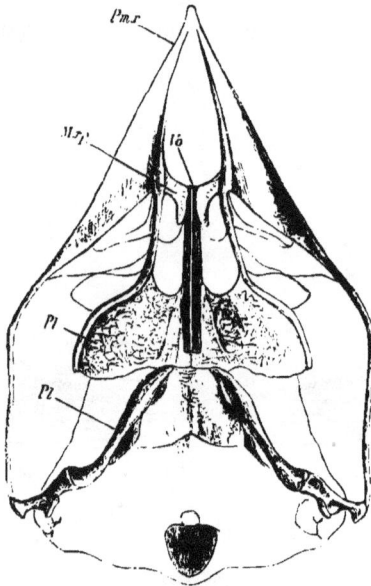

FIG. 118.—SKULL OF *Caprimulgus* (AFTER HUXLEY).
Pm.x, premaxilla ; *M.xp,* maxillo-palatine ; *Vo,* vomer ; *Pl,* palatine ; *Pt,* pterygoid.

with the hook-like maxillo-palatines. Each of the latter is connected by a ligament with the internal forward process of the palatine of its own side, the hinder part of which is largely ossified. The basipterygoid processes are well developed. The lacrymal is large and ' binds upon the zygoma.' The ectethmoids are attached to the broad outer flange of the palatines by a cartilaginous prolongation.

[1] PARKER, ' On the Structure and Development of the Bird's Skull,' *Linn. Tr.* (2), i.

In *Chordeiles*[1] the skull is much upon the same plan, but the maxillo-palatines meet in the middle line, and may even become ankylosed.

The skull is thus desmognathous, in fact. In *C. virginianus*, however, the bones do not meet.

The skull of *Nyctibius jamaicensis* (see fig. 119), described by HUXLEY, is not widely different from that of *Caprimulgus*. The ligaments which unite the inner angle of the

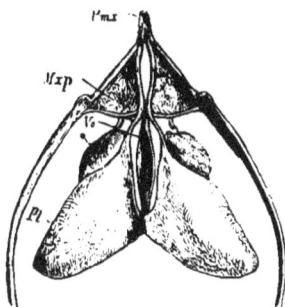

Fig. 119.—FORE PART OF SKULL OF *Nyctibius jamaicensis* (AFTER HUXLEY).
*, prefrontal process. Other letters as in fig. 118.

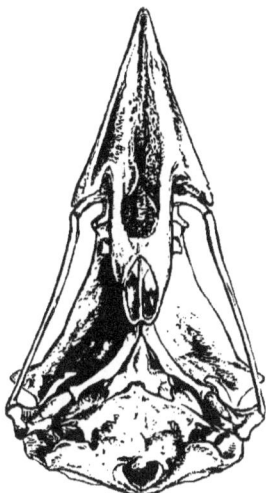

Fig. 120.— SKULL OF *Steatornis* (AFTER HUXLEY).

palatine to the maxillo-palatines are, however, completely ossified.

In *Podargus*[2] the skull is completely doubly desmognathous. The basipterygoid processes are quite rudimentary. There are two small azygous vomers. The palatines have coalesced in the middle line. The lacrymal is small, if not absent.

The skull of *Steatornis* has been described, with figures,

[1] SHUFELDT, 'On the Osteology of the Trochilidae,' &c.. *P. Z. S.* 1885, p. 891.

[2] PARKER, *loc. cit.* p. 124 (with figs. on Pl. xxiii.)

by GARROD,[1] and more recently and more fully (also with illustration) by PARKER.[2] The skull is quite owl-like in general aspect with its decurved beak, and in the equality between greatest length and greatest breadth. The lacrymal is very small and is ankylosed to the orbital wall. The

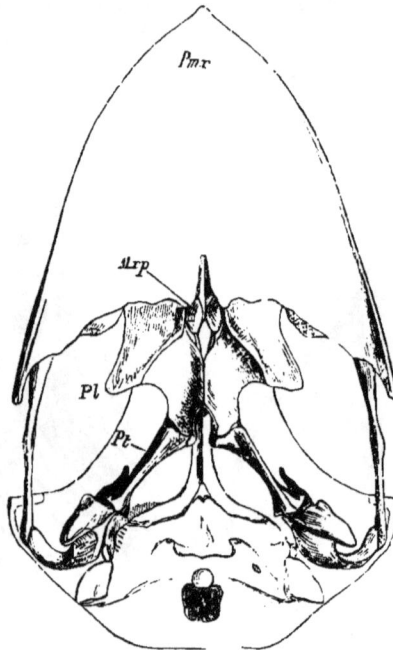

FIG. 121. -SKULL OF *Podargus* (AFTER HUXLEY). LETTERS AS IN FIG. 118.

maxillo-palatines are completely fused across the middle line, the skull being desmognathous. The palatines also in their middle part meet across the middle line. The vomer is divided into two parts, one lying behind the other. The anterior part is small, like the posterior part; they both measure 5.5 mm. The posterior vomer is probably the equivalent of the medio-palatine of *Caprimulgus* and other birds. The basipterygoid processes are well developed. The

[1] 'On some Points in the Anatomy of *Steatornis*,' *P. Z. S.* 1873, p. 526.
[2] 'On the Osteology of *Steatornis caripensis*,' *ibid.* 1889, p. 161.

ectethmoids are continued on to the quadrato-jugal bar by a
distinct ossified *os uncinatum*, as in *Todus*, *Scythrops*,
Musophagidæ, *Piaya cayana*, *Cariama*, and Procellariidæ.
The relations between the procoracoid and the clavicle vary
somewhat. In *Podargus* the process is large and reaches

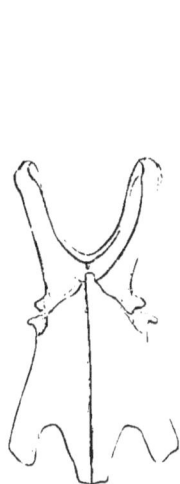

Fig. 122.—Sternum
of *Caprimulgus*
(after Sclater).

Fig. 123.—Sternum of
Podargus (after
Sclater).

Fig. 124. · Sternum
of *Nyctibius* (after
Sclater).

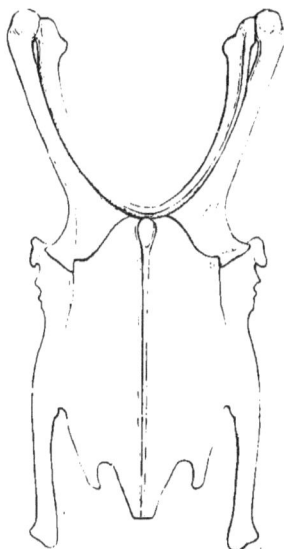

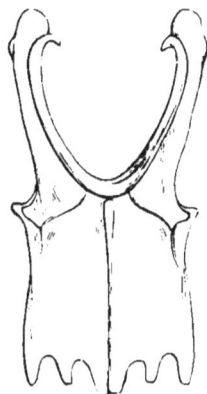

the clavicle ; it is small and does not in *Steatornis* and
Caprimulgus.

The number of *cervical vertebræ* varies from thirteen
(*Chordeiles*) to fifteen (*Steatornis*). PARKER has commented
upon the fact that in *Steatornis* the atlas, instead of being,
as is the rule among birds presumably allied to it, perforated
below for the reception of the odontoid process, is merely
deeply notched for the same. In *Steatornis* the dorsal
vertebræ are opisthocœlous, as among the parrots alone
among probable allies. It is the rule among the Caprimulgi
for four ribs to reach the sternum. The *sternum* itself is
one-notched on each side in *Steatornis*, &c. It has four
foramina in *Ægotheles*, and is doubly notched on each side

R

in *Podargus*. The spina externa is developed and slightly bifurcate in *Steatornis*. There is no spina interna.

In view of the considerable variation in structure exhibited by the group, the following tabular statement may be of use.

—	Carotids	Oil Gland	Biceps Slip	Exp. Sec.	Thigh Muscles	Glutaeus	Caeca	Gall Bladder	Skull	Basipt. Pr.	Sternum	Powder Downs	Syrinx
Podargus . .	1	−	−	.	AXY	Not over biceps	+	+	Desm.	−	2 notches	+	Bronch.
Steatornis .	2	+	−	+	NY	,,	+	+	,,	+	1 notch	−	,,
Ægotheles .	2	+	−	−	AXY		−	+	,,		2 for.		,,
Batrachostomus .	+						1	+	,,		2 notches	+	,,
Nyctibius . .									Schiz.		,, ,,	+	
Nyctidromus .	2	+	−		AXY	Over biceps	+	+	,,	+	1 notch	−	Trach.-Bronch.
Caprimulgus .	2	+	+	−	AXY	,, ,,	+	+	,,	+	,, ,,	−	,,
Chordeiles .	2	+	+	−	AXY	,, ,,	+	−	,,	+	,, ,,	−	,,
Antrostomus .	2	+	+	−	AXY	,, ,,	+	+	,,	+	,, ,,	−	,,

There can be little doubt, from a consideration of the above table, of the naturalness of a family Caprimulgidæ to include the last four genera. In these forms, in all of them, the toes are aberrant in that the last has only four phalanges, and further that the claw of the middle one is serrated.

It will be noticed from the table that the amount of structural variation among the Caprimulgidæ (*s.s.*) is exceedingly small, the only character, indeed, of those selected which varies being the gall bladder, which is absent in *Chordeiles*, and shows signs of commencing disappearance by its small size in some of the others. We might, perhaps, add the desmognathism of *Chordeiles*; but this is obviously but a slight exaggeration of the ægithognathous palate of the others. The enormous length of the second primary of *Macrodipteryx* and *Cosmetornis* is a variation which does not appear to be of great importance from a classificatory point of view.

The remaining genera are by no means so uniform a group as that which those that have been already considered form. In all of them, however (so far as is known), the biceps slip is absent, the glutæus primus is of limited extent,

---

[1] See BLYTH in *Ibis*, 1866, p. 357.

the skull is desmognathous, the syrinx is bronchial (least
marked in *Ægotheles*), and the outer toe has five phalanges,
while the middle toe has no serration. This is a fair
assemblage of identical characters. It is usual for *Steatornis*
to be separated as a distinct family from the Podargidæ, as
has been done by GADOW. It differs from *Podargus* in
seven out of the thirteen characters made use of in the
above table. It is often supposed that the Guacharo is
peculiar among goatsuckers by reason of its vegetable diet;
but NEWTON states in his ' Dictionary of Birds ' that the
Podargidæ also partly nourish themselves on fruit. Moreover *Ægotheles*, which is by universal consent placed in the
immediate neighbourhood of *Podargus*, differs from that
genus in four out of the thirteen characters, and from *Steatornis* in exactly the same number. It appears possible to
place all these genera in one family, which, on account of its
greater antiquity, has had time to vary more than the Caprimulgidæ. It is also among members of this family that the
greatest number of points of affinity to the owls is met with,
a further argument in favour of their basal position.

The relationship of the **Caprimulgi** to other groups is a
puzzle hard of solution. This is partly, perhaps, due to the
fact that the goatsuckers are probably a somewhat ancient
group. That they are an ancient group seems to be shown
by the considerable amount of specialisation of structure
which they exhibit, by the primitive character of the
intestinal folds—the cæca being at the same time well
developed—by the double carotids, and by their wide distribution, with a restriction in range of some peculiar types,
such as *Steatornis*.

The remarkable series of modifications of the syrinx is
one of the most striking facts in the anatomy of the group.
They share this with the **Cuculi**, and, though to a less extent,
with the **Striges**. It is, indeed, with this latter group that
the goatsuckers seem to be most nearly allied. This conclusion, which is in harmony with much recent opinion, is
curious in view of the external likenesses [1] which bind

[1] For instance, the ' ears ' of certain Podargidæ and of *Lyncornis*. It is a

together the two groups of birds, likenesses which might fairly be put down to similarity of habit. These superficial resemblances are, however, enforced by more deep-lying structural similarities. MITCHELL has found that of the various groups which may be supposed reasonably to be allied to the **Caprimulgi** the owls come nearest to them in the primitive character of the gut, while the cæca, swollen at the ends, are alike in both. The owls too are nearly the only other Coraciiform birds besides the **Caprimulgi** which have well-developed basipterygoid processes. The trogons, it is true, possess them, but then they differ in many other important particulars.

STRIGES

Definition. - Oil gland nude.[1] Aquincubital. Both carotids present. Cæca well developed, ending in a dilatation. Skull desmognathous and holorhinal, with basipterygoid processes. No ambiens, semitendinosus, accessory femoro-caudal, biceps slip, or expansor secundariorum.

The owls, formerly associated with the **Accipitres** and termed 'Accipitres nocturnæ,' or 'Nyctharpages,' are now generally placed by themselves away from the hawks and in the neighbourhood of some of the birds comprised under the term 'picarian.' The group itself is characterised by a great uniformity of structure, and by the possession, so to speak, of so many negative characters. The resemblances to the hawks are really only in habits and in beak and claw. These, however, will be dealt with more fully later.

The owls comprise a considerable number of genera, of which *Strix* stands rather apart from the rest, having as near neighbours the Eastern *Photodilus*[2] and the Madagascar *Heliodilus*.[3]

curious coincidence that the term 'morepork' is applied to *Podargus* in Australia, and in New Zealand to an owl, *Spiloglaux Novæ-Zelandiæ* (*fide* NEWTON, *Dict. of Birds, sub voce* 'Morepork ').

[1] Except *Strix* and *Asio otus*.

[2] BEDDARD, 'On *Photodilus badius*,' *Ibis*, 1890, p. 293.

[3] MILNE-EDWARDS, 'Observations sur les Affinités Zoologiques,' &c., *Nouv. Arch. Mus.* (2), i. 1878, and in *Hist. Nat. de Madagascar.* See also R. B. SHARPE, 'A Note on *Heliodilus*,' *P. Z. S.* 1879, p. 175.

It is often given as a character of the owls (and as a bond of union with *Pandion*) that there is no *aftershaft*. There is, however, a small one in *Strix*. As a rule the *oil gland* is nude, but NITZSCH gives as a constant character of *Hybris* (= *Strix*) the presence of two minute feathers upon the apex of that gland ; in *Asio otus* too there are two or three small down feathers upon the apex of the oil gland—all of which facts seem to indicate a comparatively recent loss of the apical tuft so often found in birds.

The *rectrices* are invariably twelve in number, except in *Micropallas*, where there are but ten.[1] A singular external character of the owls, carefully gone into by KAUP,[2] concerns the asymmetry of the ears; with this is sometimes correlated an asymmetry of the skull in the region of the ear.[3]

The *tensores patagii* of the owls [4] are on the whole simple. In none of them is there a *biceps slip*. Very rarely is there a recurrent tendon uniting the insertion of the brevis with the longus (patagial fan). This occurs, however, in *Strix Novæ Hollandiæ* (? always), *S. flammea* (occasionally), and *S. pratincola* (? always). Something of the same kind is found in *Syrnium aluco*, where, however, the connection is between the tendon of the *longus*, just at its origin, and the *inner* of the two branches of the *brevis*. In most owls (*Bubo maculosus, Scops leucotis, Pulsatrix torquata, Athene noctua, Strix, Syrnium nebulosum*) the *tensor brevis* sends off a wristward slip not far from insertion of main tendon ; the latter alone crosses the fore arm to be inserted on to ulna.

In *Scops Lempiji, S. Asio, Asio otus, Ketupa javanensis, Otus vulgaris, Bubo ascalaphus*, and *Nyctea nivea*, there is an additional posterior brevis tendon, arising separately from the muscle. This latter tendon sometimes in *Otus vulgaris* is connected with the middle one.

[1] *Vide* CLARK, 'On the Pterylography of certain American Goatsuckers and Owls,' *Proc. U. S. Nat. Mus.* xvii. p. 551.

[2] A monograph of the Strigidæ, *Zool. Trans.* iv.

[3] Cf. COLLETT, 'On the Asymmetry of the Skull in *Strix Tengmalmi*,' *P. Z. S.* 1870, p. 739.

[4] D'ALTON, *De Strigum Musculis Commentatio*, Halis, 1837 ; HEUSINGER, *Arch. f. Phys.* vii. 1822.

In *Asio otus* there is a fleshy slip from pectoral to longus tendon, and a fibrous slip from humerus to patagial muscle just at origin of innermost brevis tendon. In *Scops Lempiji* there are muscular slips to both longus and brevis from the pectoral, and a fibrous slip from the humerus to the longus tendon just at its origin.

The humeral origin of the *anconæus* is always present.

There is never an *expansor secundariorum*, so far as is known.

The *pectoralis primus* is not at all double.

The only one of the *leg muscles* used by GARROD in classification that is present is the *femoro-caudal*, the formula thus being A—. Glutæus I. is absent, glutæus V. small. In *Ketupa javanensis*, however, V. is absent and I. not quite so. Only one *peroneal* muscle is present, which is attached to head of metacarpal. The tendon of insertion of the *tibialis anticus* is divided; in *Pulsatrix* I found it to be trifid at insertion, and even the muscle itself was divided into two for a short distance.

The *deep flexor tendons* are of type I. In *Ketupa ceylonensis* the two tendons blend a quarter of the way down the tarso-metatarsus; though blended the fibres can be recognised, and it may be seen that those of *flexor hallucis* mainly supply digits I., II., a small part only to rest of common tendon.

	Small Int.	Large Int.	Cæca
	Ins.	Ins.	Ins.
Asio otus . .	20		2·5
Syrnium aluco .	22	2	3·5
„ nebulosum	25	2	3
Bubo ignavus .	40	2·25	4·5
„ virginianus .	34·5	3·5	4
„ capensis .	30·25	2·8	3·8
Ketupa javanensis	24	1·75	4
„ ceylonensis	30	3·5	3·5, 3·25
Athene noctua .	14	1	2·5
„ passerina .	14	1	2
Pulsatrix torquata	21	2	2·5
Speotyto cunicularia	15		2
Nyctea nivea .	42		4
Surnia funerea .	16·5	1·5	2
Gymnoglaux nudipes	9·9	2·2	1·4

The colic cæca are always (fig. 125) dilated at the blind end. The liver lobes are subequal, and the gall bladder appears to be always present.[1]

The *skull* of the owls [2] shows some differences in *Strix* from the characters which distinguish the majority of the group.

In *Strix* the skull is elongate, the proportions being in *Strix flammea* 56 length : 36 breadth. In another species (*Strix* sp. inc.) 62 : 37·5.

On the other hand in other owls the differences between breadth and length show gradually progressive series, culminating in *Speotyto cunicularia*, in which the proportions are nearly equal, viz. 38 : 37. The skull of *Strix* further differs from that of other owls in the swollen character of the prefrontal processes, which are thin, almost paper-like sheets of bone in other owls. The interorbital septum of *Strix* is thick, while in the remaining genera it is reduced to a thin dividing lamina, as is the case with most birds. The skull characters of the genus *Photodilus* are to some extent

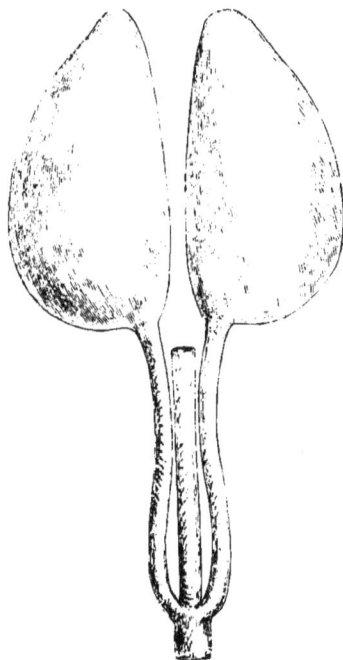

FIG. 125.—COLIC CÆCA OF *Photodilus* (AFTER BEDDARD).

intermediate between *Strix* and the remaining genera of the Striges. The interorbital septum of *Photodilus* is not so thick as in *Strix*, but, on the other hand, not so thin as in other owls, as, for example, *Bubo*. The prefrontals, although not so swollen as in *Strix*, are not nearly so flattened as they

---

[1] Absent in *Speotyto*; cf. SHUFELDT, 'Notes on the Anatomy of *Speotyto*,' *Journ. Morph.* iii. 1889, p. 122.

[2] See for certain details of skull structure PARKER, *Linn. Trans.* (2), i. p. 138.

are in *Bubo*, where, as already explained, they are thin plates, hardly thicker than a piece of paper. So far *Photodilus* agrees with *Strix*; but there are points in which the skull of this aberrant owl is nearer to the bubonine section of the order. It has not the occipital convexities which are so striking a feature of the skull of *Strix*. Finally *Photodilus* is strigine in the non-extension over the occipital region of the temporal fossæ, which do so extend in many of the Bubonidæ. In *Strix* there is but one notch on either side

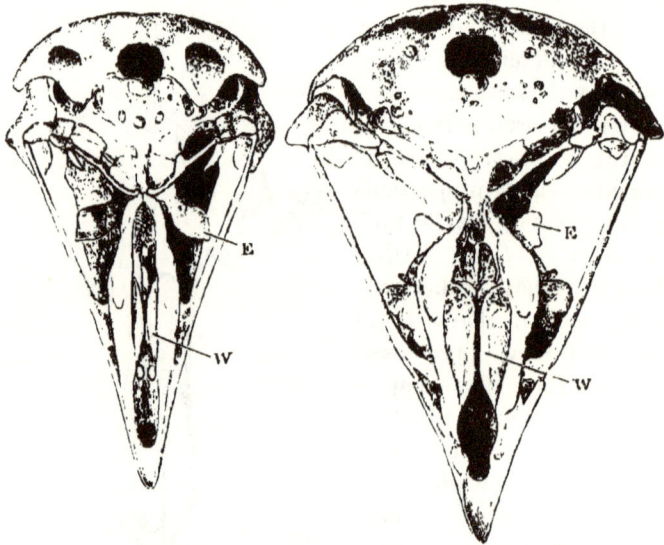

FIG. 126.—SKULLS OF *Strix* (LEFT-HAND FIGURE) AND *Bubo* (AFTER BEDDARD).

E, prefrontal process; *W*, maxillo-palatines.

of the sternum; in other owls, including *Photodilus*, there are two. The vomer of the owls is not large,[1] and behind it there is a medio-palatine, at least occasionally present.

The lacrymals, like the maxillo-palatines, with which they come into contact, are swollen and spongy. The nostrils are often partly covered by ossified alinasals, and there is a largely bony internasal septum.

[1] Said by SHUFELDT to be absent in *Speotyto*.

In the foot of *Strix* (see fig. 127) the first phalanx of digit III. is much less than the second in length ; in most other owls (fig. 128) these two phalanges are small and subequal ; *Photodilus* is intermediate. The latter genus has the peculiarity that the last digit has only four phalanges instead of five, the two basal ones being fused.

The number of cervical vertebræ does not appear to vary.

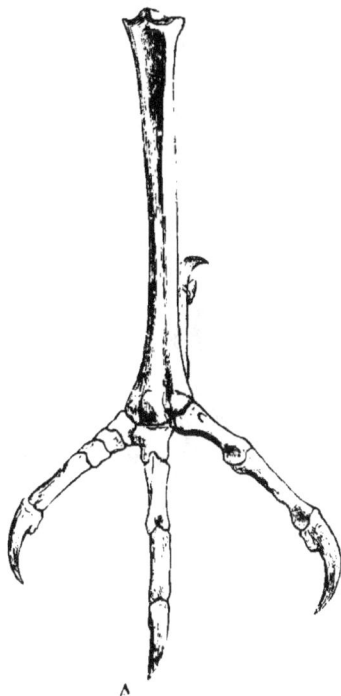

Fig. 127.—Right Foot of *Strix* (after Beddard).

I have found fourteen in *Strix*, *Photodilus*, *Ketupa*, and other genera which I have examined. In *Bubo bengalensis* the ring of the atlas is incomplete above ; the significance of the occurrence of this same deficiency in *Pandion* is largely lost, owing to the fact that *Cariama* is similarly characterised. The hæmapophyses in *Ketupa javanensis* commence as single processes on C11, and extend to D3 ; on C10 is a bifid hypapophysis, and on C9 the catapophyses nearly form a canal. Posterior catapophyses begin on C12, whence they gradually climb the hypapophyses. *Bubo bengalensis* is much the same, save that there is not a bifid hæmapophysis

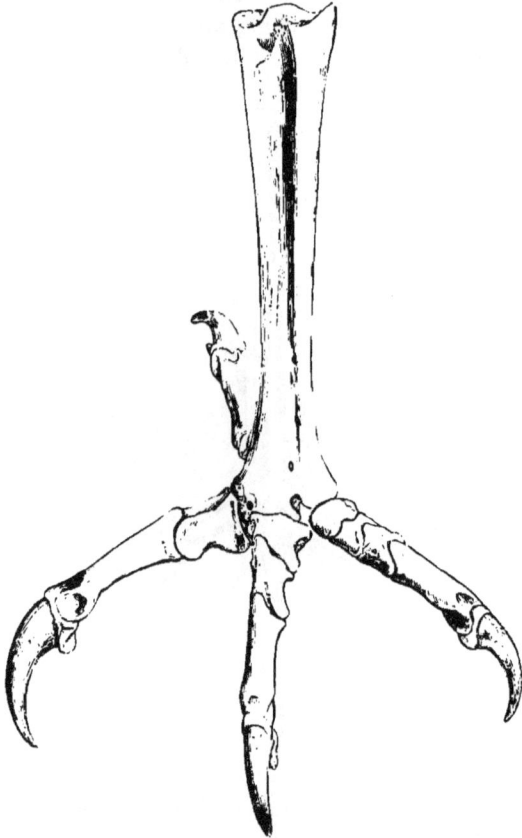

FIG. 128. LEFT FOOT OF *Bubo* (AFTER BEDDARD).

on 10. In *Strix* the catapophyses on C9 are distant. *Photodilus* is like *Ketupa*.

To the sternum, which is one-notched in *Strix* and two-notched in other owls, five ribs are attached. Dr. COUES [1]

---

[1] *Key to N. American Birds.*

has mentioned the existence among the Striges of a 'scapula accessoria.' The coracoids are not in contact at their articulation with sternum; the procoracoids are moderately large, and the clavicle reaches both them and the scapula.

If it were not for *Photodilus*, it might be possible to divide the Striges into two families, Strigidæ and Bubonidæ. As it is, it may perhaps be permissible to regard the order as containing but one family, but two sub-families, viz. Striginæ

FIG. 129.- SYRINX OF *Scops leucotis* (AFTER BEDDARD).

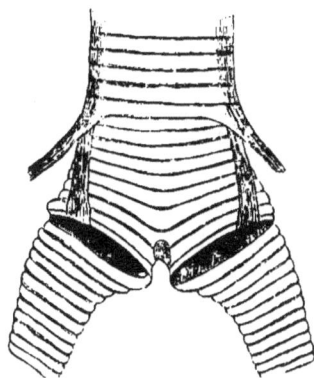

FIG. 130.—SYRINX OF *Bubo* (AFTER BEDDARD).

and Buboninæ, to which possibly a third, Photodilinæ, might be added.

The syrinx of the Striges has been chiefly described by WUNDERLICH[1] and by myself.[2] This group is one of the few that present the remarkable variety of the voice organ which has been termed the bronchial syrinx. All the owls, so far as they have been examined, possess one pair of intrinsic muscles and the usual one pair of extrinsic muscles. *Scops leucotis* has the most modified syrinx. In this bird (see fig. 129) the intrinsic muscles are attached so far down the bronchus as to the tenth bronchial ring, and, as will be

[1] 'Beiträge zur vergleichenden Anatomie und Entwicklungsgeschichte des unteren Kehlkopfes der Vögel,' *Nov. Act. Leop. Akad.* xlviii. 1884. p. 1.

[2] 'On the Classification of the Striges,' *Ibis* (5), vi. p. 355.

seen from the figure, the bronchial rings in front of this attachment are complete rings, with no membranous inter-space left.   In *Strix*, on the other hand, and in *Bubo* and *Syrnium*, the intrinsic muscles are inserted on to the first bronchial semi-ring.   In *Asio* seven complete rings intervene between the bifurcation of the trachea and the first incom-plete bronchial semi-ring, to which the muscles are attached. In *Photodilus* the intrinsic muscles are inserted on to the second bronchial semi-ring.

Until lately the owls have been almost invariably placed in the immediate neighbourhood of the diurnal birds of prey. Latterly, however, the opinion has been gaining ground that it is to the picarian birds (in a wide sense) that they are most nearly allied.   This opinion, more than hinted at by GARROD and NEWTON, has been given a practical shape in the classi-fications of FÜRBRINGER and GADOW.   The latter has in-geniously pointed out that it is impossible to imagine that the **Striges** have been derived from the **Accipitres**, since, although without an ambiens, they have much the same structure of foot as the **Accipitres** *with* an ambiens.   Hence it is difficult to believe that they would have lost it ; he con-cludes that they are derived from some bird without an ambiens, and the failure of MITCHELL to find the last trace of the missing ambiens—obvious in some birds which are clearly the descendants of birds with an ambiens—still further supports that way of looking at the matter.   Even in the skull, where the principal likenesses between the Accipitres diurnæ and nocturnæ (as the two groups in question have been called) have been seen, there are really many differences.   It is only, for example, in the skulls of those **Accipitres** to which the owls have been supposed to have the least resemblance, *i.e.* the Cathartidæ and Serpen-tariidæ, that there are basipterygoid processes.   The owls are decidedly not desmognathous (in the sense of a maxillo-palatine union), and their lacrymal is quite different from that of the hawks and eagles.   The palate, too, is incom-plete in front of the maxillo-palatines, not solid, as in the **Accipitres**.   As to other anatomical features, it is harder to find likenesses than differences. The **Accipitres** have rudimentary cæca, a biceps slip, the expansor secundariorum, a tufted oil gland, an aftershaft (except *Pandion*) ; the deep flexor tendons are different, and, in short, the differences are as great as those which separate any two groups of carinate birds.[1]

PSITTACI

Definition.—Twelve rectrices;[2] aftershaft present; aquintocubital; zygodactyle. Skull desmognathous, holorhinal, without basipterygoid processes. Biceps slip and expansor secundariorum absent.[3] Muscle formula, AXY + or −. No cæca ; a crop present.

The parrots are an almost cosmopolitan group, being most abundant, however, in the tropics. Count SALVADORI, in his British Museum catalogue of the group, allows five hundred species, distributed among seventy-nine genera. The parrots are a very sharply defined group, there being no dubious outlying forms. They are usually brilliantly coloured, and lay white eggs in hollows of trees. With the exception of the owl parrakeet (*Stringops*) of New Zealand the parrots are arboreal birds, as, indeed, the zygodactyle feet denote. As to external characters, the exaggeratedly hawk-like bill is well known. The almost universal twelve rectrices distinguish the group, but in other external and internal characters the parrots show considerable diversity of structure, as is sometimes the case with large and widely distributed groups ; compare, for instance, the pigeons, which present many other analogies to the parrots.

The *oil gland*[1] is a structure which may be wanting or developed. The table on p. 268 indicates some of the genera

---

[1] See also under 'Caprimulgi,' p. 243.
[2] With the sole exception (cf. GADOW) of *Oreopsittacus Arfaki*.
[3] See below, p. 261.
[1] External characters and many other points in the anatomy of parrots are dealt with by GARROD, 'On some Points in the Anatomy of the Parrots,' &c., *P. Z. S.* 1874, p. 247, and 'Notes on the Anatomy of certain Parrots,' *ibid.* 1876, p. 691 ; see also FORBES, 'On the Systematic Position of the Genus *Lathamus*,' *ibid.* 1879, p. 166.

in which it is present or absent. When present it is invariably tufted, and generally of fair size. In *Cacatua sulphurea*, however, the oil gland, though present, is small, and has but a single small down feather upon it.

The parrots are a group of birds which agree with the **Accipitres** in that some genera have *powder-down patches* while others have not. The table already referred to indicates the facts so far as they have been ascertained. The degree of development of the powder-downs, however, differs considerably, though in no parrot is there more than a single pair of definite powder-downs which are lumbar in position. In *Cacatua sulphurea*, for instance, there are a pair of such patches, one on either side of the dorsal tract. These send up a few scattered powder-down feathers as far forward as the neck, and a few to carinal spaces and between the branches of the ventral tract.

In *Calopsitta Novæ Hollandiæ* there are the same lumbar patches of a reniform outline; but the powder-downs are entirely confined to this region of the body.

Calyptorhynchus stellatus is more like *Cacatua*, but the lumbar patch is not so well developed.

In *Psittacula passerina* there are lumbar patches more elongated but narrower than those of the parrots already referred to; there are also scattered powder-downs not aggregated into patches.

Brotogerys tirica has no defined patches, but simply a few scattered powder-downs, which, however, are more numerous in the lumbar regions. In *Coracopsis*, *Chrysotis*, and *Pionus* there are the same generally diffused powder-down feathers not aggregated into definite tracts. The same may be said generally of *Psittacus*, though such powder-downs as there are are limited to the lumbar region.

The general *pterylosis* of the parrots is as follows: From the general covering of the head arises a narrow dorsal tract, which bifurcates in the interscapular region. Between the arms of this fork are the arms of another fork, which unite near the oil gland to form the single straight, short posterior part of the dorsal tract. On the ventral surface the tract is single, or double on the neck, and where it widens out on either side of the carina sterni a strongly feathered lateral branch is given off.

The variations shown in the pterylosis are not great. They concern the more or less definite separation of the lateral ventral tract and the slighter or more pronounced feathering of the anterior end of the posterior dorsal **Y**. Thus in *Lathamus* and *Platycercus* the lateral ventral tracts are well marked, and the posterior fork of the dorsal tract does not diminish much in width where it comes into contact with the anterior fork. In *Trichoglossus* the exact reverse of these conditions obtains, the lateral ventral tracts being but obscurely delimited from the main tract, and the dorsal tracts of the hinder part of the body almost ceasing before their junction with the anterior half.

Psephotus, Cyanorhamphus, Pyrrhulopsis, Agapornis, &c., agree with *Platycercus.* In *Ara* the outer branch of the pectoral tract is not definitely separable, but the dorsal tracts are more like those of *Platycercus.* *Conurus* is much the same.

The *syrinx* of the Psittaci [1] seems to show two main varieties.

a. In the following species there is a syrinx of the type which will be described immediately :—

> *Cacatua cristata.*
> „ *triton.*
> „ *Philippinarum.*
> *Microglossa aterrima.*
> *Calyptorhynchus Banksi.*
> *Stringops habroptilus.*

The syrinx is in these species remarkable for the fact that the first semi-rings of the bronchi are weak and cartilaginous, and are usually separated from each other by considerable tracts of membrane. *Cacatua* itself represents the most ex-

---

[1] The syrinx has been chiefly studied by GIEBEL, 'Zur Anatomie der Papageien,' *Zeitschr. f. d. ges. Wiss.* xix. p. 133, and by PARSONS and myself, 'On certain Points in the Anatomy of Parrots,' &c., *P. Z. S.* 1893, p. 507.

treme type ; in *Cacatua cristata*, for example, when the syrinx
is seen on a lateral view the membrane occupies a great deal
of the outer lateral region of the commencement of the
bronchus.

The semi-rings of the bronchus are at first very small, and
do not extend across the side of the bronchus ; they gradually increase in length, until at the sixth or seventh they
come to extend right across the syrinx. In *Microglossa
aterrima* the syrinx is in certain respects less abnormal ;
the rings are still feeble, but on a lateral view of the organ
they extend completely across, and there is on such a
view no bare tract of membrane, such as we have figured
in *Cacatua*.

Calyptorhynchus Banksi is intermediate between the
two extremes ; the first semi-ring only is incomplete, inasmuch as it does not reach from one side of the syrinx to
the other—or rather we should say from the anterior to the
posterior side.

Stringops habroptilus has the same weak cartilaginous
bronchial semi-rings ; but on a lateral view of the syrinx they
are seen to extend right across.

b. The second group contains the following genera :—

Chrysotis.	*Tanygnathus.*
Pyrrhulopsis.	*Eos.*
Trichoglossus.	*Polyteles.*
Lorius.	*Platycercus.*
Pionus.	*Pæocephalus.*
Psittacus.	

These genera are differentiated from those of the first
division by the fact that the bronchial semi-rings are as a rule
ossified, and are frequently more or less fused together; at
the same time the first ring is commonly concave upwards,
whereas in the parrots of the first-mentioned group the
bronchial semi-rings are straight.

The most extreme type is perhaps offered by *Chrysotis* ;
of this genus I have seen the following species :—

Chrysotis versicolor.	*Chrysotis Bodin.*
„ *erythrura.*	„ *viridigenalis.*
„ *leucocephala.*	„ *Levaillanti.*

In all these species the first two rings of the bronchus are closely fused together, and form a bowed piece of bone forming with the last tracheal ring a semicircular outline ; the space between the two is, of course, occupied by membrane. In *Chrysotis Levaillanti*, for instance—and there is no great difference in the other species—the double character of the apparently single first bronchial semi-ring is only to be seen at the two ends. In a number of other parrots the first bronchial semi-ring is larger than that which follows, though not fused with it ; this is the case with *Trichoglossus*, *Pyrrhulopsis*, and *Chalcopsitta* ; the genera *Eos*, *Polyteles*, *Platycercus*, and *Tanygnathus* have syringes which are constructed on the same plan. In *Conurus* there is a little difference ; here the first two rings of the bronchus are equisized ; this at any rate applies to the two species *Conurus aureus* and *Conurus cruentatus*, which are the only two that we have examined from this point of view. The genus *Ara* (species *Ara Leari*, *Ara militaris*) agrees with *Conurus*. *Psittacus* is like these genera ; but *Pionus* agrees more closely with *Chrysotis*.

It will be obvious that no hard and fast line can really be drawn between the two groups of parrots ; if it were thought desirable to draw such a line, it would be between the genus *Cacatua* on the one hand and all the remaining parrots on the other. *Cacatua* alone has a syrinx in which the first bronchial semi-rings are incomplete, leaving a bare tract laterally which is easily visible when the syrinx is viewed from the side : but in this genus there is another peculiarity—the intrinsic muscle of the syrinx ends in a very narrow point, which passes into a fine tendon of attachment ; in *Chrysotis*, *Eos*, &c., the muscle is comparatively broad down to its actual attachment. In this particular *Microglossa* and *Stringops* agree with *Cacatua*, although they do not show the incomplete rings that have been mentioned as

S

characteristic of the latter genus. These genera, in fact, are to this extent intermediate between *Cacatua* and the more normal (at any rate more usual) form of syrinx in the parrots ; the rings are still, however, soft and cartilaginous, thus different from *Conurus*, which is a further step in the direction of *Chrysotis* ; *Chrysotis* seems to represent the opposite extreme to *Cacatua*. *Ara* is a genus which is also intermediate in the characters of its syrinx ; it has weakish and straight rings, as in *Stringops*, for instance ; but the muscles are as in the second group of parrots, and the general aspect of the syrinx is more in accord with this placing of it.

Finally it should be added that occasionally (e.g. *Polyteles melanurus*) the extrinsic muscles are attached not to the sternum, but to the membrane covering the lungs, being continued there by thin tendons. In *Platycercus Barnardi* there would seem to be no extrinsic muscles at all.

Parrots are very much alike in their *myology* ; there are, however, a few points in which they show differences, and which may be useful for the purposes of classification.

The *tensores patagii* of the parrots are like those of many homalogonatous birds in the broad aponeurotic character of the tendon of the tensor brevis, which, however, has two or three thickened bands in it corresponding to the discrete tendons of most other birds (*c.g.* Charadriidae). Of these thickened bands the anterior commonly gives off a wristward slip ; but there appears to be never any patagial fan. The aponeurosis is inserted, as usual, on to the tendon of the extensor metacarpi radialis, and is continued over it by two tendinous slips, of which the posterior runs obliquely to the elbow joint, like the 'passerine slip' of many birds. The common tensor patagii muscle is usually very large, and often completely covers the posterior deltoid (*d. major*). GARROD dissected away the anterior thickened tendon of the *brevis* in *Deroptyus accipitrinus*, and found it to arise from a distinctly separate slip of the patagial muscle attached to clavicle. The tendon in question is inserted on to the lower external humeral process, and may represent with its muscle a primitive passerine *tensor patagii brevis*, to which has been subsequently added an extension of the deltoid.

With the general structure that has been described the parrots show much difference in the details of the patagial tendons.

The arrangement of the tendons of the tensor patagii is very much the same in *Nestor, Stringops,* and *Calyptorhynchus*; in all three the tendons are relatively very long when compared with the fleshy part of the muscle, and they are all close to one another, so as to give the appearance in *Stringops* of one tendon. In *Calyptorhynchus* the anterior tendon leaves the others in the lower part of the patagium and runs forwards after its usual fashion, so that the main distinctive point of these three genera, as far as the tensor patagii goes, is that the middle and posterior tendons are close together. In *Coracopsis* these tendons are separated by a slight interval, but closely correspond to the arrangement in the birds last named.

Eos, Lorius, Psocephalus, and *Caica* have a characteristic and almost uniform arrangement of the patagial tendons. In them the three tendons are very difficult to distinguish, because the fibrous membrane between them, of which they are only specialised parts, is as thick as they are. The result is that in these birds the patagial muscle seems to be inserted by a broad, short, membranous-looking tendon.

Conurus shows a transitional stage between these last genera and the typical arrangement; the three tendons are more distinct, and they are equally short and show the same mode of attachment to the fleshy part of the muscle.

Lathamus is remarkable for having the anterior tendon separate in its whole length from the middle one, instead of being fused with it in the upper part of its course.

Chrysotis and *Bolborhynchus* have a small extra tendon between the middle and posterior ones; in *C. Guildingi* this was only present on one side, but in *C. leucocephala* it was found on both.

Psittacus has three tendons which are completely separate in the whole of their course, and in this respect it corre-

s 2

sponds to *Lathamus*. The anterior tendon may represent the
fused anterior and middle tendons of *Lathamus*, and the
middle tendon may be an extra one, as in *Chrysotis*. Our
reasons for this are that there is a considerable interval
between the two tendons, and that they do not diverge, as
in all other cases. If this view is correct, the patagial
tendons of *Psittacus* closely resemble those of *Chrysotis*,
while they also agree in having the anterior deltoid larger
than the posterior, in the absence of a lower head to the
anconæus, and in having the deltoid completely covered by
the tensor patagii.

The two *deltoids* are but small muscles, and are largely
covered by the relatively enormous tensor patagii. It is
better to use the terms ' anterior ' and ' posterior ' for the deltoids, since their relative dimensions vary considerably. The
major is by no means always the larger. Sometimes the
two deltoids are entirely covered by the tensor patagii, sometimes the posterior is partly exposed. Thus in *Nestor* the
muscle is exposed, in *Deroptyus* and *Chrysotis* it is covered.
In *Nestor* and *Stringops* the anterior deltoid is the smaller,
in *Caica* it is the larger. In *Tanygnathus*, *Bolborhynchus*,
and *Eclectus* the deltoids are narrow and equisized. In *Eos
cardinalis* the anterior deltoid (which is the larger) is divisible into two distinct parts.

As regards the relative sizes of the two deltoid muscles,
where they differ, such genera as are known may be arranged
as follows :—

A. Delt. Larger

*Deroptyus, Psittacula, Aprosmic-
tus, Lorius, Caica, Eos, Pyrrhulopsis,
Lathamus, Palæornis, Loriculus,
Psephotus, Pæocephalus, Cyanorhamphus, Psittacus, Melopsittacus.*

Post. Delt. Larger

*Nestor, Stringops, Calyptorhynchus,
Cacatua.*

In some parrots the *anconæus longus* has an accessory
head from the humerus, which is especially broad in *Stringops*
The table on p. 268 shows the distribution of this accessory
head among the genera.

The *expansor secundariorum* is stated by GARROD to be

absent from the Psittaci. This is certainly almost universally the case. But Fürbringer speaks of a rudiment—a short length of tendon—in *Platycercus palliceps*.

No parrot has a *biceps slip*. A muscular *cucullaris patagialis* is generally, if not always, present.

It is well known that the *ambiens* muscle is present in some parrots, and absent from the leg of others. The actual occurrences of this muscle are shown in table (p. 268). *Stringops* is peculiar in that the muscle is sometimes complete and quite normally developed, and sometimes ends in a thin tendon on the capsule of the knee joint. This recalls *Œdicnemus*.

Of the other muscles of the leg used by GARROD in classification A, X, and Y are nearly always present, the only exception, so far as I am aware, being *Chrysotis Guildingi*, in a specimen of which I failed to find Y.

Sometimes (as in *Ara chloroptera*) the semitendinosus gives off a tendinous slip to the gastrocnemius, but in *Chrysotis* there is no such slip. The *tibialis anticus* is usually inserted by a single tendon. This Mr. PARSONS and I found to be the case in the majority of parrots which we examined. But in *Chrysotis* the tendon is distinctly double. In *Deroptyus*, *Caica*, *Pæocephalus*, *Platycercus*, and a few others, there are more or less evident indications of a double tendon.

The *deep flexor tendons* of the parrots are gallinaceous, with a vinculum such as is illustrated in fig. 54 (p. 100). There are some inconsiderable variations of this ground plan ; for instance, in *Platycercus Barnardi* the vinculum is divided into two parts, one to digit II., the other to III. and IV.

Peroneals.—The peroneus longus and brevis are, as far as we have observed, always present in parrots, but the origin of the former differs somewhat in different genera.

In *Stringops* and *Nestor* the peroneus longus rises from the front of the bony fibula and its membranous continuation for about the upper half of the leg. The muscular belly overlaps that of the peroneus brevis very much near its origin, and the muscle is large and well marked.

In *Chrysotis*, on the other hand, the peroneus longus is very small and only rises from the membranous continuation of the fibula in the lower part of the leg; it is so small that it does not overlap the peroneus brevis at all, but lies behind it.

The parrots have a well-developed *crop* and a zonary *proventriculus*. In the *liver* the right lobe is the larger; rarely are they subequal. The *gall bladder* is as a rule absent ; but it is present in *Cacatua* and in *Calopsitta*, though 'small and easily overlooked' in the latter.

The *intestinal* measurements in a series are as follows :—

	Inches.		Inches.
Stringops . .	73	*Tanygnathus Mulleri*	57
Eclectus polychlorus .	93	*Psittacus erithacus* .	48
Calyptorhynchus		*Conurus Petzii* . .	12
Banksi .	61	*Palæornis Alexandri*	30
Ara ambigua	62	*Aprosmictus erythro-*	
„ *ararauna* .	50	*pterus* . .	41
Cacatua sulfurea	33·75	*Lorius lori* . .	33
„ *cristata*	37	*Loriculus chrysonotus*	12
„ *triton* .	51	„ *galgulus* .	16
Pionus senilis . .	50·5	*Chalcopsitta scintillata*	37
„ *Maximiliani*	49	*Psittinus malaccensis*	12
Eos reticulata .	25·5	*Euphema pulchella*	12
„ *indica* . .	26	„ *splendida* .	12
Platycercus Barnardi	29·5	*Deroptyus accipitrinus*	31
„ *pallidiceps*	20	*Pionopsitta pileata* .	39
Nestor meridionalis	38	*Lathamus discolor* .	18
Microglossa aterrima	34	*Coracopsis Barkleyi* .	33
Pyrrhulopsis splendens	43	*Dasyptilus Pecqueti* .	17·25
Chrysotis collaria	45	*Brotogerys tirica*	31
„ *festiva* .	36	„ *tovi* .	21
Geopsittacus occiden-			
talis	15		

The most obvious comment upon the above list is to draw attention to the very great length of the gut in *Eclectus*, whose relations in this particular to other parrots are almost those of *Didunculus* to other pigeons.[1]

The most recent and elaborate essay upon the *osteology* of the parrots is by MIVART,[2] who has described the entire skeleton of *Lorius* and *Psittacus*; some of his illustrations are reproduced here. Fourteen is the prevalent number of *cervical vertebræ* (e.g. *Lorius, Psittacus, Platycercus, Caica*). *Stringops*, however, has fifteen. The atlas is notched (*Ara militaris*) or perforated (*Pyrrhulopsis*) for the odontoid process. Five (*Ara, Psittacus*) or six (*Platycercus, Pyrrhulopsis*) ribs articulate with the sternum. The *sternum* has as a rule an entire posterior margin, which in *Licmetis* is entirely unnotched and unfenestrated. Most parrots have a pair of fenestræ which in *Deroptyus* and *Microglossa* are converted into notches. The sternum has a spina externa, slightly forked occasionally (e.g. *Psittacus erithacus, Callocephalon*), but no spina interna. The carina is deep—deeper in *Platycercus* (without furcula) than in *Caica* (with furcula), the species being approximately of the same size. The furcula is sometimes present and sometimes rudimentary.

It is present, and forms a complete U, in *Nestor, Conurus, Caica, Licmetis, Microglossa, Ara, Palæornis*, &c.

An intermediate condition is observable in *Eos*, where the furcula thins much towards its sternal end. A still further reduction is seen where the two clavicles are separate below and only bound by cartilage. Finally there are those parrots with a quite rudimentary pair of clavicles, consisting only of a small piece of bone at the coraco-scapular end. This is the case, for example, with *Pyrrhulopsis* and *Platycercus*.

The following table shows the number of cervical vertebræ and the position of the first and last hæmophyses in a number of parrots :—

---

[1] For structure of tongue see CIACCIO, 'Nota preventiva sull' interna struttura della lingua dei Papagalli.' *Rendic. Sess. Acc. Ist. Bologna*, 1877-8, p. 157.

[2] 'The skeleton of *Lorius flavopalliatus* compared with that of *Psittacus erithacus*,' *P. Z. S.* 1895. There is no account of the bones of the limbs

—	No. of C.V.	First Hæm.	Last Hæm.	Catap.
Pyrrhulopsis personata	14	C9	D3	C13-D1
Conurus hæmorrhous .	14	C9	D2	C14
Platycercus Pennanti .	14	C9	D1	C12-14
Ara militaris . .	14	C8	D3	C13-D1
Eclectus polychloros .	14	C8	D1	C11-D1 [1]
Nestor notabilis . .	14	C8 [2]	D2	C12-D1
Caica melanocephala .	13	C8	D3	C13-D1
Callocephalon galeatum . .	14	C9	D1	C13-14
Calopsitta Novæ Hollandiæ .	14	C9	D1	C12-14
Conurus cruentatus	14	C8	D2	C12-13

The *humerus* of parrots is peculiar, and, as GARROD [3] has pointed out, there are features of resemblance to the **Columbæ** and to the Alcidæ. This peculiarity will be found described and figured in the chapter dealing with the Columbæ. The *skull* is very uniform in its structure throughout the group. It is desmognathous, holorhinal, and without basipterygoid processes.

The front part of the face (nasals, maxillæ, and premaxillæ) articulates by a transverse joint with the frontals, which is movable. The mobility of the anterior part of the face is aided by the movable articulation to it of the palatines and the jugals. The palatines have a peculiar form; for the most part they are laterally flattened plates of great depth and considerable extent. The quadrate of parrots too is peculiar in the great length of the neck, which bears the squamosal articulation. In many parrots the lacrymal bone joins the forward process of the squamosal, thus completely encircling the orbit with bone.

The *hyoid* has been extensively studied by MIVART; [4] in

[1] On D1 the median part of the hæmapophysis has vanished, leaving only the lateral.

[2] On this vertebra is a double hæmapophysis, forming a canal.

[3] See also for osteology of parrots BLANCHARD, 'Des Caractères Ostéologiques chez les Oiseaux de la Famille des Psittacides,' *Compt. Rend.* xliii. p. 1097. and xlix. p. 518 ; MILNE-EDWARDS, 'Observations sur les Caractères Ostéologiques,' &c., *Ann. Sci. Nat.* (6), vi. p. 91 ; L. VON LORENZ, 'Über die Skelete von *Stringops habroptilus* u. *Nestor notabilis*,' *S.B. k. Ak. Wien*, lxxxiv. 1882, p. 624.

[4] 'On the Hyoid Bone of certain Parrots,' *P. Z. S.* 1895, p. 162.

this paper references will be found to previous figures and descriptions. It has features which absolutely distinguish this group of birds.

The entoglossal has a considerable median foramen, or

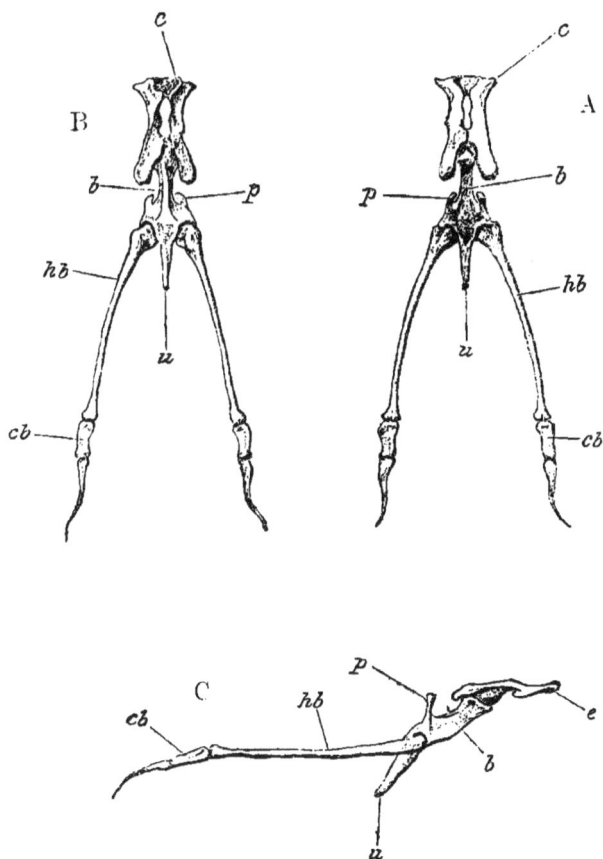

Fig. 131. Hyoid of *Stringops* (after Mivart). A. Dorsal Aspect. B. Ventral. C. Lateral.

b, basihyal; e, entoglossum; p, parahyal process; u, urohyal; hb, hypobranchial; cb, ceratobranchial.

more usually is composed of two separate pieces united in front by cartilage. The basihyal is broad, and it develops on either side a forwardly directed piece (figs. 131-3), for which Dr. Mivart has suggested the name of parahyal piece. This latter is merely a short process in *Ara*, *Psittacus*, and *Stringops* (fig. 131); in *Lorius*, *Eos*, and *Trichoglossus* the two parahyals (figs. 132, 133) unite and form a single Y-shaped

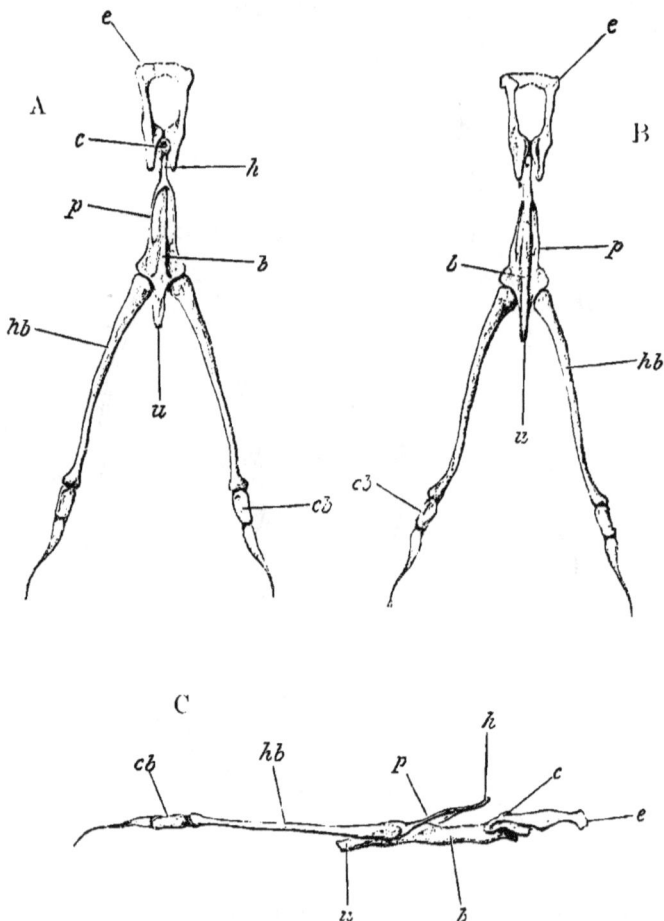

FIG. 132.—HYOID OF *Lorius flavopalliatus* (AFTER MIVART).
A. DORSAL ASPECT.   B. VENTRAL.   C. LATERAL.
c, cup-like excavation; *h*, symphyses of parahyals.   Other letters as in fig. 131.

bone inclined obliquely upwards.   The only bird which seems to present much resemblance to the parrots is the eagle, which, according to the figure in BRONN's 'Thierreich,' [1]

[1] *Aves*, Pl. xxxi. fig. 23.

has a broad basihyal with the short angular processes which
suggest the more elaborate parahyals of the parrots.

The classification of the parrots has been attempted by
more than one naturalist ; but, as GADOW has justly observed, 'our knowledge of the anatomical structure of these
birds is at present too incomplete in relation to their large
numbers.'

GADOW himself has practically divided them by the
structure of the tongue into two families, Trichoglossidæ and
Psittacidæ. In all the Trichoglossidæ the orbital ring is

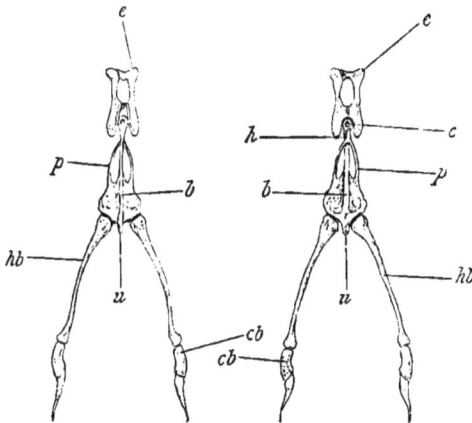

FIG. 133. HYOID OF *Lorius Comicella* (AFTER MIVART). LETTERS
AS IN FIG. 131.

incomplete, and it is possible that the remarkable structure
of the hyoid, described above, may serve to distinguish this
family. The first family contains only *Nestor*, the lories,
Cyclopsittacus, and *Lathamus* ; but the two latter are very
imperfectly known. The remaining genera are relegated to
the second family.

GARROD'S arrangement of the group, anterior to that of
GADOW'S in point of time, is based upon the variations of
the ambiens, oil gland, furcula, and carotids. The facts,
with a few others added, are displayed in the following
table :—

—	Oil Gland	Ambiens	Humeral Head of Ancouæus	Carotids	Furcula	Distribution
Stringops .	+	x	+	2	—	New Zealand
Palæornis .	+	—	—	2	+	Africa, India, China, East Indies
Aprosmictus	+	—	—	2	+	Australia, Austro-Malaya
Polytcles .	+	—		2	+	Australia
Eclectus .	+	—	—	2	+	Moluccas, Papua
Tanygnathus	+	—		2	+	Philippines, Celebes, New Guinea
Prioniturus	+	—		2	+	Philippines to Celebes
Eos . .	+	—	—	2	+	Moluccas, Papua, Solomon Islands
Trichoglossus	+	—		2	+	Australia, Celebes, Papua. Timor, New Caledonia
Lorius .	+	—		2	+	Moluccas to Solomon Islands
Loriculus .	+	—	—	2	+	India to Papua
Coriphilus .	+	—		2	+	Society and Marquesas
Chalcopsitta	—	—		2	+	Papua
Psittinus .	+	—		2	?	Malay Peninsula to Borneo
Agapornis .	+	—		2	—	Ethiopian region
Euphema .	+	—		2	—	Australia and Tasmania
Melopsittacus	+	—		2	—	South Australia
Geopsittacus	+	—		2	—	South Australia
Eolophus .	+	—		2	+	Australia to Philippines
Calyptorhynchus	+	—	+	2	+	Australia
Calopsitta .	+	—		2	+	Australia
Licmetis .	+	--		2 or l.	+	Australia
Microglossa	—	—		2	+	Papua, North Australia
Cacatua .	+	—	+	2	+	Australia, Austro-Malay, Philippines
Callocephalon	+	—		2	+	Australia
Psittacus .	+	+	—	2	+	Tropical Africa
Pœocephalus	+	+	—	2	+	Tropical Africa
Nestor .	+	+	+	2	+	New Zealand
Nasiterna [1] .	+	—		2	—	New Guinea
Ara . .	+	+	+	2	+	Mexico to South America
Conurus	+	+	—	2	+	Mexico to South America
Bolborhynchus	+	+		2	+	Mexico to South America
Caica . .	+	+	—	2	+	Guiana and Amazons
Pyrrhula .	+	—		2	+	Costa Rica, South America
Deroptyus .	+	—	—	2	+	Guiana, North-East Brazil

[1] FORBES, 'On some Points in the Structure of *Nasiterna*,' &c., *P. Z. S.* 1880, p. 76.

	Oil Gland	Ambiens	Humeral Head of Anconeus	Carotids	Furcula	Distribution
Pionopsitta	+	–	–	2	+	Central and South America
Lathamus .	÷	–	–	2	÷	Australia and Tasmania
Coracopsis .	÷	–	–	2	⊣	Madagascar
Pyrrhulopsis	÷	··	+	2	÷	Fiji
Dasyptilus .	+	–		2	+	New Guinea
Chrysotis .	–	–	–	2	+	Mexico to South America
Pionus	–	–	–	2	+	Mexico to South America
Brotogerys .	–	–		2	+	Central to South America
Platycercus	+	–	–	2	–	Australia, Tasmania, Norfolk Island
Psephotus .	+	–	–	2	–	Australia
Cyanorhamphus .	+	–	–	2	–	New Zealand, New Caledonia, Society Islands
Psittacula .	+	–	–	2	–	Mexico to South America
Nymphicus .	+	–		2	+	New Caledonia to Loyalty Islands

From these facts and a few others may be derived the
following scheme :—

Sub-Order **Psittaci** : —

Fam. I. Palæornithidæ. Two carotids. Ambiens absent. Oil gland present.

Subf. (1.) Palæornithinæ. *Palæornis, Eclectus, Aprosmictus, Eos, Tanygnathus, Prioniturus, Psittinus, Loriculus, Trichoglossus, Lorius.*

Subf. (2.) Cacatuinæ. Orbital ring complete.
Calopsitta, Calyptorhynchus, Licmetis, Eolophus, Cacatua.

Subf. (3.) Stringopinæ. Furcula lost.
Stringops, Euphema, Geopsittacus, Melopsittacus, Agapornis.

Fam. II. Psittacidæ. Left carotid superficial.

Div. *a*. Ambiens present.

Subf. (4.) Arinæ. *Ara, Conurus, Bolborhynchus, Caica, Psittacus, Pæocephalus, Nestor.*

Div. *b.* Ambiens absent.

Subf. (5.) Pyrrhurinæ. *Pyrrhura, Lathamus, Coracopsis, Pyrrhulopsis.*

Subf. (6.) Platycercinæ. Furcula lost.
Platycercus, Psephotus, Cyanorhamphus, Psittacula.

Subf. (7.) Chrysotinæ. Oil gland lost.
Chrysotis, Pionus, Brotogerys.

A phylogenetic tree accompanies GARROD's scheme, in which it is assumed that the ancestral parrot possessed the normally running carotids, an ambiens, an oil gland, and a complete furcula. From this the main stems are given off, in one of which the carotids remained normal, while in the other the left became superficial. The loss of the other characters leads to further branching of both main branches.

The Stringopinæ, especially *Stringops* itself, are the nearest living representatives of the ancestral stem.

FÜRBRINGER [1] also argues for the low position of *Stringops*, in contradiction to MARSHALL, who holds that it is an extremely modified form. Its owl-like plumage, defective carinal keel, and associated loss of the power of flight are undoubtedly modifications, but it seems more probable that they are modifications of an ancient than of a modern type of parrot. FÜRBRINGER's views are chiefly based upon the *flexibility* of its anatomical characters. I have already referred to the variability of the ambiens : the sternum offers another fact of the same kind ; sometimes it is entire, without notches or foramina, sometimes there are one or two upon one or the other side, and occasionally two incisuræ, one upon each side. This variability must not be associated, as variability may often be associated, with a rudimentary structure ; the xiphosternum is not rudimentary, though the keel is.

---

[1] 'Einige Bemerkungen über die Stellung von *Stringops*.' &c., *Journ. f. O.* 1889, p. 236.

Mr. Parsons and I have pointed out certain likenesses between *Stringops*, the Cacatuinæ of Garrod, and *Nestor*, which he places in an altogether different family. These partly concern the syrinx, to which attention has been already directed, partly the muscular system. In those birds the posterior deltoid is larger than the anterior. It may be noted also that powder-down patches are best developed and more universal among the Cacatuinæ, while it is in that family only that the gall bladder exists.

Of extinct parrots among the most remarkable is *Lophopsittacus mauritianus*,[1] characterised, as was also *Necropsittacus rodericanus*, by its enormous jaws.

The principal interest attaching to other remains of parrots is the light that they throw upon the former distribution of the group; for *Psittacus* has been found in the lower Miocene of France.

The determination of the affinities of the parrots to other groups of birds is one of the hardest problems in ornithology. They have been likened to the **Accipitres** (mainly, perhaps, on account of the hooked beak and its cere), and to the gallinaceous birds, in the neighbourhood of which they were placed by Garrod. It seems to me that the parrots, like the cuckoos, are a group of birds which are on the borderland between the Anomalogonatæ and the higher birds. It is remarkable what a number of points there are in which they show resemblances to the **Passeres** the complicated musculature of the syrinx, the absence of biceps slip and expansor secundariorum, the presence of a cucullaris propatagialis, found in the Passeres and in the somewhat passeriform *Upupa* and **Pici**, the small number of cervical vertebræ, the total want of cæca, allying them not certainly to the **Passeres** but again to the **Pici** and many Anomalogonatæ, the reduced clavicles of some genera. Zygodactyle feet, moreover, are not found among the higher birds except in the **Cuculi** and

[1] Sir E. Newton and H. Gadow, 'On Additional Bones of the Dodo and other Extinct Birds of Mauritius,' &c., *Tr. Zool. Soc.* xiii. p. 281. See also for a figure and account of this bird Newton's *Dict. Birds*, sub voce 'Extermination.'

the **Musophagi**, which are, similarly to the parrots, on the border line between the Anomalogonatæ and higher birds. It is noteworthy also that of the Anomalogonatæ which present a catapophysial canal (found at any rate in one parrot) it is the **Pici** and the passerine alone. But while it is not so difficult to point out likenesses to the Anomalogonatæ it is much harder to indicate resemblances to any of the higher groups of birds. It must be held, in my opinion, that they have emerged from a low anomalogonatous stock at a time not far removed from that at which the **Cuculi** and **Musophagi** also emerged, but that there is not a common starting point of the three groups.

CUCULI

Definition.—**Feet zygodactyle by reversion of fourth toe. Skull desmognathous, without basipterygoid processes.[1] Oil gland nude. Quintocubital. Two carotids. Cæca longish. Ambiens present.**

The family which is defined by the above characters is a large one, comprising, according to the recent catalogue of Captain SHELLEY, 165 species, which are distributed by that ornithologist into forty-two genera.

The family is almost world-wide in range, being most abundant, however, in the tropics. Correlated with its numerous genera and species and wide range we find a certain amount of structural variation in the family, which permits of its division into several subfamilies, concerning the number and extent of which there is some divergence of opinion. It may be convenient, however, to consider the general anatomy of the family before dealing with its major subdivisions.

Apart from the facts used in the definition of the family the cuckoos are characterised by the absence of, or the presence of only a rudimentary, *aftershaft* to the contour feathers. The number of *rectrices* is not, as was stated by NITZSCH, constantly ten; for in *Saurothera*, *Guira*, and *Croto-*

---

[1] Well-marked rudiments in *Rhinococcyx* and *Eudynamis*.

phaga there are only eight. The feather tracts are somewhat diverse in their disposition. The *pterylosis* of the European cuckoo (*Cuculus canorus*) has been described by NITZSCH in his 'Pterylography.' The feathering on the throat completely occupies the intermandibular space. The ventral pteryla is divided upon the neck into its two halves, which are not again divided ; each passes backwards, gradually

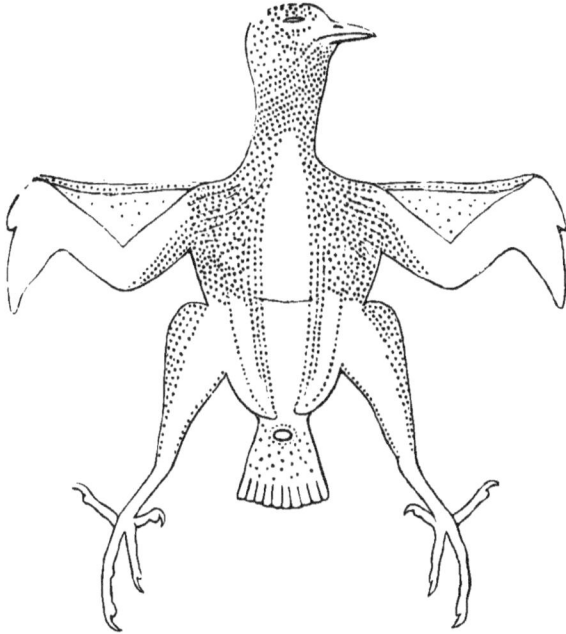

FIG. 134. PTERYLOSIS OF *Eudynamis orientalis*. VENTRAL VIEW. (AFTER BEDDARD.)

diminishing in extent, until it ends in a single row of feathers in the neighbourhood of the cloaca. Over the sternum this pectoral tract is very wide ; later its three rows of feathers become separated by a slight interval, two on one side and one on the other, which, however, reunite before ending at the cloaca. The spinal tract is narrow in the neck region. It bifurcates on the shoulder to enclose a lanceolate space.

T

Of other cuckoos whose pterylosis has been studied [1] *Caco-mantis, Piaya, Saurothera, Diplopterus, Coccyzus, Chrysococcyx,* and *Coccystes* agree in most points with *Cuculus.* But in the American genera *Piaya, Diplopterus, Saurothera,* and *Coccyzus,* the ventral tract is double from the very first —that is to say, in the mandibular region.

A more complicated pterylosis characterises certain other

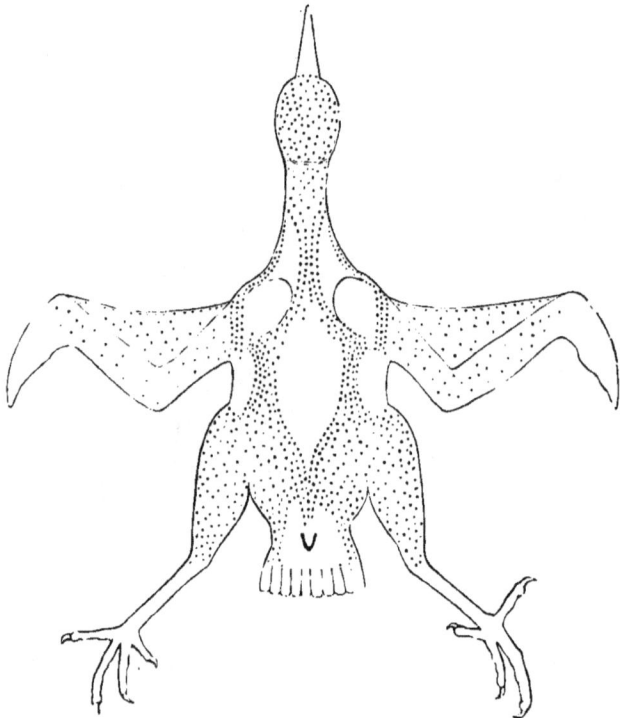

Fig. 135.   PTERYLOSIS OF *Piaya cayana.*   DORSAL VIEW.   (AFTER BEDDARD.)

cuckoos.   In the genus *Centropus* the feathering upon the throat is close and continuous, the two ventral tracts diverging at about the junction of the neck and trunk. Each of these,

[1] BEDDARD, ' On the Structural Characters and Classification of the Cuckoos,' P. Z. S. 1885, p. 168; SHUFELDT, ' Contributions to the Anatomy of Geococcyx,' *ibid.* 1886, p. 966.

again, divides into two separate tracts, of which the inner
is at first two feathers wide, which number is reduced to one
just before the termination of the row a little way in front of
the cloacal aperture. The outer branch consists of one row
only, and terminates some way in front of the end of the inner
branch, without, however, showing any signs of being fused
with it. *Pyrrhocentor, Geococcyx, Crotophaga, Eudynamis,*

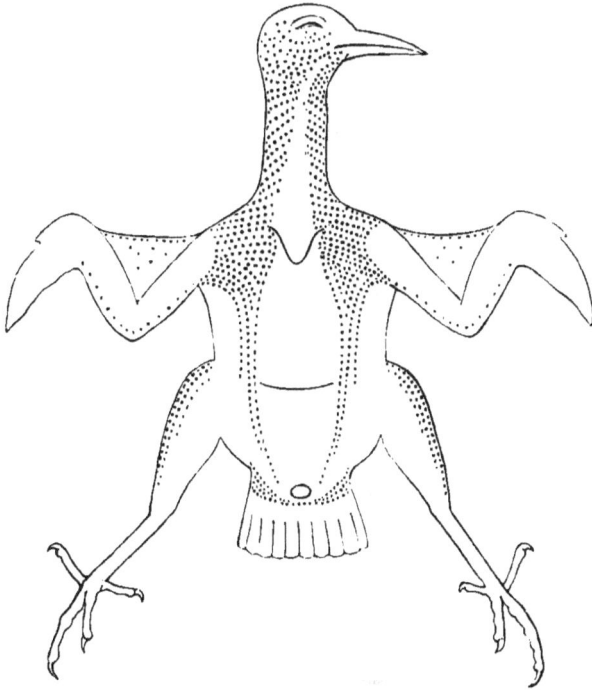

FIG. 136.—PTERYLOSIS OF *Piaya cayana*. VENTRAL VIEW.
(AFTER BEDDARD.)

Scythrops, and *Phœnicophaes* have a pterylosis which is much
like that of *Centropus.* The two types of pterylosis observable in the cuckoos are illustrated in the accompanying woodcuts. Unfortunately the pterylosis of some important genera,
such as *Coua,* is not known.

All cuckoos possess the *ambiens.* In all cuckoos we also
find the *semitendinosus,* the *accessory semitendinosus,* and

the *femorocaudal*; in *Centropus* and its allies, both in the Old World and in the New, the *accessory femorocaudal* is also present. In many groups of birds the arrangement of the tendons ending in the patagium is very complicated. This is not the case with the **Cuculi**, where the disposition of these tendons is very uniform. For the most part the simplicity is suggestive of the perhaps allied picarian birds. In *Cuculus canorus*, for example, which has been figured by GARROD, the tensor patagii brevis is inserted on to the fore arm without any bifurcation. So too in *Piaya, Saurothera*. The only exception to this which has been noted occurs in *Geococcyx*, where the said tendon bifurcates just before its insertion, the anterior branch being inserted on to the extensor metacarpi radialis a little way in front of the main attachment. This is also the case with *Guira* and *Phœnico-phaes*.

No cuckoo has any *biceps slip*. In some genera, *e.g.* in *Saurothera, Coccyzus, Pyrrhocentor*, there is an attachment between the anconæus and the humerus. In *Guira* there is none. The *expansor secundariorum* is what GARROD (see p. 85) has termed 'ciconiiform. The *glutæus primus* is extensive in most cuckoos;[1] its origin in them reaches behind the head of the humerus as well as in front. There is no glutæus V. The deep plantar tendons of *Pyrrhocentor* and *Centropus* are peculiar in that no branch is sent to the hallux.

The *syrinx* in the cuckoos shows greater variability than in any other group of birds excepting the goatsuckers. We meet with the typical tracheo-bronchial syrinx in a considerable number of genera, while in others is the much-modified bronchial syrinx. The latter was first described many years ago in *Crotophaga* by JOHANNES MÜLLER. In that genus the syrinx closely resembles that of *Steatornis*, which has been already described in the introductory chapter (see p. 69). There are, however, as might be imagined, differences of detail. In *Crotophaga* the membrana tympaniformis commences at the seventh ring of the bronchi, the rings in front

[1] Apparently absent or very slight in *Cuculus*.

being perfectly complete rings, the trachea dividing, 'as in the mammalia.' From the seventh onward all the bronchial rings are semi-rings, the intrinsic muscles being attached to the tenth. This is one extreme of the series, the other being offered by such a type as *Piaya*. In *Piaya* (see fig. 137) there is a purely tracheo-bronchial syrinx. The third bron-chial semi-ring is of compara-tively speaking enormous size, and to it are attached the in-trinsic muscles of the syrinx. In *Saurothera* we have a syrinx which is quite similar save for the fact that the third bron-chial ring is not enlarged. *Diplopterus* is much the same as the last. *Cuculus* has also a perfectly typical tracheo-bron-chial syrinx. In *Eudynamis* there is a cuculine syrinx, the last tracheal and the first three bronchial semi-rings being ossi-fied; the intrinsic muscles are attached, as in *Piaya*, to the third bronchial semi-ring. *Phœnicophaes* is much the same.

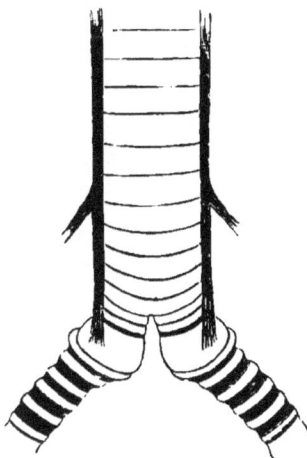

Fig. 137.- Syrinx of *Piaya cayana* (after Beddard).

The remaining genera of cuckoos whose syrinx is known are nearer akin to *Crotophaga*, though in them the bronchial syrinx is not quite so typical. In *Centropus ateralbus*, for instance, the first fifteen rings of the bronchi are incomplete internally and are closed by membrane, but the membranous area is narrow; this area widens out at the sixteenth ring, which with the following is much stronger than the pre-ceeding and succeeding rings of the bronchus; to the sixteenth ring are attached the intrinsic muscles of the syrinx. *Pyr-rhocentor* and *Geococcyx* have a similar syrinx. The syrinx of *Guira* is in many respects very remarkable. On a superficial view it is not unlike that of *Cuculus*. The voice organ in this genus is placed further forwards than in the genera just

considered.   But the first two or three rings of the bronchus
are complete rings with no membrane internally.   From the
fourth onwards the rings are semi-rings.   Upon the sixth
are inserted the syringeal muscles.   *Coua* is somewhat intermediate.   The first seven bronchial semi-rings have their
inner extremities separated by a narrow area of membrane.
To the seventh are attached the intrinsic muscles of the
syrinx.   From this point commences the tympaniform mem-

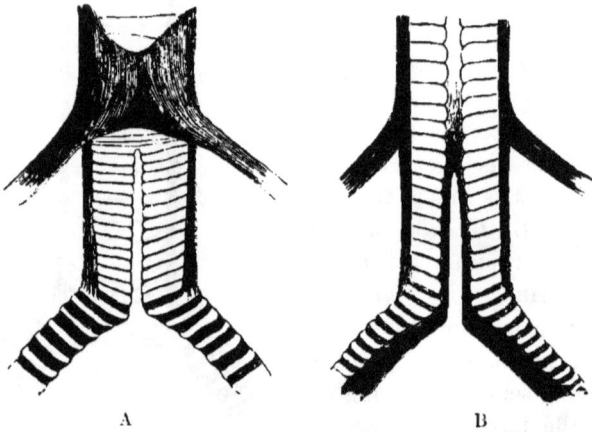

Fig. 138.—Syrinx of *Centropus ateralbus* (after Beddard).
A. Front View.   B. Back View.

brane.   The accompanying woodcuts will serve to illustrate
the varying form of the syrinx among the **Cuculi**.

As with some other large groups of birds, such as the
pigeons and parrots, the *gall bladder* is present in some
cuckoos and absent in others.   The gall bladder exists in
the genera *Saurothera, Coccyzus, Guira, Pyrrhocentor,
Scythrops*, and *Cuculus* ; it is absent in *Crotophaga* and in
some species of *Centropus*.   *Coua* has a gall bladder ; *Eudynamis* appears not to have one.

In *Guira* the right lobe of the *liver* is five or six times as
large as the left.

The main *artery of the leg* is the femoral in *Piaya, Centropus*, the sciatic in *Diploptera, Saurothera, Coccyzus,
Pyrrhocentor*.

The following are some intestinal measurements, principally of the cæca :—

—	S. I.	L. I.	Cæ.
	Inches	Inches	Inches
Cuculus canorus .	18		1·1
Piaya cayana .		1·8	1·4
Diplopterus nævius .	7	1	·8
Saurothera dominicensis .		2	1·7, 1·5
Coccyzus americanus .		1·3	1·2, 1·1
Pyrrhocentor celebensis .		2·5	1·5
Centropus ateralbus .		2·1	1·75, 1·5
„ phasianus .	16	2	2·25
Phœnicophaes sp. .		1·75	2·25
Geococcyx affinis .	13		2
Crotophaga sulcirostris .	13		1
Guira piririgua .	12	1·5	1·5
Scythrops .	20	4·5	6·5
Chrysococcyx .	8		·9

As for the skull of the cuckoos, that of *Scythrops* has been described by PARKER [1] and by myself. [2] SHUFELDT has studied the skeleton of *Geococcyx*. [3] *Scythrops* is doubly desmognathous, the maxillo-palatines being united for their whole lengths, and the palatines also being fused with each other posteriorly. In other cuckoos the maxillo-palatines diverge posteriorly for a short space, and there is no union between the palatines. *Scythrops* has, according to PARKER, two small vomers, situated one behind the other. [4] In cuckoos the ectethmoid processes are large, and the lacrymal has often (e.g. *Scythrops*, *Crotophaga*) a large descending process nearly touching the jugal. *Scythrops* and *Eudynamis* have an os uncinatum lying between the descending process of the lacrymal and the ectethmoid.

The holorhinal nostrils are much obliterated by bony growths, the degree varying. As a consequence the nostrils are as a rule impervious; but in the dried skulls of *Pyrrhocentor* and *Cuculus* there is a considerable foramen.

No cuckoo has more than fourteen *cervical vertebræ*, and some have only thirteen. Three or four ribs only articulate

Trans. Linn. Soc. (2), i.   [2] *P. Z. S.* 1898.
J. Anat. Phys. xx. 1886, p. 244.
I could only find one lying entirely between the palatines.

with the sternum.   The sternum may be fenestrated or
marked by posterior incisions.

The atlas of *Rhinococcyx*[1] is perforated by the odontoid
process ; C11–D1 have hypapophyses in addition to the axis,
and two or three following, C12–14, have also catapophyses,
which on 14 ascend on to the hæmapophysis.

In *Saurothera* C10 has closely applied paired hæmapophyses.

The modifications of the syrinx, the pterylosis, and the
leg muscles permit the family to be subdivided thus, the
subdivisions, as will be observed, corresponding to the
geographical range of the birds :--

Subfamily I. Cuculinæ.   Syrinx tracheo-bronchial.   Ven-
     tral feather tract single.   Muscle formula AXY + .
  a. Ventral tract single at commencement.
     Cuculus
     Chrysococcyx ⎫
     Cacomantis   ⎬ Old World.
     Coccystes (?) ⎭
  b. Ventral tract double at commencement.
     Saurothera   ⎫
     Diplopterus (?) ⎬ New World.
     Piaya   ⎬
     Coccyzus   ⎭

Subfamily II. Phœnicophainæ.   Syrinx tracheo-bronchial.
     Ventral feather tract bifurcate.   Muscle formula
     ABXY + .
     Scythrops   ⎫
     Eudynamis   ⎬ Old World.
     Phœnicophaes   ⎭

Subfamily III. Centropodinæ.   Syrinx bronchial.   Ven-
     tral feather tract bifurcate.   Muscle formula
     ABXY + .

_____

[1] In *Scythrops* there is a notch nearly completely converted into a foramen ;
in *Eudynamis* a notch less nearly converted into a foramen, also in *Guira* and
Diplopterus. In *Cuculus* and *Saurothera* there is a foramen.

a. Ventral tract occupying whole of space between jaws.

Pyrrhocentor)
Centropus } Old World.
Coua (?))

b. Ventral tract only occupying median region.

Geococcyx)
Crotophaga } New World.
Guira)

The question of the affinities of the cuckoos is a difficult one.

By Gadow they are placed nearest to the **Musophagi** and next nearest to the **Psittaci**. Fürbringer's views do not greatly differ. There seems to be no doubt that these birds are an archaic group not far from the point where the Anomalogonatæ and Homalogonatæ of Garrod diverge. They are, like the **Musophagi**, quintocubital; their intestines are simple; and they have the complete muscle formula (B being in some forms absent). These characters are found in others among the more primitive of the higher birds.

The likenesses which the cuckoos show to the Pico-Passeres are mainly in the structure of the foot, in the simple character of the tendons of the patagium, and the marked resemblance in the syrinx to that of the **Caprimulgi**, and in a less degree to the **Striges**. As has been pointed out, precisely the same series of modifications between the extreme bronchial syrinx of *Crotophaga* and the purely tracheo-bronchial syrinx of *Cuculus* are to be seen among the **Caprimulgi**. The syrinx is really the only salient point in the anatomy of the group that can be laid hold of for purposes of comparison, and, considering the dissimilarities in the voice and habits of cuckoos and goatsuckers, it is particularly noteworthy.

As to fossil cuckoos, the two most interesting facts are, perhaps, the occurrence of *Centropus* and *Phœnicophaes*[1] in

[1] See Milne-Edwards in *Comptes Rendus* for 1891.

Europe, especially the latter, as it has some claims to represent the most ancient form of cuckoo, with complete muscle formula and tracheo-bronchial syrinx.[1]

MUSOPHAGI

Definition.—Oil gland tufted. Aftershaft present. Quintocubital. Rectrices, ten. Muscle formula of leg, ABXY+. Expansor secundariorum present. Biceps slip absent. Cœca absent. Both carotids present. Skull holorhinal, desmognathous, without basipterygoid processes.

This group of birds, purely African in range,[2] is divisible into three genera, *Corythaix*, *Musophaga*, and *Schizorhis*. These genera do not show a large amount of structural variation.

As to the *pterylosis*, the two ventral tracts are double upon the neck (in *C. albocristata*); they remain separate until just in front of the cloaca, being especially weak and narrow in the breast region. Longitudinally arranged rows of feathers connect the pectoral tracts above with the humeral. The other important external characters are stated in the definition.

In *Schizorhis* the normal arrangement of the leg arteries obtains.

In *Corythaix* and *Musophaga* the femoral artery is the one developed. The *right jugular* is the largest, and in *Corythaix albocristata* seems to have entirely disappeared.

In the *liver* the right lobe is the larger,[3] sometimes considerably so. The *gall bladder* is present, and sometimes is elongated in form. The *tongue* is short and triangular; the proventriculus is zonary, the gizzard weak. The *intestines* are capacious and short, without cœca. The following are a few measurements :—

[1] Cf. also partial persistence of basipterygoid processes in Phœnicophainæ.
[2] The extinct *Necrornis* of French Miocene may be a Touraco.
[3] The viscera are described by OWEN for *Corythaix porphyreolopha*, P. Z. S 1834, p. 3, and by MARTIN for *Corythaix Buffonii*, *ibid.* 1836, p. 32.

Corythaix erythrolophus .   . 17 inches
      ,,   albocristata   . 18   ,,
      ,,   persa   .   . 18   ,,
Musophaga violacea .   . 18   ,,
Schizorhis africanus   . 20   ,,

The *windpipe* of *Corythaix persa* is slightly swollen along its course, narrowing again at the bifurcation. It is much ossified. Counting as the last tracheal ring that to which the pessulus is attached in front, the intrinsic muscles are inserted on to the third in front of this. The first two bronchial semi-rings are ossified; the third, between which and the second there is a considerable membranous interval, is the first of the purely cartilaginous series. The extrinsic muscles are stout, and arise seven or eight rings from the end of the trachea, and pass at once to

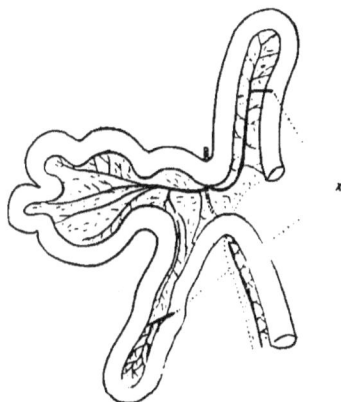

Fig. 139.—INTESTINES OF *Corythaix chlorochlamys* (AFTER MITCHELL).
x, short-circuiting vessel divided.

their insertion; they run no distance along the trachea, as is so common.

Musophaga has no intrinsic muscles; *Schizorhis* has.

As to muscles, the *tensores patagii* are very simple, and the *biceps slip* is entirely absent.

The tensor brevis sends off a wristward slip just before its insertion in both *Musophaga* and *Corythaix* ;[1] it is reinforced by a pectoral slip and by a fibrous slip from the humeral crest.

The *anconæus longus* has not, at any rate in *Corythaix albocristata*, a humeral head. *Glutæus I.* is large, covering the biceps; *glutæus V.* is absent (*Corythaix erythrolophus, Musophaga*) or present (*Schizorhis africanus*). The muscle

[1] Absent —perhaps as an individual variation—in *C. albocristata*.

formula of the leg, as stated in the definition, is ABXY+.
Both *peroneals* are present. The *deep flexor tendons* are
bound by a vinculum, which is single in *Corythaix* and
double in *Schizorhis*.

The *skull* of *Corythaix* is barely desmognathous, and
by no manner of means especially like that of a cuckoo,
to which group the **Musophagi** have been often compared.
The hinge in the middle of the face is nearly complete, but
there is a bridge on each side, formed by an ankylosis
between the frontal and nasal. The holorhinal nostrils are
situated very far forwards, and each has, as PARKER [1] has
pointed out, an osseous fold upon the ossified internarial
septum. The maxillo-palatines diverge from each other
posteriorly for a much longer space than in any cuckoo.
The ascending laminæ of the palatines come into contact
for a very brief space over the rostrum in front, and are
continued forward for a short distance as a sharp spike.
Between them lies a minute interpalatine splint (or vomer).
The interorbital septum is moderately fenestrate. The
lacrymal bones are of some size, and the descending process
is closely applied to, but does not fuse with, the square
ectethmoid process ; connected with both is a small os
uncinatum, which reaches the palatine.[2]

	Cuculi	Musophagi
Bony nostrils .	Close to fronto-nasal hinge.	Near end of bill.
Ectethmoid	Continuous with dorsal wall of orbit.	Not so continuous.
Palatines	Ascending laminæ come into contact *posteriorly* for a *considerable* space.	Ascending laminæ come into contact *anteriorly* for a *short* space.
Maxillo-palatines .	More completely fused.	Less completely fused.
Jugal bar . .	Expanded where it joins maxilla on, or close to, ventral surface.	Not expanded ; junction with face higher up, very much as in parrots.

In view of the general opinion as to the nearness of the

---

[1] In his paper upon *Opisthocomus*. See below. p. 286, footnote [2].
[2] J. T. REINHARDT, *Vid. Medd.* Kjöbenhavn, 1871, p. 326.

alliance between Musophagidae and Cuculidae it may be useful to tabulate the principal divergences in the skull.

Corythaix has 14 *cervical vertebrae* ;[1] the *atlas* is notched, not perforated by the odontoid process. Four ribs reach the *sternum*, which is doubly notched and has a strong spina externa.

In the pelvis the prepubic process is very markedly developed, as in *Geococcyx*.

The *clavicle* comes into contact both with the scapula and with the moderately large procoracoid; the latter is fused with the acrocoracoid,[2] making thus a complete bridge over the sulcus supracoracoideus. The coracoids slightly overlap at their articulation with the sternum.

The haemapophyses are characteristic. In *Corythaix albocristata* C11 has paired processes; on C12 and C13 the haemapophyses form a continuous ventral keel to those vertebrae. The three following vertebrae have haemapophyses. The extremities of those of D1 and D2 are expanded with a median ridge, owing to the catapophyses having descended them.[3]

OPISTHOCOMI

Definition.—**Aftershaft present. Oil gland feathered. Rectrices, ten. Quintocubital Muscle formula, ABXY+. Biceps slip and expansor secundariorum present. Carotids, two. Cæca present. Skull holorhinal, schizognathous, without basipterygoid processes. Sternum peculiar in form, wider behind than in front; the spina externa ankylosed with furcula.**

There is no doubt that this group, consisting of but a single genus and species, *Opisthocomus cristatus*, the hoatzin of British Guiana, forms a well-marked group of birds.

The external characters of the adult and young have been chiefly described in recent years by NITZSCH, PYCRAFT,[4] and

[1] In *Corythaix persa* the atlas ring is incomplete.
[2] Cf. *Galli*, in which the same fusion occurs.
[3] E. BLANCHARD, 'Remarques sur l'Ostéologie des Musophagides,' *Compt. Rend.* xlv. p. 599.
[4] 'On the Pterylography of the Hoatzin,' *Ibis*, 1895, p. 345.

myself,[1] the internal anatomy of the soft parts by PERRIN,[2] GARROD,[3] GADOW,[4] MITCHELL,[5] and myself.

The *pterylosis* of *Opisthocomus* shows a less sharp differentiation into pterylæ and apteria than in many birds; many of the spaces are covered with a sparse feathering of semiplumes. This condition may be, as GARROD has suggested, immediately derived from a continuous feathering. Nevertheless there is a sharply marked apterion upon the carina sterni, which is necessarily, therefore, of limited extent. The dorsal apterion is but feebly marked. In the young chick the ventral apterion was as clear as in the adult, but I could discover no trace of a dorsal bare space. In the young unhatched chick the ventral feathering was closer than the dorsal.

The skin lying upon the carina sterni is dense and thickened, a state of affairs which appears to have some relation to its habit of squatting close to the branch upon which it is resting.

The principal recent papers upon the *osteology* of this bird are those of HUXLEY[6] and the more recent and more elaborate treatise of PARKER.[7] The *cervical vertebræ* are nineteen in the adult; but PARKER found only eighteen in the unhatched young. The *atlas* is notched for the odontoid process. Hæmapophyses are very feeble. The last two or three cervicals are ankylosed with each other, and form part of the dorsal series of ankylosed vertebræ. There is some

[1] 'Contributions to the Anatomy of the Hoatzin,' &c., *Ibis*, 1889, p. 283. See also C. G. YOUNG, 'On the Habits and Anatomy of *Opisthocomus cristatus*; ' *Notes Leyden Mus.* x. p. 169; and J. J. QUELCH, 'On the Habits of the Hoatzin,' *Ibis*, 1890, p. 327.

[2] 'On the Myology of *Opisthocomus cristatus*,' *Tr. Zool. Soc.* ix. p. 353.

[3] 'Notes on Points in the Anatomy of the Hoatzin,' *P. Z. S.* 1879, p. 109.

[4] 'Description of the Modification of certain Organs,' &c., *Zool. J.B.* v. Abth. Syst. v. 1891, p. 629, and *Proc. R. Irish Ac.* (3). ii. p. 147.

[5] See below, p. 288, footnote, for reference. See also a recent paper by GOELDI in *Ornith. M.B.* May 1895, and more fully in *Bol. Mus. Para.* 1895.

[6] 'On the Classification and Distribution of the Alectoromorphæ,' *P. Z. S.* 1868.

[7] 'On the Morphology of a Reptilian Bird, *Opisthocomus cristatus*,' *Tr. Z. S.* 1891.

variation as to the number of cervical vertebræ, which bear
long rib stylets. As a rule five complete ribs exist, of which
all bear uncinate processes. The *sternum* (see fig. 140) is exceedingly remarkable in its form. It is wider behind than in
front, with a pair of notches, and outside of these a pair of
foramina ; the keel is shorn away anteriorly, but well developed posteriorly. The furcula, which is shaped like a fork
with nearly straight lines, is completely ankylosed on the
one hand with the coracoids, and by its median region with

FIG. 140.—STERNUM OF *Opisthocomus*.
SIDE VIEW. (AFTER HUXLEY.)

FIG. 141. STERNUM OF *Opisthocomus*. FRONT VIEW.
(AFTER HUXLEY.)

the spina externa sterni. The region of the furcula, however,
which comes into contact with the sternum was found by
PARKER to be a separate ' needle of bone,' which he regarded
as the interclavicle (see p. 131). The *scapula* is provided in
the young with a distinct suprascapula, segmented off from
the scapula.

The *pelvis* is especially compared by HUXLEY with that
of *Coturnix* : it has no ileo-pectineal processes.

In the *skull* the rostrum is articulated with the frontal region by a well-marked hinge. The skull is holorhinal, schizognathous, and there are no basipterygoid processes. The vomer is expanded and bifid in front in a fashion that recalls the ægithognathous skull and that of certain of the Charadriiformes (see below).

In the young skull PARKER figures basipterygoid processes, not, however, articulating with the pterygoids; they appear to be not unlike those of *Aptornis* (see below). The holorhinal nostrils are partly obliterated by an ossified alinasal, as in so many picarian and passerine birds.

The *alimentary canal* of *Opisthocomus* is remarkable in more than one way. There is, in the first place, the enormous *crop*, which has been most recently and most fully described by GADOW. This organ is very large, and rests upon the furcula and the fore part of the sternum, for the abortion of the anterior part of whose keel GADOW thinks the crop is by its pressure responsible. The crop too is exceedingly muscular, and has numerous parallel folds in its interior, some of which are continued into the œsophagus below. The gizzard is much reduced in size. Probably the crop is not a mere storehouse, but a compartment where at least trituration of the food (chiefly leaves) takes place.

The most remarkable feature about the *intestine* [1] is the long and coiled rectum, a feature which is also found among the struthious birds and in the archaic *Chauna*. The general arrangement of the coils of the small intestine is intermediate between those of *Pterocles* and pigeons. There is in the middle loop a faint trace of the spiral found in the corresponding loop of the pigeon's gut. There are also likenesses to the form of gut in the cuckoos. The *cæca* are fairly developed.

The peculiarities of the *muscular system* mainly concern the hind limb, and chiefly characterise the *ambiens*. The muscle formula is complete, *i.e.* ABXY+. The ambiens, however, is subject to variation. GARROD found that in all

[1] P. CHALMERS MITCHELL, ' A Contribution to the Anatomy of the Hoatzin (*Opisthocomus cristatus*),' *P. Z. S.* 1896, p. 618.

of six knees that he examined the ambiens was present, though small ; but in only one knee did it cross the knee to be inserted in the usual fashion in connection with the flexors of the leg. MITCHELL dissected this muscle in two specimens ; in one the ambiens was completely absent above the knee, but in each case (see fig. 53, p. 96) a ligament left the fibula, and, dividing into three, joined each of the three perforated flexors in the way in which, as has been already described, it occurs in birds which have this ambiens rudiment. In the second case there was an ambiens above the knee, but it became lost upon the fascia of the knee, and not connected with the ambiens rudiment springing from the fibula, which was there present. So in this bird there are many stages in the reduction of this characteristic muscle, which is clearly in them on the wane.

It is apparently the rule among birds for there to be a vinculum

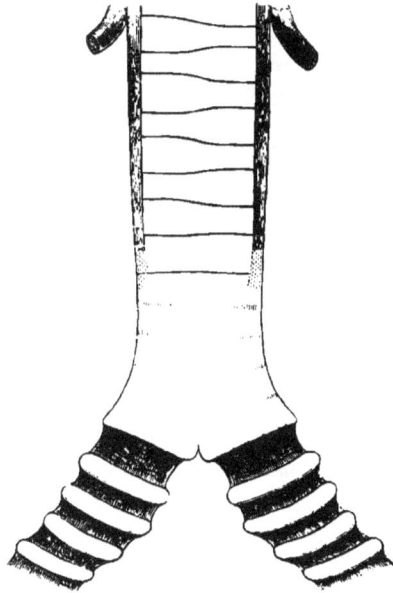

FIG. 142.—SYRINX OF *Opisthocomus*. FRONT VIEW. (AFTER GARROD.)

between the two superficial flexors of digit III. This slip is wanting in *Opisthocomus*, as it is, according to MITCHELL, in *Asio otus* and *Rhytidiceros plicatus*. The *deep flexor tendons* are connected by a strong vinculum.

The *syrinx* has been described by GARROD and myself. The accompanying figure is from GARROD's paper. The last few rings of the trachea are solidified into a tracheal box, and the intrinsic muscles do not reach this box, being only continued on to it by a ligamentous continuation. There is

U

some variation in the number of rings which coalesce to form
the box, while the fibrous continuation of the intrinsic muscles
may reach the first bronchial semi-ring. This muscle is
evidently decaying in *Opisthocomus.*

GALLI [1]

Definition.—**Quintocubital birds with an aftershaft. Muscles of leg
generally ABXY +. Expansor secundariorum present. Entepicondylo-ulnaris present. Cæca large; a crop present. Skull
schizognathous, holorhinal, with sessile basipterygoid processes.
Palatines without internal lamina.**

This very large group of birds, universal in range, shows
a considerable amount of structural variation.

The *oil gland* is generally tufted; [2] but it is nude in the
Megapodes and absent altogether in *Argus.*

The *pterylosis* of the **Galli** is, according to NITZSCH,
singularly uniform. He figures *Gallus bankiva, Pavo
cristatus,* and *Meleagris gallo-pavo,* describing also a few
other types. There are lateral neck spaces in all; the dorsal
tract is single in *Gallus,* widening out on the back; in the
peacock it widens out in a more pronounced fashion and
further back than in *Gallus.* In the turkey there is a
narrow space in it between the shoulder blades.

The ventral tract divides early upon the neck, and each
tract gives off on the breast a wider, denser outer branch;
the two median tracts then continue nearly to the cloaca,
where they unite.

In *Perdix* and *Tetrao* there is a dorsal space, as in *Meleagris.*

Among the Cracidæ there may or may not be a space in
the dorsal tract.

The *pectoral* muscles of gallinaceous birds, like those of
the tinamous, meet over the keel of the sternum; this is at
least the case with *Euplocamus Vieilloti* and some others.

[1] H. SEEBOHM, 'An Attempt to Diagnose the Sub-Orders of the Great Galli-
naceo-Gralline Group of Birds by the Aid of Osteological Characters alone,' *Ibis,*
1888, p. 415.

[2] *Callipepla californica* has a small tuft; in *C. squamata* I have observed
both the complete absence of a tuft and the presence of a very small one.

The *deltoid* may or may not possess a special tendinous slip from the scapula. This slip is absent in *Mitua tomentosa*, *Excalfactoria chinensis*, and *Callipepla*, but present in *Ortalis albiventris*, *Crax Sclateri*, *C. Daubentoni*, *Crossoptilon mantchuricum*. It is evidently, therefore, not of great use in classifying the group.

The *biceps slip* is generally present, but absent in *Ortalis albiventris*, *Crax*, *Mitua* ; it is present in *Megapodius* and *Megacephalon* ;[1] absent in *Talegalla*, *Numida*, and *Meleagris*. The same remark may, therefore, be made about this muscle. The humeral head of the *anconæus* is not always present.

The *tensor patagii brevis* of gallinaceous birds has a thin, wide, diffused tendon, as in the tinamous; there is no patagial fan.

The entepicondylo-ulnaris is another muscle which they share with the last-mentioned group.

The *expansor secundariorum* is a muscle which appears to be invariably present among the **Galli**, but to have varying relations at its scapular insertion.

'In the majority of the gallinaceous birds,' wrote Professor GARROD, ' the *expansor secundariorum*, with the normal origin from the secondary quills, has a different method of insertion, which has led M. A. MILNE-EDWARDS to describe the muscle in the common fowl as a part of the *coraco-brachialis (brevis)* in his superb work on fossil birds.

'In the genera *Tetrao*, *Francolinus*, *Rollulus*, *Phasianus*, *Euplocamus*, *Gallus*, *Ceriornis*, and *Pavo*, the muscle, instead of being inserted into the scapulo-sternal fibrous band, above referred to, after blending to a certain extent with the axillary margin of the *teres*, ceases by becoming fixed to a fibrous intersection about one-third down the *coraco-brachialis brevis* muscle.

'In *Francolinus Clappertoni* from among the francolins, *Coturnix*, *Odontophorus*, *Ortyx*, *Eupsychortyx*, and *Numida*, the tendon does not go so far as the short *coraco-brachialis*, but ends either by simply joining the axillary margin of the

---

[1] Absent, according to FÜRBRINGER.

teres or by at the same time sending a tendinous slip
behind it to the scapula. In *Argus giganteus* the tendon,
running from the elbow, turns round the axillary border of
the *teres* to end by joining a triangular muscular fasciculus,
attached by its base to the upper portion of the thoracic
surface, which appears to be nothing but a differentiation-off
of the upper portion of the last-named muscle. In the
Cracidæ this insertion into the scapula is also found, but it
is tendinous, like the upper element of the thoracic band
above described in the storks and *Chauna* ; and in them
there is also a second tendinous slip from the axillary margin
of the *coraco-brachialis longus* (not the *brevis*).   In the Megapodidæ also the attachment to the *coraco-brachialis brevis* is
wanting, the tendon ending either by blending with the *teres*
margin or running on to the scapula.'

The *glutæus primus* is a large muscle covering the
biceps. *Glutæus V.* appears to be always present, but is
sometimes (*Thaumalea picta*) quite tendinous.[1]

Most gallinaceous birds have the complete muscle formula
ABXY+.   The femoro-caudal, however, varies in size,
and is quite absent in *Pavo* and *Meleagris*. It is very
slender in *Crax* and *Ortalis*.[2]

The deep flexor tendons belong to type I., and are illustrated in fig. 54 (p. 100).

There are two *carotids* in all but the Megapodes, where
the left only is present. A *gall bladder* is present.

Some *intestinal measurements* are given on p. 293.

The trachea has in a few gallinaceous birds two pairs of
extrinsic muscles, *thus resembling*, it will be observed, *the*
Anseres and **Palamedeæ**.

Thus in *Crax Daubentoni*, besides the usual sterno-tracheales, which arise in the ordinary way from the costal processes,
there are a pair of cleido-tracheales, springing from the

[1] For the tail muscles of the peacock see HEMMING, *Proc. Linn. Soc.* 1844,
p. 212.
[2] GARROD has figured (in MS.) an "abnormal *Gallus domesticus* with a
peculiar additional muscle springing by tendinous slips from femur, femoro-
caudal, accessory ditto, and semitendinosus, and running to gastrocnemius.'
It was the same on both sides.

Name of Bird	Small Intest.	Large Intest.	Caeca
	Inches	Inches	Inches
Pavo cristatus	56	4	9
„ nigripennis	50	4	8 and 9
„ „	39	3·5	6·5
„ muticus	44	3	9
„ spicifer	46 — —	3·5	7·5
Caccabis chukar		30	4·5
Argus giganteus	70 to 84	5	5·5 to 7·5
„ „	66	4·5	6·5
Ithaginis Geoffroyi	31 ————	3·5	7·5
Polyplectron chinquis		37	3·5
♂ „ bicalcaratum . .	37	4	3·75
Rollulus coronatus	25	2·5	3
Arboricola torquecola	37	3	5·25
Coturnix communis	————		2·5
Ortyx virginianus		22	4
„ cristatus	22	2·5	3
„ Gambelii	18	3	4
Odontophorus dentatus . . .	28	3·5	3·5
Perdix cinerea	————		
Phasianus versicolor		43	5·75
Thaumalea Amherstiae . . .	46	3	5
Euplocamus Swinhoii . . .	53	3·5	7
„ erythrophthalmus . .	30	3	6
„ Vieilloti	66	3·5	7
„ nycthemerus . . .	56	4	7
„ cristatus	42	3	5·5
„ albo-cristatus . . .	42	3	8
„ Andersoni . . .	34	3·5	6
„ nobilis	47	3·5	5
Ceriornis Temmincki	64	3·5	8
♂ „ satyra	59	3·5	8
Lophophorus impeyanus . . .	61	5	6·5
Crossoptilon mantchuricum . . .	39	4	10·5
Lobiphasianus Bulweri . . .	54	4	6
Gallus bankiva	61	5	6·25
„ Sonnerati	35	2·5	4
Numida meleagris		3	5
„ ptilorhyncha . . .	29	3	5·25
„ cristata	33 ————	3	6
„ Edouardi . . .		34	5·5
„ vulturina	39	4	9 and 10
Tetrao urogallus	78	8	30·5
„ tetrix	59	6	2·75
„ phasianellus . . .			16
„ cupido	48	6	17
Meleagris ocellata	66	5·5	13·5
Francolinus afer	31 ————	2	6
„ gularis		30	4
Crax globicera	126	4 (2)	7 and 8
„ Sclateri	85	4	4·5
„ Daubentoni	104	4	6·5
„ Alberti	118	4	6
„ globulosa	121 ————	5	4·5
Mitua tuberosa		123	6·5
„ tomentosa	90	4	5 5
Penelope cristata	38	2·5	

Name of Bird	Small Intest.	Large Intest.	Cæca
	Inches	Inches	Inches
Penelope cujubi . .	48	3	3
„ cujubi . .	29	2·5	2·5
„ pileata . .	36	3	3·5
„ jacucaca .	42		2·75
Pipile jacutinga . .	51	4	4
„ cumanensis .	56		4
Aburria carunculata .	40	2·25	4
Ortalis albiventris .	24		2·5
Talegalla Lathami .	72		5
Megacephalon maleo .	51		5·5

anterior end of the sternum and from fibrous septum between it and the pessular process ; these muscles run up the sides of the trachea, reaching further than the sterno-tracheales.

Where the trachea is convoluted it sometimes happens that the extrinsic muscles are quite abnormal in their attachments ; thus in the males of *Penelope pileata* and *Ortalis albiventris* the muscles in question do not enter the thoracic

FIG. 143.—SYRINX OF *Pavo spicifer.* FRONT VIEW. (AFTER GARROD.)

FIG. 144.—SYRINX OF SAME. BACK VIEW. (AFTER GARROD.

cavity, but pass close to the carina sterni and are inserted at the very end of the sternum.

The *syrinx* of gallinaceous birds has been chiefly studied by GARROD.[1] From his paper the following account of this

[1] 'On the Conformation of the Thoracic Extremity of the Trachea in the Class Aves, Pt. 1, The Gallinæ,' *P. Z. S.* 1879, p. 354.

organ has been mainly drawn. The simplest form of the
syrinx is seen in the peacock (fig. 143), where the modification
of rings at the bifurcation is of the slightest. The last two
tracheal rings are partly fused behind. The accompanying
series of illustrations (figs.143–147) show some of the principal forms of syrinx among the Alectoropodes, in which it will

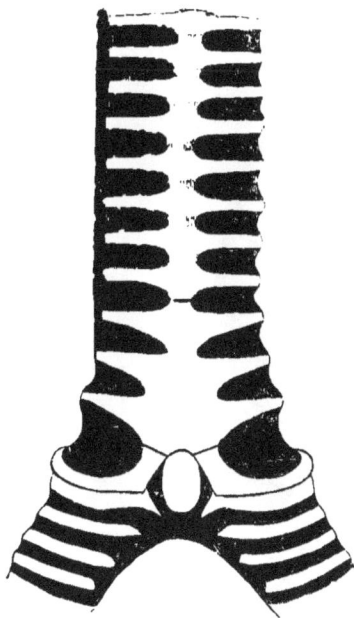

FIG. 145.—SYRINX OF *Callipepla califor-*   FIG. 146.—SYRINX OF SAME. BACK
nica. FRONT VIEW. (AFTER GARROD.)   VIEW. (AFTER GARROD.)

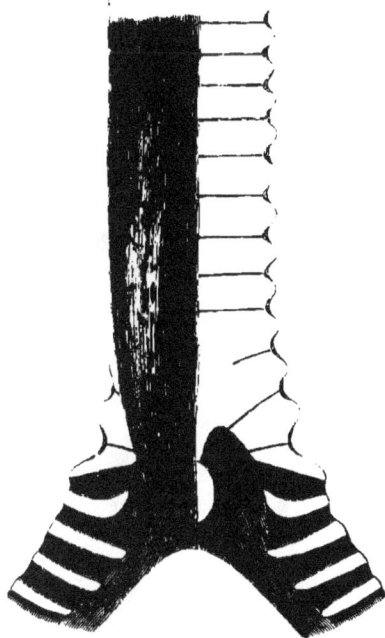

be noticed that intrinsic muscles are but occasionally present.
When present they do not descend to the bronchi, but cease
upon the trachea some way in front of the bifurcation. The
most remarkable modification of the intrinsic muscles is in
Callipepla californica (figs. 145, 146), where the muscles
descend the trachea posteriorly, and are inserted on to the
bronchidesmus. This state of affairs is not unlike what is
found, and will be described (see below), in the condor. It is
uncertain whether these muscles may be not more accurately

referred to the cleido-tracheal extrinsic pair (see below).   A
very singular syrinx is that of the male *Tetrao tetrix* (fig. 147).
On each side of the trachea at the lower end is an 'immense
irregular tumefaction, connected with its fellow by a bridge
of fatty tissue.'   It appears to be mucous in its chemical
nature ; but it may possibly have some relation to the tracheal

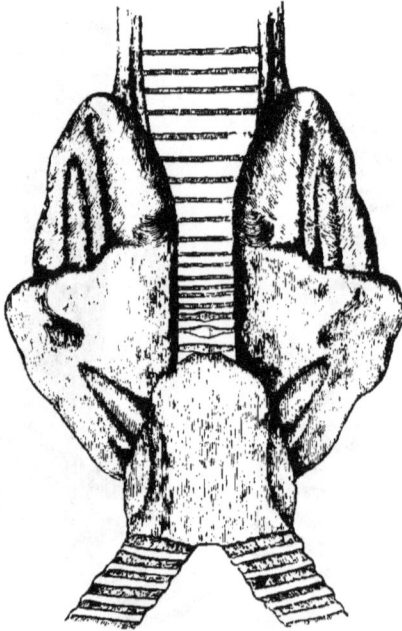

Fig. 147.— Syrinx of Male *Tetrao tetrix*.   Front View.
(After Garrod.)

box of the male ducks, and be thus another of those many
unexpected resemblances between the two groups.

The Cracidæ (fig. 148) generally possess the intrinsic muscles, which are, however, short, as in the other **Galli**.   The
syrinx itself has no salient characters by which it may be
distinguished from the Alectoropodes.

GARROD [1] has also described and figured (fig. 150) the
syrinx of the megapode *Megacephalon maleo*.   It is rather

[1] 'On the Anatomy of the Maleo,' *P. Z. S.* 1878, p. 629.

peculiar in form, but has a pair of intrinsic muscles, which reach the first bronchial semi-ring ; in this point the syrinx is more primitive than that of other **Galli**.

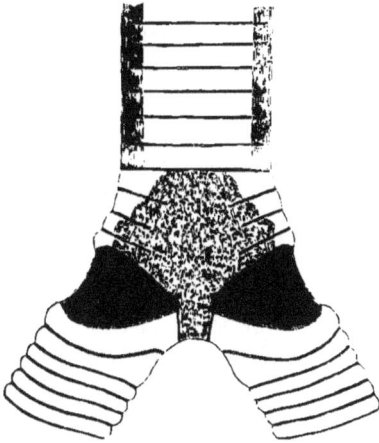

FIG. 148.—SYRINX OF *Aburria carunculata*. FRONT VIEW. (AFTER GARROD.)

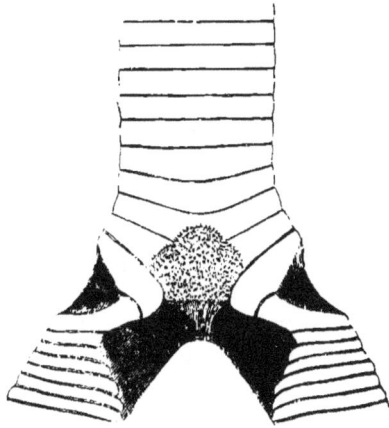

FIG. 149.—SYRINX OF SAME. BACK VIEW. (AFTER GARROD.)

The skull in gallinaceous birds is in many respects remarkably duck-like.[1] The basipterygoid processes—a little,

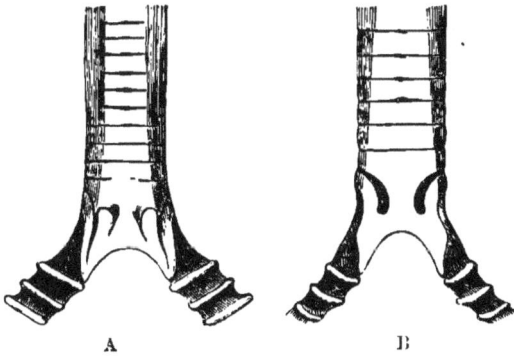

A   B

FIG. 150.—SYRINX OF *Megacephalon maleo*. A. FRONT VIEW. B. BACK VIEW. (AFTER GARROD.)

but not much, more pronounced in the Megapodes—are oval sessile structures, with which again, as in the ducks, the ptery-

[1] As pointed out by PARKER.

goids articulate by their anterior ends. The palatines too are devoid of an internal lamina, and the angle of the mandible is recurved and produced ; it is enormous, extended upwards, in *Tetrao*.

In the typical gallinaceous birds the maxillo-palatines are generally small or even obsolescent. In *Gallus bankiva*[1] they are triangular plates of fair size; in *Tetrao urogallus* (fig. 152) they are small, narrow, backwardly projecting plates, not quite so long and thin, and not so curved as those of *Talegalla*. In *Ptilopachys* they are somewhat intermediate ; in *Callipepla californica* they are still longer. In *Numida* and *Meleagris* they are much the same.

The lacrymals are not large,[2] and have a feeble or aborted descending process. It is very general for the post-frontal and zygomatic processes to fuse and enclose a more or less triangular foramen, and sometimes, as in *Tetrao*[3] and *Crossoptilon*, the zygomatic bar extends forwards a considerable way in front of its junction with the other.

The interorbital septum is sometimes (*Coturnix, Callipepla, Perdicula, Ptilopachys*) considerably fenestrated.

The vomer, in gallinaceous birds generally, is thin and splint-like.

As to the Megapodes, there are some differences in the skulls of the two genera *Talegalla*[4] and *Megacephalon*. The latter has the well-known hammer-shaped projection of the back part of the skull. In both genera the palatines are slender, and there is some ossification of the nasal septum. The interorbital septum is not much fenestrate, but it is deficient in front. In *Talegalla* the maxillo-palatines are thin

---

[1] SHUFELDT, 'Observations upon the Morphology of *Gallus bankiva*,' &c. *Journ. Comp. Med. Surg.* ix. p. 343.

[2] WOOD-MASON has described (*Ann. Mag. Nat. Hist.* xvi. 1875, p. 145) supraorbital bones in certain partridges. Cf. as to this point Tinamous, *Psophia*, and *Menura*.

[3] SHUFELDT, 'Osteology of the N. American Tetraonidæ,' *Bull. U. S. Geol. Surv.* vi. p. 309.

[4] The skeleton of this bird is described by PARKER, 'On the Osteology of Gallinaceous Birds and Tinamous,' *Tr. Z. S.* v. p. 160. See also W. K. PARKER, 'On the Structure and Development of the Skull of the Common Fowl,' *Phil. Trans.* 1870, p. 159.

plates, ending in a curved point very much like those of some passerines (e.g. *Pteroptochus*). Each is vacuolate posteriorly in *Talegalla*. They do not nearly come into contact in the middle line. In *Megacephalon* these bones are spongy plates, which do nearly come into contact; the palatines, slender in both birds, are more bowed in *Talegalla*, and thus enclose a wider interpalatine vacuity. The lacrymals of *Talegalla* are small and ankylosed to the skull wall; the ectethmoids are thin plates. A curious difference in the skulls of these two birds concerns the nasals and premaxillaries. In *Megacephalon* there is nothing worthy of special remark except the tumid outer part of the nasals; in *Talegalla* the premaxillary process of the nasals approach each other in the middle line, and cut the nasal process of the premaxillary into two, an anterior and a posterior portion.

In the Cracidae the maxillo-palatines are largish plates, concave inferiorly and convex above, which in *Crax globicera* actually come in contact for a short space, and fuse with each other and with the median septum. In *Pauxi galeata* this fusion (perhaps owing to the great casque) is even better marked. There is nearly a fusion in *Aburria carunculata*; in *Ortalis albiventris*, on the other hand, the maxillo-palatines are well apart.

In *Ortalis* and *Aburria* there is no ossification of the nasal septum; in *Crax* there is a median piece, which expands below to become attached to (*C. Sclateri*) or fused with (*C. globicera*) the maxillo-palatines. In *Pauxi galeata* the nasal septum, as might be expected, is quite complete and very strong.

There are also in this group of gallinaceous birds a series of stages in the development of the zygoma and the post-orbital processes. In *Ortalis* they are short; in *Aburria* longer and convergent; in *Crax Sclateri* and *Pauxi galeata* they meet distally, and enclose a triangular foramen; finally in *Crax globicera* they are completely fused throughout, and form a stout triangular process.

The lacrymals in this family are large, with a large descending process. The ectethmoids are but slightly ossified.

The interorbital septum is more fenestrate than in the Mega-podes, but still not markedly so.

The **Galli** have sixteen cervical vertebræ. In *Gallus bankiva* the catapophyses of vertebra 10 nearly enclose a canal; on the next vertebra the two processes have almost fused into a single one, the two processes being closely

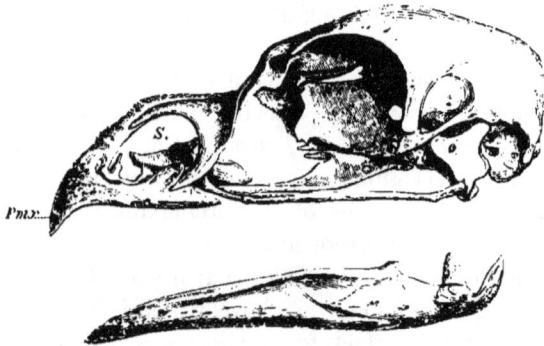

FIG. 151.—SKULL OF *Crax globicera*. SIDE VIEW. (AFTER HUXLEY.)
Pm.x, premaxilla; *S*, nasal septum.

soldered together for their whole length, and not, as is so often the case, divergent at the end. All the remaining cervical vertebræ have strong median hæmapophyses, those of the sixteenth being fused with the two following at their ex-tremities (cf. *Musophaga*, p. 285). The last cervical vertebra and the three anterior dorsals are themselves fused.

It will be simpler to compare the vertebræ of *Gallus* with those of some other **Galli** by means of the following table:—

—	First Hæm-apoph.	Last Hæm.	Hæm. fused.	Vertebræ fused.
Crossoptilon mantchuricum	C11	D4	C16–D3	C16–D3
Numida cristata	C11	D3	C16–D2	C16–D3
Talegalla Lathami	C11	D2	C16–D2	C16–D3
Megacephalon maleo	C10 [1]	D1	C15–D1	C15–D2
Callipepla californica	C13	D2	C16–D2	C16–D3
Ptilopachys ventralis	C11	D3	D1–D3	D1–D3
Aburria carunculata	C12	D3	D1–D2	C16–D3

[1] Bifid at end.

It is evident that not much of classificatory importance is deducible from the above facts. HUXLEY has used with more success the remaining parts of the skeleton.

The gallinaceous birds are divided by HUXLEY into two main subdivisions, Peristeropodes and Alectoropodes.

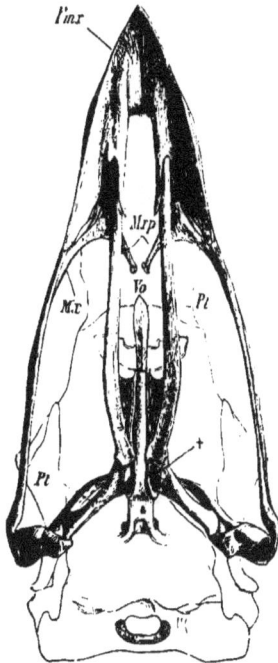

FIG. 152.—SKULL OF *Tetrao urogallus*. VENTRAL VIEW. (AFTER HUXLEY.)

Mxp, maxillo-palatines; *Vo*, vomer; *Mx*, maxilla; *Pl*, palatines; *Pt* pterygoids; +, basipterygoid facets; *Pmx*, premaxilla.

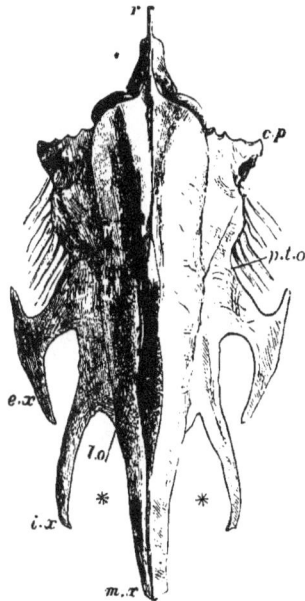

FIG. 153.—STERNUM OF *Crax globicera* (AFTER HUXLEY). LETTERS AS IN FIG. 72, P. 128, WITH WHICH THIS FIGURE IS TO BE COMPARED.

The former contains the Cracidæ and Megapodidæ, the latter the remaining families.

The Peristeropodes may be thus defined :—

1. Sternum with not very deep inner notches (fig. 153) and with short obtuse costal processes, the anterior edge of which is at right angles to the long axis of sternum.

2. Hallux on a level with other toes.

The Alectoropodes in this way :—

1. Sternum with very deep inner notches (fig. 72, p. 128), and with long costal processes, whose long axis corresponds with the long axis of sternum.

2. Hallux attached above the level of other toes.

The former group, which undoubtedly, so far as the above characters are concerned, is the more primitive, consists of the curassows and the Megapodes, limited respectively to Central and South America and to Australia and some of the Indian islands.

These two families are distinguished by HUXLEY mainly on account of the differing form of the hallux ; but there are other points of dissimilarity, not known at the time when he wrote. The two families may be thus differentiated :—

Cracidæ	Megapodidæ
1. Hind toe not so long in proportion to rest.	Hind toe longer in proportion.
2. Oil gland feathered.	Oil gland nude.
3. Biceps slip never present.	Biceps slip sometimes present.
4. Two carotids.	Left carotid only.
5. Trachea generally coiled.	Trachea always straight.

The Alectoropodes may be readily divided into three groups, which may be thus differentiated :—

The NUMIDIDÆ.

1. Second metacarpal without backward process.

2. Costal processes outwardly inclined (thus forming a transition between Alectoropodes and Peristeropodes).

The remaining families, which will be distinguished from each other immediately, agree with each in differing from the Numididæ in both the points mentioned.

MELEAGRIDÆ.

1. Postacetabulum longer than preacetabulum.

2. Postacetabulum longer than broad.

3. Furcula weak and straight (viewed laterally), with straight rod-like hypocleidium.

In the remaining gallinaceous birds the preacetabular length is greater than the postacetabular (equal in *Tetrao cupido*) ; the postacetabular area is broader than it is long ; the contour of the furcula is curved, with an expanded hypocleidium. The series of birds which have these characters may be divided into the galline and the tetraonine type.

In the former the postacetabular region is only moderately broad ; the hypocleidium is oval in contour ; the tarso-metatarsus is more than half as long as the tibia, and there are a number of smaller osteological marks.

In the grouse-like birds the postacetabular region is very broad ; the hypocleidium has a triangular form, and the tarso-metatarsus is not half as long as the tibia. I do not give HUXLEY's characters in detail, since he has pointed out that the two series meet among the partridges and quails, and cannot thus be sharply marked off.

The Alectoropodes have a range which is related to their anatomical differences, as have the Peristeropodes. The Meleagridæ are confined to America, the Numididæ to Africa, the Phasianidæ to the Oriental region just encroaching upon the Palæarctic, while the Tetraonidæ are Palæarctic and Nearctic.

The **Galli** seem to be undoubtedly an ancient group of birds, a view which is upheld by their points of likeness to many diverse groups.

That they are an ancient group is also shown by their quintocubitalism, the primitive character of the gut convolutions, the often complete muscle formula, and the existence of basipterygoid processes. The existence of these structures has led to their comparison with other groups of birds. Less weight, however, appears to be due to these more general points of resemblance than to other slighter but equally constant similarities. The existence of the entepicondylo-ulnaris muscle is an example to the point.

The **Galli** share this muscle with the **Tinami** only; and no one will doubt on other grounds that the gallinaceous birds and tinamous are distinctly related. It is, however, with the **Anseres** that the greatest number of special resemblances of this character exist. These have been admitted by HUXLEY and PARKER; and more recently, in one of his alternative schemes, SEEBOHM united the two groups. In these two groups the palatines have the peculiar character of wanting the internal lamina, which is at most indicated by a slight ridge; in both of them the basipterygoid processes can hardly be described in those words, as they are but oval facets for the articulation of the pterygoids. The two pairs of extrinsic muscles of the syrinx are one of the strong reasons for uniting the **Anseres** and the **Palamedeæ**, and we have among the **Galli** forms in which there are the same extra pair of muscles present. The general habit of a gallinaceous bird is, it is true, somewhat remote from that of an anserine bird; but *Palamedea* might with truth be described as a goose-like bird with external likeness to a curassow, or as a gallinaceous bird which had put on the characters of the **Anseres**. Its likeness to both is considered on another page.

The existing genera *Tetrao*, *Lagopus*, and *Francolinus* are known from Pleistocene deposits in countries which they at present inhabit, and the species from which these few remains have been described are existing species. Similarly *Coturnix Novæ Zealandiæ* and *Talegalla Lathami* have been met with in the Pleistocene of Australia and New Zealand, both of them being species now living in the localities whence their fossil remains have been extracted. *Phasianus* is represented by a number of extinct species from Europe as old as the Miocene. Three totally extinct genera described by MILNE-EDWARDS are *Palæortyx*, *Palæoperdix*, and *Taoperdix*. These are all Eocene or Miocene and European. *Taoperdix* presents affinities to *Meleagris* and *Numida*.

COLUMBÆ [1]

Definition.—**Aftershaft absent. Oil gland absent or nude. Crop present. Cæca nipples. Syrinx with asymmetrical extrinsic muscles. Of the leg muscles ABX always present; deep plantar tendons of type I. Biceps slip present and peculiar in form. Skull schizognathous, schizorhinal. Basipterygoid processes present, except in Didus.**

This group of birds contains something like 458 species, which are divided into sixty-eight genera. Pigeons are cosmopolitan in range, and show a considerable amount of structural variation.

The absence of an *aftershaft* and the often rudimentary condition of the *oil gland* are among the most important variable external characteristics of the pigeons. The latter is never feathered, and is sometimes totally absent. But its presence or absence cannot be made use of as a fact of great systematic importance; for we find in the same genus *Ptilopus* two species, *Pt. coronulatus* and *Pt. superbus*, with a minute oil gland, and two others, *Pt. assimilis* and *Pt. puella*, from which the gland has entirely disappeared.

The *rectrices* vary in number from 12 (e.g. *Columbula*) through 14 (e.g. *Phaps*), 16 (*Goura*), to 20 (*Otidiphaps*). But the variations do not invariably coincide with the limits of genera. *Phaps chalcoptera* has 16 rectrices, P. *elegans* only 14. *Phlogœnas Stairi* has 12 rectrices, *P. cruentata* 14. There are other examples. Though for the most part aquincubital, it is remarkable to find in *Columbula* a quinto-cubital bird (MITCHELL).

I have by me a careful manuscript account of the *pterylosis* of *Didunculus* by Mr. FORBES. The dorsal tract is strong and broad upon the neck, narrowing a little below; where it narrows it becomes stronger. Below the shoulder it bifurcates and becomes suddenly weak; below this point the whole dorsal region is covered with a weak and diffused feathering, which is especially weak over the head of the femur. In the

---

[1] C. J. TEMMINCK, *Histoire Naturelle Générale des Pigeons et des Gallinacées.* Amsterdam et Paris, 1813-1815.

X

middle line posteriorly the dorsal tract is again strongish. The lumbar region is also strongly feathered, and the thigh is covered by a strong tract which ends very abruptly below.

The ventral tract is much weaker, and on the neck gradually merges into the lateral space which extends from a short way below the head to the shoulder. Below this the ventral tract is still weaker, and does not bifurcate until the upper end of the carina sterni. The median apterion is oblong and narrow, and reaches the cloacal aperture. In the abdominal region the two tracts get stronger. The pterylosis of *Columba livia* as figured by NITZSCH hardly

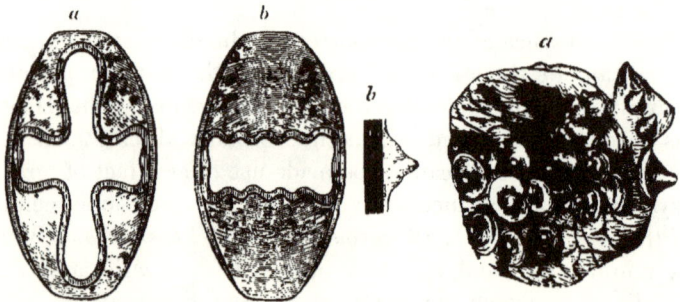

FIG. 154.—HORIZONTAL SECTIONS OF GIZZARDS OF, *a*, *Ptilopus jambu*, *b*, *Treron calva* (AFTER GARROD).

FIG. 155.—*a*, GIZZARD OF *Carpophaga latrans*. *b*, ONE OF HORNY TUBERCLES IN SECTION. (AFTER GARROD.)

differs, and is typical of the pigeons in general; there is no down.

All pigeons have a well-developed *crop*, the presence of which organ is presumably related to their fruit- and grain-eating mode of life.

In *Carpophaga*, a fruit-eating pigeon, the *gizzard* is very weak; but in other pigeons this organ is very strong, its walls being even ossified in *Calœnas nicobarica*.[1] In *Ptilopus* a cross section of the gizzard has the peculiar form shown in the accompanying drawing, where it is compared with a

---

[1] VERREAUX and DES MURS describe an exaggeration like this in *Phœnorhina goliath*, where also the tubercles are ossified (*Rev. Mag. Zool.* 1862, p. 138). See also FLOWER, 'On the Structure of the Gizzard of the Nicobar Pigeon,' *P. Z. S.* 1860, p. 330.

pigeon showing a more normal state of affairs. There are four muscular pads in its walls, so that in transverse sections the lumen is cruciform. In *Ptilopus coronulatus* the lumen of the gizzard is not regularly cruciform, like *Pt. jambu*, but irregular and asterisk-shaped; so with *Pt. superbus*. *Chrysœna viridis* has also a ptilopine gizzard. In *Carpophaga paulina* the transverse section shows a close approach in the structure of its gizzard to that of *Ptilopus*. In *C. latrans* (fig. 155) the gizzard is lined with extraordinary conical horny processes.[1] The right lobe of the *liver* is larger than the left, and the *gall bladder* may be absent or present.

'In the Columbæ which I have examined (*Columba* of several species (fig. 156), *Phlogœnas cruentata*),' remarks Mr. Mitchell, 'it is tempting to regard the gut as a simple derivative of the type seen in *Pterocles*. The duodenum is longer and narrower. The circular loop is enormously expanded, but the three subsidiary loops seen

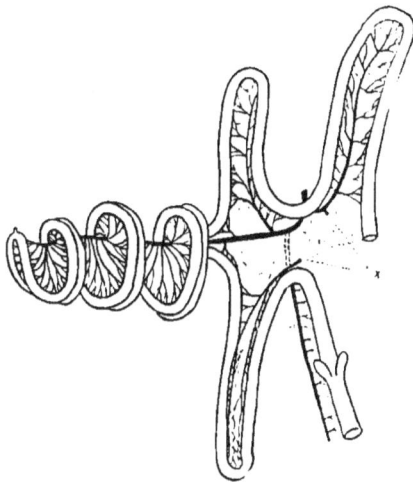

FIG. 156.— INTESTINES OF *Columba livia* (AFTER MITCHELL).

in *Pterocles* remain. The first of these is somewhat shortened; the second, that bearing the yolk-sac vestige at its end, is enormously lengthened; the mesentery is folded along the line of the median mesenteric vessel, so that the two limbs of the loop are brought in contact with each other, and, finally, the whole folded loop is rolled into a rough spiral. The third subsidiary loop of the circular part of the gut has

[1] R. Germain, 'Note sur la Structure du Gésier chez le Pigeon Nicobar,' *Ann. Sci. Nat.* (5). iii. p. 352; Garrod, 'Note on the Gizzard and other Organs of *Carpophaga latrans*,' *P. Z. S.* 1878. p. 102.

x 2

the same arrangement and veins as in *Pterocles* ; but the cæca no longer run along it, but occur as very short stumps upon the rectum.'

The *cæca* are small and nipple-like ; they may be entirely absent. In *Tympanistria bicolor* I found one, an especially minute one, on the left side.

The *intestines* of the **Columbæ** are very short and voluminous in some of the fruit-eating forms, moderately long in the majority of forms, and extraordinarily long in *Didunculus*. The following are a few measurements :—

	Ft.	Ins.		Ft.	Ins.
Carpophaga ænea	1	6	*Chrysœna viridis*		7
Chalcophaps chryso-chlora	1	8½	*Macropygia lepto-grammica*	2	10¼
Phlogœnas Stairi	2	6	*Peristera Geoffroyi*		10
„ *cruentata*	2	4¼	*Ianthœnas leuco-læma*	3	11
Columba maculosa	2	8			
Geopelia cuneata		8½	*Goura coronata*	5	1
„ *humilis*	1	6	„ *Victoriæ*	4	0
Ptilopus jambu	1	0	*Phaps chalchoptera*	2	6
„ *melanoce-phalus*		9	„ *elegans*	1	10¼
Ptilopus coronulatus		6½	*Didunculus strigi-rostris*	7	0

In *Goura Victoriæ* the *tensor brevis* muscle is bordered on the patagial side for the last half of its course by a strong tendon, which arises from the pectoralis muscle. Below this tendon forms the outer and stronger part of a thin and rather wide tendon, in which the muscle itself ends. A wristward slip is given off, but there is no patagial fan. The *biceps slip* arises tendinously from the biceps, has a short muscular belly, and ends tendinously upon tensor patagii longus tendon.

Goura coronata is much the same, but the biceps slip is (? exceptionally) digastric, a second tendon springing from tensor patagii longus muscle, and becoming muscular before it joins the muscular belly derived from biceps itself.

In other pigeons—and this is one of the most salient

features of the group—the tensor brevis muscle is often very extensive—in fact, overlapping the extensors of the fore arm. This is the case, for instance, with *Geophaps scripta*, where I could find no deltoid beneath it. In *Ptilopus* and *Phaps*, on the other hand, a good deltoid is present, and though mainly attached to the humerus is also at its extremity inserted on to tendon of tensor brevis. This latter insertion is wanting in many pigeons which have a long deltoid. In *Erythrœnas* the deltoid reaches nearly to elbow. There is very often a wristward slip from tendon of tensor patagii brevis, but never, so far as I have ascertained, a patagial fan.

The *biceps slip* appears to be always an independent muscle, arising, as described above, by a thin tendon from the biceps, and it frequently has a second tendon of origin, as in *Goura coronata*. This is the case, for example, with *Carpophaga*, *Didunculus*, and *Phaps*. It is absent in *Erythrœnas*.

The *expansor secundariorum* is very constantly present ; but there are indications that it is on the wane in these birds. In *Phlogœnas cruentata* the tendon has not the characteristic T-shape that it has in other pigeons, but blends with a triangular fibrous slip arising from the scapula near the scapulo-coracoid joint. In *Carpophaga paulina* there is much the same arrangement. In *Phlogœnas Stairi* the muscle appears to be totally absent.

Of the two *latissimi dorsi* muscles the posterior is sometimes wanting in pigeons.[1] HASWELL[2] first directed attention to this point, which was denied by GADOW[3] and FORBES, but reaffirmed by FÜRBRINGER. According to the last-mentioned observer its occasional absence is a matter of individual variation.

The *anconæus* has a tendinous insertion on to the humerus.

---

[1] And also (? occasionally) in *Otis*, *Pterocles*, and various passerines, according to FÜRBRINGER.

[2] 'The Myological Characters of the Columbidæ,' *Proc. Linn. Soc. N. S. W.* iv. 1880, p. 303 ; 'Note on the Anatomy of Two Rare Genera of Pigeons,' *ibid.* vii. 1883, pp. 115, 397.

[3] In his memoir upon *Pterocles*, quoted below, p. 315, footnote.

The second pectoral is inserted, in an unusual way, on to the anterior face of the humerus, and not on to the pectoral ridge, as in other birds.

In the hind limb nearly all pigeons have the complete formula ABXY, the ambiens being sometimes present and sometimes absent. Only in *Lopholæmus* apparently is the accessory femoro-caudal absent.

The *deep flexor tendons* are gallinaceous, either blending or with a vinculum. In *Lopholæmus antarcticus* and *Erythrænas* there is a vinculum, and also a special slip to tendon of digit II.

Fig. 157.—SYRINX OF *Carpophaga latrans* (AFTER GARROD).

Considering the variability of so many muscles and organs among the Columbæ, it is remarkable to note what slight variations there are in that, as a rule, rather variable organ, the *syrinx*. The accompanying illustration will serve to show the form of the syrinx among the Columbæ; but in the species illustrated the origin of the sterno-tracheales is not so highly asymmetrical as is sometimes the case; they occasionally arise more markedly from the right side of the trachea. The intrinsic muscles are always present, and generally attached, as shown in the figure, to the membrane between the penultimate and antepenultimate tracheal rings. They are sometimes continued a little further by ligamentous tissue. The last three tracheal rings are united by median bony or cartilaginous pieces. Posteriorly the tracheal rings are weak, or even defective in the middle line. These are the general characters of the windpipe and syrinx in the Columbæ. A few of the genera which show some slight divergencies may now be mentioned. In *Calœnas nicobarica* the intrinsic muscles are continued by ligament as far as the penultimate tracheal ring; the first four or five bronchial semi-rings are connected posteriorly by a cartilaginous bar, which borders the membrana tympaniformis and is continued up as far as the sixth tracheal ring from the end of the series.

In *Goura* there is no union posteriorly between successive tracheal rings, and the last two or three are quite discontinuous in the middle line posteriorly.

In *Didunculus* also the last few tracheal rings do not meet in the middle line posteriorly.

As to the skull, the pigeons are schizognathous birds with a slender vomer and basipterygoid processes, absent only in *Didus*. They are also schizorhinal, but *Goura*, like *Cursorius*, &c., among the Charadrii, is pseudo-holorhinal. The lacrymal fuses below with the ectethmoid, and, indeed, forms with it a nearly solid and often rather massive plate of bone. In *Goura*, at any rate, the descending process of the lacrymal is perforated in front by a largish foramen, as in the *Rhea*. Some pigeons—*e.g. Lopholæmus*—have a median small circular foramen above the foramen magnum; in *Macropygia*, &c., this becomes a notch upon the upper border of the foramen magnum. In *Goura* the foramen is totally absent.

The skull of *Didunculus* is exceptional. The basipterygoid processes are very large. The palatines, instead of widening out posteriorly, are narrow, solid bars throughout their whole extent. HUXLEY states that the internal lamina of these bones is 'altogether obsolete.' I find, however, in my specimen a pair of small downwardly directed hooks arising from where the palatines come into contact posteriorly, which I take to be the homologues of these structures. Owing to the shortened and curved bill the bony nostrils are much reduced in extent. There is no supraoccipital foramen. There is, as in gallinaceous birds, a fusion between the postfrontal process and the zygoma.

There are 15 *cervical vertebræ* in *Goura, Carpophaga*, &c., 14 in *Columba, Phaps*, &c. Vertebræ 15-17 appear to be nearly always ankylosed.[1] The *atlas* is notched for the odontoid process. The hypapophyses begin in *Goura* upon the eleventh cervical and end upon the first dorsal. Four ribs reach the sternum in *Goura Victoriæ*, of which the three first bear uncinate processes. Only three reach the sternum in some

[1] *Fide* NEWTON and GADOW (*loc. cit.* on p. 314). In *Leucosarcia picata* I found four fused, and in *Geotrygon violacea* only two.

pigeons. The *sternum* has both spina externa and interna ; it has two pairs of notches (*Goura*), or one pair of notches and one pair of foramina (*Phaps, Carpophaga*).

The furcula is U-shaped without hypocleidium. The coracoids may meet, but they do not overlap.

The following table gives the variations of the principal variable organs in a number of genera :—

—	Ambiens	Caeca	Oil Gland	Gall Bladder	Distribution
Columba	+	+	+	0	Cosmopolitan
Turtur .	+	+	+	0	Old World
Leptoptila	+	+	+	0	South America
Macropygia	+	+	+	0	Australia, East Indies
Ectopistes .	+	+	+	0	North America
Ianthœnas .	+	+	+	0	Japan, Fiji, Samoa
Chamæpelia .	+	0	+	0	America
Geotrygon .	+	0	+	0	South America
Metriopelia .	+	0	+	0	South America
Peristera .	+	0	+	0	South America
Zenaida . .	+	0	+	0	America
Zenaidura .	+	0	+	0	North America
Calœnas .	+	0	+	0	New Guinea, China, Nicobar
Chalcopelia .	+	0	+	0	Africa
Chalcophaps .	+	0	+	0	Australia, East Indies
Tympanistria .	+	0 or 1	+	0	Africa
Ocyphaps .	+	0	+	0	Australia
Leucosarcia .	+	0	+	0	Australia
Phaps .	+	0	+	0	Australia
Didunculus .	+	0	0	0	Samoa
Carpophaga .	+	0	+	+	East Indies, Pacific Islands
Ptilopus .	0	0	0	+	Australia, Fiji, Celebes
Erythrœnas . .	0	0	+	+	Madagascar
Lopholæmus .	+	0	+	+	Australia
Treron . .	0	0	0	0	India, East Indies, Africa, Madagascar, Celebes, New Guinea
Goura .	0	0	0	0	New Guinea
Œna .	0	0	+	0	South Africa
Geopelia .	0	0	+	0	Australia, East Indies
Starnœnas .	0	+	0	0	America
Phlogœnas .	0	+	+	0	East Indies
Columbula .	+	0	+	0	America
Turacœna .	+	?	0	?	Celebes, Timor
Chrysœna .	0	0	?	+	Fiji

GARROD included with the **Columbæ** *Pterocles*, which I treat of separately. Accordingly, in quoting his classification of the group,[1] I term families what he termed subfamilies. For him the group contained two families, viz. Columbidæ and Pteroclidæ. The following is, with the alterations referred to, GARROD's scheme of division of the group :—

Family I. **Columbidæ.** Columbæ with ambiens, cæca, oil gland, no gall bladder, and 12 rectrices.

Genera : *Columba, Turtur, Macropygia, Ectopistes, Leptoptila, Ianthœnas.*

Family II. **Phapidæ.** Columbæ with ambiens and no cæca.

Division A. Oil gland present, no gall bladder.
Genera : *Chamœpelia.*
Metriopelia.
Zenaida, Zenaidura.
Geotrygon.
Peristera.
Calœnas.
Chalcopelia.
Tympanistria.
Ocyphaps.
Leucosarcia.
Phaps.
Division B. The oil gland and gall bladder present.
Genus *Carpophaga.*
Division C. The accessory femoro-caudal absent ; oil gland and gall bladder present.
Genus *Lopholœmus.*
Division D. Oil gland and gall bladder absent.
Genus *Didunculus.*

Family III. **Treronidæ.** Columbæ without ambiens.

Division A. With cæca and oil gland ; no gall bladder.
Genus *Phlogœnas.*

[1] 'On some Points in the Anatomy of the Columbæ,' *P. Z. S.* 1874, p. 249, and ' Notes on Two Pigeons,' &c.. *ibid.* 1875, p. 367.

Division B. With cæca, no gall bladder, no oil gland.
Genus *Starnœnas*.
Division C. With oil gland, without gall bladder and
cæca.
Genera *Geopelia*, *Œna*.
Division D. Without oil gland (or rudimentary) and
cæca ; tarsi scutellate.
Genera *Treron*, *Ptilopus*, *Erythrœnas*.
Division E. With cæca, oil gland, and gall bladder ;
tarsi reticulate.
Genus *Goura*.

To these must in any case be added another family to
include the flightless dodo (*Didus*) and the equally flightless
solitaire (*Pezophaps*),[1] the former from Mauritius, the latter
from Rodriguez.

The dodo at the time of its description by Messrs.
STRICKLAND and MELVILLE[2] was only very imperfectly
known. Subsequently OWEN[3] gave an account of and figured
the greater part of the skeleton. Later still Sir E. NEWTON
and GADOW[4] supplemented this account by further details,
and published a figure of 'the first correctly restored and
properly mounted skeleton.' There are naturally many
other notices of this much-written-about bird.

The skull has the median supra-occipital foramen of
some pigeons, but not the basipterygoid processes. The
nostrils are schizorhinal. The palatines have, contrary to
what is found in *Didunculus*, with which Sir R. OWEN
specially compares *Didus*, an internal lamina. Neither do
the postfrontal process and zygoma meet and fuse, as they
do in *Didunculus*. The interorbital septum is thick and
complete.

There are 15 cervical vertebræ, and the atlas is notched for
the odontoid process. The last cervical and first two dorsals

---

[1] A. and E. NEWTON, 'On the Osteology of the Solitaire,' &c., *Phil. Trans.*
1869, p. 327 ; E. NEWTON and J. W. CLARK, 'On the Osteology of the Solitaire
(*Pezophaps solitarius*),' *ibid.* vol. clxviii. 1879, p. 438.
[2] *The Dodo and its Kindred.* London, 1848.
[4] 'On the Osteology of the Dodo,' *Tr. Zool. Soc.* vi. p. 49.
[ ] 'On Additional Bones of the Dodo,' &c., *ibid.* xiii. p. 281.

are fused. Four ribs reach the sternum, of which the last belongs, as in *Pezophaps* only, to the first pelvic vertebra.

The scapula and coracoid are, as in the Ratites,[1] ankylosed, and the angle between them approximates to that of the Ratites in its wideness.

The sternum has a fair keel, neither spina interna nor externa, and the coracoids do not nearly meet. The clavicles have no hypocleidium.

I discuss the affinities of the **Columbæ** under **Pterocletes.**

PTEROCLETES [2]

Definition.—Aftershaft small; aquincubital. Oil gland nude. A crop present. Cæca long. Muscle formula of leg, ABXY+. Biceps slip and expansor secundariorum present; plantar tendons of type IV. Skull schizognathous, holorhinal, with basipterygoid processes. Both carotids present.

This group contains but two genera, *Pterocles* and *Syrrhaptes*, both of which are Old-World in range. *Pterocles* is more widely spread than *Syrrhaptes*, extending southwards to Africa and Madagascar. *Syrrhaptes* is confined to Central Asia. These birds have a pigeon-like aspect, though NEWTON has pointed out that *Syrrhaptes* has a plover-like flight.

The *feet* are peculiar for the feathering, which extends to the claws; and in *Syrrhaptes* the three toes (the first is altogether aborted) are encased in a common 'podotheca,' which presents the appearance of 'a fingerless glove.' This is not the case with *Pterocles*, which, moreover, has the first toe. The sand grouse have a small *aftershaft* and a nude *oil gland.* The *rectrices* vary in number from fourteen to eighteen. Contrary to what is found in the pigeons the newly hatched sand grouse is covered with down, and in the

[1] MOSELEY ('On the Structure and Arrangement of the Feathers in the Dodo,' *Rep. Brit. Ass.* for 1884, p. 782) notes that the feathers are disposed in threes, a feature which is, he says, apparent in pictures of the bird.

[2] M. BOGDANOW, 'Bemerkungen über die Gruppe der Pterocliden,' *Bull. Soc. Imp. St. Petersb.* xxvii. 1881, p. 164; H. GADOW, 'On some Points in the Anatomy of Pterocles,' &c., *P. Z. S.* 1882, p. 312; D. G. ELLIOT, 'A Study of the Pteroclidæ, or Family of the Sand Grouse,' *ibid.* 1878, p. 233.

adult there is down upon the apteria. The *pterylosis*, as figured by NITZSCH, is almost exactly like that of the pigeons, but there are no neck spaces. The beak has no soft 'cere,' such as exists among the pigeons.

GADOW has contrasted the *crop* of the sand grouse with that of the pigeon. In the former it is a simple dilatation of the anterior and lateral walls of the œsophagus, without any constriction in the middle line. In the pigeons, on the other hand, the crop consists of two symmetrical swellings of the œsophagus, between which is continued the œsophagus. The intestinal coils have, according to MITCHELL, an 'extremely primitive character.' The resemblance to the intestine of the pigeons is great. The **Pterocletes** differ from the pigeons in the large size of the *cæca*. The lining of the cæca is marked by about six longitudinal folds, according to GADOW, but according to PARKER no less than twelve. In *Pterocles* the right lobe of the *liver* is about three times the size of the left. A gall bladder is always present. Figs. 17 and 18 (p. 33) show some variations in the positions of the liver and pancreatic ducts in the sand grouse, which are taken from GADOW's paper upon the anatomy of this group.

The *syrinx* (of *Pterocles*) is not in the least like that of the **Columbæ**. The extrinsic muscles are perfectly symmetrical, and the intrinsic muscles are enormously developed. The ordinary pair present in the **Columbæ** are attached to what I regard as the first bronchial semi-ring, and are not specially large; the second pair[1] are only visible on the posterior aspect of the windpipe; they are two large fusiform muscles which are inserted in common into the middle line of the trachea, near to its termination.

The *muscles* of the hind limb have been described in some detail by GADOW.

The muscle formula is complete, being thus expressible, on GARROD's notation, by the letters ABXY+. There is only one *peroneal*, the *longus*, which has the usual attachments to the ankle and to the *flexor perforatus*.

---

[1] *Pterocles arenarius*; it appears that *P. alchata* has not the second larger pair.

The *deep flexor tendons* fuse at the ankle, but no branch is given off to the small hallux ; this digit has, however, a special *flexor hallucis brevis*. GADOW concludes his survey of the muscles of the bird with the following remarks : ‘ Of all the other muscles of the leg (excluding ambiens, peroneus, and absent flexor hallucis slip) there is none that shows any practical difference between sand grouse, pigeons, and even (if we include them in our comparison) the plovers. On the whole, however, the myology of *Pterocles* indicates that it is more nearly allied to the pigeons than to any other group of birds.'

The general aspect of the *skull* of *Pterocles arenarius* is much like that of a similarly sized pigeon. The nostrils, however, are more distinctly holorhinal, thus leading towards the gallinaceous birds. They end on a level with the ends of the nasal processes of the premaxillaries, and do not narrow at all at their broadly rounded terminations. As is the case with *Goura*, *Opisthocomus* (a fact of possible importance), and some other holorhinal birds, a plate of bone underlies the extremity of the nostrils, reducing the extent of the orifices.

As in pigeons the ectethmoids are very solid plates of bone which fuse with the lacrymals, and nearly reach the jugal bar : only the minutest foramen perforates this plate above. *Syrrhaptes* [1] agrees so far with *Pterocles*, but has a rather more vacuolated interorbital septum.

The post-orbital and post-frontal processes are long, and nearly (*Pterocles*) or quite (*Syrrhaptes*) fuse at their extremities, as in some gallinaceous birds.

The maxillo-palatines are not like those of pigeons ; they are, as in gallinaceous birds, slender curved hooks.

The basipterygoid processes are well developed.

There are fifteen or sixteen cervical vertebræ.

‘ In almost all those respects,' remarked HUXLEY, ‘ in which the grouse differ from the fowls they approach the pigeons ; and an absolute transition between these groups is effected by the Pteroclidæ, whose popular name of “ sand grouse ” might fitly

[1] See PARKER, ‘ On the Osteology of Gallinaceous Birds,' &c., *Trans. Zool. Soc.* vol. v.

be exchanged for that of " pigeon grouse." ' There is, in my opinion, much to be said for this view, which, however, is not now so generally held.   HUXLEY's view was based almost exclusively upon osteological characters, with but slight reference to the anatomy of the soft parts, which were indeed—when he wrote (in 1868)—scarcely known.   The several regions of the vertebral column in the sand grouse have the same number of vertebrae as in the 'Alectoromorphae,' 'and ankylosis takes place in the same manner.'   The skull is dove-like for the most part; but in certain ways it approaches the **Galli**.   For instance, the maxillo-palatines are alike in both groups of birds; the union between the squamosal and the post-frontal process is gallinaceous; the holorhinal nostrils, which I must term those of *Pterocles*, indicate a likeness to all the members of the group **Galli**. The remainder of the skeleton is, in HUXLEY's opinion, ' peristeromorphous,' but the pelvis is partly grouse-like. Attention may be directed to the likeness of the sand grouse humerus to that of pigeons.   The osteological characters, however, are not quite so intermediate in some respects as might be inferred from HUXLEY's paper.   The at least ' pseudo-holorhinal ' nostrils have their counterpart among the **Limicolæ**, in *Thinocorus*, and in some others (see below).   The solid ectethmoids too are also seen in that group, while GARROD's remark that the **Alcæ** have a humerus like that of **Columbæ** and *Pterocles* is suggestive in the light of the unquestionable likeness of the **Alcæ** for the **Limicolæ**, though the actual weight of this character may be thought by some to be discounted by the fact that it is met with in the **Psittaci**.

Moreover *Otis*, which is to be placed somewhere near the **Limicolæ**, has the gallinaceous union between the squamosal and the post-frontal process, to which I have referred as possibly affining the **Pterocletes** to the **Galli**.   Other characters too, which appear at first sight to be arguments in favour of the position taken up by HUXLEY, may be interpreted fairly as marks of affinity with the **Limicolæ** (and their immediate allies).   Such are, for example, the long cæca (with folds in the bustards), the crop (present in *Thinocorus*), the gall bladder, &c. MITCHELL distinctly places both **Pterocletes** and **Columbæ** in the neighbourhood of the **Limicolæ** by reason of the arrangement of the intestinal coils.

It is at any rate clear that the **Pterocletes** occupy a lower place than the **Columbæ**—that *they* have given rise to the **Columbæ**, and not *vice versa*. The justice of this view is shown by the long cæca, the existence of an aftershaft, the complete muscle formula of the leg, and by a few other equally unmistakable characters. On the whole it seems not unreasonable to look upon the **Pterocletes** as not far from the stock which produced the **Limicolæ**, which itself was possibly not far again from the primitive gallinaceous stock.

TURNICES

Definition.—Rectrices, twelve. Aftershaft present. Oil gland tufted. Cæca long. Muscle formula of leg, A(B XY+. Skull ægithognathous, schizorhinal, with basipterygoid processes. Cervical vertebræ, fifteen. Sternum one-notched.

This group of birds consists of the genera *Turnix* and *Pedionomus*.[1] It has been confounded with the gallinaceous birds; but the discovery of PARKER that the skull is ægithognathous, and further investigations into the structure of the group—of which the most important is a recent paper by GADOW[2]—have rendered it necessary to remove the two genera from close association with the **Galli**.

Of the two genera *Turnix* (*Hemipodius*) is European, African, and Indian in range: *Pedionomus* is Australian.

Besides the external characters mentioned in the definition, which are common to both genera, *Pedionomus* wants the fifth cubital, which is present in *Turnix*; there are four toes in *Pedionomus*; *Turnix* has lost the small hallux of the former genus.

[1] LEGGE (*P. Z. S.* 1869, p. 236), from a consideration of some external characters and habits, was impressed by possible charadriine affinities of *Pedionomus*.

[2] 'Notes on the Structure of *Pedionomus torquatus*,' &c., *Records Austral. Mus.* i. 1891, p. 205.

As to the *pterylosis*, there is a long spinal apterion, which begins on a level with the shoulder joint and reaches to a little beyond the level of the hip joint.   Thence the two dorsal tracts are continued on as a single tract to the feathered oil gland.   On the neck below there is no apterion ; the two tracts then divide, leaving a bare interclavicular space ; they divide again on a level with the anterior end of the carina sterni into a lateral thick patch and a median thinner one ; this latter swells out in its course and then again dwindles, being continued to the cloacal aperture by a few scattered feathers.

The *pectoralis I.* (at any rate in *T. Sykesi*) is two-layered. The *tensor patagii brevis* tendon gives off a wristward slip, but there is no patagial fan.

There is no *biceps slip*, but the *expansor secundariorum* is present (? as to both these structures in *Pedionomus*). The *muscle formula of the leg* is the complete one ABXY + in *Pedionomus* ; *Turnix* has lost the accessory femoro-caudal for the most part—not, however, in *T. Kleinschmidti*, where it is present.   It is remarkable that in *Pedionomus* it is not *B* but *A* which is on the wane.

Both *carotids* are present in *Pedionomus*, only the left in *Turnix* ; but in *Pedionomus* the left is the weaker and not the right, as might perhaps have been suspected.

The *alimentary canal* has no *crop*, ' but the upper half of the œsophagus is very dilatable ' (in *Pedionomus*).   The *cæca* are well developed ; the *liver* in both genera is split into three nearly equisized lobes.   The *gall bladder* is present.

The *syrinx* (of *Turnix lepurana*) is not at all gallinaceous in its characters.   The tracheal rings are weak and cartilaginous.   The intrinsic muscles are thick and originate in close contact from the anterior face of the trachea : they are inserted some way down the bronchi on to the opposite face of the tubes.   In *Hemipodius tachydromus* the windpipe is very soft, and is much dilated in front of the origin of the intrinsic muscles, which, as in the last species, are large.

Our knowledge of the *osteology* of the Hemipodes is

chiefly due to PARKER,[1] who has described in detail the entire skeleton of *Hemipodius varius*, and also the skull of *Turnix rostrata*. The atlas is perforated for the odontoid. There are 15 *cervical vertebræ*, of which Nos. 10–15 bear hæmapophyses. None of the dorsals are ankylosed. The sternum is reached by three or four ribs, and has one pair of deep lateral incisions cutting off two long thin postero-lateral processes.

The *skull* is ægithognathous in its vomer, broad in front, and double posteriorly, and in the slender hook-like maxillo-palatines ; the latter, however, are not unlike those of many gallinaceous birds, while HUXLEY has compared the vomer with that of *Tetrao urogallus*. The nostrils are pseudo-holorhinal, and, as in pigeons, there is a considerable alinasal ossification, reducing the long nares, which are perfectly pervious. As is also the case with the pigeons, the ectethmoids are large and solid, and have fused with the lacrymals. There are well-marked basipterygoid processes.

RALLI

Definition.—Aftershaft usually present. Carotids, two. Muscle formula of leg, ABX(Y)+. Expansor secundariorum always present. Tensor patagii brevis without recurrent slip to tensor patagii longus. Cæca long. Skull schizognathous and holorhinal.

The rails are a group of birds of very uniform structure. They have as a rule a tufted *oil gland*, but *Porzana carolina* is an exception. The *aftershaft* is present. The *rectrices* vary in number from 10 (*Aramides cayennensis*) through 12 (*Porzana carolina*) to 14 (*Ocydromus Earlei*).

The *spinal tract* encloses a long narrow apterion, which commences earlier in *Rallus aquaticus* than in *Fulica atra*. The latter bird has almost a gap between the anterior and posterior parts of the spinal tracts. The pectoral tract of

---

[1] In his papers on the osteology of gallinaceous birds and of the ægithognathous skull in *Zool. Trans.* vols. v. ix. x.

each side is double in *Rallus* and *Ocydromus*, single in *Fulica*.

The following are the *intestinal measurements* of a series of species :—

	Small Int.	Large Int.	Cæca
	Inches	Inches	Inches
Rallus aquaticus . .	15 ——	1·5	1·25
Ocydromus sylvestris .	27		3
O. lefresnayanus . .	42		3·5
O. Earlei . . .	31	3·5	2·5
Aramides cayennensis .	23	1·5	1·75
Porzana carolina . .	18	1·5	2·25 [1]
Crex pratensis . . .	?	2	1
Porphyrio madagascariensis	24	3	2
Gallinula chloropus . .	39 ——	3	·5
Fulica atra . . .	74		14
F. ardesiaca . . .	39	3	6·5
Tribonyx Mortieri . .	40·5	2·75	6, 6·75

The folds of the intestine (fig. 159) are remarkably like those of the cranes (fig. 158), so much so that on intestinal characters only the two groups could not be separated.

There is no *crop*; the *proventriculus* is zonary; the stomach a '*gizzard*.' The right lobe of the *liver* is larger than the left, and the *gall bladder* is always present.

The *atlas* is notched for the odontoid process; it has no lateral canals. The number of *cervical vertebræ* is 15 in *Fulica ardesiaca*, in which there are 7 complete ribs (6 in *Ocydromus*). On the eleventh cervical (of *Fulica ardesiaca*) the catapophyses nearly unite; the hæmapophyses, up which the catapophyses do not climb, extend as far as the second dorsal. The *sternum* has very long lateral processes, with a larger or smaller spina externa. The clavicle comes into near relations with both procoracoid and scapula. In the *skull* [2] there are no basipterygoid processes, and the lacrymals (in *Fulica, Ocydromus,* and *Aramides*) do not join the

[1] In *Porzana notata* the cæca are minute—·8 inch in length. Cf. Parridæ among Limicolæ.

[2] C. G. GIEBEL, 'Osteologie der gemeinen Ralle,' &c., *Zeitschr. ges. Naturw.* v. (1855), p. 185. SHUFELDT, 'Osteology of certain Cranes, Rails, and their Allies,' *J. Anat. Phys.* 1895, p. 2; and 'Osteology of *Porzana carolina,*' *Journ. Comp. Med. Surg.* 1888.

ectethmoids, as they do in all charadriiform birds. The ectethmoids themselves in all rails that I have examined send a process upwards, which joins the frontal bone and leaves a foramen for the passage of nerves. The interorbital septum is widely fenestrate.

The *pelvis* in the rails has a longer preacetabulum than postacetabular portion. The ilia are vertical in their plane anteriorly, and in *Tribonyx* and *Fulica* are completely sepa-

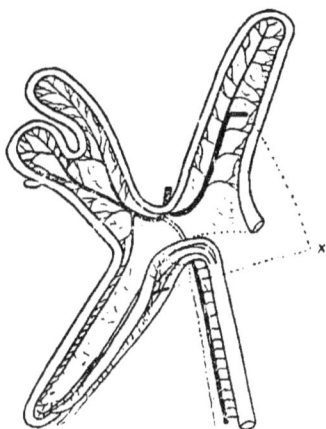

FIG. 158. -INTESTINES OF *Cariama cristata* (AFTER MITCHELL).

x, short-circuiting vessel.

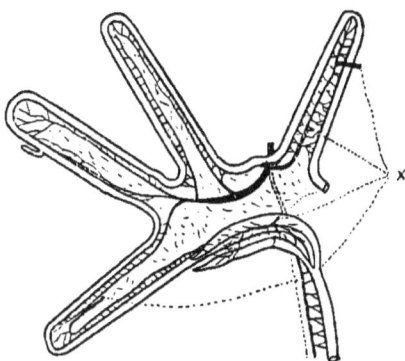

FIG. 159.—INTESTINES OF *Crex pratensis* (AFTER MITCHELL).

x, as in fig. 158.

rated from each other by the fused neural spines of the vertebræ. In *Aramides* and *Ocydromus*, on the other hand, the ilia reach the summit of those neural spines. In all these rails the pubes are fairly strong bones, which are not ankylosed anywhere with the ischia.

Nearly all the Rallidæ have a *biceps slip*. *Ocydromus Earlei* and *Rallus maculatus* are the only exceptions known to me. The tensor brevis is simpler than in many birds; in *Ocydromus Earlei* it consists of only a single tendon. In *Crex*, as in most others, this tendon gives off a wristward slip. In *Aramides* and *Porphyrio martinicus* the hinder branch of the tendon is very feeble, and in the latter does not reach the fore arm. In no rail is there any distal *patagial fan*, a fact justly emphasised by FORBES [1] in discriminating from the rails the somewhat rail-like *Parra*. In some (e.g. *Ocydromus Earlei*, *Aramides*) the tendon of the tensor brevis does not run over the arm to the ulnar side, but in others (e.g. *Crex*) it does.

The *anconaeus* has a humeral attachment.

FÜRBRINGER figures in *Porphyrio* an interesting condition of the *biceps* and of the *biceps slip*. The biceps slip arises by a tendinous head close to, but apparently independently of, the humeral head of the biceps. Close to the coracoidal head of the biceps springs a ligament which is inserted on to the humerus just in front of the origin of the humeral head. This ligament seems to be a detached portion of the biceps, since in the **Steganopodes** (*q.v.*; cf. also FÜRBRINGER, pl. xxvi.) it is perfectly continuous with the biceps.

In the leg both *peroneals* are present, with the usual insertions.

The deep flexor tendons are often of No. I. type. But in *Aramides* there is a modification of this in the shape of a second vinculum, attached partly to the tendon just before its trifurcation and partly to the special tendon of digit II. In *Ocydromus Earlei* the second vinculum is also present, but feebler, not having the second attachment to the flexor of digit II.

The *syrinx* in the rails is of a quite typical tracheo-bronchial form, except for the fact that the intrinsic muscles are attached rather far down the bronchi to the fourth bronchial semi-rings in *Ocydromus*, *Aramides*, and some others. In *Ocydromus sylvestris* none of the tracheal rings are fused; the first two bronchial semi-rings are ossified, and there is an ossified pessulus. In *Aramides* the last three tracheal rings are partly fused; the last two of these and the first two bronchial semi-rings are ossified; there is no ossified pessulus. In *Fulica* (*ardesiaca* and *leucoptera*) the bronchidesmus, which is incomplete, is at first strengthened

[1] 'Notes on the Anatomy and Systematic Position of the Jaçanas,' *P. Z. S.* 1881, p. 639.

by two yellow elastic pads of tissue springing from the membrana tympaniformis.

The genera *Heliornis* and *Podica*—the former American, the latter Old-World in distribution—seem to require a separate family for their reception. The structure of these two birds has been mainly investigated by myself;[1] the skeleton, however, has been described by Brandt also.[2]

In neither bird is there an *aftershaft*, though the *oil gland* is tufted. Unlike other rails they are quintocubital. The *pterylosis* is essentially ralline. In *Heliornis* the neck is nearly continuously feathered, there being only a short ventral apterion. The dorsal tract is strong between the shoulder blades, and is forked; the hinder parts of the tracts scarcely join the anterior; they become fused some way in front of the oil gland. In *Podica senegalensis*, but not in *Heliornis*, the ventral tracts are undivided. The less degree of specialisation is seen in other features of the anatomy of the smaller American finfoot.

The *muscle formula* of the leg in both genera is ABX +, the Y of the rails not being developed. The chief peculiarity of the leg muscles, however, concerns the *biceps*. This is a very large muscle; in *Podica* it has no less than three separate insertions on the leg. First there is the ordinary insertion through a perfectly normal sling; just before this tendon a branch is given off which is inserted independently on to the leg some way further down. In addition there is an extensive insertion on to the fascia covering the calf of the leg. In *Heliornis* the muscle is somewhat simplified, only the first and third insertions being present. The complications of the biceps may have some relation to swimming; for in certain auks (*q.v.*) there is a similar gastrocnemial attachment.

---

[1] 'On the Anatomy of *Podica senegalensis*,' *P. Z. S.* 1890, p. 425; 'On the Osteology, Pterylosis, and Muscular Anatomy of the American Finfoot (*Heliornis surinamensis*),' *Ibis*, 1893, p. 30.

[2] 'Beiträge zur Kenntniss der Naturgeschichte der Vögel,' &c., *Mém. Ac. Sci. St. Petersburg*, 1810, p. 81. See also Giebel, 'Zur Naturgeschichte des surinamischen Wasserhuhns *Podoa surinamensis*,' *Zeitschr. ges. Naturw.* xviii. 1861, p. 421, and Nitzsch's *Pterylography*.

The *deep flexor tendons* of *Heliornis* are remarkable (see fig. 160) for the fact that both tendons split into three branches for the three digits before they unite, the slip to the hallux being given off from the flexor hallucis previously. The conditions which characterise *Podica* are unknown. In the fore limb some of the muscles are characteristic, and differ from those of the rails. The *patagialis brevis* consists in

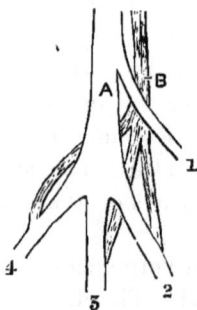

FIG. 160.—DEEP FLEXOR TENDONS OF *Heliornis* (AFTER BEDDARD).
A, fl. hallucis ; *B*, fl. communis ; 1–4. slips to digits.

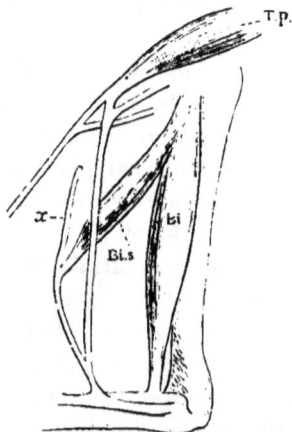

FIG. 161.—PATAGIAL MUSCLES OF *Heliornis* (AFTER BEDDARD)
Tp. tensor patagii ; *Bi.* biceps ; *Bi.s,* biceps slip ; *x*, tendinous slip.

both of a simple undivided tendon, which has not more than an indication of a patagial fan. This indication is seen only in *Heliornis* (fig. 161) in the shape of an upwardly directed tendinous slip, to which, a little before its termination—apparently upon the patagium—a well-developed *biceps slip* is attached. In *Podica* the *biceps slip* ends freely upon the patagium. Both these conditions are different from what is met with in the rails, but are to some extent paralleled among the grebes. In *Podiceps* the biceps slip ends freely upon the patagium, as in *Podica*. In *Æchmophorus* (p. 387) the biceps slip is connected directly with the patagial fan. The likeness between that bird and *Heliornis* appears to me to be unmistakable.

The *expansor secundariorum* is present, and the *anconæus*

is anchored to the humerus. The finfoot agrees with the divers in the extensive origin of the posterior *latissimus dorsi* from the front end of the ilium. The largely tendinous origin of the *rhomboidei* appears to point in the same direction. As to the *alimentary viscera*, there are long cæca ; the right lobe of the liver is the larger ; a gall bladder is stated by GIEBEL to be present in *Heliornis*. I did not find one in *Podica*, but the matter doubtless requires re-examination.

FIG. 162.—SYRINX OF *Podica senegalensis* (AFTER BEDDARD).

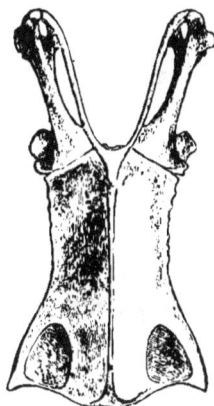

FIG. 163. —STERNUM OF *Heliornis*. VENTRAL VIEW. (AFTER BEDDARD.)

The *syrinx* is typically tracheo-bronchial, and in no way remarkable.

The *osteology* of the Heliornithidæ is not very decisive as to their affinities. In *Podica* there are seventeen *cervical vertebræ*, an advance upon the fifteen of the rails and an approach to the twenty-one of *Podiceps*. Six *ribs* reach the sternum in both genera of Heliornithidæ. The *sternum* has but one pair of notches, and in *Podica*, at any rate, the spina externa is well developed. The *skull* is, on the whole, rail-like, bearing, perhaps, a greater resemblance to *Aramides* than to any other genus of rails. The clavicles, contrary to what is met with in the rails, reach, and are firmly attached to, the carina sterni.

The *pelvis* (fig. 164) of the Heliornithidæ is in some respects

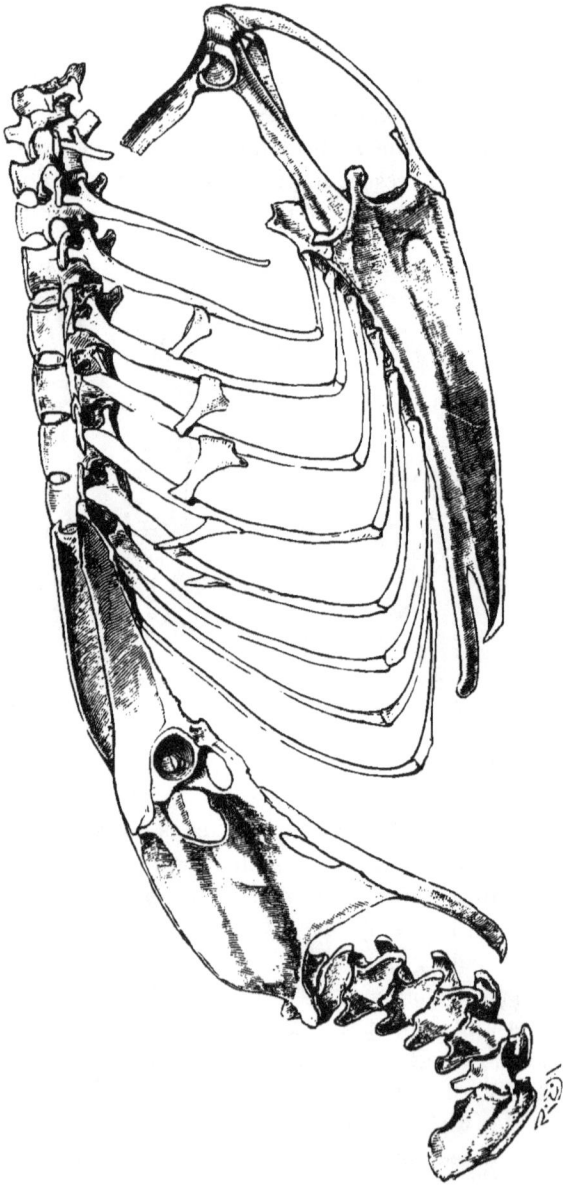

FIG. 164.—LATERAL VIEW OF VERTEBRAL COLUMN, PELVIS AND STERNUM OF *Podica senegalensis* (AFTER BEDDARD).

unlike that of the typical rails. As in *Fulica*, the ilia are widely separated by the fused neural spines of the dorsal

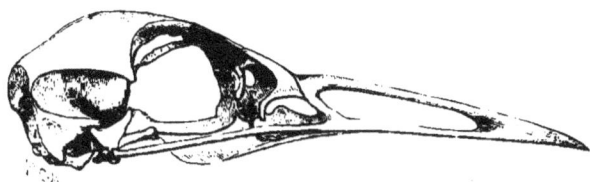

FIG. 165.—SKULL OF *Podica*. LATERAL VIEW. (AFTER BEDDARD.)

vertebræ concerned; but the ischia are broader and directed more downwards (their position is, in fact, more primitive)

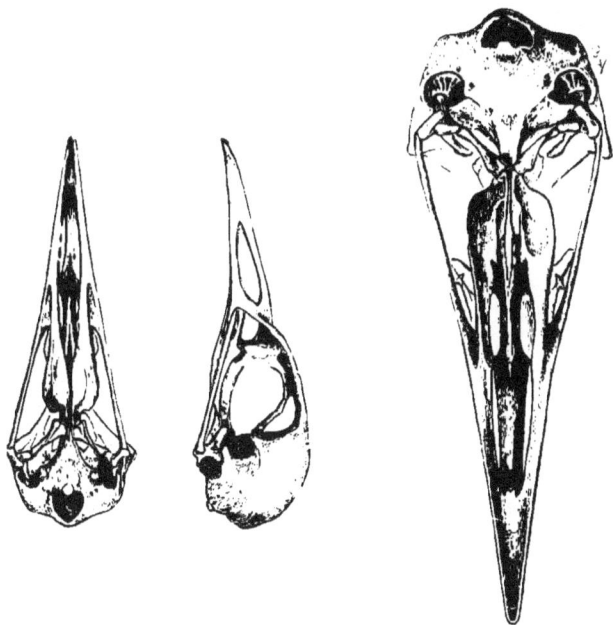

FIG. 166. SKULL OF *Heliornis*. VENTRAL AND LATERAL VIEWS. (AFTER BEDDARD.)

FIG. 167. SKULL OF *Podica*. VENTRAL VIEW. (AFTER BEDDARD.)

than in the rails, while the pubes are ankylosed at least at one point with the ischia.

There are a considerable number of extinct rails, many of which were flightless, thus showing an exaggeration of a tendency of many existing rails which either do not fly much or are but feebly fitted for flying.

It is particularly upon small islands that these flightless rails have been discovered, both living and fossil, and in islands where loss of flight may be regarded as having been of less importance as a disadvantage in the struggle for existence. Thus in New Zealand there was (until recently) the large *Notornis*, whose skeleton has been described by T. J. PARKER.[1] The last living specimen was taken in 1879. From the Chatham Islands are known *Palæolimnas chathamensis* and *Nesolimnas Dieffenbachii*. The skeletons of these rails have been described by H. O. FORBES,[2] MILNE-EDWARDS,[3] and ANDREWS;[4] the latter by ANDREWS. *Nesolimnas* appears to be not yet extinct; the former species is. Another form from the Chatham Islands was originally described under the generic name of *Aphanapteryx*, and supposed to be congeneric with *A. Broecki* of Mauritius. Both of these birds have been dealt with by MILNE-EDWARDS and ANDREWS. *Diaphorapteryx Hawkinsi* was a largish rail, with the keel of the sternum much reduced, being about half the height of the keel of the flying *Hypotænidia celebensis* and slightly less than that of *Ocydromus*. The scapula and coracoid make an exceedingly wide angle, as in all flightless birds— about 130 degrees. The resemblances of *Diaphorapteryx* to *Aphanapteryx* are set down by ANDREWS and GADOW to parallelism of development, and not to real affinity.

Palæolimnas chathamensis is not, as was at one time thought, identical with *Fulica Newtoni* of Mauritius; it may, however, be the same as *Fulica prisca* of New Zealand. The bird is much like *Fulica* in osteological characters, the principal difference being the large size of the impressions

---

[1] 'On the Skeleton of *Notornis Mantelli*.' *Tr. N. Zeal. Inst.* xiv. (1881), p. 245.

[2] In *Nature*, xlv. (1892), p. 416; *ibid.* p. 580; and *Ibis*, 1893, p. 254.

[3] 'Sur les Ressemblances qui existent entre la Faune,' &c., *Ann. Sci. Nat.* (8), ii. 1896, p. 117.

[4] 'On the Extinct Birds of the Chatham Islands,' parts i. and ii.; *Novitates Zool.* iii. p. 73 *et seq.*, p. 260 *et seq.*; and 'Note on the Skeleton of *Diaphorapteryx Hawkinsi*,' *Geol. Mag.* 1896, p. 337.

for the supra-orbital glands. The keel is reduced as compared with living coots, its height being 12 mm. as compared with 17 mm. and 15 mm. in *F. atra* and *F. cristata*. The wing is short in proportion to the leg, shorter than in *F. atra* ; but ANDREWS thinks that the bird 'may still have been capable of heavy flight for short distances.'

Nesolimnas is a more aberrant form in some particulars. It may be still living, but the only specimen was obtained in 1840. The most striking feature in the osteology of this bird appears to be the schizorhinal nostrils, which do not occur elsewhere in the rails (as defined in the present volume). The wings are reduced, but the scapula and the coracoid do not make a wide angle (forty-five as against sixty for *Ocydromus*).

Of the extinct European rails described by MILNE-EDWARDS, from the Eocene and Miocene, a number of species have been described and referred to the genus *Rallus*. *Gypsornis* is believed to be most nearly akin to *Aramides*.

OTIDES

Definition.—**Three-toed birds. Oil gland absent. Aftershaft present. Aquincubital. Skull schizognathous, holorhinal, without basipterygoid processes. Muscle formula, BXY + . Cæca long. No biceps slip. Expansor secundariorum present. Syrinx without intrinsic muscles.**

THE bustards are undoubtedly a much-specialised group, not (in my opinion) distinctly nearer to the charadriiform birds, where they are placed by FÜRBRINGER, than to the cranes, with which they are associated by GADOW.

They are distinguished from all their allies by the total absence of an *oil gland*. The feathers have an *aftershaft*. There are twenty *rectrices* in *Otis* and *Houbara*, sixteen in *Eupodotis Denhami*, eighteen in *Tetrax*.

In both *Otis* and *Tetrax* the lateral neck spaces are reduced to a rudiment on each side close to the shoulder. The dorsal tract is divided high up on the back of the neck : the two halves come nearer together, and at the same time get broader in the lumbar region; they finally completely fuse to form a broad and uninterrupted tract.

The ventral tract is undivided in the neck in *Otis* ; it is very broad in the pectoral region, where it divides into two narrow bands, with an indication of a third in the form of a slightly divergent outer group of particularly strong feathers. The two pairs of narrow tracts unite round the cloaca.

In *Eupodotis australis* the dorsal tract is divided high up on the neck, the ends of the tracts at ends of scapula are particularly strongly feathered, and the two halves of the posterior region of the dorsal tract run in anteriorly between the anterior forks. The ventral tract has a fainter indication than in *Otis* of the outermost branch. It does not appear again to divide into two.

A striking peculiarity of the bustards is the variability of the *carotids*. In *Otis* and *Houbara* there are two ; in *Eupodotis* only the right; in *Tetrax* only the left.

The following are intestinal measurements :—

	Small Int.	Large Int.	Cæca
	Inches	Inches	Inches
Otis tarda	46	--	10·5
Eupodotis australis	28·5	4·75	13·5 and 14
E. kori	42	11	14·5
E. arabs	33	9	9·5 and 8·75
E. Denhami	29	4·5	7·5 and 9
Houbara undulata	22	—	7
	18	4	8
H. Macqueeni	36	4·5	8
Tetrax campestris	30		7
	31		6·25

The *cæcum* of *Otis tarda* is highly remarkable. The median third of the gut is much dilated, and is lined by a smooth mucous membrane, which is marked by about seven slightly raised longitudinal folds, which are visible externally, but connected with no sacculation. Scattered about are numerous circular glands, of the size of hemp seed. The terminal part of each cæcum, which is some two inches in length, has villi, like intestine. The apical region has not, but there is a close retiform disposition of mucous membrane, which gradually passes into longitudinal folds of the middle region. The cæca of *Eupodotis Denhami* appear to be much the same.

In *Eupodotis australis* the cæca are not dilated only in the middle, as are those of *Otis tarda*, but for the apical eleven or twelve inches or so. The dilated region is lined with irregular folds. *E. Denhami* agrees with *O. tarda.* The *liver* lobes are equal, or (*Tetrax*) the right is the larger. A *gall bladder* is present.

The bustards (at any rate *Eupodotis Denhami*) are remarkable for possessing a rudimentary penis, as does the perhaps nearly allied *Œdicnemus*. It is a short blunt cone, grooved above, with a row of glandular pores on each side below. Internally there are two oval spongy bodies, attached to anterior part of sphincter muscle, and external to these on each side are retractor muscles, attached to back of cloaca.

The bustards have a tracheo-bronchial *syrinx*, but the intrinsic muscles are either absent or, if present, are but feebly developed.

Eupodotis australis has perhaps the least modified syrinx. The first two or three bronchial semi-rings seem really to belong to the tracheal series, on account of their greater depth and slighter dividing membranous intervals than those which follow. The intrinsic muscles are reduced to a narrow ligament, fanning out somewhat below.

In *E. kori* the ligament representing the intrinsic muscle of each side is even feebler, and in *E. Denhami* it has absolutely vanished.

In all the above species the rings and semi-rings preserve their independence, and are not fused, except one or two ventrally to form the pessulus, which is strong and ossified.

In *Otis*, on the other hand, the pessulus is slender and cartilaginous, being formed by one ring only. There are no traces of intrinsic muscles.

The genus *Houbara* has a rather peculiar syrinx, which, however, like the last, is without intrinsic muscles. It is compressed from side to side just before the bifurcation. But the 'waist,' thus formed does not correspond to the

boundary line between trachea and bronchi; it lies between
the penultimate and antepenultimate tracheal rings.

In front of the last three tracheal rings there is no particular modification of the trachea. The antepenultimate ring
is strongly ossified in front, where it is convex downwards,
thus leaving a considerable membranous interval between
itself and the ring in front. The next ring is of the same
size, and also ossified in front; the tough and elastic membrane uniting the two can be easily stretched. The last
tracheal ring is much narrower, but also ossified in front; it
passes into the cartilaginous pessulus. Posteriorly these
rings are incomplete, but are joined by a particularly tough
membrane. The bronchial semi-rings are delicate, and not
so long (from before backwards); they naturally diminish
successively in length. The above description refers to *H.
Macqueeni*, but *A. undulata* hardly differs.

The bustards exhibit a phenomenon known as 'showing
off,' which is associated with certain anatomical peculiarities.
The appearance of the male bird, when indulging in this
display, is illustrated by a plate which accompanies Dr.
Murie's paper [1] upon the subject. The neck is immensely
puffed out, so as actually to trail upon the ground. This
singular behaviour on the part of the cock bird during the
breeding season is not confined to the European *Otis tarda*;
it has been observed in both *Eupodotis australis* and *E.
Denhami*. It is curious that, though the result to all outward appearance is much the same, the mechanism which
produces the inflation of the neck differs in the two cases.
In *Eupodotis* the anterior section of the oesophagus becomes
dilated. In *Otis* there is a special pouch [2] developed between
the two halves of the lingual frenum, which extends for a
considerable way down the neck.

The *tensores patagii* are fairly characteristic. No bustard

---

[1] *P. Z. S.* 1868, p. 471 ; Sir W. ELLIOT, 'Notes on the Indian Bustard,' &c.,
P. Z. S. 1880, p. 486. See also FLOWER, *P. Z. S.* 1865, p. 747 ; NEWTON, *Ibis*,
1862, p. 107; MURIE, *P. Z. S.* 1869, p. 140; GARROD, *ibid.* 1874, pp. 471,
673; FORBES, *ibid.* 1880, p. 477.

[2] Cf. similar pouch in duck **Biziura**, v. p. 458.

has a biceps slip. In *Eupodotis Denhami* the brevis tendon is a broad fibrous band spreading out after the ulnar muscles and inserted on to humeral tubercle. In *Eupodotis australis* and *Houbara Macqueeni* there is, in addition, a broad wrist-ward slip which does not cross the fore arm. In *Otis tarda* the extreme degree of complication is reached, for there is, in addition to the structures described, a slight patagial fan joining the longus tendon in the usual way.

The *anconæus* has a tendinous humeral head (at least in *Eupodotis australis* and *Otis tarda*). *Otis* has no *latissimus dorsi posterior*.

As in other three-toed birds, the *deep flexor tendons* are completely blended.

The *glutæus maximus* is large and quite covers the biceps.

The number of *cervical vertebræ* is sixteen, seventeen (*fide* GADOW and FÜRBRINGER), or eighteen (*Houbara Macqueeni*). The atlas is notched. In *Houbara Macqueeni* at any rate the eleventh to thirteenth cervicals have closely approximated hæmapophyses. The last hæmapophysis is on the D1; in the two vertebræ in front these processes are trifid.

Five ribs reach the sternum, all of them with uncinate processes. The *sternum* has two notches. There is neither spina externa nor spina interna.

In the *skull* the margins of the orbit are very sharp, as in *Œdicnemus* and *Rhinochetus*. The interorbital septum is not greatly fenestrate. The descending process of the lacrymal just comes into contact, but does not ankylose, with the prefrontal process of the ethmoid. The maxillo-palatine processes are curved and shell-like. In *Houbara Macqueeni* at any rate this bone reaches the jugal arch. The temporal fossa is guarded by two long and spine-like processes of the squamosal bones, as in gallinaceous birds, and much more marked than in *Rhinochetus*.

The procoracoid is of moderate size and does not reach the clavicle. The two coracoids are not in contact at their articulation with sternum.

I place the bustards in a group by themselves, largely

on account of the fact that they are in several respects much
altered by modification from their allies. They show
evidence of degeneration in the loss of the oil gland, in the
occasional loss of one of the two carotids, in the absence of
the biceps slip, and in the reduced muscle formula of the
leg. GARROD associated with the bustards the Cariamidæ,
Œdicnemidæ, *Serpentarius*, and possibly *Phœnicopterus*.
There is, in my opinion, more to be said in favour of
associating the first two families with the bustards than the
last two. But, as I have pointed out elsewhere, *Serpentarius*
shows more than one hint of a crane-like origin. As to the
first two groups, they agree with the bustards in the muscle
formula BXY, in the holorhinal nostrils, in the absence of the
biceps slip (*Cariama*), the absence of basipterygoid processes,
the absence or feeble development of intrinsic muscles to the
syrinx ; the oil gland too, absent in the bustards, is nude in the
cariamas, and thus shows a commencing reduction. But these
various cases of reduction cannot be held to be necessarily
indications of relationship. I should, however, lay some stress
upon the holorhinal nostrils, the leg muscles, and the syrinx ;
in this case the same conclusion as that advanced by
FÜRBRINGER is arrived at, viz. that the **Otides** come nearest
to the Œdicnemidæ. The very difficulty of associating the
Otides with either gruiform or charadriiform birds is evidence
of the common descent of all three divisions of the class.

LIMICOLÆ [1]

Definition. Oil gland feathered. Aftershaft present. Aquincubital.
Skull schizognathous. Both carotids present. Cæca nearly
always large. Ambiens [2] always present. Biceps slip to patagium
nearly always present.

This is a large group of birds which are cosmopolitan in
range and embrace a variety of types, which may perhaps
be arranged in six families. The type family is that of the

---

[1] SEEBOHM, *The Geographical Distribution of the Charadriidæ* &c., London,
1887. A monograph of all the species (excl. gulls).

[2] *Rhynchops* is alone exceptional in having no ambiens.

Charadriidæ, which contains the largest number of genera ; the remaining families are not separated from it by very numerous points of difference, and the group as a whole is very near to the gulls, which I only divide as a family. The birds of this group, though they are generally good flyers, are mostly found upon the margins of the sea or of marshes and pools ; and their long bills are apparently constructed with a view to probing the mud and sand of such localities for their food, which is, with the exception of the vegetable-feeding Thinocoridæ, animal. The bill is usually long, and, in the woodcock, soft at the extremity, reminding us of the bill of *Apteryx*, being used, indeed, for the same purpose, to extract earthworms. In the curlew (*Numenius*) the bill is curved downwards, as in the ibis. In the avocet (*Recurvirostra avocetta*) it is curved upwards ; in *Eurynorhynchus* it is spatulate at the extremity, and, finally, in the crooked-billed plover it is bent sideways. The legs are often long, and the toes moderately or very much so (Parridæ). There are either four toes, the hallux being small, as in the whimbrels, pratincole, &c., or the hallux and the remaining toes also are of enormous length, as in the Parridæ only ; in many forms, such as the stilt plover, the hallux is absent. In *Recurvirostra* and *Himantopus andinus* the feet are well webbed. In the phalaropes the feet are lobate. The colour of these last-mentioned birds is suggestive of that of the mature gulls, just as the markings of the immature gulls is suggestive of the coloration of many **Limicolæ**, such as the dunlin, knot, &c. The number of *rectrices* varies from ten in *Rhynchæa* and twelve in *Eurynorhynchus* to as many as twenty-six in *Scolopax*. The face in *Lobivanellus* is adorned with fleshy lobes, so often found in birds.

The *pterylosis* of the **Limicolæ** has been chiefly studied by Nitzsch,[1] who figures *Scolopax* and *Charadrius*. The dorsal tract, single on the neck, bifurcates between the shoulders into two strong bands, which either are (*Scolopax*) or are not (*Charadrius*) continuous with the anteriorly bifurcate posterior section of the dorsal tract. The ventral tract divides

---

[1] See also Anderson as quoted on p. 343.

Z

at the beginning of the neck ; on the breast each half gives off
a strong lateral branch.

All the Charadriidæ, and indeed all the **Limicolæ**,[1] are schizognathous; but many of them have not the typical condition
of the vomer which accompanies as a rule the schizognathous
skull.   In the woodcut (fig. 169) a few exceptions to this are

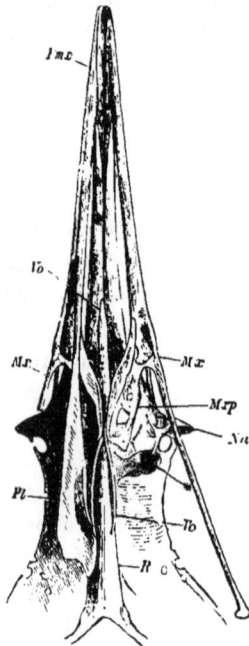

given, which range from the typical
condition observable in *Sterna* to an
excavated extremity, such as characterises *Recurvirostra*. In *Chionis* the
vomer ends in the typical manner, *i.e.*
in a point ; but it is exceedingly broad
before its termination, and therefore
quite unusual.

In *Thinocorus* and *Attagis* the
vomer is short and broad, and almost
passerine in form.

The maxillo-palatines are, as a
rule, thin and scroll-like plates, which
are bent downwards and often deficient in ossification, leaving holes here
and there.   The palatines have a
spout-like process, extending upwards
towards the base of the skull, which
is especially well marked in *Œdicnemus*.   The **Limicolæ** are nearly all of
them schizorhinal, the delicate bar of
the premaxilla being inserted at a
different plane from the attachment
of the nasals to the frontals.   In the
Charadriidæ proper there are a pair of

Fig. 168. —UNDER VIEW OF
SKULL OF *Charadrius pluvialis* (AFTER HUXLEY).

Pmx. premaxilla ; *Vo.* vomer ; *Mx.*
maxilla ; *Pl.* palatine ; *R.* rostrum ; *Na.* nasal ; *Mxp.* maxillo-
palatines ; *, ectethmoid.

largish occipital vacuities, one on either side of the foramen
magnum.   These same birds have basi-occipital processes,

---

[1] SHUFELDT ('Contributions to the Comparative Osteology of Arctic and
Subarctic Water Birds,' *J. Anat. Phys.* 1889 and 1890) figures a few skulls of
Limicolæ.   See also the same, ' On *Aphriza virgata*,' *Journ. Morph.* ii. p. 311.
See also 'Osteology of *Numenius*,' &c., *Journ. Anat. Phys.* 1885, p. 51 ;
' Observations on the Osteology of *Podasocys montanus*,' *ibid.* 1884, p. 86.

which vary somewhat in the degree of their development. The lacrymal bone articulates with the ectethmoid, and makes a complete arch of bone in the anterior region of the orbit. The bones are particularly slender in *Himantopus*, leaving, therefore, a large vacuity. In one or two types the foramen of the arch is almost obliterated by the thickness of the ectethmoid. The upper margin of the orbits in the Charadriidæ is marked with conspicuous grooves for the supra-orbital glands.

There are fifteen *cervical vertebræ* in *Hæmatopus*, *Numenius*, &c.

The *atlas* is perforated by the odontoid (*Numenius*,

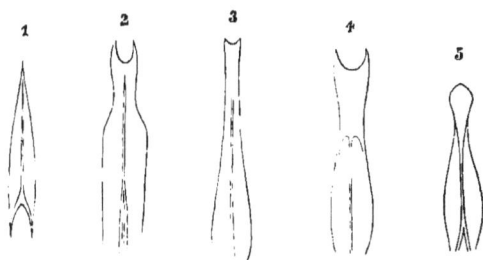

FIG. 169. —VOMERS OF VARIOUS *Limicolæ* (AFTER GARROD).
1. *Sterna hirundo.*   2. *Hæmatopus ostralegus.*
3. *Numenius arquatus.*   4. *Recurvirostra avocetta.*
5. *Chionis alba.*

Limosa). The **Limicolæ** differ from the **Grues** in the fact that only one or two vertebræ (the ninth in *Limosa*) are furnished with two fairly closely approximated hæmapophyses for the reception of the carotids. In *Numenius* these exist on the ninth and tenth, and there are traces on the eleventh. The first dorsal vertebra, or the first two, has a large distally expanded hæmapophysis, an indication of a state of affairs which is carried much further in the allied group of **Alcæ** (*q.v.*) In these points the gulls may be contrasted with the more typical **Limicolæ**. There are no paired hæmapophyses borne by the centra.

The unpaired hæmapophyses extend (in *Lestris*) from C10 to C15 ; on D1–3 there are slightly bifid hæmapophyses. The atlas is notched.

In *Chionis* C10 has a slightly excavated hæmapophysis,

z 2

a trace of the double one of other birds. The hæmapophyses of D1, D2 are the longest, and the former is slightly trifid; this trifidity is very marked in the case of the two last cervicals. The atlas is notched.

In *Parra*, as in the typical **Limicolæ**, the atlas is perforated by the odontoid process. On the eleventh vertebra only do the two processes, which form an incompletely closed canal for the carotids, approach each other markedly in the middle line. The first dorsal vertebra has the largest hæmapophysis, which is flattened slightly distally.

Œdicnemus, with a notched *atlas*, has paired ventral outgrowths for the carotid, closely approximated only on C10. The three following have median blade-like hæmapophyses. On the fourteenth to sixteenth there are lateral outgrowths of these. The first dorsal has the last and the strongest hæmapophysis.

In *Attagis* the atlas is perforated; the hæmapophyses are very feeble.

In *Limosa* and in other genera the clavicle is attached by ligaments to the acrocoracoid, procoracoid (which is moderately developed and curved upwards), and scapula. The two coracoids are not in contact at articulation with *sternum*; the latter is two-notched and has the spina externa only; six ribs reach it. The same statements may be made about *Hæmatopus* and *Numenius*, *Eudromias* (all examined by myself), as well as other genera.

The *pelvis* of *Numenius* may serve as a type for that of the Limicolæ.

The preacetabular portion of the ilium is about equal in length to the postacetabular portion of that bone; the two bones are excavated horizontally, and are just prevented from coming into contact by the fused neural spines of the vertebræ. The pubes are strongish bones and not fused with the ischia. The ischia end in long thin processes which extend back beyond the ilia and nearly as far as the ends of the pubes. *Hæmatopus* and *Limosa* are much the same. *Chionis* hardly differs. In *Parra* the pelvis has rather more the look of that of a rail. The pelvis of *Attagis* is wider than that of other **Limicolæ**, and the ilia are rather further apart. In *Œdicnemus bistriatus*, but not in *Œ. grallarius*, the ischium has a well-marked pubic process, which reaches the pubis.

As to the muscular anatomy, there is great uniformity in the *tensores patagii* of this group. *Charadrius pluvialis* may serve as a type. In that bird there is a *biceps slip*; the tensor brevis early divides into two, of which the anterior is again divided not far from its ending; a recurrent slip runs to the longus.

The same disposition of tendons is found in *Glareola*, *Numenius*, *Scolopax*, *Himantopus*, *Vanellus*, *Machetes*, *Parra*, *Recurvirostra*, *Totanus*, *Limosa*, the only differences being that in some (e.g. *Parra jacana*) the middle only of the three tendons which are inserted upon the fore arm is continued over the muscles of the fore arm to the lower border of the ulna, while in others (e.g. *Numenius*) both the principal tendons are thus continued. In *Glareola* the middle and wristward tendons are thus continued.

Vanellus cristatus has been recorded with two separate biceps slips, which both run to the tendon of the longus. In *Tringa canutus* I found a second biceps slip, largely but not entirely tendinous, which is attached to the outer of the two main tendons of the tensor brevis, this latter tendon indeed only dividing into two near its insertion on to the ulna.

In *Gallinago* the recurrent slip connecting the two tensores seems to be absent.

In *Gambetta flavipes* GARROD found no biceps slip at all, and it seems also to be absent in *Metopidius africanus*.[1]

Thinocorus rumicivorus has the typical pluvialine arrangement of the tensores tendons that has been already described, but the biceps slip is remarkable for the fact that it has a tendinous band running along it. The characters of the tensores patagii in this group are fairly distinctive and at any rate serve to distinguish the **Limicolæ** from the **Ralli**. They do not, however, permit of the enforcement of any

[1] FORBES, *P. Z. S.* 1881, p.

views respecting the families into which the **Limicolæ** have been divided.

An *expansor secundariorum* is almost universally present, but is often feebly developed. It appears to be absent in *Tringa canutus* and *Chionis*.

The *biceps* is so far noteworthy that only the coracoidal head is present in *Rhynchæa* and in *Parra sinensis*.[1] The condition of this muscle in *Himantopus* is extremely interesting. It is stated by Dr. GADOW (on the authority of MECKEL) to be a double muscle. In *H. nigricollis* I find the following arrangement : There are two distinct portions— (1) a part which may perhaps correspond to the entire biceps of other birds, with two heads, a coracoidal and a humeral ; from the former of these arises the biceps slip : (2) in addition there is a distinct coracoidal portion, with a fleshy belly, which has, however, a common origin from the coracoid with the coracoidal head of the main muscle. Something of the same kind appears to occur in *Chionis* and *Scolopax* ; it may obviously be compared with the gulls (*q.v.*) In *Cursorius* the biceps was also double, though the division only commenced a little below the level of the humeral attachment.

In *Lobivanellus* there were indications merely of the same division in the lower part of the belly of the muscle. *Glareola* has a biceps which is double for the greater part of its course.

The *anconæus* appears to invariably possess the tendinous humeral head.

My remaining notes upon the myology of the group are scanty. In *Lobivanellus atronuchalis*, the *semimembranosus* and *semitendinosus* are inserted by a common tendon, and the latter gives off a branch to *gastrocnemius*. There is but one peroneal muscle (the longus). The latter is alone present in *Chionis alba* and *Himantopus nigricollis*. The *pectoralis primus* in these birds does not appear to be divided into two layers.

There is some variation in the *deep flexor tendons* of this group.

---

[1] In this bird the biceps slip arises (as figured by FÜRBRINGER) from the humerus itself.

In *Totanus calidris* there is a slender vinculum, and, in addition, the flexor hallucis gives off a special slip to the branch of the flexor communis which supplies the second digit. The arrangement, in fact, is like that in *Scopus*, in many **Accipitres**, &c. In the Parridæ, on the other hand, the tendons blend early upon the ankle, and in those that have been examined no branch to the hallux has been discovered.

Chionis alba has the deep flexor tendons of *Totanus*.

In all Charadriidæ the *ambiens* is present. The genera *Charadrius*, *Calidris*, *Gambetta*, *Gallinago*, *Limosa*, *Machetes*, *Scolopax*, *Strepsilas*, *Totanus*, *Tringa*, and *Vanellus* have the reduced formula AXY +.

In *Hæmatopus*, *Himantopus*, *Recurvirostra*, *Ægialitis*, *Numenius*, there is the full formula ABXY + ; so, too, in the representatives of the remaining families of the Charadrii, with the sole exception of the Australian thickknee, *Burrhinus*, and some *Œdicnemus*, which have the formula BXY +.

With the exception of the Parridæ (*q.v.*) the charadriiform birds have well-developed cæca. The intestinal measurements of a few types are as follows:—

	Small Int.	Large Int.	Cæca
	Inches	Inches	Inches
Hæmatopus ostralegus	34	1·6	2·75
Œdicnemus crepitans	22 (32)	3	2½ (3) [1]
" *grallarius*	16 (17)	2½ (3)	2½
" *superciliaris*	16	2	2
" *bistriatus*	25	2½	2½
Himantopus brasiliensis	18	2¼	1½
Numenius arquatus	30	2	2½
Strepsilas interpres	13	1	2
Gambetta flavipes	18	1½	1
Glareola pratincola	8½	1¼	1¼
Scolopax rusticola	48	3	½
Gallinago gallinula	12	1	1
Tringa canutus	18½	1¼	1¼
Limosa rufa	33	1¼	⅔
Numenius phæopus	20	2	1¾
Vanellus cayennensis	18		2¾
Recurvirostra avocetta	41 (30)	–	3 (2½
Hydrophasianus chirurga	12	–	⅙
Thinocorus rumicivorus	12·6		2½ and 2¼
Chionis alba	33	1·75	8·25 and 9
Attagis	12·5	—	3
Eurynorhynchus pygmæus [2]	8·75	·88	·7

[1] The brackets contain measurements of a second specimen.
[2] ANDERSON, 'On the Pterylosis, &c., of the Spoon-billed Sandpiper,' *Tr. Linn. Soc.* (2), i. p. 213.

Comments upon the facts set forth in the above table must obviously be discounted by the variations (quite considerable in amount) which occur in one or two of the species. The table given by GADOW[1] increases the number and extent of the individual differences in intestinal length.

In the *liver* the right lobe is, as a rule, larger than the left. Sometimes it is only slightly so; but in *Charadrius pluvialis* the right lobe is twice as large. In *Scolopax rusticola* and in *Œdicnemus crepitans* the lobes are equal. The *gall bladder* is nearly always present. GADOW did not find it in a specimen of *Numenius arquatus* and of *Tringa arenaria* and *T. alpina*. This, however, appears to be individual.

In all the Charadriidæ the *syrinx* is of the tracheo-bronchial pattern.

In *Vanellus cayennensis*[2] (fig. 45, p. 66) the last twenty-five tracheal rings are narrower than those which precede them, and of equal diameter throughout. The first two bronchial semi-rings are like split tracheal rings; the next two are very closely applied together; the remainder are normal bronchial semi-rings with no modification. The most remarkable fact about the windpipe is the enormous size of the intrinsic muscles, of which, however, there is only a single pair. The muscles end in a tendon, which is inserted on to the second, third, and fourth semi-rings. In *Vanellus cristatus* the intrinsic muscles are certainly large, but not so abnormal as in the other species of the genus. Only the last four tracheal rings are modified, and in front they are all fused in the middle line to form a bony box; behind the last three tracheal rings are semi-rings, the pessulus being attached to the fourth. The muscles are attached to the first bronchial semi-ring.

In *Himantopus nigricollis* there was no trace whatever, that I could detect, of these muscles; nor in *H. brasiliensis*. *Charadrius pluvialis*, *Hæmatopus ostralegus*, and *Squatarola helvetica* are also without these muscles. On the other hand they are present in *Totanus*

[1] In Bronn's *Thierreich*, ' Aves.' p. 624.
[2] GARROD, ' On the Trachea of *Tantalus loculator* and of *Vanellus cayennensis*,' *P. Z. S.* 1878, p. 625.

canutus, *Tringa cinclus*, *Numenius arquatus*, *Ægialitis hiaticula*, *Limosa rufa*, *L. ægocephala*, *Machetes pugnax*, *Strepsilas interpres*, and *Scolopax rusticola*. But although the muscles are present in the birds included in the second list they do not, in all of them at least, reach as far as the bronchi, though they may possibly be continued by fibrous tissue to a more normal point of attachment. Thus in *Limosa ægocephala* the muscles stop three or four rings before the end of the trachea. In *Scolopax rusticola*, on the other hand, the rather broad intrinsic muscles reach as far as the first bronchial semi-ring.

The windpipe of the Indian painted snipe (*Rhynchæa capensis*) is peculiar in that it is convoluted slightly in the female, *not* in the male, as might be expected in view of this frequent difference between the sexes in other birds. As has been pointed out, the female is the larger and more richly coloured of the two, a fact which is in harmony with the more complicated trachea.

This is not seen in young females. The same condition is stated by GOULD to characterise the Australian *R. australis*.[1]

The large group of the **Limicolæ** has been variously divided. I follow GADOW in the families, but include also the Laridæ, which I separate from the auks. I should define the **Charadriidæ**, to which most of the foregoing refers, as schizorhinal birds, with occipital fontanelles, furrows for supra-orbital glands and basipterygoid processes, and fifteen cervical vertebræ.

The family **Œdicnemidæ** has been instituted for the genus *Œdicnemus*, which includes the Norfolk plover and a number of other species closely allied; these range widely, being only absent from North America, Central Asia, and New Zealand. The Australian *Œ. grallarius* has been separated as a distinct genus, *Burhinus*, which, as also *Œ. crepitans*, instead of possessing the complete muscle formula (ABXY+), as in

[1] WOOD-MASON, 'On the Structure and Development of the Trachea in the Indian Painted Snipe (*Rhynchæa capensis*),' *P. Z. S.* 1878, p. 745.

the other members of the genus, has the formula BXY+.
The *pterylosis* is as in the Charadriidæ, but the number of
rectrices may be as numerous as fourteen.   There is no
hallux.   In the *skull* the absence of basipterygoid processes
(sometimes indications of them are present, according to
GADOW) and the holorhinal character of the nostrils distinguish this family from the Charadriidæ.   The depressions
for the supra-orbital glands are well marked.   The lacrymal
bones are nearly (occasionally quite, though by suture)
united with a process of the frontals, and form a canal,
through which the gland apparently passes ; this is seen in a
more exaggerated way in *Chionis* and in *Vanellus* (*v.* description of Chionididæ).   The post-orbital angle is not
distinct from the post-frontal process.   There are no occipital foramina, as in Charadriidæ.   There are sixteen *cervical
vertebræ*, the last three of which have ribs of progressively
increasing length ; five or six ribs articulate with sternum.
Contrary to what is found in the Charadriidæ, the coracoids
slightly overlap at their external articulation.

Some of the visceral characters have been already dealt
with above in the general description of the **Limicolæ**.

There is a tendency for the *ambiens* not to cross the knee ;
this occurs individually with specimens of several species.

The *syrinx* has not, except as an occasional variation,
any intrinsic muscles.

Œdicnemus bistriatus has on the anterior wall of the
cloaca two hardly elevated ridges, which end by slightly free
points, and seem to represent a rudimentary penis.

The surgeon birds and jaçanas of the tropical regions of
the Old and New Worlds, with their enormously elongated
feet and their somewhat rail-like aspect, are now known to
belong to the **Limicolæ** (and not to the **Ralli**), of which they
may be regarded as forming a distinct family, **Parridæ**.
Their anatomy has been chiefly studied by FORBES.[1]   As
with the Thinocoridæ so with the present family there is a
character of the alimentary canal which immediately distin-

[1] 'Notes on the Anatomy of the Jaçanas,' *P. Z. S.* 1881, p. 639.

guishes them from all their allies. In this family the cæca
are mere passeriform nipples, measuring from ·15 to ·2 of an
inch in length.

The muscular anatomy has already been to some extent
treated of in connection with the structure of the entire group
of birds of which the present genera form a family. In all
of them, *Parra*, *Hydrophasianus*, and *Metopidius*, the muscle formula is complete, *i.e.* ABXY+. The condition of the
deep flexor tendons of the foot is very singular. As Forbes
justly pointed out, the peculiarly large size of the hallux (as
of all the digits) of the foot in these birds seems to be un-
reconcilable with the entire absence of a special slip from the
conjoined tendon of the long flexors. 'This fact,' he thinks,
'seems to indicate that the Parridæ may have been developed
from some form with a more normal-sized foot, and a small
hallux which had no special long flexor, the great size of
their feet having been developed in accordance with their
peculiar habits.'

The *syrinx* has a pair of intrinsic muscles.

The *skull* has well-formed basipterygoid processes, but
no occipital foramina or supra-orbital impressions. In
Metopidius the *radius* is extraordinarily enlarged (see fig. 70,
p. 125). In the remaining genera there is no such modification of the bone, but there is a metacarpal spur, which may
be of the same use, *i.e.* for fighting. In *Parra* the clavicle
is at its articulation further from the procoracoid than in the
Charadriidæ, and the *sternum* has only one pair of notches.
Five ribs reach it.[1]

The single genus *Chionis*,[2] of antarctic range and somewhat gull-like form, makes up the family **Chionididæ**.

There are twelve *rectrices*.

The *skull* is peculiar in that the grooves for the supra-
orbital glands end in a large foramen on each side, which is

---

[1] The bones of *Parra albinucha* are described and figured by Milne-Edwards, *Hist. Madagascar*.

[2] The peculiar sheath which covers the base of the bill and the nostrils (whence 'Sheathbill') is declared by Steien to be developmentally different from the tube of the Tubinares.

formed by the union of the lacrymal with a process of the
frontal, forming a continuous bony bar.   As has been already
mentioned, *Œdicnemus* shows a very considerable approach to
this state of affairs.   So too, as I interpret it, do the quite
typical charadriid *Vanellus* and *Eudromias*.   In these birds
the grooves for the supra-orbital glands do not, as they
do in *Limosa* and *Recurvirostra*, border the margin of the
orbit.   They are situated at some distance from it, and each
ends in a very small foramen, bordered in front by the
ankylosed lacrymal, which I take to correspond to the large
foramen of *Chionis*.

There are no occipital foramina or basipterygoid processes.

The *cervical vertebræ* are fifteen, of which the last three
bear discrete ribs ; six ribs reach the two-notched sternum,
and there is one behind.   The *clavicles* have no hypocleidium,
and end a long way in front of carina.   The *coracoids* are
not in contact at sternal articulation.

The *muscle formula* is ABXY + .

The *syrinx* of *Chionis* is not widely different from that
of other **Limicolæ**, and exhibits, as will be seen from the following description, no particular resemblance to the **Galli**.   As
is the case in so many **Limicolæ**, the intrinsic muscles end as
such some way in front of the bifurcation of the windpipe,
though they are continued on to the bronchi by fibrous tissue.
They end in *Chionis* upon the fifth tracheal ring counting
from the last.

The last four tracheal rings are more or less closely united
to form an ossified box.   The first bronchial semi-ring to
which the fibrous continuation of the intrinsic muscles is
attached is the widest (from before backwards) of the rings
of the windpipe, and is deeper than the bronchial semi-rings
which follow.[1]

[1] The following are the principal memoirs dealing with the anatomy of this
bird : EYTON, ' Note on the Skeleton of the Sheathbill,' *P. Z. S.* 1858, p. 99 ;
A. REICHENOW, ' Osteologie von *Chionis minor*,' &c., *J. f. O.* xxiv. 1876, p. 84 ;
SHUFELDT, ' The Chionidæ : a Review of the Opinions on the Systematic
Position of the Family,' *Auk*, 1893, p. 158, and in ' Contributions to Comparative
Osteology,' &c., *J. Anat. Phys.* 1891, p. 509 ; KIDDER and COUES, ' A Study of

It is usual to separate the two genera *Thinocorus* and *Attagis*, both South American birds, into a family, **Thinocoridæ**. In contradistinction to their allies they are grain-eating birds, connected with which habit is the presence of a *crop*, an absolutely distinctive character so far as the present group is concerned. Their anatomy has been chiefly studied by GARROD.[1] The differences which distinguish them from

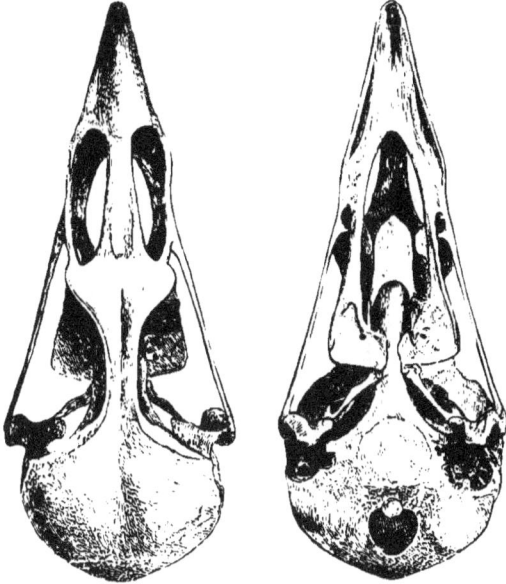

Fig. 170.—SKULL OF *Attagis Gayi* (AFTER GARROD).

other **Limicolæ** are neither great nor numerous. In the *skull* the basipterygoid processes are absent, and the vomer is broad and rounded in front as figured by PARKER;[2] the skull, in fact, as has been mentioned on a previous page, is ægithognathous rather than schizognathous. There are no occipital foramina,

Chionis minor,' *Bull. U.S. Nat. Mus.* iii. 1876, p. 85; R. O. CUNNINGHAM, ' On *Chionis alba*,' *J. Anat. Phys.* 1870, p. 87; BLAINVILLE, ' Mémoire sur la Place que doit occuper le Genre *Chionis*,' *Ann. Sci. Nat.* 1836, p. 97; STUDER, *Forschungsreise S. M. S.* ' *Gazelle*,' Bd. iii. ' Zoologie u. Geologie,' p. 107.

[1] ' Notes on the Anatomy and Systematic Position of the Genera *Thinocorus* and *Attagis*,' *P. Z. S.* 1877, p. 413.

[2] ' On Ægithognathous Birds,' *Zool. Trans.* vol. x.

but the supra-orbital impressions are present. The maxillo-palatines are very far apart, and besides being short are swollen instead of being leaflike plates. The same pseudo-holorhiny that characterises the Glareolidæ is also found in the present family. Five ribs reach the sternum. The coracoids are quite separate at their insertion on to sternum. The *muscular formula* of the two genera is the complete one ABXY +.

In *Thinocorus rumicivorus* the *syrinx* has a pair of lateral muscles, which are attached to the fourth incomplete ring when seen from in front. This is probably the first bronchial, the last three tracheal rings being thus incomplete.

Glareola, Cursorius, Pluvianus, and *Dromas,* all Old-World genera, are included in a separate family, **Glareolidæ**, which GARROD regarded as very near akin to the Thinocoridæ.[1] As in them, basipterygoid processes and occipital fontanelles are absent, and impressions of supra-orbital glands present. The *muscle formula* too is complete, and the *syrinx* has intrinsic muscles.

Except in *Cursorius* there is a hallux. FORBES, with some reason, has united into a group Pluviales, equivalent to the rest of the **Limicolæ**, this family, together with the Thinocoridæ and Chionididæ, mainly on account of the above combination of skull characters found in no other Limicolæ.

As in them also there is pseudo-holorhiny, the bony nostrils, though extending back beyond the nasal process of the premaxilla, being distinctly rounded off.

Pluvianus, however, has typically holorhinal nostrils.

In *Cursorius* the grooves for nasal glands are converted into elongated foramina by a fusion between adjacent processes of the skull.

The gulls form another distinct family, **Laridæ**, containing the genera *Larus, Lestris, Sterna, Rhynchops, Anous,* and *Gygis.*

The *skull* is schizorhinal and schizognathous, without

---

[1] LINNÆUS placed *Glareola* in genus *Hirundo,* SUNDEVALL in Caprimulgidæ.

basipterygoid processes or occipital fontanelles,[1] but with well-marked grooves for supra-orbital glands. In *Lestris antarcticus* at any rate there is a foramen formed round anterior end of supra-orbital groove, as in *Chionis* (*q.v.*) In *Lestris* there is a distinct tendency towards the pseudo-holo-rhinal condition of the Thinocoridæ and Glareolidæ.

In *Larus* the relations of the clavicles, scapula, and cora-coids are as in the Charadriidæ; but the clavicles provided with a hypocleidium come into nearer relations with the carina sterni, to which they are attached by a ligament. The cora-coids are in contact behind the spina (externa) sterni. The same statement may be made of *Lestris*.

The *cervical vertebræ* are fifteen in number. The dorsal vertebræ are, as in Alcidæ and plovers, opisthocœlous. Six (*Lestris antarcticus*) or seven ribs articulate with the sternum. The first phalanx of the second is commonly perforated.[2]

As to the *pterylosis*, the plan is that of the **Limicolæ**, but the ventral tract does not divide until some way down the neck. The feet are webbed, the hallux is small or absent, and there are twelve *rectrices*.

The *muscular formula* of *Rissa tridactyla* is AX + ; of *Larus*, *Stercorarius*, and *Gygis*, AXY +. The other genera have the complete formula ABXY +. *Rhynchops* has no *ambiens*.

In *Lestris crepidatus* and *L. antarcticus* the semi-membranosus is sometimes two-headed, one arising from ischium and one from postacetabular ridge of ilium.

The *tensores patagii*[3] (see fig. 171) are on the plan of

---

[1] These are present in the young: see PARKER, *Linn. Trans.* (2), i. p. 142.

[2] This does not, however, as it has been stated to do, distinguish the Laridæ from other Limicolæ ; though apparently universal in the gulls (including *Anous* and *Gygis*), the same perforations are found in *Glareola*. The value of this anatomical fact may be judged by the perforation of the same bone in such varied types as *Pterocles*, *Coracopsis obscura*, (not in *C. nigra*), *Psittacula passerina*, *Machærhamphus Andersoni*, *Heliodilus*, *Caprimulgus*, *Phaethon candidus* (not *P. rubricauda*), and *Fregata minor*. These instances are taken from the osteological plates illustrating MM. GRANDIDIER and MILNE-EDWARDS's *Histoire Naturelle de Madagascar*.

[3] For muscular anatomy of Laridæ see GIEBEL, 'Beiträge zur Anatomie d. Möven,' &c., *Zeitschr. f. d. ges. Naturw.* x. (1857), p. 20; BEDDARD, 'A Contribution to the Knowledge of the Anatomy of *Rhynchops*,' *P. Z. S.* 1896, p. 299.

those of other limicoline birds, but are apt to be a little more complicated. In *Larus argentatus*, of which the tendons are represented in the annexed cut, the anterior stronger branch of the *brevis* tendon gives off a forwards and downwards slip to the extensors of the fore arm, from which arises the usual connection with the *longus* tendon. There is a second connection between the two tendons. At the origin of the

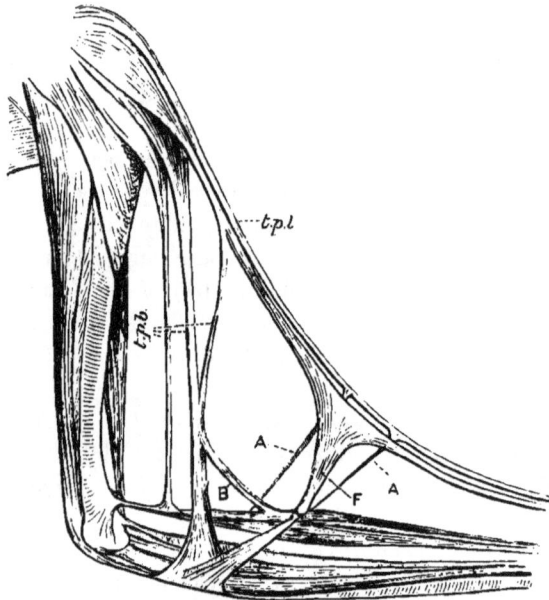

FIG. 171.—TENSORES PATAGII OF *Rhynchops* (AFTER BEDDARD).
t.p.l, tensor longus; *t.p.b*, tensor brevis; *B*, wristward slip; *F*, patagial fan;
A, tendons to ulnar side of arm.

patagial fan is a bony nodule, as in the petrel. The tensor longus tendon also gives off a slip (A, fig. 172) to the inside of fore arm which is also present in the auks (*q.v.*) In *Rhynchops* (fig. 171, A) there are two such tendons.

Lestris antarctica, *Sterna*, and *Larus marinus* (according to FÜRBRINGER's figure, Pl. xix.) are much the same, but are without the additional slips A and B. Of these A is present in *Rissa tridactyla*.

The *expansor secundariorum* is frequently absent, but it

is present in *Larus argentatus, marinus,* and *glaucus, not* in *fuscus.* It is absent in *Sterna* and *Lestris*; present in *Anous*; absent again in *Rhynchops.*

The *biceps* is peculiar in some members of this family. In *Larus, Sterna,* and *Anous* the *biceps* has, as usual, the two heads, humeral and coracoidal; but they form two perfectly distinct muscles, of which one, the coracoidal, soon divides into two distinct muscles again, which are inserted

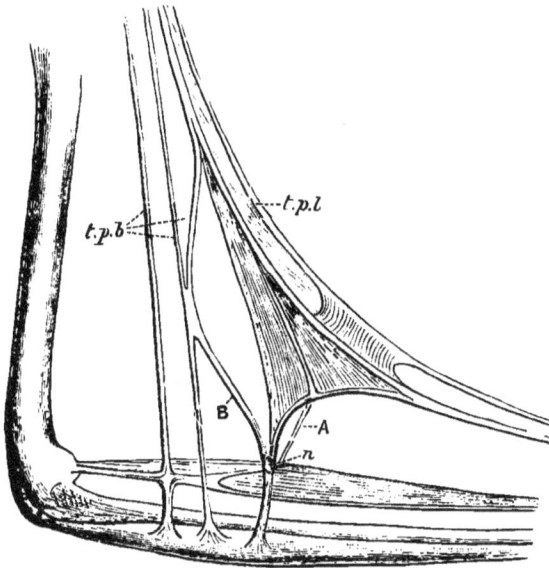

FIG. 172.—TENSORES PATAGII OF *Larus argentatus* (AFTER BEDDARD FROM FORBES).

n, osseous nodule. Other letters as in fig. 171.

respectively upon the radius and ulna. The biceps head goes to the radius, and as a rule gives off the biceps slip to the patagium; but in *Larus ridibundus,* according to FÜR-BRINGER, this slip arises from the coracoids, an anomaly observable also in the petrel, *Thalassiarche. Rhynchops* has no biceps slip at all.

The *anconæus* is generally attached by a tendon to the humerus.

A A

The *cæca* vary considerably in their development, as the following table of measurements shows :—

	Small Int.	Cæca	Large Int.
Larus argentatus .	56·25 (36)	·5 (·25)	3·25 (1)
„ *glaucus* .	66	·5	2·75
„ *ridibundus* .	26	·25	.
„ *Jamesoni* .	12·9	·12	1·75
Lestris pomatorhinus	30·5	3	3·5
„ *antarcticus*	38	3·4	2
„ *crepidatus* .	18·5	2·25	2·5
Sterna cantiaca .	24	·25	—
Rissa tridactyla .	25	·25	·75

The lobes of the liver are subequal in *Sterna, Anous, Gygis,* and *Rissa.* In *Larus* and *Lestris* the right is the larger. A *gall bladder* is always present.

In *Anous* the *cæca* are quite short; in *Gygis* long and charadriine.

The *syrinx* of this family is typically tracheo-bronchial (at any rate in *Larus*), with well-developed muscles, which always reach the bronchi. The family thus differs from other **Limicolæ** where there is a tendency towards a retrogression of the intrinsic muscles, sometimes culminating in actual disappearance.

In *Larus marinus* the last six or seven tracheal rings are rather narrower from above downwards than those which precede them, and are more or less firmly attached (except the last ring, which is incomplete both in front and behind) to form a box. When the syrinx is viewed from behind, a broad three-way piece is seen, into the formation of which the penultimate tracheal ring and the four or five in front of it appear to enter. This piece, however, is only really solid at the edges, the bars being a continuation of the penultimate tracheal rings. In the middle it is so thin as to be little more than a membrane. The first bronchial semi-ring (to which the intrinsic muscles are attached) is bow-shaped and in close contact with the last tracheal ring. The remaining semi-rings are narrower and run in a straight direction across the bronchi.

The membrana tympaniformis is distinguished by its

thinness and transparency from the thick yellowish membrane which unites the edges of the greater number of the bronchial semi-rings.

In *Larus fuscus* and *L. glaucus* the differences are but slight, and chiefly concern the greater solidity of the three-way piece.

In *Larus argentatus* the edges of the three-way piece are most solid; but they are connected by a series of four or five bars which divide up the central region of the three-way piece into alternate thicker and thinner portions. This is a peculiar specialisation of the three-way piece which I have not observed elsewhere. There is, furthermore, a thin bony curved rod, closely applied to the outside of the three-way piece, which arises from, or is at least connected with, the last tracheal ring.

Lestris antarcticus has a syrinx which is rather different from that of *Larus*, and which points in the direction of the Charadriidae, owing to the fact that the intrinsic muscles do not reach the bronchial semi-ring except as a fibrous band. The three-way piece, which is solid, is formed by two or three tracheal rings; there is no differentiation in membrane closing bronchial semi-rings.

It is evident from the foregoing account that, while there are a few differences between the various genera of Laridæ, *Rhynchops* is quite the most anomalous form in structure as well as in external appearance, as seen in its remarkable bill, with its scissor-like edge and projecting mandible. Still, the differences are, in my opinion, not sufficient to place *Rhynchops* in a family by itself opposite to the remaining Laridæ. It seems that the usually received division of the family, making an additional one for *Rhynchops*, will serve to divide naturally the Laridæ : we may term these divisions subfamilies. They will be thus characterised :

Subfamily 1. **Larinæ.**

> *Muscle formula of the leg, AXY+. Caeca rudimentary. Biceps slips and expansor secundariorum present.*

Subfamily II. **Stercorariinæ**.

> *Muscular formula, AXY+. Cœca long. Biceps slip present. Expansor secundariorum absent.*

Subfamily III. **Sterninæ**.

> *Muscular formula, ABXY+. Cœca rudimentary. Biceps slip present. Expansor secundariorum absent.*

Subfamily IV. **Rhynchopinæ**.

> *Muscular formula, ABXY—. Cœca rudimentary. Biceps slip and expansor secundariorum absent.*

Gygis and *Anous* require further investigation before they can be placed in this system; they are usually regarded as terns—by HOWARD SAUNDERS, for example. *Anous* has the complete muscle formula. *Gygis* has the formula of the gulls; but then the tern, *Sternula* (sp.), wants the accessory femoro-caudal, and in this approaches the gulls. *Anous* has the expansor secundariorum.

Of undoubted or reputed extinct limicolous birds a number have been described. If *Palæotringa* (with three species), from North America, is rightly referred by MARSH to this group, it goes back to Cretaceous times. *Milnea*, from French Miocene, is known by the humerus. *Elorius*, known by an imperfect coracoid, a tarso-metatarsus, and parts of the skull, seems to belong here. *Ægial-ornis* is considered by LYDEKKER to have been a gull-like bird, largely on account of the perforated first phalanx of the second digit.[1] *Halcyornis* has been described from the extremity of a humerus and the back of the skull.

—	Atlas	Cerv. Vert.	Coracoids	Sternum	Nostrils	Basipt. Pr.	Occipital Foramina	Muscle Formula of Leg
Charadriidæ .	Perf.	15	Distant	2-notched	Schiz.	+	+	A(B)XY+
Œdicnemidæ .	Notch	16	Overl.	2-notched	Holorh.	—	—	(A)BXY+
Parridæ .	Perf.	16	Overl.	1-notched	Schiz.	+	—	ABXY+
Chionididæ .	Notch	15	Distant	2-notched	Schiz.	—	—	ABXY+
Thinocoridæ .	Perf.	15	Distant	1-notched	Ps.-hol	—	—	ABXY+
Glareolidæ .		15, 16	Distant	2-notched	Ps.-hol.	—	—	ABXY+
Laridæ . .	Notch	15	In cont.	2-notched	Schiz.	—	—	A(B)XY(+)

[1] A character which we have seen (*supra*, p. 351, footnote) to be of no account in fixing affinities.

The foregoing table shows some of the principal points in which the several families of the **Limicolæ** differ from each other, and will afford a justification for the divisions adopted in the present work. Whatever is the relation between the other families, we can clear the ground by removing the Laridæ from competition for the basal place in the series. As was discovered by PARKER, the young of these birds have basipterygoid processes and occipital foramina, the persistence of which, therefore, in the Charadriidæ and Parridæ (basipterygoids only) places those two groups lower in the series than the more specialised gulls. That the gulls are rightly placed here, and therefore as rightly removed from a closer association with the **Alcæ**, can hardly be disputed. GADOW, who does the reverse in his scheme, enumerates only the following points in which the gulls differ from the **Limicolæ** :—

In the Laridæ—

Down feathers are thicker. Coracoids in contact. Hæmapophyses *mostly* (*not* in *Lestris*) wanting to the dorsal vertebræ. Hypotarsus simpler. In muscle formula of leg disappearance of B instead of Y. Webbed feet.

As a matter of fact, the crossing of the coracoids in *Œdicnemus* destroys the second of these, at best very slender, grounds, and, as GADOW admits, the webbing is almost as well developed in *Recurrirostra*.

On the other hand the differences from the **Alcæ** are more pronounced.

These latter birds have --

A much longer sternum.

Largely developed dorsal hæmapophyses, of which indications only are to be found in the gulls and in other **Limicolæ**.

The biceps slip is peculiar.

The leg muscles are always reduced, the formula being in *Phaleris* only AX − .

It may be mentioned in addition that the expansor secundariorum is always absent in the **Alcæ** and only sometimes in the Laridæ. The auks are, in fact, so far as we

can see, a further modification of the gull type, but further
from them than they are from the **Limicolæ**.

We may, therefore, probably regard the gulls as a slightly
modified offshoot of the typical **Limicolæ** (*Chionis*, perhaps,
being slightly intermediate), leading towards the **Alcæ**. The
arrangement *inter se* of the remaining families brings us to
the broader question of the affinities of the **Limicolæ** as a
group. It appears to me that the only other groups which
need be specially considered in this connection are the
Grues, **Otides**, and **Ralli**. It may be convenient to preface the
discussion with a tabular statement of the actual points of
likeness between these several groups.

	Limicolæ	Grues	Otides	Ralli
Cerv. vertebræ .	15–16	14–20	16–18	14, 15
Basipt. pr. . .	− or +	−	−	−
Nares . . .	Schiz. or hol.	Schiz. or hol.	Hol.	Hol.
Occip. foram. .	+ or −	+ or −		−
Sternum .	1- or 2-notched	1- or 2-notched	2-notched	1-notched
Leg muscles .	A(B)XY +	(AB)XY +	BXY +	ABXY +
Patagial fan .	+	+	+	−
Biceps slip .	+	+ or −	−	+

The very difficulty of finding any characters, greatly
noteworthy, in which the groups in question vary is an
index of how closely allied all four are. There can, to my mind,
be no doubt of their common origin. The **Limicolæ** on the
whole come nearest to the **Grues**, and especially to the true
cranes, whose distinctive characters are a little swamped in
the above table, owing to the aberrant cranes (e.g. *Rhinochetus*,
Psophia), whose peculiarities have naturally modified that
table. Taking this point of view, it seems to follow that of the
Limicolæ the most primitive section is that of the Charadriidæ proper; for it is among them that the forms with
the greatest number of points of resemblance to the cranes
occur. I should consider these birds to be slightly lower in
the scale than the cranes.

ALCÆ

Definition.—Oil gland tufted.   Aftershaft present ; aquincubital.
Skull schizognathous and schizorhinal.   Occipital fontanelles
present, but no basipterygoid processes.   Two carotids.[1]   Tensores patagii and biceps slip distinctive.   No expansor secundariorum.   Dorsal vertebræ opisthcocœlous.

This group of birds comes nearest to the **Limicolæ**, but
differs more from any of them than they do among themselves.   The group is entirely confined to the northern
hemisphere, and is mainly Arctic.

The *oil gland* is invariably tufted and the feathers have an
aftershaft.

In the *pterylosis* the dorsal tract divides between the
scapula, and there is (? universally) a well-marked spinal
space.   But there is no break between anterior and posterior
parts (as in **Limicolæ**).   The ventral tract, contrary to what
we find in the **Limicolæ** (incl. Laridæ), does not divide early
in the neck.   The *rectrices* (see table, later) vary in number
from twelve to sixteen.   The Great auk (*Alca impennis*) is
said to have possessed eighteen.

The *oil gland* has often many apertures ; there are only
two in *Brachyrhamphus marmoratus*, but four in *Lunda
cirrhata*, six in *Synthliborhamphus antiquus*, and eight in
Uria columba.[2]

The *skull* is schizorhinal [3] and schizognathous, with well-
marked occipital fontanelles [4] and impressions for supra-orbital
glands ; the latter nearly meet in the middle line, leaving but

[1] *Synthliborhamphus antiquus* has only one (the left).

[2] The remarkable shedding of the beak of the puffin (stated also to occur in
the penguin ; cf. *P. Z. S.* 1880, p. 2) has been described by Bureau in *Bull.
Soc. Zool. Fr.* ii. 1877, pp. 377, 432.   See also *ibid.* iv. 1879, p. 1, for the same
phenomenon in other auks.

[3] For osteology of auks see Owen, ' On *Alca impennis*,' *Tr. Z. S.* v. p. 317 ;
Shufeldt, *J. Anat. Phys.* vol. xxiii.; and Parker, ' On the Morphology of the
Duck Tribe (Anatidæ) and Auk Tribe (Alcidæ),' *Cunningham Memoirs R. Irish
Ac.* No. 6, 1890.

[4] These are sometimes obliterated with age.   I find them present in a young
Uria troile, absent in an old one.   In a specimen of *Fratercula arctica* there
was only one present.   They also may be present or absent in *Brachyrhamphus*
and *Synthliborhamphus.*

a thin median line. The interorbital septum is very imperfect. Basipterygoid processes are absent in the adult, but
present as rudiments in the young. The lacrymals are
firmly united with the prefrontals, as in gulls and **Limicolæ**.
The vomer is distinctly double in the young. *Alca torda*
has an os uncinatum. In *Uria troile* and in *Alca torda* and
A. impennis there is on each side a foramen at anterior end
of supra-orbital grooves, as in *Chionis*, &c. There are fifteen
cervical vertebræ ; the first few dorsal vertebræ have very
conspicuous hæmapophyses, bifid at their free ends. Seven

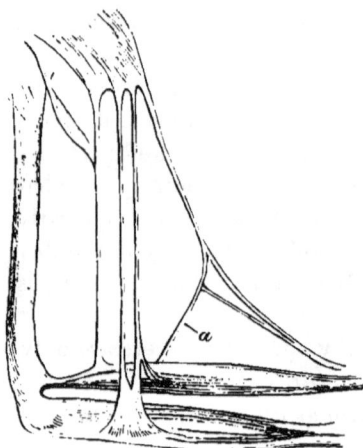

FIG. 173.—TENSORES PATAGII OF *Lunda cirrhata* (AFTER BEDDARD FROM FORBES).
a, slip to ulnar side of fore arm.

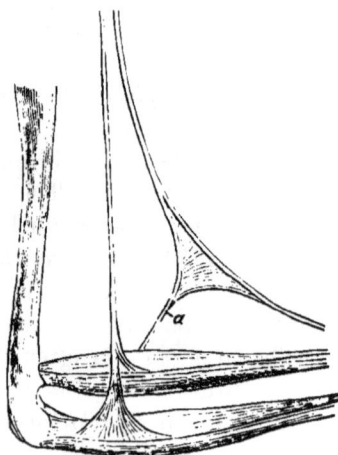

FIG. 174.—THE SAME OF *Synthliborhamphus antiquus*.

ribs reach the very long and narrow *sternum*, which has one
notch on each side posteriorly (*Uria*), or in addition a fenestra
on each side and a median notch (*Fratercula*) or no notches
at all (*Alca impennis*).

As to the *muscular anatomy* of the auks, the simplest
form of the *tensores patagii* tendons [1] is seen in *Alca torda*,
where there are two *brevis* tendons, both of which pass over

[1] For muscles of *Alca* (and *Spheniscus*) see A. CARLSSON, 'Beiträge zur
Kenntniss d. Anatomie d. Schwimmvögel,' *Bih. K. Svensk. Vet. Ak. Handl.*
ix. 1884, No. 3.

the extensors of the fore arm to be inserted on to the ulna ; the anterior of these bifurcates in the limicoline fashion just in front of its insertion, and here a very faintly marked fan [1] (FÜRBRINGER) connects it with the longus.

Synthliborhamphus antiquus is in some respects even more simple.[2]

There is but one brevis tendon, from which—just as it passes over the extensor of the fore arm—the merest apology for a forward branch exists ; from this branch arises a special muscular belly of the *extensor metacarpi* (cf. Petrels). There is no patagial fan, but a special slip, found in all the other members of the family (and also occurring in *Larus argentatus*), runs from the longus tendon to the opposite side of the fore arm to that upon which the patagial fan, when present, is inserted.

In *Brachyrhamphus marmoratus* three separate and parallel tendons arise

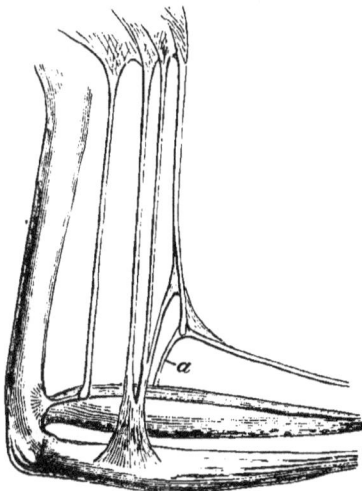

Fig. 175.—TENSORES PATAGII OF *Ceratorhina monocerata* (AFTER BEDDARD FROM FORBES).

from tensor patagii brevis muscle, of which the anterior is the strongest and alone passes to the ulna. From a small wristward slip arise a few fibres of the extensor metacarpi, as in the last species. There is no patagial fan, but an ulnar slip, which gives off a branch running back to the humerus. In *Uria columba* there is the same.

. In *Lunda cirrhata* there are but slight differences ; the two most anterior of the brevis tendons cross the extensor

---

[1] Not figured at all by GARROD in a MS. sketch.

[2] For various details in anatomy of soft parts see BEDDARD, ' On the Anatomy of a Grebe (*Æchmophorus major*), with Remarks upon the Classification of some of the Schizognathous Birds,' *P. Z. S.* 1896, p. 538.

muscles and reach ulna, but they cross it as a single diffuse band formed by their fusion.

Ceratorhina monocerata has a slight patagial fan, as well as ulnar slip ; otherwise it is like the last.

Fratercula arctica has only two *brevis* tendons, upon the anterior of which, at the origin of the patagial fan, is an ossicle.

The *biceps slip* is present in Alcidæ, but is generally, if not always, peculiar. Thus in *Alca torda* it is inserted partly on to patagial membrane, partly on to inner of two *brevis* tendons (*not longus* tendon).

In *Lunda cirrhata* it arises tendinously and joins inner of three patagialis brevis tendons. The biceps slip (as in some petrels) is all that is left of the humeral head of the biceps.

The biceps slip of *Fratercula arctica* is inserted on to middle of three tendons.

In *Brachyrhamphus marmoratus* the biceps slip ends in a long and fine tendinous thread, which is inserted on to the innermost of the three tendons of the brevis.

In *Phaleris psittacula* the biceps slip is firmly adherent to the single brevis tendon ; but from it just at the lower end of the line of the brevis tendon an obliquely running strand is found, which reaches the tendon of the longus.

The *biceps*, as already mentioned, consists merely of the coracoidal head, the humeral head being represented only by the biceps slip.

The muscle is not large, and in *Phaleris* its muscular belly is largely divided into two.

The humeral head of the *anconæus* seems to be nearly always present.[1]

There is no *expansor secundariorum*. The *deltoid* has, as a rule, no scapular slip, but there is one in *Uria*.

The muscles of the leg which are invariably present are the *femoro-caudal* and the *semitendinosus*. The accessory head of the latter is never present. The *ambiens* and the *accessory femoro-caudal* may be present, and, except in

---

[1] It is (? individually) absent in *Brachyrhamphus marmoratus*.

Phaleris psittacula, are never *both* absent; the formulæ are the three following :—

ABX— *Uria.*
AX+ *Ceratorhina.*
AX— *Phaleris.*

The *biceps femoris* of *Brachyrhamphus* gives off a fleshy slip to the outside of the thigh, to the *gastrocnemius* (cf. *Podica*). This slip is not to be found in *Phaleris*.

The least modified form of *syrinx* is seen in *Alca torda,*

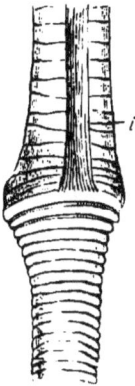

FIG. 176.—SYRINX OF *Lomvia troile* (AFTER BEDDARD).
i, intrinsic muscle.

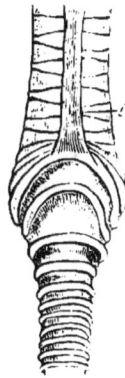

FIG. 177.—SYRINX OF *Ceratorhina monocerata* (AFTER BEDDARD).
i, intrinsic muscle.

Lomvia troile, Synthliborhamphus antiquus, and *Uria columba.*

In *Lomvia troile* the last four or five tracheal rings enter into the formation of a three-way piece, which is deeply excavate medianly. The intrinsic muscles are attached to the first bronchial semi-ring, which hardly differs from those that follow; it and the one which immediately succeeds are slightly more broad than the rest.

Alca torda is rather more gull-like, there being no marked depression in the pessulus medianly, and the transparent membrana tympaniformis being sharply marked off from a thicker yellowish region behind.

Synthliborhamphus antiquus has also a perfectly typical tracheo-bronchial syrinx, the first bronchial semi-ring (to which intrinsic muscles are fixed) being longer and deeper than those which follow. *Phaleris* is similar.

Uria columba distinctly differs from *Lomvia troile* (with which it is often considered to be congeneric), and is an approach towards a type of syrinx to be described immediately, but with certain peculiarities of its own.

A dozen tracheal rings in front of the last are very thin (more particularly in front), and have, therefore, wide membranous intervals. The last tracheal ring, however, is stout and ossified; it appears to be composed of two closely adjoined; posteriorly three rings enter into the formation of the tracheal box. The first bronchial semi-ring is very much arched, so much so that laterally it conceals the last tracheal rings. To it the intrinsic muscles are attached.

In *Ceratorhina monocerata* this state of affairs is exaggerated. Not only the first, to which muscles are attached, but the second bronchial semi-ring is very convex upwards, forming, indeed, the half of a rather elongated ellipse, as shown in the figure (fig. 177). The last two tracheal rings are ossified and closely connected. The last twelve tracheal rings are shallow vertically and leave considerable membranous interspaces.

In *Lunda cirrhata* there is an almost identical syrinx, but the last tracheal rings are not particularly thin in front.

Fratercula arctica and *F. corniculata* are sufficiently similar to need no special description.

The *tongue* is generally fleshy, elongated, triangular, and spiny only at the base.

The relative proportions of the *liver* lobes not only vary, but the absolute size of the organ varies greatly.

A *gall bladder* is always present. There is no crop.

The *cœca* are usually mere nipples, ·35–·25 inch in length, but in *Alca torda* one inch. The length of the *small intestine* in *Fratercula arctica* is 28·5 inches, in *Alca torda* 49·5 inches.

From the characters displayed in the accompanying

table it seems possible to divide the group into two families, Uriidæ and Fraterculidæ, which may be thus defined :—

Uriidæ.

> *Rectrices, twelve or fourteen. Lobes of liver equal, or left larger than right. Muscle formula, ABX— (or AX—).*

Fraterculidæ.

> *Rectrices, sixteen. Lobes of liver equal, or right larger than left. Muscle formula, AX+. Syrinx peculiar.*

—	Rectrices	Accessory Femoro-caudal	Ambiens	Liver Lobes	Syrinx
Uria columba . . .	14	Good	0		
Synthliborhamphus antiquus	14	Good	0	L>R	
Ceratorhina monocerata	16	0	Good	L=R	With U-shaped first bronchial semi-ring
Lunda cirrhata . . .	16	0	Good .	R L	With U-shaped first bronchial semi-ring
Brachyrhamphus marmoratus	14	Broad	0	L>R	
Phaleris psittacula . .	14	0	0	L=R	
Fratercula arctica . .	16	0	Slender	R>L	With U-shaped first bronchial semi-ring
Alca torda .	12	Present	0	L>R	
Uria troile .		Good	0	L>R	

There appears to me to be no doubt that the **Alcæ** are best placed in the neighbourhood of the **Limicolæ**, though, as FÜRBRINGER justly states, 'at first sight the relations between the two groups do not appear to be intimate.' These differences, however, merely concern outward form, in which it is perhaps reasonable to compare the auks with the grebes. But an anatomical study shows plainly that the grebes are much further away from the auks than are the **Limicolæ**. Such points of likeness as there are with the **Colymbi** are largely, if not entirely, due to the similar life ; thus the elongated sternum, which is also shared by the aquatic ducks, and possibly the muscle formula ABX+. With the **Limicolæ** are many positive points of likeness, to which no such explanation seems to be applicable. In the skull it is hard to find points of difference ; but the most remarkable point of similarity is the presence in both groups of those additional tendinous slips upon the patagium on the

ulnar side (in *Charadrius* as well as Laridæ) which have been
duly described in the foregoing pages. Nor are there any
salient facts, save such as are evidently associated with loss
of the power of flight, which contradict such a placing.

GRUES

Definition.— Oil gland present and tufted; [1] feathers with an aftershaft.
Rectrices, twelve. A quintocubital or quintocubital.[2] Ambiens,
semitendinosus, and accessory always present. Expansor secundariorum present. Cæca large.[3] Skull schizognathous, schizorhinal, without basipterygoid processes. Two carotids.

Among the typical cranes of the family **Gruidæ** I include
not only the nearly cosmopolitan *Grus* and the African
Balearica, but also the South American *Aramus*.

There are no particular remarks to be made about the
pterylosis, which NITZSCH states to be precisely like that of
Psophia (see below, p. 374).

The muscular system is fairly uniform in its characters, as
will be seen from the length of the above definition.

The *tensores patagii* of the demoiselle crane (*G. virgo*)
are furnished with a muscular *biceps slip*, which is reinforced
by a tendon springing from the biceps below the origin of the
biceps slip. There is also the usual fibrous junction with
the deltoid crest of the humerus.

From the pectoralis [1] springs a broad flat tendinous slip,
which joins the undivided tensor patagii. The tensor brevis
divides at once into two thin broad diffuse tendons, of which
the anterior sends forward a wristward slip, from whose
junction with extensor metacarpi a slight patagial fan pro-
ceeds to the longus tendon.

In *Grus leucogeranos* the tensor brevis tendon widens out
shortly after crossing biceps slip into a wide diffuse band,
composed of many strands, but not distinctly separable into
two or three tendons. There is a patagial fan.

[1] Except in *Mesites*, *Cariama*, and *Rhinochetus*.
[2] *Rhinochetus*, *Cariama*, *Psophia*.   [3] Not in *Eurypyga*.
[4] The *pectoralis I.* is usually stated to be single. It appeared to me to be
distinctly double in *Grus carunculatus* and in *Balearica pavonina*, especially
in the latter.

Aramus scolopaceus has the same thin diffused tendons; but they are distinctly divided below into a main tendon and a wristward slip. There is no patagial fan.

The *anconæus* has generally, if not always, a well-marked broad humeral slip.

The typical formula of the *leg muscles* for the cranes is ABXY+. This is the case with all the members of the genus *Grus*, excepting *G. leucogeranos*, where I could find neither A nor B. In *Aramus* and *Balearica pavonina* the formula is BXY+, and in *B. regulorum*, as in *G. leucogeranos*, XY+ only. In *G. americana* the femoro-caudal is minute and has but a feeble accessory.

The *deep flexor tendons* are united by a strong vinculum.

Both *peroneals* appear to be present; but the only notes at my disposal on this matter refer to *G. leucogeranos*.

The left lobe of the *liver* is much smaller in *B. pavonina*, a little smaller in *G. antigone* and *G. virgo*. The proportions are reversed in *Aramus*.

The *gall bladder* is present; there is a good *gizzard*; the *proventriculus* is zonary. The following are intestinal measurements:—

–	Small Int.	Large Int	Caeca
	Inches	Inches	Inches
Grus antigone .	69	3·5	7 and 8
„ *leucogeranos* .	78	6	6·5
„ *carunculata* .	80·5 (♂) 86 (.)	7 ♂ 4 ?	8 ♂ 5 ?
„ *americana* ♂ .	76	3·5	7·5
„ *canadensis* ♂ .	72	3·5	4·25
„ *australasiana* ♂	84 (73)	6 (3·5)	7·5 (7 and 8·25)
„ *virgo* . .	51·5	3	2·5 and 3·25
Balearica pavonina ?	54	2·5	6·8 and 6·1
„ *regulorum* ♂	64		5·5
Aramus scolopaceus .	40		2, 2¼

The intestinal coils in the crane tribe are very characteristic and quite unlike those of any other birds except the rails and bustards. The figure of *Cariama* shows the characters of the **Grues** generally and may be compared with that of *Crex* on p. 323.

The genus *Grus* has the most typical syrinx. In *G. leucogeranos* the first bronchial semi-rings are firmly attached to each other, and the first two are ossified and somewhat arched. To the first of these are inserted on each side the *two* flat, rather broad intrinsic muscles, which run side by side, and which appear to be continued by fibrous tissue on to the second semi-ring. There is a normal pessulus. The membrana tympaniformis gets narrower from above downwards (having, therefore, a triangular form), and finally ends opposite the thirteenth semi-ring ; but the rings remain semi-rings after this point, though their ends are very closely approximated, until close to their opening into the lung. *G. australasiana* shows no special differences. In *G. canadensis* the two muscles, though distinct above, appear to fuse below ; they do not quite reach the bronchial semi-ring as muscle, but are attached to it by a short ligamentous ending. *Grus carunculata* agrees with the last.

A peculiarity found in many cranes is the convoluted trachea.[1] This state of affairs is not found in *Balearica* or *Aramus*.

In both males and females of the following species the trachea is convoluted : *G. cinerea*, *G. antigone*, *G. carunculata*, and *G. leucogeranos*. The males of *G. australasiana* and *G. canadensis* are known to be the same, and the female of *G. americana*. In the female of *G. leucogeranos* and in the male of *G. carunculata* the trachea, though convoluted more or less, does not enter the substance of the sternum, as it does in the others. This too holds good for *Tetrapteryx* and *Anthropoides*.

The trachea has the usual pair of extrinsic muscles, which in *Balearica pavonina* arise not from the costal processes, as is the rule, but from the angle of the first rib.

I have myself examined syringes of the following species : *Grus canadensis*, *G. australasiana*, *G. leucogeranos*, *G. carunculata*, *Balearica pavonina*, and *B. regulorum*.

The syrinx of *Balearica* is rather different and less typical.

The two intrinsic muscles are present, but they end in a

<hr>

[1] See 'A Natural History of the Cranes,' by W. B. Tegetmeier, and Forbes, *P. Z. S.* 1882, p. 353.

fibrous band fourteen rings above the end of the trachea. The first tracheal ring is not so strongly modified as in *Grus*.

B. pavonina hardly differs.

There are nineteen *cervical vertebræ* in *G. carunculata*, twenty in *Balearica*.

Seven ribs reach the sternum in both. The *clavicles* in the former and in *Tetrapteryx* are ankylosed with the sternum, but not in *Balearica*. Some of the dorsal vertebræ are partly ankylosed.

The *skull* has occipital fontanelles, as in most charadriiform birds. This holds good also of the slightly aberrant *Aramus*. The impressions for the supra-orbital glands are slight, and largely concealed when viewed from above. The lacrymal bones do not blend with the ectethmoid. The interorbital septum is much fenestrated, but not so much so as in the rails. In *Tetrapteryx* and *Balearica* [1] the palatine bones do not appear to come into contact posteriorly, and at any rate the inner lamina is continued right to the end of the bone. This is not the case with *Grus*, where the bones do come into contact posteriorly and the inner laminæ are not continued to the end.

The *pelvis* of the typical cranes (*Grus*, *Balearica*, *Tetrapteryx*) hardly differs from that of such a rail as *Aramides*.

An outlying member of this group is usually included in the family **Rhinochetidæ**. This family is represented by but a single species, the kagu (*Rhinochetus jubatus*), of New Caledonia. The bird is not unlike a heron in appearance; but BARTLETT, who made a careful study [2] of the habits of specimens at the Zoological Society's Gardens, compared its quick active movements rather with those of a crane than with the slow motions of a heron. The anatomy of the bird has been chiefly studied by PARKER (osteology),[3]

[1] So too apparently in *Anthropoides stanleyanus* (PARKER, *Tr. Z. S.* x. pl. liv. fig. 6).

[2] *P. Z. S.* 1862, p. 218.

[3] 'On the Osteology of the Kagu,' *Zool. Trans.* vi. p. 501.

MURIE ('dermal and visceral structures'),[1] and myself (syrinx and muscular anatomy).[2]   Others, however, particularly FÜRBRINGER and GARROD, have contributed details of importance to our knowledge of this bird.

The *powder-down patches*, which were originally discovered by BARTLETT, exist as scattered groups of feathers of the kind; there are not the regular patches found in the

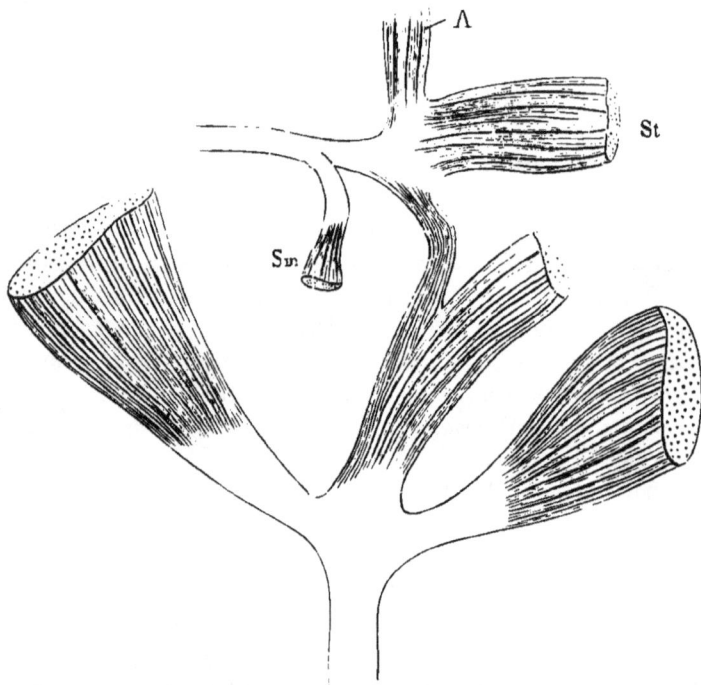

FIG. 178.—CERTAIN LEG MUSCLES OF *Rhinochetus* (AFTER BEDDARD).
St, semitendinosus; *A*, its accessory; *Sm*, semimembranosus.

near ally of *Rhinochetus, Mesites*.   The *oil gland* is present but nude.   The feathers have an *aftershaft*.   There are twelve rectrices.   The *pterylosis*, imperfectly described by

[1] 'On the Dermal and Visceral Structure of the Kagu,' *Zool. Trans.* vii. p. 465.

[2] 'Contributions to the Anatomy of the Kagu,' *P. Z. S.* 1891, p. 9.   See also W. MARSHALL, 'Quelques Observations sur la Splanchnologie de *Rhinochetus jubatus*,' *Arch. Néerl.* 1870, p. 402.

MURIE, has been rather more fully dealt with by FORBES.
The dorsal tract is double on the neck, and continues so
until its termination about on a level with the scapulæ.
The posterior portion of the dorsal tract is not continuous
with the anterior portions; it terminates with a slight bifurcation anteriorly and is widely dilated mesially.   The ventral
tract is broken into two by the intervention of powder-downs,
and the pectoral branch is perfectly separated from the
main tract, a unique feature, save for *Mesites*.   It is the
scattered powder-downs which are apparently responsible
for much of the breaking up of the pterylæ of *Rhinochetus*.[1]

. The *semitendinosus*, its *accessory*, the *femoro-caudal*, and
the *ambiens* are all present in the kagu.
As in *Psophia* and some other birds, the
semitendinosus is inserted in common with
the *semimembranosus*.   The relations of the
last muscles and of the *gastrocnemius* are
illustrated in the accompanying figure (fig.
178), which will explain itself.   Both *peroneals* are present, and have the typical
arrangement seen when both muscles are
developed.   The *deep plantar tendons* are as
shown in the figure (fig. 179).   The *flexor
hallucis* supplies the hallux alone, and is
tied to the *flexor communis* by a strong
vinculum before the trifurcation of the

FIG.   179.--DEEP
FLEXOR TENDONS
OF   *Rhinochetus*
(AFTER BEDDARD).

latter.   The mode of insertion of the *tensor patagii brevis*
is complicated; the tendon divides into three branches,
the two inner of which are prolonged some way beyond the
tendon of the extensor metacarpi radialis longi, to which
they are first of all attached.   There is a *biceps slip*.[2]
The *anconæus longus* has a flat tendon of origin from the
humerus, as well, of course, as its scapular head.   A muscle
apparently peculiar to *Rhinochetus* (see fig. 180) is what
has been termed by me an 'accessory biceps.'   This arises

---

[1] In his paper on *Mesites*, P. Z. S. 1882, p. 267.

[2] I wrongly asserted the absence of this in my paper upon the anatomy of
the bird.

from the humerus just below the insertion of the deltoid,
and is inserted near to the insertion of the biceps. The
expansor secundariorum is present.

The *syrinx* of *Rhinochetus* is tracheo-bronchial, and
presents us with no features of special interest. The
accompanying drawing (fig. 181) shows its lateral aspect.
The bronchidesmus is incomplete; the intrinsic muscles are
attached to the third bronchial semi-ring.

The number of *cervical vertebræ* is sixteen. Four of the

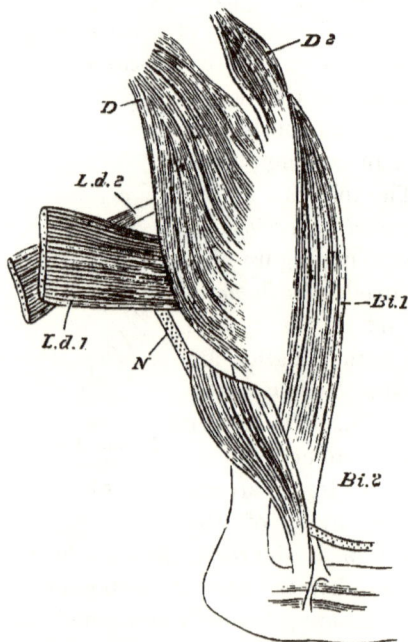

FIG. 180.—MUSCLES OF FORE LIMB OF
Rhinochetus (AFTER BEDDARD).
D, D2, deltoid; *Ld1, Ld2*, latissimus dorsi; *Bi1*, biceps;
Bi2, accessory biceps; *N*, nerve.

FIG. 181.—SYRINX OF
Rhinochetus (AFTER
BEDDARD).

last *dorsal vertebræ* are ankylosed. Five ribs articulate with
the sternum. The *sternum* is unnotched. The *skull* is
schizorhinal; there are no basipterygoid processes. There
is a partial bony internasal septum not to be found in the
cranes. The interorbital septum is more fenestrate than in

them. The palatines are abruptly truncated posteriorly, as in the herons.[1] There are small occipital foramina.

An abnormal member of the crane[2] group is the South American seriema, of which it is usually considered that there are two genera, *Cariama* and *Chunga*, of the family **Cariamidæ**. These birds agree with the cranes in possessing an aftershaft and in the number of their rectrices (twelve). The *oil gland*, however, is nude. In the *pterylosis* (which has been described by NITZSCH) there is a marked break between the posterior forks of the anterior section of the dorsal tracts and the anterior fork of the posterior section of the same tracts.

The dorsal tract is single on the neck and divides interscapularly. The posterior parts of the ventral tract are formed of two rows about two feathers wide. Each joins the outer branch above by one row of feathers merely.

The *skull* (fig. 182) is desmognathous, but the two maxillo-palatines, though they come into contact in the middle line, are not fused. The nasals are apparently holorhinal, really schizorhinal (see p. 144), and there are no basipterygoid processes. There are

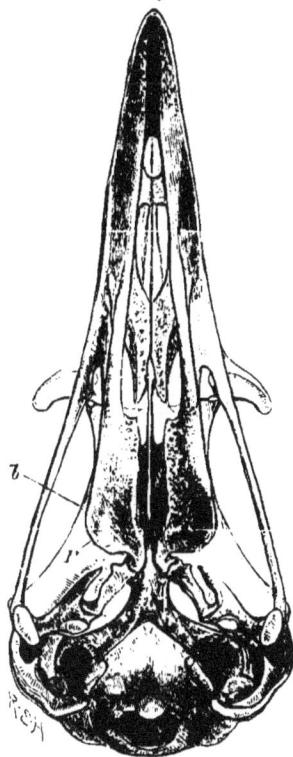

FIG. 182.—SKULL OF *Chunga*
VENTRAL VIEW. (AFTER BEDDARD.)
P, palatine ; *b*, supraorbital ridge.

fifteen *cervical vertebræ* ; five ribs articulate with the *sternum*, which is one-notched ; it has the spina externa.

[1] Stress has been laid upon this fact and comparison, but a posterior truncation of the palatines, nearly as marked, is to be seen in *Fratercula arctica* and *not* in some other auks.

[2] The skull is described by PARKER, *Tr. Linn. Soc.* (2), i. p. 128 ; the general osteology and to some extent the visceral anatomy by BURMEISTER,

The *muscle formula* of the leg is BXY+, as in bustards. The *accessory femoro-caudal* muscle (of *Chunga*) is peculiar in that it becomes reduced in the middle to a thin tendon, being muscular at both extremities. Both *peroneals* are present.

The *tensores patagii* spring from a single muscle. There is no *biceps slip*, another point of likeness to bustards. The brevis tendon spreads out into a broad aponeurosis, but there is no patagial fan. The *anconæus* has a humeral head. The intestinal measurements of the two birds are as follows :-

Cariama cristata.	*Chunga Burmeisteri.*
Small intestines, 33 inches.	33 inches.
Large intestine, 3 ,,	3·5 ,,
Cæca, 8·75 ,,	8·5 and 10·5 inches.

The family **Psophiidæ** is represented by the single South American genus *Psophia*, including some four species. These birds have the outer aspect of a rail rather than of a crane, and PARKER has commented upon their ' phasianine ' expression of face. Nevertheless their nearest alliance seems to be with the crane tribe, and perhaps more especially with *Cariama*.

The *pterylosis* has been described and figured by NITZSCH. There are apparently ten rectrices (not twelve, as NITZSCH stated), and the *oil gland*, as in *Grus*, is tufted. The dorsal tract is single on the neck and forms a strong bifurcation between the shoulders ; from the two ends of the fork a single row of feathers descend and unite to form a weakly feathered but widish posterior part of the dorsal tract. The ventral tract bifurcates early in the neck, and each in the pectoral region gives off a strong band on the outside ; the main portion of each tract is continued on to the cloaca by

' Beiträge z. Naturgeschichte der Seriema,' *Abhandl. nat. Ges. Halle*, i. (1851), p. 17 : the viscera also by GADOW, *J. f. O.* xxiv. (1876), p. 445, and by MARTIN, *P. Z. S.* 1836, p. 29. See also BEDDARD, ' On the Anatomy of Burmeister's Cariama,' *P. Z. S.* 1889, p. 594, and literature there quoted.

a very narrow band of feathers, which is only one feather wide to begin with, and afterwards only two feathers wide.

The *tensor patagii* muscles are distinctly gruine ; the *biceps slip* is present.

In the hind limb the *muscle formula* is BXY+, as in *Cariama* and the bustards.

The *syrinx*, shown in the accompanying woodcut (fig. 381), presents no remarkable features. It is quite typically tracheo-bronchial, and has, as will be observed, an incomplete bronchidesmus. It has been stated (by TRAIL) that the windpipe communicates with an air space, apparently after the fashion of the emu. But there is no doubt that this statement was based upon some imperfection of the example studied. It has been also stated that the windpipe in the male is convoluted ; this requires confirmation also.

The *skull* of *Psophia* [1] is schizognathous and holorhinal (fig. 81, p. 143). As PARKER first observed, the orbital margin is furnished with about five smallish supra-orbital bones, a feature which reminds us of certain archaic birds, as the tinamous, *Arboricola*, and *Menura*. The lacrymal has a descending process, which is swollen and nearly comes into contact with the ectethmoid. The maxillo-palatines are comparatively large and swollen bones ; as in *Cariama* these bones are convex on the outer side, and not concave— as in *Grus*. There are no occipital foramina. It may be remarked that the holorhinal nostrils of this bird show no such approach to schizorhiny as is displayed by *Chunga*.

From the anterior part of the maxillo-palatines, on a level with a point just in front of the commencement of the bony nostrils, a stoutish knob of bone [2] projects inwards on either side. Of this there are traces in the cranes, particularly in *Tetrapteryx*. If these processes were to be increased in size and to meet a bony internasal septum, we

FIG. 183.—SYRINX OF *Psophia leucoptera* (AFTER BEDDARD).

---

[1] F. E. BEDDARD, 'On the Structure of *Psophia*,' &c., *P. Z. S.* 1890, p. 329.
[2] Duly referred to by PARKER, 'Osteology of the Kagu,' *Tr. Z. S.* vi. p. 507.

should have the 'desmognathous' skull of the American vultures.

Psophia has seventeen *cervical vertebræ*, of which the

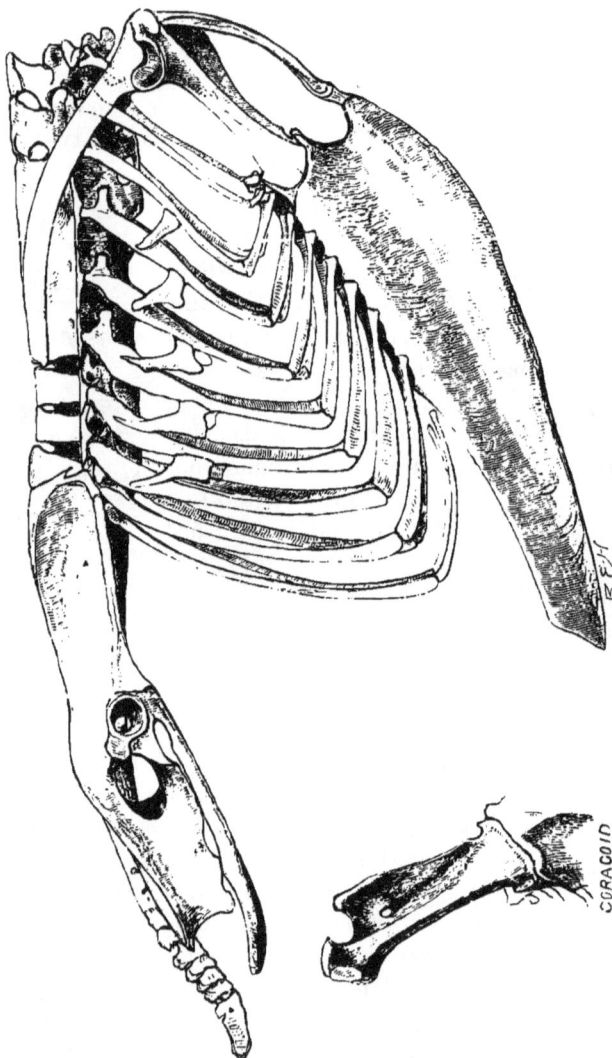

FIG. 184.— STERNUM, PELVIS, &c., OF *Psophia leucoptera* (AFTER BEDDARD).

last bears a rudimentary rib. Five *dorsal vertebræ* are ankylosed, there being two free ones behind. The *sternum* (fig. 184) is entire and unnotched; eight ribs articulate with it.

The *atlas* is notched for the odontoid process. From the fourteenth cervical vertebra to the third dorsal there are blade-like median hypapophyses. In front of the fourteenth the catapophyses nearly enclose a canal; they get further apart and die away anteriorly. The following table shows the number and character of the hypapophyses in various **Grues** :—

—	Chunga	Cariama	Psophia	Rhinochetus	Grus	Balearica
Catapophyses	Last on C11	C12	C13	C11	C15	C16
Hypapophyses	C12–D1	C13–D1	C14–D3	C12–D3	C16 C19	C17 C19

The family **Eurypygidæ** contains but one genus and species, *Eurypyga helias,* native of South America. It has an *oil gland,* which is generally nude but occasionally tufted, and twelve rectrices. *Eurypyga,* like *Rhinochetus* and *Mesites,* has powder-down patches, but their arrangement is very different from those of *Rhinochetus.* Dorsally there is on either side of the dorsal tract a compact dense triangular patch; in front it continues over scapula as a band which runs on to the sternal surface, and there forms a sparsely feathered patch more or less continuous with pectoral tract of contour feathers. There are a few scattered powder-downs on axilla and along neck.

The *tensor patagii brevis* is broad and rather diffused, stronger at the two edges; it sends off a wristward slip. The tensor longus is reinforced by a strong *biceps slip.*

The *expansor secundariorum* is strong and 'ciconiine.' The *anconæus* has a humeral attachment. The insertion of the *deltoid* extends halfway down the humerus. I have noticed in the *pectoralis primus* a vertical septum dividing the muscle into a right and left half.

The *muscle formula* of the hind limb is complete, *i.e.* ABXY+. The *glutæus I.* extends well over the biceps. Both *peronæals* are present.

The *liver* is equilobed, with a gall bladder. The intestines are 18 inches long, the short cæca (¾ inch) being 1½ inch from cloaca.

Both carotids are present.

The *skull* of the sun bittern has been described [1] and figured by PARKER.

It presents several points of likeness to that of *Rhinochetus*, notably in the ardeine character of the palatines,[2] which are cut off squarely behind and are of approximate length throughout; each palatine, moreover, has a fenestra, as in *Tigrisoma leucolophum* (and also in *Numenius phæopus* and *Anous stolidus*). The interorbital septum is widely fenestrate; there are no occipital foramina. As in *Psophia* (*q.v.*), *Rhinochetus*, and cranes, there is a rudimentary 'snag' from the anterior part of maxillo-palatine.

The nostrils are schizorhinal, and the curves of the various surfaces of the bones are such that if the very narrow anterior chink were closed a well-rounded and quite typical holorhinal skull would be the result.

Eurypyga has a one-notched sternum with well-developed spina externa. There are eighteen cervical vertebræ, and three dorsals are fused.

The pelvis is a little less rail-like than in *Grus*, *Rhinochetus*, *Psophia*, &c., in being wider, and in the more horizontal plane of ilia, which do not meet.

Family **Aptornithidæ**.[3] The two species of *Aptornis*, *A. defossor* and *A. otidiformis*, from New Zealand quaternary deposits, were originally referred to the Dinornithidæ, and more lately to the rails. FÜRBRINGER has, however, advanced certain reasons for relegating them to the neighbourhood of *Rhinochetus*, and I follow him in placing them in the present group. The chief reason which persuaded FÜRBRINGER to this conclusion was the schizorhinal nostrils, quite evident

---

[1] 'On the Osteology of the Kagu,' *Tr. Z. S.* vol. vi., and 'On Ægithognathous Birds,' *ibid.* vol. x. p. 307, pl. liv. figs. 7, 8, 9.

[2] See, however, footnote, p. 373.

[3] OWEN, 'On *Dinornis*,' pt. xv. *Tr. Z. S.* vii. p. 353.

in OWEN's plates,[1] and showing the inward curvature so often found in the schizorhinal nostril, and quite apparent in *Grus* (though not in *Rhinochetus*). A special point of resemblance to *Rhinochetus* among the crane-like birds seems to me to be in the partial ossification of the nasal septum. The solidity, posteriorly at any rate, of the interorbital septum is like *Psophia* so far as gruine birds are concerned, while the spout-like process upwards of the palatines is quite in harmony with FÜRBRINGER's views of the affinities of *Aptornis*. The junction of the zygoma with the post-frontal process is not crane-like ; it occurs among gallinaceous birds, and there is a near approach to it in *Otis*.

In the view given by OWEN of the under surface of the skull is a bone described, though not figured, which appears to me to correspond to the desmognathous palate of *Cariama*. The union of the bones and their divergence posteriorly are precisely like what is to be seen in *Cariama*.

Large basipterygoid processes are present, but OWEN failed to find upon them an articular surface.

So that while the outline of the skull of *Aptornis* is very like that of some of the large rails its affinities have been probably more correctly diagnosed by FÜRBRINGER.

Besides the Aptornithidæ already mentioned other forms referable to the **Grues** have been obtained from Tertiary strata. Of these *Aletornis* (with a number of species) is placed among the Gruidæ and *Geranopsis* of LYDEKKER. The latter is known only by the coracoid, which differs somewhat from that of *Grus*. LYDEKKER does not admit the genus *Palæogrus* of PORTIS.

The family **Mesitidæ** is represented by the Madagascar *Mesites*, a genus containing but a single species, which has been investigated anatomically by MILNE-EDWARDS[2] and by FORBES.[3] One of its principal characteristics was originally

---

[1] *Loc. cit.* pl. xl. figs. 1, 2.

[2] 'Remarques sur le Genre *Mesites*,' &c., *Ann. Sci. Nat.* (6), vii., and in *Hist. Nat. de Madagascar*.

[3] 'Description of the Pterylosis of *Mesites*,' &c., *P. Z. S.* 1882, p. 267.

discovered by E. BARTLETT,[1] who found and described briefly the powder-down patches.

The bird has sixteen rectrices, and apparently—but there is some little doubt about the matter—a nude *oil gland*. The contour feathers have no aftershaft. There are five pairs of *powder-down patches*. The most anterior pair lie in the interscapular region, and are enclosed by the dorsal tracts. The second pair are upon the rump, the third pair at the commencement of the pectoral region; the fourth pair lie also on the ventral region, but posteriorly; the fifth pair, finally, are axillary. The number of these pairs is greater than in any known bird, and their definition and complete separation as distinct patches contrasts with the diffused arrangement characteristic of *Rhinochetus* and *Eurypyga*.

There are four apteria on the neck, since both dorsal and ventral tracts divide early. The dorsal tracts converge interscapularly, and then become much feebler, and are continued on to the Y-shaped posterior part of the tract. The ventral tracts cease altogether at the commencement of the pectoral region, but recommence behind the powder-downs. The outer branch is present, but is quite unconnected with the main stem.

The *muscle formula* is complete, ABXY +. Both *carotids* are present.

As FÜRBRINGER removed *Aptornis* from the rails and placed it in the present group largely on account of its schizorhinal nostrils, it is remarkable that he did not also do so with the present bird. The bony nostrils are, in fact, of the type that has been termed pseudo-holorhinal. They are rounded at their end, but elongated and curved inwards; they are exactly like those of *Glareola*.

Mesites is schizognathous, with delicate maxillo-palatines. The descending process of the lacrymal abuts upon, but does not fuse with, the very stout square ectethmoid. This part of the skull, again, is more like *Glareola* than any gruine form; but it is also like *Pterocles* and various other birds.

There are seventeen *cervical vertebræ*, and four *ribs* reach

---

[1] 'Remarks on the Affinities of *Mesites*,' *P. Z. S.* 1877. p. 292.

the one-notched sternum. The dorsal vertebræ with complete ribs are ankylosed. The furcula is quite degenerate.

	Aftershaft	Oil Gland	Powder-downs	Fifth Remex	Nostrils	Occipital Fontan.	Supra-orbital Grooves	Interorbital Septum	Cervical Vertebræ	Dorsal Vertebræ fused	Sternum	Muscle Formula of Leg	Biceps Slip
Gruidæ .	+	Tuft	0	0	schiz.	+	+	Incompl.	19,20	2 ?(1)	Solid	ABXY+ BXY+ XY+	+
Cariamidæ .	+	Nude	0	+	schiz.	0	,0	Compl.	15	None	One incis.	BXY+ XY+	0
Rhinochetidæ.	+	N.	+	+	schiz.	R.	0	Incompl.	16	1 4	Sol.	AXY+	+
Eurypygidæ .	+	N., t.	+	0	Schiz.	0	0	Incompl.	18	1-3	One incis.	ABXY+	+
Mesitidæ	0	N.	+	?	schiz.	0	0	Incompl.	17	1-4	One incis.	ABXY+	?
Psophiidæ .	+	T.	0	+	Hol.	0	0	Compl.	17	2-4	Sol.	BXY+	+
Ralli .	+,0	T.	0	+,0	Hol.	0	0	Incompl.	15,14	None	One incis.	ABXY+	+
Otides .	+	0	0	0	Hol.	0	0	Incompl.	16,17	None	Two incis.	BXY+	0
Œdicnemidæ.	+	T.	0	0	Hol.	0	0	Incompl.	16	None	Two incis.	ABXY+ BXY+	+

¹ Not completely fused, and not always.

It is clear from the accompanying table that the seven families which I here include with the **Grues** are a tolerably divergent series of birds. Yet it does not appear to me possible to locate any one of them elsewhere. The bird concerning whose position I am most doubtful is naturally *Mesites*. It is placed near the hemipodes among the gallinaceous birds by GADOW, and by FÜRBRINGER near the hemipodes but among the rails. SHARPE takes the view which is urged here, while FORBES and some other recent writers are impressed by its likenesses to *Eurypyga* and *Rhinochetus*, FORBES, indeed, having associated all three in a separate group. There are, unfortunately, so many lacunæ in our knowledge of this form that a strict comparison is as yet hardly possible. Allowing the characters of the deep flexor tendons, not mentioned by FÜRBRINGER, the agreements with the **Turnices** do not appear to me to be more numerous than those with *Eurypyga*, while the powder-down patches are unknown in either rail or gallinaceous bird.

There are difficulties too with other genera of crane-like

birds.   For PARKER the inclusion of *Cariama* in the present
group is impossible.   Yet I find myself in this matter in
accord with most recent writers.   If *Cariama* is not allied
to the cranes, where are we to put it ?   The only alternative
seems to be the rails, which are in any case not far removed
from the present group.   MITCHELL has pointed out the
very close resemblance in the intestinal tract of cranes, rails,
Cariama, Otis, and (as yet unpublished) *Psophia.*   PARKER,
on the other hand, has emphasised the likeness between
Cariama and the **Accipitres**, a likeness which has even impressed itself upon their external physiognomy.   FORBES
went so far as to include in the same group with *Cariama*
the secretary bird.[1]   It appears to me, in fact, that the
origin of the **Accipitres** is to be traced to some crane-like form,
and that the very varied characters of the **Grues** point to
their being a basal form connected with more groups than
one.   As is pointed out elsewhere, unless we are to regard
the **Accipitres** as here treated as diphyletic, or even triphy-
letic, we must assume that the Cathartæ and *Serpentarius*
are the most primitive forms whence the typical Falconidæ
have been derived by a loss of two muscles of the leg.   In
these birds the formula BXY or AXY must come before A
alone.   But little change is required to convert *Cariama* into
Gypogeranus; and if it be objected that this is because
Cariama is an accipitrine, the quite similar skull, so far as
the desmognathism is concerned, of *Aptornis* may be pointed
to ; it will hardly be contended that *Aptornis* is anything but
a crane or a rail.   Along another line probably the peculiar
desmognathism of the American vultures may be derived
from conditions obtaining in the crane group.   PARKER has
pointed out the ossification and fusion in the middle line of
alinasals in *Cariama* ; but in that bird the fused bones, though
quite analogous with, cannot be the exact homologues of,
the fusion of bones which has occurred in *Cathartes*, &c. ;
for the position is totally different, being much further
forward in *Cariama*.   There is, however, in *Psophia* and in
others of its allies - particularly well developed in *Psophia*—

[1] 'A raptorial isomorph of the Cranes ' (PARKER).

a small snag of bone, already referred to as projecting out-wards from the maxillo-palatine in precisely the spot required ; if this projection were to grow more extensively, we should have a palate exactly like that of *Cathartes* and *Gypagus*.

To consider another quite different group of birds, the **Herodiones**, I have urged on another page (p. 441) the low position of *Platalea* among the **Herodiones**. In more than one particular these birds recall the cranes. The occipital fontanelles, the complete muscle formula are among these characters. FÜRBRINGER has hinted at, but not accepted, the diphyletic origin of the **Herodiones** ; were it not for *Scopus* and *Balæniceps*, this might be fairly assumed indeed ; and in view of this possibility the strongly ardeine character of the palatines in *Rhinochetus* and *Eurypyga* may be commented upon.[1] This character, too, is coupled with the existence of powder-down patches.

Though the charadriiform birds form, in my opinion, a group distinct from the **Grues**, there are many points of simi-larity between them. This is practically shown by the fact that *Œdicnemus* has been much bandied about between the two groups. The occipital fontanelles, the supra-orbital impressions are among the points of likeness. In view of these facts the likeness in some matters of the skull of *Mesites* to that of *Glareola* is perhaps significant. It is really mainly the form of the intestinal coils as described by MITCHELL that leads me to dwell upon the separateness of the two groups. But this question is dealt with more fully under **Limicolæ** (on p. 358).

As an appendix to the **Grues** we may perhaps consider the

STEREORNITHES

This singular group of birds was originally discovered in the lower Tertiary strata of Patagonia by the well-known palæontologist AMEGHINO. Of the several genera assigned to the family, many of which are very imperfectly known

---

[1] See, however, p. 373 footnote, where a similar conformation of these bones in the auk, *Fratercula*, is described.

and doubtfully referable to the group,[1] *Phororhacos* is the
best known, almost the entire skeleton being now in the
museum at Buenos Ayres. A convenient summary of AME-
GHINO's work, with criticisms, has lately appeared in the 'Ibis,'[2]
by Mr. ANDREWS, from the whose paper the information here
has been extracted. When first discovered (the anterior portion of the lower jaw) the creature was referred to the mammalia. As with the leg bone of the *Dinornis* its ornithic
nature was doubted, though the lesson afforded by the
Dinornis might have tempted naturalists to be bolder than
was perhaps reasonable in the forties. The bird was formerly

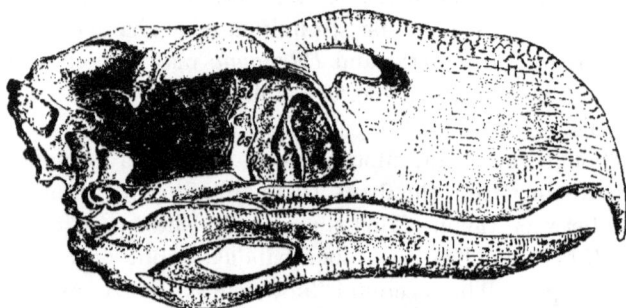

FIG. 185.—SKULL OF *Phororhacos*. LATERAL ASPECT.
(AFTER ANDREWS.)

held to be ratite. But LYDEKKER, who examined the quadrate, found that, as in carinates, it articulated by two heads
instead of the one which characterises that bone in the
ratites.[3]

The skull of *Phororhacos longissimus* is two feet in
length ; the hooked beak has two or three serrations at the
commencement of the hook, which remind one of the Eocene
Odontopteryx, in which bird, however (see p. 418), the serrations extend along the entire length of both jaws. Seen
from above certain resemblances between this skull and that
of *Phalacrocorax* or *Plotus* will be apparent. ANDREWS
thinks that the lacrymal was united with the jugal by a

[1] E.g. *Brontornis, Dryornis, Pelecyornis*.
[2] Jan. 1896.   [3] Except *Apteryx*.

separate bone which exists in *Cariama* (and of course
Chunga), a bone which is found in other birds, though its
articulation with the palatines often makes its identification
with that of the seriema doubtful.

The ventral surface of the skull is crane-like.[1]

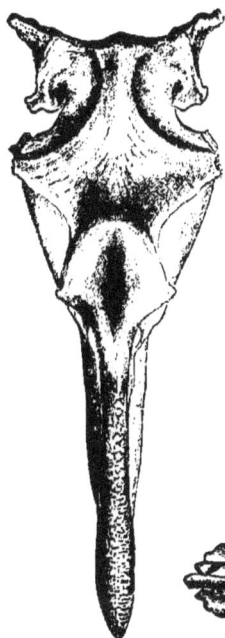

The vertebral column is interesting on account of the
fact that many of the dorsal and all
of the caudals have their centra perforated for the remains of the notochord. There is no pygostyle. The
pelvis is remarkably like that of *Hesperornis* and the grebes. It has the
length and narrowness of that limb
girdle in the birds mentioned. It
has, however, been more satisfactorily perhaps compared with the
pelvis of *Cariama*. The slenderness
and length of the coracoid, together
with the absence of a procoracoid,
prevent a comparison in this particular with the ratites. It is typically

Fig. 186.—Skull of *Pho-
rorhacos*. Dorsal Aspect.
(After Andrews.)

Fig. 187.— Pelvis of *Phororhacos*. Dorsal
Aspect. (After Andrews.)

carinate in fact, and shows perhaps special resemblances to
Cariama.

The ulna shows well-marked tubercles, which indicate the
insertion of the secondaries. Though reduced in size, and
therefore possibly useless for purposes of flight, it seems

[1] See letter from Dr. Gadow, *Ibis*, Oct. 1896, p. 586, where it is pointed out
that *Stereornis* is synonymous with *Phororhacos*, and that therefore the name
Stereornithes cannot stand. It is, however, difficult to compose a name out of
Phororhacos, and it is not proved that the birds in question are definitely
cranes. I therefore leave the name.

likely that the wings of this bird were more efficient than those of the ratites on account of the apparently well-developed remiges.

COLYMBI

Definition.—**Oil gland tufted. Aftershaft present. Aquincubital. Accessory semitendinosus absent. Biceps slip present. Glutæus maximus large, extending behind acetabulum. Cœca long. Skull holorhinal, without basipterygoid processes, schizognathous. Tibial crest strongly developed.**

This group of birds contains two very well-marked families, the divers (Colymbidæ) and the grebes (Podicipedidæ). In view of their numerous and important points of similarity I have not thought it desirable to separate these two families quite so widely as has GADOW. The Colymbidæ contain but one genus, *Colymbus*, with four species. Of the grebes there is perhaps also only one well-marked genus, *Podicipes*. But the Central American *Centropelma* (of SCLATER and SALVIN) has some claims, on account of the complete loss of flight, to generic distinction, while in the course of the following pages it will be seen that there are certain, if small, reasons for distinguishing *Æchmophorus* and *Tachybaptes*. *Podilymbus*, another alleged genus, has not been dissected.

Our knowledge of the anatomy of this group of birds is chiefly due to NITZSCH,[1] BRANDT,[2] COUES,[3] SHUFELDT,[4] and myself.[5]

As to *external characters*, there is a close agreement among the **Colymbi**. The number of *rectrices* in *Colymbus glacialis* I find to be twenty. NITZSCH gives eighteen to twenty for the genus. Specialised rectrices are not recognisable among the grebes. The inferior tract of feathers is

---

[1] *Loc. cit.*

[2] 'Beiträge z. Kenntniss d. Naturg. d. Vögel,' *Mem. Ac. St. Petersb.* 1840, p. 197.

[3] 'On the Osteology of *Colymbus*,' &c., *Mem. Bost. Soc. Nat. Hist.* i. 1866, p. 131.

[4] 'Concerning the Taxonomy of the N. American Pygopodes,' &c., *J. Anat. Phys.* 1892, p. 198.

[5] 'Notes upon the Anatomy of a Grebe (*Æchmophorus major*),' &c., *P. Z. S.* 1896, p. 538.

divided into two about halfway down the neck; the two tracts are not again divided upon the trunk, where they are broad. The spinal tract divides high up on the neck (*Podiceps cristatus*) or only between the shoulders (*Colymbus glacialis*). The anterior part of the spinal tract is stronger than the posterior part, and is separated from it; the latter is solid, enclosing no space.

The *patagial tendons* of *Colymbus glacialis* are rather simple. The tendon of the brevis is a rather broad undivided band. There is no patagial fan, but, as in *Æchmophorus*, a delicate tendon arises from the fore arm near the insertion of the tensor brevis and ends in the biceps slip. In *Colymbus arcticus*, according to FÜRBRINGER's sketch, a broad diffuse tendon arises from biceps slip, and ends freely upon the patagium. The first description given is from my own dissection of *C. glacialis*, and agrees with a manuscript note of Mr. FORBES upon *C. septentrionalis*. But *C. glacialis* apparently sometimes approaches *C. arcticus*. I have a manuscript sketch by Professor GARROD showing a broad band of fibres arising from the biceps slip, but ending on the fore arm.

The *anconæus* has a humeral head. The *expansor secundariorum* appears to be absent. The *biceps* is single-headed.

The *leg muscle formula*, in contradistinction to what we find in the grebes (where it is BX −), is ABX +.

There is a peculiarity in the *gastrocnemius* of *C. septentrionalis* which is worth calling attention to: one of the heads of that muscle arises from the tendinous end of the glutæus maximus.

The *femorocaudal* is tendinous for half its length; its accessory is a very large muscle. The *ambiens* in *Colymbus glacialis* has two heads of origin.

The combined plantar tendons give off a small slip to the hallux in *C. septentrionalis*, as in *Podicipes minor*.

The divers have two *carotids*, the more modified (cf. leg formula) grebes only the left.

The lobes of the *liver* in the Colymbi are equal, and there is a *gall bladder*.

The following are a few intestinal measurements:—

	S. I.		L. I.	Cæca
	Ft.	Ins.	Inches	Inches
Colymbus septentrionalis	3	11	3·7	1·7, 1·8
„ glacialis .	5	4	2·5	2
Podicipes cornutus .	2	5	1·5	1·1, 1·6
„ minor . .	2	5	1·5	1·5

The *cæca* of *C. septentrionalis* are conical and saccular in form, with irregular but distinct transverse rugæ.

The *syrinx* in the divers (*Colymbus septentrionalis*) is not in any way remarkable in form. The last tracheal rings and

FIG. 188.—SYRINX OF *Æchmophorus*.
i, intrinsic muscles.

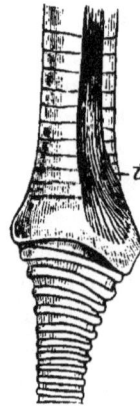

FIG. 189.—SYRINX OF *Tachybaptes*.
i, intrinsic muscles.

the first bronchial are ossified and firmly attached to each other. The pessulus is also ossified. The succeeding bronchial rings are soft and cartilaginous. The powerful intrinsic muscles are inserted partly on to the last tracheal ring and partly on to the first bronchial, the line of insertion being oblique.

The syrinx of *Æchmophorus* (fig. 188) has a very incomplete bronchidesmus, a very wide space between the two bronchi existing above its anterior edge. The last two tracheal rings are fused to form a long box, into the composition of which it appears to me that the first bronchial semiring enters. In any case, if that be not so, the first bronchial semi-ring has the unusual relations shown in the drawing, which are perfectly consistent with the belief that the ring is the second bronchial. The intrinsic muscles are attached to the third tracheal ring in front of the tracheo-bronchial box. The bronchial semi-rings are fairly ossified, but have rather wide membranous interspaces.

In *Podicipes cristatus* there is the same failure of the intrinsic syringeal muscles to reach even the end of the trachea. A box is formed by fusion at the end of the trachea, into which it appears to me the first bronchial semi-ring does not enter. The bronchial semi-rings are deeper and closer together, and the whole bronchus is more ossified, than in the last genus. The bronchi, too, are longer.

In *Podicipes cornutus* the syrinx is much the same, but of course smaller. The first free semi-ring of the bronchus seems to be No. 2. There is a wider membranous interval between it and the antecedent tracheo-bronchial box than in the last species.

Tachybaptes fluviatilis (fig. 189, p. 388) has a different syrinx. The last three tracheal rings are only fused in front, though they are closely united laterally. These rings are much ossified. The insertion of the intrinsic muscles is remarkable. They run obliquely forward, converging, to be inserted into the last three tracheal rings. The first bronchial semi-ring is arched, and ossified in front, where it is fused with the tracheal box ; otherwise it and the succeeding rings are cartilaginous. It is clear, therefore, that the syringeal characters justify the generic distinction here adopted.

The *cervical vertebræ* are more numerous in the grebes than in the divers; they are only fourteen or fifteen in the latter, twenty-one in *Æchmophorus* and *Podicipes cornutus*. Cervical vertebræ 10 to 16 in *Æchmophorus*[1] have a catapophysial canal; on 17 and 18 the hypapophyses are blade-like and enormous ; these processes extend to the very end of the dorsal series. The last cervical and the first three dorsals are fused. There is no catapophysial

[1] A canal is nearly formed in *Podiceps*.

canal in *Colymbus* ; but the hypapophyses are greatly developed, and in the dorsal region Y-shaped, with widely divergent flattened and expanded limbs. None of the dorsal
vertebræ are fused. Six ribs reach the sternum in the two
grebes mentioned ; eight or nine in the divers. The *sternum*
is one-notched ; in the grebes it has in addition a median
triangular notch. There is no anterior spine to the sternum
in the grebes. The procoracoid is moderately large and
hooked in the divers, absent in grebes. The *pelvis* is very
elongated and compressed, as in *Hesperornis*, but the ischia
are not free from the ilia, as in that bird.

In the skeleton of the leg the most conspicuous feature
is the highly developed cnemial crest. The patella too is
very large in the grebes, but not in Colymbidæ, where, indeed,
it is not ossified. In these particulars the grebes resemble
Hesperornis. The principal *skull* characters have been
already mentioned in the definition of the group. In addition
to these matters the strongly marked temporal fossæ may be
mentioned, which nearly meet on the upper surface of the
skull. They are not so well marked in the small *Podiceps
minor*. In the divers are strongly marked furrows for the
supra-orbital glands ; this is not the case with the grebes.
The former have also a single median occipital foramen
above the foramen magnum. The ectethmoids exhibit the
bullate form so characteristic of the **Anseres**.

It may be useful to state in a tabular form the principal
characters of the grebes and divers.[1]

	Muscle Formula	Carotids	Cerv. Vert.	Dors. Vert.	No. of Compl. Ribs	Semi-membra-nosus
Grebes	BX –	1	21	Ankylosed	4–6	–
Divers	ABX +	2	15	Not ank.	8, 9	+

These characters appear distinctly to point to the more
modified structure of the grebes. There are reasons (p. 165)
for regarding a small number of cervical vertebræ and a

[1] The extinct *Colymboides* (*C. minutus*, M.-ED. ; *C. anglicus*, LYDEKKER), of
which the former is known by a humerus, the latter by a coracoid, and possibly
a piece of sternum, is said to combine the characters of grebes and divers.

large number of complete ribs as more archaic characters than the reverse ; the other features as evidence of degeneration require no comment.

In considering, then, the affinities of the **Colymbi** the Colymbidæ are to be chiefly taken into account. But any comparisons bristle with difficulties. The late Mr. FORBES in his final scheme of classification definitely associated the **Colymbi** with the Heliornithidæ ; in this course I supported him by reason of certain muscular characters of both groups of birds. The outward appearance, too, of both birds is not at variance with such an affinity. The muscular formula of the leg is the very unusual one of ABX+. This reduced formula is found in *Podoa* and in the **Sphenisci**, Anatidæ, **Tubinares**, *Phalacrocorax*, and certain Alcidæ. It is to be noted that all of these are at least largely aquatic in their habits, a fact which must, of course, discount the value of the character ; but still ABX is an unusual formula, and there are other grounds for regarding the birds mentioned (with the exception, perhaps, of the Alcidæ) as having some relation to each other. The insertion of the biceps slip on to the patagium allies *Colymbus* and *Podoa* ; though this also occurs in other birds, it is, again, in some that are presumably not far from the **Colymbi** : for example, of the forty characters selected by GADOW for the comparison of the various groups of birds the Colymbidæ agree with the **Sphenisci** in twenty-eight ; with the **Tubinares** in twenty-seven ; with the **Steganopodes** in twenty-six ; with the Anatidæ in twenty-three ; with the Heliornithidæ in twenty-three.

On the other hand there are twenty-five points of likeness to the Laridæ, as deduced from the same tables, and no less than twenty-nine to the **Alcæ**. As might indeed be imagined, it would be rash to lay much stress upon the proportions of these numbers. It is not to my mind clear with which of the groups mentioned the **Colymbi** are most nearly allied. Their undoubted relationship to the **Hesperornithes** is treated of on another page (p. 395) ; and it is perhaps this very fact which prevents us from detecting likenesses to more modern (?) groups.

HESPERORNITHES

Definition. **Large extinct birds of a diver-like form. Skull holorhinal, with supra-orbital impressions. Vomers paired. Quadrate single-headed. Sternum without keel. Shoulder girdle platy-coracoidal ; clavicles fully developed. Fore limb represented by humerus only.**

The principal source of information concerning the remarkable genus *Hesperornis,* from the Cretaceous of North America (to which three species, viz. *H. regalis, H. crassipes,* and *H. gracilis,* are assigned), is naturally the magnificent treatise on this bird (and on *Ichthyornis*) by MARSH.[1]

The bird stood about six feet high, and presented the general form of the diver, to which bird it is now held to come nearest. MARSH, however, compared it more especially with the **Struthiones**, and it is spoken of by him and by others as an ' aquatic ostrich.' The former view now holds the field.

There are, nevertheless, various struthious features in the skull, and in other parts of the skeleton, some of which may be held to be due to its loss of the power of flight ; others are not so explicable from the point of view of our actual knowledge of bird structure.

The *skull* has the general contours of that of the diver, and, like that bird, has very well-marked furrows for the supra-orbital glands. The nasal bones produce a holorhinal nostril. The figure given by MARSH of the upper surface of the skull is a little suggestive of there having been an appearance of the ossified ethmoid on the surface of the skull, as in some struthious birds and tinamous. The interorbital septum is fairly ossified with a large foramen. The lacrymal is large with a large descending process, which nearly, if not quite, reaches the jugal bar. The pitting of the premaxillaries led MARSH to the inference that the beak was present, in addition to the teeth, which will be presently referred to.

The skull has been described as ' saurognathous,' on

[1] *Odontornithes.* Washington, 1880.

account of the paired vomers. These were, however, not
particularly like the presumed vomers of the woodpeckers
(see p. 187). Each bone is broad behind, where it may, as
in struthious birds, have articulated with the pterygoids, and
tapers in front to almost a point. It is not clear whether
the bird really had, as Marsh is disposed to infer, a 'dromæognathous' palate. In any case the basipterygoid processes
are present, and the articulations on the pterygoids are
towards the posterior end of these bones, as in the skull of
the **Struthiones**. The palatines are longish bones, and are
compared to those of the ostrich. They taper in front. As
in most **Struthiones** the articular head of the quadrate, for
articulation with the skull, is not divided into two facets.
The rami of the lower jaw do not appear to have been
ankylosed together, but to have been connected by a possibly
merely chondrified or ligamentous tract, which would have
allowed a gaping of the mandibles—seen partly in the
pelican, and obviously useful to a fish-eating bird, as we
may presume *Hesperornis* to have been. Both lower and
upper jaws have *teeth*, which are implanted in a continuous
groove, widened at the implantation of each tooth. In the
upper jaw the teeth are limited to the maxillæ, and there
were fourteen to each maxilla. The lower jaw had teeth
along its entire length, and the number given is thirty-
three.

The *vertebræ* are saddle-shaped. The number of cervical
vertebræ is seventeen. The entire vertebral column consisted of forty-nine vertebræ. None are ankylosed, except, of
course, the sacral series, and some at the end of the tail.
Pneumatic openings were not discoverable in any of the
vertebræ. The *atlas* has not been found. In no vertebra
do the catapophyses seem to have united to form a ventral
canal, a state of affairs which is occasionally met with in
the **Struthiones** and is characteristic of the **Herodiones** and
Steganopodes. The fourteenth cervical is extraordinary by
reason of the enormous size of the two catapophyses, which
are approximated and nearly parallel. The following vertebra
has the first median hypapophysis, which is bifid at the free

tip; the hypapophyses die away on the fourth dorsal vertebra.

With regard to the caudal vertebræ, the most remarkable fact is that while there is no pygostyle there is no rudimentary state of affairs observable; for the last few vertebræ have greatly expanded transverse processes, which would not move independently. MARSH thinks that 'the end of the tail would move mainly as a whole. This would give great power, similar to that in the beaver's tail or the flexible blade of an oar.'

As to the *shoulder girdle*, the *scapula* is in the same straight line, or nearly so, with the coracoid; it belongs, in fact, to what FÜRBRINGER has termed the 'platycoracoidal' type, seen also in the **Struthiones**. But, contrary to what we find in those birds, there is not a fusion between the coracoids and scapula in *Hesperornis*, and there is, moreover, a complete pair of clavicles. The *coracoid* has a strong procoracoidal process, and also a supra-coracoidal foramen, as in various **Struthiones**.

The two coracoids are widely removed at their articulation with the sternum. The *clavicles*, though complete, appear to have joined each other ventrally by a joint; they arise from the procoracoid. The *sternum*, which has articular surfaces for five ribs, has no keel. It is notched in the middle line posteriorly, and is wider in front than behind. The elongated form of the sternum is compared by MARSH to that of *Uria*. The ribs have uncinate processes.

The *fore limb* of *Hesperornis* appears to have consisted only of the humerus, as no other bones were discovered, and as there were no distal facets for articulation with a radius and ulna.

The *pelvis* of *Hesperornis* in its general form resembles that of *Podiceps*. The constituent bones are, however, entirely free distally. The acetabulum is closed by bone, a state of affairs only seen in the emu and to some extent in *Tinamus*.

The ischium has no processes tending upwards to the ilium and downwards to the pubis, as in **Struthiones**.

The prepubic process is large, but not larger than in certain recent birds, and not so large as in *Geococcyx* and *Tinamus*.

'The *femur* of *Hesperornis* is remarkably short and stout, more so than in any known bird, recent or fossil.' The *tibia* has an enormous cnemial crest, as in divers, and the *patella* is a huge bone; the latter is perforated by a foramen for the ambiens muscle. The feet have four toes.

Nothing is known about the soft parts of this bird, save the feathers, which have been lately[1] stated to be quite ostrich-like. But casts of the brain have brought to light the interesting fact that it has smaller cerebral hemispheres than are found in any existing bird.

It seems clear that, as D'ARCY THOMPSON and others have argued, the nearest affinities of *Hesperornis* are with the **Colymbi**. That they are more nearly related to the ratites was the opinion of MARSH. The likeness to the ratites, however, seems mainly to be based upon the degenerate structure of the wings in both. The degeneration of the wings, however, has not proceeded along precisely similar lines. Although the angle between the scapula and the coracoid is nearly as open as in the ratites, the two bones have not become ankylosed; and moreover the clavicles are retained. The fact that of the fore limb only the humerus remains (at least in an ossified condition) may be compared with the fact that in the Dinornithidæ traces of the same bone have been met with. On the other hand there are more positive likenesses to the **Colymbi**, which are not so clearly due to the immediate action of environment. The long and narrow pelvis, and the huge cnemial crest with the relatively enormous patella, are among the more salient points of resemblance.

The characters in which the **Hesperornithes** approach the **Colymbi** are divided between the two families of the latter group; in some points *Hesperornis* is more of a diver; in others it comes nearer to the grebes. The following table of comparison may help to make its 'mixed' characters clear:—

[1] In a letter to *Nature*, April 8, 1897, by Prof. MARSH.

	Hesperornis	Divers	Grebes
Cerv. vert. .	17	14, 15	17-21
Catapoph. canal	0	0	+
Hæmap. .	With expanded ends	With expanded ends	Not so expanded at ends
Dors. vert. fused .	None	None	Several
Patella . . .	Large	Very small	Large
Supra-orbit. grooves	Large	Large	Not marked

With the grebes *Hesperornis* agrees in only two of the characters, and in the remaining four with the divers. It presents in fact, as might be expected of so ancient a form, a compound of the grebe and the diver.

We cannot however, in my opinion, put it down definitely as the ancestral form whence both divers and grebes have branched off; but it seems to approach that form, agreeing as it does in most points with the more generalised divers.[1]

SPHENISCI [2]

Definition.—**Moderately sized to large birds, with wings modified to form a swimming paddle. Aftershaft present. Feathering continuous. Oil gland tufted. Muscle formula of leg, ABX +. No biceps brachii or expansor secundariorum. Two carotids. Skull schizognathous, holorhinal, without basipterygoid processes. Dorsal vertebræ markedly opisthocœlous. Scapula flattened and expanded. Metatarsals short and incompletely fused.**

The anatomy of the penguins has been mainly investigated by MENZBIER [2] and WATSON.[2] The group, which is entirely antarctic, contains the genera *Eudyptes, Aptenodytes, Pygosceles,* and *Spheniscus.*

[1] For discussion of the affinities of the *Hesperornis* see, besides MARSH and FÜRBRINGER and GADOW in their large works. FÜRBRINGER, ' Über die systematische Stellung der Hesperornithidæ,' *Orn. Monatsschr. d. deutsch. Ver. z. Schutz. Vögel,* xv. 1890, p. 488 ; F. HELM, ' On the Affinities of *Hesperornis,' Nature,* xliii. 1891, p. 368 ; SHUFELDT, ' On the Affinities of *Hesperornis,' ibid.* p. 176 ; D'A. W. THOMPSON, ' On the Systematic Position of Hesperornis,' *Stud. Mus. Univ. Coll. Dundee,* i. 1890.

[2] Since WATSON'S ' Report on the Penguins collected by the *Challenger* ' the following papers have appeared : M. MENZBIER, ' Vergleichende Osteologie der Penguine,' &c., *Bull. Soc. Imp. Nat. Mosc.,* 1887 ; H. SCHAUINSLAND, ' Zur Entwickelung des Pinguins,' *Verh. Ges. Deutsch. Naturf. Leipzig,* 1891, p. 135 ; STUDER in *Zoology of S. M. S. ' Gazelle.'*

The penguins have a continuous feathering, the feathers acquiring upon the paddle-like wing a scaly aspect; [1] they have an *aftershaft*. The *oil gland* is tufted.

In the *myology* of the fore limb the most remarkable fact is the entire absence of the *biceps*, a muscle which is wanting nowhere else among birds. There are also absent the *expansor secundariorum, scapulo-humeralis anterior*, and *serratus metapatagialis*. The *tensor patagii longus* is present, and its tendon is inserted on to the whole length of the bones of the arm as far as the extremity of the last phalanx. The *latissimus dorsi* is peculiar in that its two parts, ending in thin tendons, pass side by side through a pulley arising from the scapula. The *pectoralis major* (of *E. chrysocome*) meets its fellow in the middle line over the carina sterni, as in tinamous, &c. In the hind limb the *muscle formula* is ABX +. The *semimembranosus* is remarkable for the fact that it has, besides the usual head of origin, a second from the aponeurosis of the abdominal muscles. The *accessory femorocaudal* (in *E. chrysocome*) sends a muscular slip to the tendon of the femorocaudal. The *tibialis anticus* divides in *Spheniscus mendiculus* into two heads of insertion; in other species its insertion, as is usual with birds, is single. There is but one *peroneus* present, the longus. The deep flexors are as in birds with but three well-developed toes, *i.e.* they blend. In *S. demersus* there is a slip to the hallux; in *Pygoscelis papua*, as in *Heliornis*, the flexor hallucis splits into three tendons, one for each branch of flexor profundus, there being no slip to hallux. The glutaeus maximus is absent in *Eudyptes*, limited in extent in the other genera. The *ambiens* grooves the patella.

In the *vascular system* the most remarkable fact is the breaking up of the brachial artery into a rete in the arm. There are two carotids, which do not fuse. In the leg the sciatic artery is practically absent.

[1] It is stated that when moulting the scale-like feathers of the wing are detached in a continuous piece, as with reptiles (BARTLETT, *P Z. S.* 1879, p. 6).

The *tongue* and the roof of the mouth are covered with papillæ.

The *proventriculus* has a patch of glands upon the right side, heart-shaped in *Eudyptes chrysocome* and *Spheniscus demersus*; the *gizzard* is small and not gizzard-like. In *Pygosceles* the proventricular gland, however, is zonary, and the same state of affairs was found in one of four examples of *Aptenodytes*.

In *Eudyptes chrysocome* the gut is thrown into a vast number of primitive irregular folds. The duodenal loop is excessively complicated, more so than in *Haliaetus*, to which it bears some (probably a convergent) resemblance.

The following are the intestinal measurements of a series of species :—

—	Small Int.	Large Int.	Cæca
	Ft. In.	Inches	Inches
Eudyptes chrysocome	11 8 (14 ft. 8 in., 23 ft.)	3	1½ ¾
„ *chrysolophus* .	21 3 (20 ft.)	3½ (3 in.)	¾ 1
Spheniscus magellanicus	30 6 (19 ft.)	4	1½
Aptenodytes longirostris .	21 8 (17 ft. 10 in.)	4½	1¾

From these data, which are extracted from WATSON's memoir already referred to, it is clear that there is some individual variation in the length of the small intestine among the penguins, which is, indeed, greater than anything that has been recorded in any other group of birds. That the cæca are so small is a curious fact in the structure of a fish-eating bird.

In the *liver* the right lobe is larger than the left in all species. The *gall bladder* is large and extends a long way down the abdominal cavity, as in the toucans.

The *syrinx* is not especially divergent in structure. The *trachea* has a septum down the middle, as in certain petrels (*q.v.*) The intrinsic muscles are attached to the tracheal rings a considerable distance above the bifurcation of the tube, the distance varying from species to species. There is but little fusion between the last rings of the trachea. In *Eudyptes chrysocome* there is a thick fibrous

pad at the commencement of the membrana tympaniformis.
There is a tendency, as in the petrels, to the formation of a
bronchial syrinx. This is especially well seen in *Aptenodytes*
and *Pygosceles*. In the latter penguin the rings, three or
four of them, after the bifurcation preserve the character of
the tracheal rings, being deeper than those which follow,
and being at the same time complete rings.

In the penguin's *skull* the sutures are not so completely
closed as in most birds; in this they resemble the ratites
among existing birds.

In *Eudyptes chrysocome*, which may be taken as a type
of the **Sphenisci**, the skull is schizognathous; the maxillo-
palatines are thin plates which are curved backwards, as in
many **Passeres** and in some other schizognathous birds; but,
instead of being flatter upon the ventral and dorsal surfaces,
they are compressed laterally. These bones gently nip the
vomer, which is also composed of two flattened rami partially
separated up to nearly the very anterior end. The palatines
are large and flat with a very slightly developed internal
lamina; they nearly come into contact in the middle line.
The pterygoids are unique among birds for their relatively
immense size; their shape is almost that of the human
scapula, the wider region being at their junction with the
palatines. Each pterygoid is perforated at the middle of
the wider part. The nostrils are holorhinal, and the nasal
bones at their posterior extremity are seen to overlap the
frontals. There are strongly marked impressions for the
supra-orbital glands. The lacrymals reach the jugals, where
they expand into a flattened foot; the descending limb of
each lacrymal, which is flattened out in a plane parallel with
the long axis of the skull, is there perforated by a very large
foramen. It is extraordinary that in this bone, as in the
maxilio-palatines, the plane of the bone is in a different
direction from that of the same bones in other birds. There
are no ossified ectethmoids. On either side of the foramen
magnum are occipital fontanelles of small size. The interorbital septum is largely vacuolate posteriorly. In front of
the occipital condyle, in the region occupied by the basi-

occipital and the basitemporals, is a hollowed area which suggests the impress of the potter's thumb. The thumb of the writer exactly fills it.

The symphysis of the lower jaws is of unusually limited extent.[1]

The *vertebral column* of *Spheniscus Humboldti* consists of twenty-one vertebræ in front of the sacrum, of which fifteen are cervical. The atlas is notched for the odontoid process. The catapophyses do not form a canal; hypapophyses commence on the eleventh cervical and continue to the last dorsal but one; they are large on the twelfth and thirteenth; on the fourteenth they have greatly diminished, but there are two large lateral catapophyses which on C15 form two large flanges (as in the diver); these arise from one base or D1, and gradually diminish into a single median hypapophysis again. The first dorsal is the first vertebra to be opisthocœlous, but it is heterocœlous in front. The opisthocœlous characters of the dorsal vertebræ of the penguins are better marked than in other recent birds in which the same structure of vertebra occurs (cf. p. 111). In *Œdicnemus*, for example, some of the dorsal vertebræ are opisthocœlous, but the convexity in front is by no means so clear as in the penguins.

The *scapula* is remarkable for its great breadth posteriorly, which narrows towards the neck. In *Pygosceles* it is wider actually as well as relatively than in *Spheniscus*, and has a truncated extremity; the same is the case with *Eudyptes* and *Aptenodytes*.

The *coracoid* also shows some differences. In *Spheniscus* the procoracoid fuses below with the coracoid, leaving an oval foramen about half an inch long; at the sternal end of the bone is a short upwardly directed snag of bone. In *Pygosceles* the procoracoid is not fused with the coracoid, but it is carried on by ligament to the snag at the base of the coracoid—as probably also in *Spheniscus*. The acro-

---

[1] MENZBIER describes the quadrate as single-headed. This is not accepted by FÜRBRINGER.

coracoid in both is very long. The *furcula* is strong and
U-shaped.

The bones of the anterior extremity are extraordinarily
flattened, in accordance with the paddle-like function of the
limb.

The *sternum* is roughly triangular, with a well-developed
keel. There are two lateral notches (one on each side),
which are united distally by cartilage and membrane. There
is a spina externa, but no spina interna.

The *pelvis*, unlike what is found in the majority of birds,
is remarkable for the fact that the pubes take a share in the
formation of the acetabulum. The ilia are well separated
from each other by the neural spines of the dorsal vertebrae.
The pelvis is thus perfectly free from, not ankylosed to, the
vertebral column.

The point of chief interest in the hind limb of this group
is the imperfect fusion between the short metatarsals, which
closely resemble those of the dinosaur *Ceratosaurus*. The
only recent bird which approaches the penguins in the
shortness of these bones is *Fregata*.

Fossil penguins are the genera *Palaeudyptes* and *Palae-
spheniscus*. The latter, from the Tertiaries of New Zealand,
was originally described by HUXLEY,[1] and later by HECTOR.[2]
It was a large bird standing some five feet in height, the
recent birds not being larger than three feet. The wings
were proportionately longer than in recent birds, while the
partly separate metatarsals were as is the case now ; hence
this latter character is obviously an inherited one in the
recent penguins.

The affinities of the penguins are not clear. This is due
to their antiquity, the existing characters of the group having
been apparently acquired in the Tertiary period. The main
facts of structure in which the penguins differ from other
birds are—

(1) Continuous feathering (except *Chauna*).

[1] 'On a Fossil Bird . . . from New Zealand,' *Q. J. Geol. Soc.* 1859, p. 670
[2] 'On the Remains of a Gigantic Penguin,' *Trans. New Zealand Inst.* 1871,
p. 341. 'Further Notice of Bones of a Fossil Penguin,' &c., *ibid.* 1872, p. 438.

(2) Scale-like feathering of the flattened wing.

(3) Absence of biceps brachii muscle.

(4) Presence of a vascular rete in the wing.

(5) Freedom of cranial bones (not so marked in Ratitæ).

(7) Large and flattened scapula.

(8) Short and imperfectly fused metatarsals.

The opisthocœlous character of the lumbar vertebræ is more pronounced than in other birds, but is, as has been already said, a character found in many groups.

These features, some of them, have appeared so important to MENZBIER that he has divided birds into four great groups—Saururæ, Ratitæ, Odontormæ, Carinatæ, and, finally, the Eupodornithes or penguins. This, however, seems to be a too great separation from other birds. GADOW would place them nearest to the **Tubinares** and **Steganopodes**, the **Colymbi** being only a little further removed.

STEGANOPODES

Definition.—All four toes webbed.[1] Oil gland tufted; aquincubital. Skull desmognathous, holorhinal, without basipterygoid processes.[2] Cœca present, but small.

Though this group shows much divergency of structure, its naturalness can hardly be doubted. The number of rectrices varies. In *Phalacrocorax brasiliensis* and *graculus* there are twelve, so also in *Fregata aquila* and *Plotus anhinga.*[3] *Phaeton* has twelve or sixteen. *P. carbo* has fourteen. *Pelecanus* has up to twenty and twenty-four.

The aftershaft is minute but distinct in *Fregata*, apparently absent in *Plotus* and other genera.

The skin is only slightly pneumatic in *Fregata*, not so at all in *Plotus*. It is distinctly emphysematous in *Phaeton*,[4] *Pelecanus*.

---

[1] This one feature is sufficient to define the group.

[2] Rudiments exist in *Pelecanus*; cf. *infra*, p. 409.

[3] *P. melanogaster* appears to have only ten.

[4] Some few details of the structure of the soft parts of this tern-like steganopod are to be found in BRANDT, 'Monographia Phaethontum,' *Mem. Ac*

The tufted oil gland has four orifices in *Phalacrocorax brasiliensis* and in *Plotus melanogaster*. *Phaeton* has six, the other genera apparently two, save *Pelecanus*, which has the unusually large number of twelve.

The pterylosis, on the other hand, is very uniform, and the feathering is very close. The neck and head are closely feathered, and there is a very narrow apterion on the breast. The spinal tract has a very limited apterion between the shoulder blades.

The *tongue* is small in this group, practically obsolete in *Plotus* (see fig. 5, p. 20). The proventricular glands are in two large squarish patches in *Phalacrocorax* (*brasiliensis*). The remarkable modifications of these organs in *Plotus* are described later (p. 414.)

The *gizzard* is very small in *Pelecanus* and *Sula*, the proventriculus being enormous. The glands are zonary in arrangement.

The following are measurements of the alimentary tract in a series of **Steganopodes**. The most remarkable fact to be noted in the table is the great length of the large intestine in *Plotus*.

	Small Intestine	Large Intestine	Caeca
	Inches	Inches	Inches
Fregata aquila	36	3	·25
Phalacrocorax brasiliensis	42·75	3	·2
,, *carbo*	111	4	·2
Sula bassana	57	2	·25
Pelecanus onocrotalus	93	3	1·75
,, *mitratus*	90	2·5	1·25
,, *rufescens*	65	1·5	1·25
Plotus anhinga	54	6	(One cæcum only in some specimens)
,, *Levaillanti*	24	3	·2
,, *melanogaster*	30	5·5	·2 *
Phaeton sp.	42	·75	·2 *

* These measurements are derived from *descriptions*.

The varying proportions of the *liver* lobes [1] are given in

St. *Petersburg*, 1810, p. 239 and pl. v., and in a paper by myself, *P. Z. S* 1897, p. 288.

[1] G. Alix, 'Sur l'Anatomie du Pélican,' *Bull. Soc. Zool. Fr.* ii. 1877, p.

the table on p. 415.   The *gall bladder* appears to be always present.

The *tensores patagii* of *Fregata*[1] are somewhat complicated.   The tensor muscle gives off two tendons, of which the anterior, the longus, is much the thickest.   The latter is reinforced by an elastic slip from the deltoid ridge of the humerus, whereupon it again divides into two, a branch going to fuse with the brevis ; where this branch joins the brevis tendon that tendon gives off a wristward slip, and is itself continued over on to the ulnar side of the fore arm.   There is a patagial fan and a bony nodule where it arises from the junction of the wristward slip of the brevis and the extensor metacarpi radialis.

In *Sula bassana* the tendons of the brevis are two from the very first ; the anterior one corresponds to the wristward slip of *Fregata*, and from it springs the patagial fan.

The other genera are not very different, save that in none is there an osseous nodule, and that all have a patagial fan.

The pectoralis I. is double in *Fregata*, *Plotus*, *Pelecanus*, and *Sula*, variable in *Phaeton*, single in *Phalacrocorax*.

The *biceps slip* has been occasionally overlooked.   It is present in *Plotus*, *Phalacrocorax*,[2] *Phaeton*, and *Sula*.[3]   It is absent in *Pelecanus*, *Fregata*.

Where present, however, it is slender, and is attached sometimes to the tensor longus, as ordinarily, and sometimes to the patagium itself, as in *Colymbus*, *Podica*, &c.   The deltoid has the scapular slip in *Phalacrocorax*.

The *expansor secundariorum* has been commonly said to be absent from the wing of the steganopods.   This is not, however, at least according to FÜRBRINGER, an accurate statement of the case.   In an embryo of *Phalacrocorax carbo* unmistakable traces of it were discovered, while in

287; G. L. DUVERNOY. ' Sur la Poche Maudibulaire du Pélican,' *Mem. l'Inst.* iii. 1835. p. 219.

[1] The anatomy of this genus has been described by BURTON in *Linn. Trans.* vol. xiii. p. 1.

[2] Not always.   FORBES records its absence in *P. brasiliensis*.   I did not find one in *P. africanus*.

[3] Not always; it is absent in *S. fusca* (*fide* GARROD).

Sula and *Pelecanus* a slender tendon, running from the arm-
pit and ending in unstriated (*Sula*) or striated (*Pelecanus*)
fibres for the movement of the secondary feathers, was discovered.

The *biceps* is two-headed in the **Steganopodes** ; but the
arrangement differs from what is common among birds.
Both heads, in fact, arise from the coracoid, but the outer
one, which corresponds to the humeral head of other birds, is
also attached to the humerus. The two muscular bellies are
separate in *Pelecanus* and *Fregata*, and their tendons unite

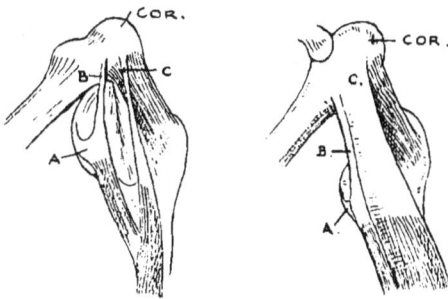

FIG. 190. - ORIGIN OF BICEPS IN *Pelecanus* (LEFT-HAND
FIGURE) AND *Phalacrocorax* (AFTER FÜRBRINGER).
Cor, coracoid ; *C*, coracoidal head of biceps ; *A*, attachment of humeral head ;
B, its prolongation to coracoid.

to divide again directly. In *Fregata*, indeed, the division of
the tendon of the coracoidal head takes place before the
junction. In *Phalacrocorax africanus* I found the coracoid
head alone, and it had but one insertion. In *Phaeton* and
Phalacrocorax and *Sula* the tendons of origin of the two
heads from the coracoid are continuous. In *Phalacrocorax*
and *Plotus*, at any rate, the *anconæus* has a humeral head.

The muscles of the leg vary greatly among the Steganopodes. The following are the formulæ for the different
genera :—

Phaeton AXY – [1]	*Pelecanus* AX –	*Phalacrocorax* { ABX + [2] / AX +
Plotus AX +	*Sula* AX +	*Fregata* A + .

[1] Perhaps the ambiens varies. GARROD, FORBES, and I did not find it.
FÜRBRINGER marks it as present, as does GADOW.

[2] ABX *P. carbo*? AX *P. lugubris*, *P. brasiliensis*.

As regards the other leg muscles *Phaeton* has quite exceptionally no tendinous loop for the *biceps* to pass through.[1]

The plantar tendons blend in *Fregata*, &c., sending slips to all four toes ; they blend in *Phaeton*, but send no branch to hallux.   In *Plotus* there is no blending, but a strong vinculum, which is attached to flexor perforatus, just as it splits up into its three branches.   This too is the case with *Phalacrocorax*, but the vinculum is not strong.

In the pelicans (at any rate in *P. rufescens* and *P. mitratus*) there is a curious relationship between the femorocaudal and the semitendinosus.   The former receives a tendinous slip not far from its insertion, which runs up from the middle of the semitendinosus at right angles to its fibres. Whether this may be regarded as a rudiment of the accessory semitendinosus or not is uncertain.

Another peculiarity shared with *Biziura lobata, Colymbus,* and the extinct *Hesperornis* is the perforating of the patella by the tendon of the ambiens in *Phalacrocorax*.   In *Plotus* there is a groove upon the ossified patella, and, remarks Professor Garrod, ' some of the fibrous ligament overlapping this groove shows traces of ossification ; so that in aged birds this groove may be converted into a foramen.'

Glutæus 1. in the **Steganopodes** is a small muscle, not extending, or hardly extending, over the biceps.   *Glutæus V.* is large.   In *Fregata* glutæus I. is absent.

The form of the syrinx of the Steganopodes varies considerably, but in all it is tracheo-bronchial ; intrinsic muscles may be present or absent.[2]

Of *Pelecanus* I have examined six species, viz. *P. mitratus, P. onocrotalus, P. rufescens, P. fuscus, P. conspicillatus,* and *P. crispus.*   In none are there any intrinsic muscles, and the bronchidesmus appears to be complete.

In *P. conspicillatus* the organ of voice is very simple. The rings are but little modified.   There is a bony pessulus, which is attached behind to the last two tracheal rings,

---

[1] This peculiarity is, however, to be found in the swifts.

[2] The anatomy of *Pelecanus rufescens* has been described by Owen, *P. Z. S.* 1835, p. 9, by Martin, *ibid.* p. 16, and by Alix, *loc. cit.* on p. 403, above.

which are fused and ossified where they pass into it; in
front the last of these and the next following are ossified and
fused in the middle line; the middle one of the three is continuous with the pessulus.

P. mitratus chiefly differs in the larger number of rings with
which the broad three-way piece comes into contact in front
and in the greater length of the membrana tympaniformis.

P. fuscus is remarkable for the degree in which the
syrinx is flattened from before backwards. This is caused
by the straightening of two of the bronchial semi-rings,
which thus come, though hardly longer than the others, to
project out considerably beyond them. At a first glance
these two semi-rings appear to be the first two of the bronchial series which are commonly among birds different from
those which follow. There are, however, five pairs of bars
in front of them, only separated in the middle line in front
by a furrow; on slitting up the windpipe it may be observed
that there is really a septum between them, and that they
are bronchial. Owing to the peculiar form of the sixth and
seventh semi-rings the membrana tympaniformis is exposed
behind, but not in front.

P. onocrotalus and *P. crispus* have syringes which are
very much alike. When removed from the body, at any
rate, the bronchi stand out at right angles to the trachea.
This is due to the very large posterior end of the pessulus,
which is broadened into a bar at right angles to and longer
than the median portion of the pessulus. It is not ossified.
There is a slight dilatation of each bronchus, which is carried
to an excess in the next species.

In *P. rufescens* the characters of the syrinx of the last
species are carried to an extreme. The pessulus is the same
in shape, but ossified. The bronchi are greatly swollen for
the space of about an inch, there being occasionally some
forking and anastomosing of the individual rings. The ends
of the rings in the swollen region are nearly in contact, being
separated only by a narrow membranous interval.

In *Fregata aquila* the syrinx has a pair of intrinsic
muscles and the bronchidesmus is incomplete. The syrinx

is flattened from before backwards, and the first two bronchial semi-rings are very prominent. To the first of them, and apparently also to the ring in front, are attached the two muscles into which the intrinsic muscle divides. There is a membranous gap separating the last of the specialised bronchial semi-rings from the first of those which follow, whose border, moreover, is concave upwards. The last few tracheal rings are ossified and firmly fused.

Phalacrocorax has a complete bronchidesmus and a single pair of intrinsic muscles. The first three bronchial semi-rings are very prominent and arched, and to the third of these the intrinsic muscles are attached. There is a membranous gap between the last of these and the first of the remaining series of bronchial semi-rings, which forms, at any rate in *P. carbo*, quite a pocket. The fourth bronchial semi-ring is curved in the same direction as that which precedes it ; both, in fact, are convex. The curvature, which is slightly more marked in *P. brasiliensis* than in either *P. carbo* or *P. varius*, suggests very much the syrinx of certain auks (cf. p. 363). In *P. varius* the intrinsic muscles are attached to the second bronchial semi-ring, as also in *P. brasiliensis*.

The syrinx of *Plotus* does not differ greatly, but it has an incomplete bronchidesmus. There are two bronchial semi-rings, which are specially increased in length and depth ; they are the second and third, and are relatively stouter than those of *Phalacrocorax* ; to the first of them the intrinsic muscles are attached.

The syrinx of *Sula* is a good deal different from that of other steganopods.

There is no ossification, except in the pessulus. A square projection is formed by a fusion between last tracheal rings ; this is continuous with the pessulus and is well shown in GARROD's figure.[1] The bronchial semi-rings are at first feeble with wide interval. Between the third and fourth of them there is, covering the insertion of the intrinsic muscle, a protuberant pad of elastic tissue about the size and shape of a pea.

[1] *P. Z. S.* 1876, pl. xxxviii. fig. 4.

In *Phaeton flavirostris* the syrinx is typically tracheo-bronchial and not flattened, as in *Fregata*.

The *skull* of the **Steganopodes** [1] is desmognathous. It is most extremely so in *Pelecanus*, in correlation, perhaps, with the long broad beak. *Pelecanus*, in fact, may be described as doubly desmognathous, for the palatines are not merely united but ankylosed behind the posterior nares, which are of limited extent. They form but one bone with a deep ventral median crest, and on the opposite side an equally pronounced dorsal crest, occupying a space left by the here deficient interorbital septum. *Fregata* is nearly at the other extreme, for the maxillo-palatines are largely free from each other in the middle line, and the palatines are only united for a short distance posteriorly. *Phaeton* [2] is most like *Fregata*, but here there is no fusion between the palatines. The strong inferior crest of *Pelecanus* is represented by two feeble ridges of limited extent. *Phalacrocorax* is intermediate. The maxillo-palatines are completely united. The palatines are fused for the greater part of their length posteriorly; they are, however, quite flat above and have below but a faint trace of the median crest. The interpalatine space anteriorly is much more capacious than in *Pelecanus*. *Plotus* agrees with *Phalacrocorax*.

The **Steganopodes** are generally (HUXLEY, FÜRBRINGER, GADOW) said to have no basipterygoid processes. In *Pelecanus rufescens*, however, I find a pair of thorn-like outgrowths in the right position (cf. *Platalea*, p. 439), which I take to be the rudiments of these structures.

The bony nostrils are holorhinal, pervious only in *Phaeton*, in others much obliterated by bony growths, as in **Herodiones**; in *Plotus*, indeed, reduced to the merest chinks. As in some **Herodiones** and **Tubinares** they are continued forward by a marked groove which runs to, or near to, the very end of the bill, absent only in *Phaeton*. In *Fregata* there is no

<hr/>

[1] All the types are described and figured by BRANDT, 'Zur Osteologie der Vögel.' *Mem. Ak. St. Petersb.* 1840 (6), iii. p. 81. See also for osteology of *Plotus* and *Phaeton* MILNE-EDWARDS in *Hist. Nat. Madagascar.*

[2] BEDDARD, 'Notes upon the Anatomy of *Phaeton*,' *P. Z. S.* 1897, p. 288.

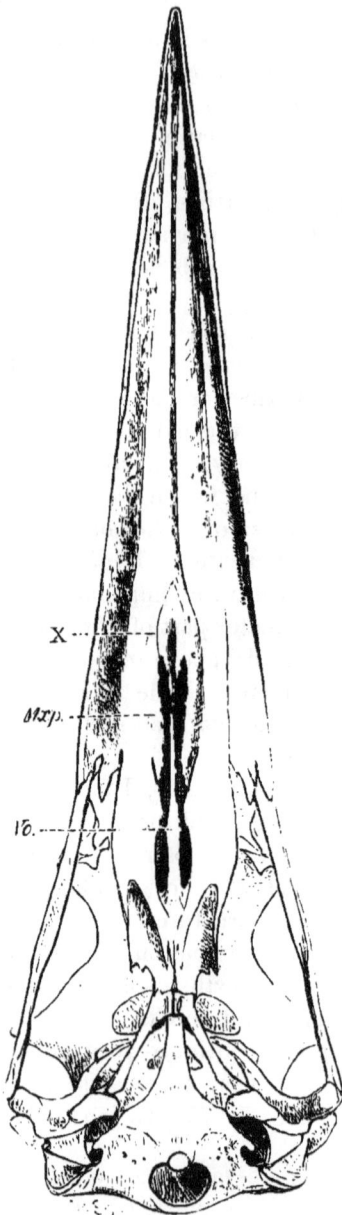

FIG. 191.—SKULL OF *Fregata*. VENTRAL
ASPECT. (AFTER BEDDARD.)

Vo, vomer ; *Mxp*, maxillo-palatines ; *X*, upwardly
directed part of maxillo-palatines.

hinge line separating the
cranium from the face.
This is present in *Pele-
canus* and *Phaeton*, while
in *Phalacrocorax* and
Plotus the existence of
fibro-cartilage allows of a
free motion.

The lacrymal in *Fre-
gata* and *Pelecanus* is large,
with a large descending
process, which in both
reaches the jugal bar ; in
addition *Fregata* has an
uncinate bone, which
reaches the palatine. Of
this there is apparently a
rudiment in *Phaeton*.

In *Phalacrocorax* the
descending process of the
lacrymal completely blends
with the ectethmoid to
form a ring of bone, as in
the **Limicolæ**. In *Plotus*
the same thing occurs, but
the lacrymal is much
smaller. The interorbital
septum as an ossified
structure is almost com-
plete in *Pelecanus* and
Fregata, largely deficient
in *Phaeton*, almost absent
in *Phalacrocorax* and
Plotus. In *Phalacrocorax
carbo* there is a small
bonelet resting upon the
jugal bar in front of the
lacrymal ;[1] this is also

[1] *Ossiculum suprajugale* of BRANDT.

well developed in *Plotus*. There are traces of it in *Fregata*. Another peculiarity shared by *Phalacrocorax* and *Plotus* is a style-like bone [1] attached to the occipital, to which the temporal muscles are partly attached. The bone varies in size, is not ankylosed to the skull, and is probably to be looked upon as an ossification in the septum between the two muscles. In these birds also the quadrate is peculiar in form in that the anterior process is short and slender and at right angles to the rest of the bone.

Sula is nearest to *Phalacrocorax*. It has the same peculiar form of the quadrate bone, and the equivalent of the small bone seated upon the jugal bone is apparently there, though ankylosed. The nostrils too are reduced to a mere pinhole, as in *Plotus*. The palatines agree absolutely with those of *Phalacrocorax* and *Plotus*, but the interorbital septum is not so completely vacuolate. It rises up, moreover, in front, as in *Pelecanus*, and a faint crest from the palatines ascends into the vacuity. The lacrymal is, however, different; the orbital part is

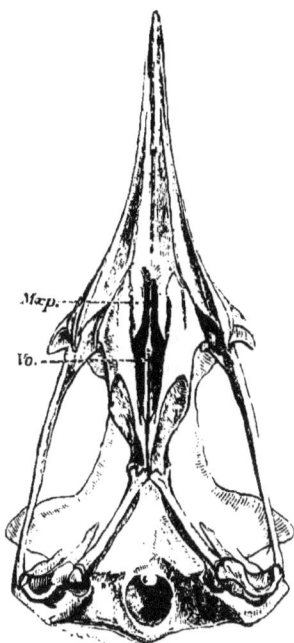

FIG. 192.—SKULL OF *Phaeton* (AFTER BEDDARD). LETTERING AS IN FIG. 191.

small, but the descending bar is large and joins the jugal; the ectethmoids appear to be deficient as bony structures.

[1] J. A. JEFFRIES, 'The Osteology of the Cormorant,' *Science*, ii. p. 739, iii. pp. 59, 274 ; GILL, 'Osteology of the Cormorant.' *ibid*. iii. p. 404 : SHUFELDT, 'Remarks upon the Osteology of *Ph. bicristatus*,' *ibid*. ii. p. 640, and 'Osteology of the Cormorant.' *ibid*. iii. p. 143 ; DOLLO, *Bull. Mus. Roy. Belg*. iii. 1884, p. 130. This bone (xiphoid, YARRELL; nuchal, MARSH ; intranuchal, DOLLO) has been wrongly compared with the post-occipitals of dinosaurs. It is merely a sesamoid. LUCAS, 'Description of some Bones of Pallas's Cormorant (*Ph. perspicillatus*),' *P. U. S. Nat. Mus*. xii. 1889, p. 88.

The lacrymal, as in *Pelecanus* and *Fregata*, is deeply notched laterally.

MIVART, from his investigations into the axial skeleton of this group,[1] set aside *Phaeton* and *Fregata* by reason of their possessing twelve or thirteen cervical vertebræ and the doubly notched (on each side) sternum of *Phaeton*, besides a number of other points.

There are seventeen *cervical vertebræ* in *Pelecanus*, eighteen in *Sula*, twenty in *Phalacrocorax*.

The *atlas* vertebra is notched for the reception of the odontoid process in *Pelecanus*, perforated in *Sula* and *Plotus* ; in *Phalacrocorax* the conditions are intermediate, a perforation being only just closed above.

The **Steganopodes** have catapophyses upon some of the cervical vertebræ which enclose a canal ; in *Pelecanus* this is found on vertebræ 8–15. In *Sula* the canal may commence on the same vertebra, but more usually on the ninth, extending to the thirteenth. In *Phalacrocorax* there are no vertebræ with a complete hæmal arch formed by union of the catapophyses. In *Plotus* the canal begins on the ninth vertebra and extends to the fourteenth. In *Phaeton* there is no canal. The dorsal vertebræ are opisthocœlous in *Phalacrocorax* and *Plotus*, not in *Sula* and *Pelecanus*.

In all these genera the *sternum* is at most only slightly notched by one notch only on each side. In *Pelecanus* the clavicles are ankylosed to its keel. They reach it in *Plotus*, *Phalacrocorax*, and *Sula*, but are only firmly connected by ligaments, not ankylosed. In *Phaeton* the clavicles are attached to the keel behind the extremity.

Both spina externa and interna are wanting in *Sula* and *Pelecanus* ; the spina externa is present, but small, in *Phalacrocorax* ; contrary to the statement of FÜRBRINGER (in his tables) I find a rudimentary spina externa in *Plotus anhinga*. *Fregata* and *Phaeton* show their divergence from the normal steganopod type by a better developed spina externa, while the latter bird may possess a rudimentary spina interna. Four to six ribs reach the sternum.

[1] 'On the Axial Skeleton of the Pelecanidæ,' *Tr. Z. S.* x. p. 315.

The relationship of the clavicles to the scapula and to the coracoid, upon which FÜRBRINGER has laid so much stress, serves to differentiate some of the Steganopodes. In *Plotus* the clavicle is connected by ligament with the scapula; this connection is nearly effected in *Phalacrocorax* and *Sula*, but not in *Pelecanus*. In *Fregata* the dilated end of the clavicle is perforated in the middle; it is, moreover, fused with the scapula.

The genus *Plotus* (consisting of the four species *P. anhinga*, *P. melanogaster*, *P. Novæ Hollandiæ*, and *P. Levaillanti*) has been investigated by BRANDT,[1] EYTON,[2] DÖNITZ,[3] GARROD,[4] FORBES,[5] FÜR-BRINGER,[5] and myself. Many of its characters have been described in the foregoing pages. I shall here direct attention to certain peculiarities of *Plotus* which it does not share with the other Steganopodes, or which it possesses in a more marked degree than its nearest ally, *Phalacrocorax*. The darters feed in a peculiar manner; they pursue fishes under water with a jerky action of the head and neck. This action, as Mr. FORBES has suggested, may be compared to that of a man poising a spear before hurling it. 'Arrived within striking distance,' continues Mr. FORBES, 'the darter suddenly transfixes—in fact, bayonets—the fish on the tip of its beak with marvellous dexterity, and then immediately comes to the surface, where the fish is shaken off the beak by jerking of the head and neck, thrown upwards, and swallowed, usually head first.' This mechanical action is associated with a mechanism in the neck.

The first eight vertebræ form a continuous curve forwards, so marked that the head when outstretched is in the same straight line with the eighth vertebra. This latter vertebra is articulated at right angles with the foregoing, and almost at right angles with that which follows; there is thus formed a conspicuous kink in the neck, which is never unbent.

<hr>

[1] *Loc. cit.* (on p. 409.) [2] *Osteologia Avium*, p. 218.

[3] 'Ueber die Halswirbelsäule der Vögel,' &c., *Arch. f. Anat. u. Phys.* 1873, p. 357.

[4] 'Notes on the Anatomy of *Plotus anhinga*,' *P. Z. S.* 1876, p. 335, and 'Note on Points in the Anatomy of Levaillant's Darter (*Plotus Levaillanti*),' *ibid.* 1878, p. 679.

[5] 'On some Points in the Anatomy of the Indian Darter,' &c., *ibid.* 1882, p. 208.

[6] 'Notes on the Anatomy and Osteology of the Indian Darter (*Plotus melanogaster*),' *ibid.* 1892, p. 291.

On the ninth vertebra there is on the dorsal surface a fibrous loop ('Dönitz's bridge'), which is fibrous only in *P. anhinga*, ossified in the three other species. Through this loop passes the tendon of the *longus colli posterior* muscle to be inserted on to vertebræ 2, 3, 4. The *longus colli anterior* is a very powerful muscle, which ends in a long tendon attached anteriorly to the

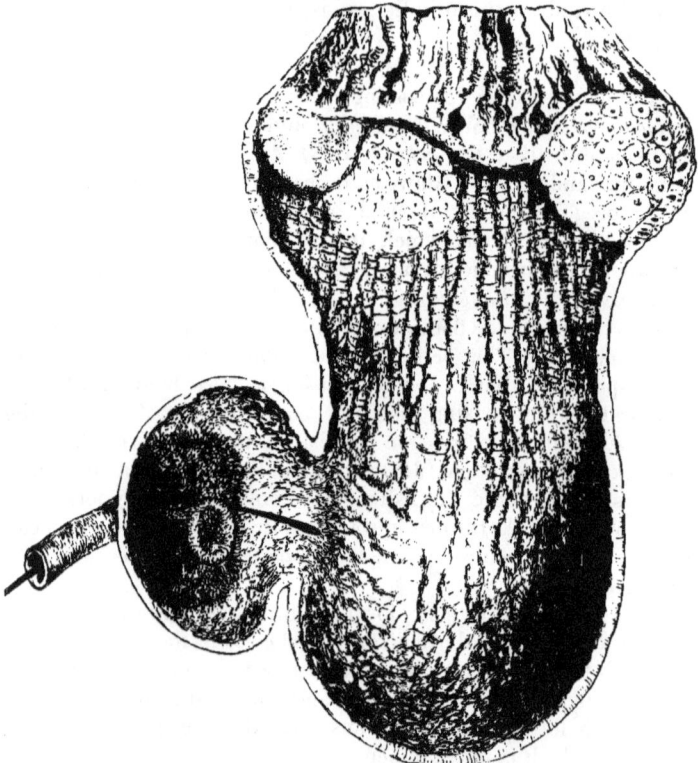

FIG. 193. —STOMACH OF LEVAILLANT'S DARTER (AFTER GARROD).

hæmapophysis of cervical vertebræ 8-10. A separate portion of the muscle is similarly attached to the eleventh.

The pulling back of the head preparatory to striking is effected by the longus colli posterior, while the bayonetting movement is produced by the longus colli anterior.

The second characteristic feature in the organisation of *Plotus* concerns the stomach. Though originally described by MACGILLI-

VRAY in AUDUBON's ornithological miscellany, it has been more fully described and illustrated by GARROD and FORBES. One great peculiarity is that in *P. anhinga* the proventricular glands, instead of forming a patch, or patches, upon the inner surface of the proventriculus, constitute a special cæcal diverticulum of the stomach, which is completely lined by the glands in question. The pyloric portion of the stomach forms a well-marked compartment, quite distinct from the gizzard region ; the opening of the pylorus into the duodenum is protected by a dense mat of hair-like processes, each of which is about half an inch in length. Microscopically these structures 'are much more like true hairs' than like any filiform papillæ which might occur in such places. In *P. Levaillanti* there is no special compartment for the pyloric proventricular glands. The hairs lining the pyloric chamber have a more complex arrangement than in *P. anhinga.* There is (fig. 193) a dense mass of them lining the distal end of the pouch, but there is also a singular conical process of the mucous membrane, covered with more hairs and serving to close the pylorus. *P. melanogaster* agrees with the last species, but the plug is less developed, being rather a well-defined ridge than a retractile plug. Mr. FORBES thinks that this hairy plug of the darters is an exaggeration of the nipple-like valve which is found to guard the pylorus in many birds. The two Old-World species thus come nearer together, as they do in the ossification of Dönitz's bridge, than does either to the New-World *P. anhinga.*

In order to facilitate a comparison of the several genera of **Steganopodes** among themselves, the annexed table, indicating the differences in some of the muscles, viscera, and bones, may be of use :—

	Muscle Formula of Leg	Exp. Sec.	Bic. Slip	Carotids	Liver Lobes	Lateral Muscles of Syrinx	Cerv. Vert.	Vomer	Maxillo-palatines
Fregata	A+	0	0	2	R2>L	+	15	Single	Free
Pelecanus	AX--	+	0	1	R2>L	0	17	Rud.	Fused
Phaeton	AXY-	0	+	2	R>L	+	15	Double posteriorly	Free
Phalacrocorax	AX+	R.	+	2	R2 L	+	20	0 ?	Absent
Plotus	AX+	0	+	1	R2>L	+	20	0 ?	Absent
Sula	AX+	+	+ or 0	2 or 1	R>L	+	18	0 ?	Absent

The Steganopodes, though allowed by all to be a natural

group), show, as is indicated in the above table, a considerable
amount of variation among themselves. And, what is even
more striking, the genera vary more than genera as a rule do
—for example, the stomach of the darters, the biceps and
carotid of *Sula*, and the muscle formula of the leg of *Phalacrocorax*. The accompanying table gives some of the chief
characters of the several genera. It is interesting to observe
from this table that there is to some extent a linear
series to be deduced from the facts of structure. If we
commence with *Phalacrocorax*, we may associate with it
certainly *Plotus* and also *Sula*. In these genera the peculiar
flat and largely fused palatines are associated with the
muscle formula AX + . *Sula* offers a slight step in the
direction of *Pelecanus*. There is a trace of the dorsal ridge
of the palatine, and of the anterior vacuity in the interorbital
septum for the reception of this edge. In *Pelecanus* these
characters are fully realised, and at the same time the
ambiens has vanished. *Fregata*, which stands apart in the
structure of its skull, is also unique in the group by reason
of its muscle formula, which is only A + . But this genus
and *Pelecanus* have lost the biceps slip, present in those
referred to before, but commencing to disappear in *Sula*.

The question is, In what direction has the modification
gone? Are we to start with *Phalacrocorax* or with *Pelecanus*,
or with some other genus?

The answer to this question naturally depends upon the
relationship of the **Steganopodes** to other groups. The group
to which they are most nearly allied appears to me (as to
FORBES and others) to be the **Tubinares**. In estimating this
affinity FORBES did not, in my opinion, lay sufficient stress
upon the 'os uncinatum.' It is true that—as he states-
this bone appears also to be present in *Chunga* and in the
touracos, not to mention other birds ; but in *Chunga* it is a
continuation of the descending process of the lacrymal, and
articulates with the jugal bar, while in *Corythaix* it articulates both with ectethmoid and lacrymal, though it ends, as
in the birds under consideration, on the palatine. In both
Fregata and *Diomedea* this bone connects the lacrymal with

the palatine, the lacrymal itself reaching (or nearly so) the jugal bar. The palate of *Diomedea* is remarkably like that of *Fregata*, which, unlike other **Steganopodes** (except *Phaeton*), is not far from being schizognathous, and represents, I am disposed to think, the nearest intermediate form. The grooves, starting from the nostrils and running towards the end of the beak, are also found in the **Tubinares** (and in the **Herodiones**, with which latter group the **Steganopodes** share the very much reduced, and yet holorhinal, nostrils). Another fact which is perhaps of importance is the much-reduced gizzard and the correspondingly enlarged proventriculus. Less important likenesses are the double pectoral, the short colic cæca, which are occasionally reduced to one in the **Steganopodes**; these points ally the group to the **Herodiones** as well as to the **Tubinares**. The very names *Fregata* and *Fregetta*, *Pelecanus* and *Pelecanoides* are an expression of these views.

FÜRBRINGER, however, and GADOW place *Phaeton* nearest to the base of the steganopod series, and there is much to be said for this way of looking at the group. There is no doubt that *Phaeton* is very different from the other genera of the group; indeed, if it were not for *Fregata* it would be difficult to avoid removing it altogether. It is really not desmognathous (in the sense of HUXLEY); for the maxillo-palatines do not fuse; in front of them there is a bony platform, forming the hard palate, but this is produced from anterior ingrowths of the maxillæ, not homologous with the maxillo-palatines, which are present and unfused. The vomer, moreover, is well developed and bifid posteriorly, being exceedingly like the bifid vomer of such schizognathous birds as the grebe, *Æchmophorus*. In *Fregata* we have a further step. The vomer has coalesced into a single rod; the palatines have united posteriorly; the os uncinatum, rudimentary in *Phaeton*, has increased (?), and the grooves running from the nostrils to the end of the beak have put in an appearance; furthermore the nostrils, pervious in *Phaeton*, have acquired the steganopodous imperviousness. The maxillo-palatines, however, are not united; but beneath them

E E

is a slanting wall of bone (cf. fig. 191 X), present in all the other **Steganopodes**, co-existing with and lying below the maxillo-palatines, and therefore *not* comparable to them. True maxillo-palatines, indeed, seem to be only present in *Pelecanus* of the remaining genera of the family, where they are small and coalesced. The other genera have only the upwardly sloping bone of *Fregata*. The vomer too appears to be at the most very small in the higher steganopods, and I have not been able to find it.

All these facts point to the basal position of *Phaeton*. It has, however, in the loss of the ambiens departed from the primitive condition. We can derive *Fregata* from *Phaeton*; but in this case the mutual relations of the **Steganopodes** and the **Tubinares**, and perhaps the **Colymbi**, become somewhat obscured.

As perhaps an appendix to the present group we may consider *Odontopteryx toliapicus* of the London clay, known [1] by a portion of a skull.

FÜRBRINGER discusses this bird very slightly under the **Tubinares** and **Anseres**; both GADOW and LYDEKKER place it near the **Steganopodes**, with which determination I associate myself. The most marked peculiarity of this bird, which has given to it its name, is the serration into longer and shorter teeth of the upper and lower jaw. The two jaws are grooved, which seems to indicate that the beak was, as in the Steganopodes (and other birds for that matter), divided into several pieces. On the right-hand side of the skull is a small notch, which has been identified with the bony nostril of that side. It has, however, in the drawings an accidental look, and the fact of the possible obliteration of the nostrils must be weighed in discussing the steganopodous affinities of the *Odontopteryx*; for in many of those birds, especially in *Sula* and *Plotus*, the nostrils have been practically obliterated. It does not seem to me that the depressed form of the skull, or, so far as we can judge them, the shape of the lacrymals, is strong evidence in favour of the steganopodous

[1] OWEN, ' Description of the Skull of a Dentigerous Bird,' &c., Q. J. Geol. Soc. 1873, p. 511.

relationships of this remarkable fossil; but the facts are at any rate not opposed to that placing.

Of birds more definitely referred to the **Steganopodes** a large number have been described from Tertiary strata. If MARSH's *Gracularus*, from the Cretaceous, really belongs here, the group is as old as any existing group, and older than most. *Argillornis* (= *Lithornis*) of OWEN is among the most interesting, inasmuch as it is known from fuller remains than others. The skull and some long bones have been found in the London clay. It is referred to the neighbourhood of the present group by LYDEKKER, but by FÜRBRINGER to the **Ichthyornithes**.

Actiornis, Pelargornis are placed here by LYDEKKER, and FÜR-BRINGER would include *Remiornis* (considered struthious by GADOW) and *Chenornis*, referred by others to the **Anseres**.

HERODIONES

Definition.—Oil gland feathered.[1] Aftershaft present.[2] Aquincubital. Skull desmognathous, holorhinal, without basipterygoid processes. Catapophysial canal nearly always present. Two carotids. Cœca present, but nearly always rudimentary. Expansor secundariorum present.

This group of birds is an extensive one, with a considerable range of structural variation. That the flamingoes form a group apart can hardly be doubted, though it is not easy to differentiate them by any very important characters from other **Herodiones**. Less easy is it to distinguish the herons from the storks. The extreme types, *e.g. Ciconia* and *Cancroma*, can be readily distinguished by the muscle formula and by the characters of the syrinx, not to mention some other points of minor importance; but between the extremes are forms like *Abdimia, Scopus,* and *Balæniceps,* which forbid so sharp a line of division. As to *Phœni-copterus*, WELDON was the first[3] to show in a convincing way its likenesses to the stork, its previous association with the duck tribe having been in large part due to the lamellated bill and the webbed feet. As to the latter character, no one

---

[1] Except in *Cancroma*. [2] ? except in *Leptoptilus argala*.
[3] 'On some Points in the Anatomy of *Phœnicopterus*.' &c., *P. Z. S.* 1883, p. 638.

would nowadays associate the gulls with the ducks for a similar reason, though it was, of course, done by the earlier ornithologists. The duck-like bill of the flamingo is not so exclusively anatiform as might be thought; for a very decided stork, *Anastomus*, has a bill which has very much the same structure as regards the lamellæ (hence, indeed, its specific name—*lamelligerus*).

The typical storks may be distinguished from the typical herons by the following table :—

	Ciconiidæ	Ardeidæ
Syrinx	Without intr. muscles. Tracheal	With intr. muscles. Tracheo-bronchial
Cæca	2, small	1, small
Powder-downs . .	0	+
Pterylæ . . .	Wide	Narrow
Pectoralis I. . .	In two layers	Single
Cucullaris dorsocutaneus . . .	0	+
Pectoralis abd. . .	0	+
Vinculum of deep flexor tendons . . .	Strong	Weak or absent
Ambiens . . .	+ or 0	0
Carotids . . .	2, separate	2, fused

The set of differences may be certainly regarded as of family value. But it must always be remembered that there are tendencies to the heron-like organisation among true storks; while *Scopus*, and possibly *Balæniceps*, are distinctly intermediate.

Family **Scopidæ**.—There is one genus only, containing but a single species, *Scopus umbretta*—African and from Madagascar. The anatomy of this stork-like heron has been principally investigated, as regards the ' soft parts,' by myself.[1]

It differs from the true herons by the absence of *powder-down patches*, in having ten primaries instead of eleven, and in possessing sixteen *cervical vertebræ*. On the other hand it differs from the storks in having an ardeiform—or at least a ' typical '—syrinx, and (from the Plataleidæ) in the

[1] 'A Contribution to the Anatomy of *Scopus umbretta*,' P. Z. S. 1884, p. 543.

absence of a *biceps slip* to the patagium. Its *leg muscle formula* too is that of a heron. Besides these facts

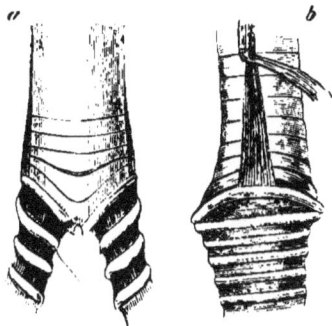

FIG. 194. SYRINX OF *Scopus* (AFTER BEDDARD). *a.* FRONT VIEW. *b.* SIDE VIEW.

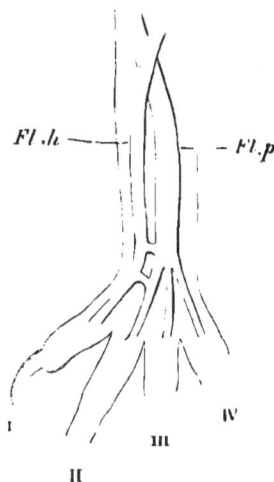

FIG. 195. DEEP PLANTAR TENDONS OF *Scopus* (AFTER BEDDARD).

Fl.h, flexor hallucis ; *Fl.p,* flexor profundus.

which incline to both the stork and the heron side, and which, perhaps, are the most important of such facts—it may be mentioned that *Scopus* shows a stork-like character in the partial division of the *pectoralis primus*, a fact which was first pointed out by FORBES incidentally, in his report upon the petrels collected during the voyage of the ' Challenger.' The *patagial muscles* and tendons are not specially distinctive of the affinities of the bird. In my dissection of *Scopus* I could not find the *expansor secundariorum* ; but as this muscle is found in most of the **Herodiones** its absence is not in any way significant, though it should doubtless be verified. The *anconæus* has a tendinous slip to the humerus. The *deep plantar tendons*

FIG. 196.—DEEP PLANTAR TENDONS OF *Scopus* (AFTER BEDDARD).

appear to vary somewhat ; in two specimens dissected by myself I met with the two conditions illustrated in the two

figures annexed. In the one the flexor hallucis was united
to the flexor communis by two distinct vincula, one before
the trifurcation of the latter tendon, the other attached to
the tendon supplying digit II. In a specimen of *Ciconia
nigra* dissected by FORBES there was an identical arrangement. In the other *Scopus* (see fig. 195) only the last of
the two vincula was present, *i.e.* that passing to tendon of
digit II.

For the skeleton of *Scopus* see MILNE-EDWARDS's account.[1]
The *skull* is on the whole more
stork-like than heron-like, but it does
not show any of the extreme modifications of the stork type. The bony
interorbital septum, as in the storks, is
not largely fenestrate. The inner
lamina of the palatines does not reach
the posterior boundary of those bones.
In front, at about the middle of the
interpalatine vacuity, the palatines are
produced into a short lateral process ;
this is well marked in many storks,
but also in *Cancroma*, to the skull of
which heron that of *Scopus* shows

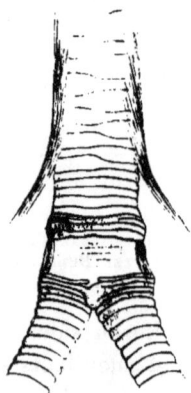

FIG. 197. SYRINX OF
Leptoptilus   (AFTER
WELDON).

another point of likeness ; in both
these birds a deepish groove runs from
the end of the nostril to the end of the
bill. This groove is also found, though it is not so conspicuous, in *Ardea* and *Butorides*, and (among storks only)
in *Platalea* ; it is suggestive of a recently closed, more
elongated nostril, like that of the cranes. The *procoracoid*
is more rudimentary than in storks, but the *coracoids* overlap
at insertion.

Family **Ciconiidæ**. I include in this family not only the
true storks but also the wood ibises (*Tantalus*). GARROD[2]

[1] *Histoire Naturelle, &c., de Madagascar,* 'Oiseaux,' p. 514.
[2] F. E. BEDDARD, ' Notes on the Convoluted Trachea of a Curassow (*No-
thocrax urumutum*), and on the Syrinx in certain Storks,' *P. Z. S.* 1886, p. 321.

contrasted *Ibis* and *Platalea* on the one hand with *Ciconia* and *Tantalus* on the other, on account of the following differences :—

Ibis and Platalea	Ciconia and Tantalus
Skull schizorhinal, angle of mandible produced.	Skull holorhinal, angle of mandible truncated.
Pectoralis major single.	Pectoralis double.
Accessory femorocaudal present.	Accessory femorocaudal absent.
Semitendinosus muscular throughout.	Semitendinosus tendinous for distal half.
Biceps slip present.	Biceps slip absent.

To which I add the form of the syrinx. These collectively appear to me to justify this separation. The true

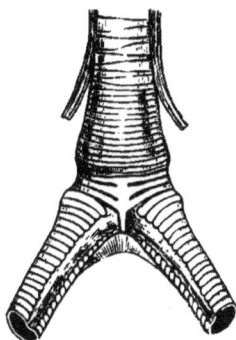

Fig. 198.—Syrinx of *Dissura episcopus* (after Beddard).

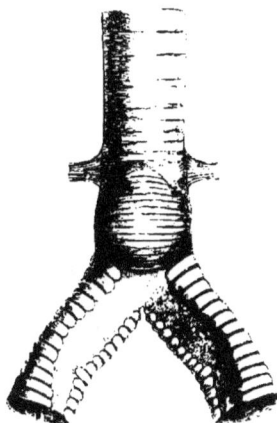

Fig. 199.—Syrinx of *Abdimia sphenorhyncha* (after Beddard).

storks, including *Tantalus*, are well characterised by the peculiar structure of the trachea and syrinx, there being, as already mentioned, an approach in these birds to the purely tracheal syrinx of the tracheophone Passeres. In the common black stork, *C. nigra*, the *syrinx* has the form illustrated in the figure (fig. 197). Its principal features are the absence of intrinsic muscles, the modification of the last tracheal rings, the existence of a rudimentary vocal process (see p. 69), and the closed character of the bronchial rings, which are *rings*, not semi-rings, the membrana tympaniformis being absent. In *Xenorhynchus senegalensis*,[1] however (fig. 200), there is some approach to the more typical tracheo-bronchial syrinx, which is further developed in *Abdimia sphenorhyncha* and *Dissura episcopus*;[1] the syringes of these two storks are illustrated herewith (figs. 199,

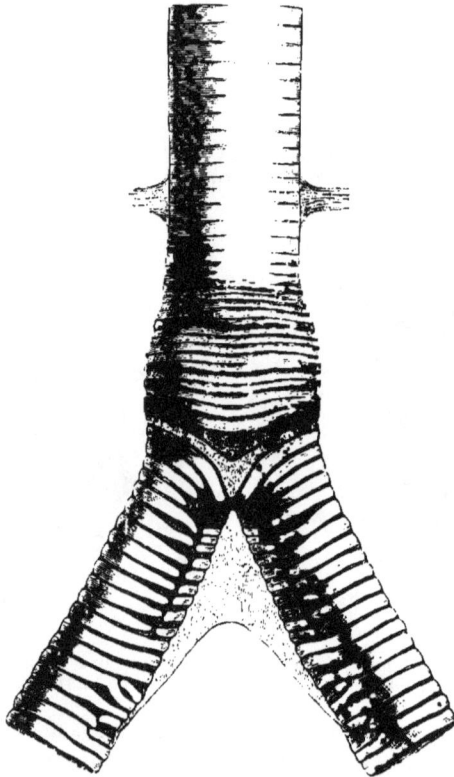

Fig. 200.—Syrinx of *Xenorhynchus senegalensis* (after Beddard).

198). In the former as well as in the latter the intrinsic muscles are still absent, and there is a considerable modification of the last tracheal rings; but in both there is a partial deficiency of cartilage where the membrana tympaniformis is developed in other birds. In *Xenorhynchus* this

[1] F. E. Beddard, 'A Note upon *Dissura episcopus*,' &c., P. Z. S. 1896, p. 231.

is but slight ; in *Abdimia* and *Dissura episcopus* it is well shown. But the latter bird has the complete bronchidesmus which marks the ciconiine as contrasted with the ardeine syrinx (cf. figs. 194 and 198). In *Mycteria americana* the bronchial rings are complete, but thinner internally, which is a hint of the otherwise absent membrana tympaniformis. The genus *Tantalus* is unique among the Ciconiidæ in having a convoluted trachea. This, however, is now known to occur only in the male of *T. ibis* ; in both sexes of *T. loculator* the trachea is unconvoluted.[1] In the former bird the tube makes several intrathoracic loops, as shown in the figure (fig. 201). The syrinx is essentially stork-like.

In most storks the muscle formula of the leg is AXY + ;[2] the only exceptions to this yet known are *Xenorhynchus senegalensis, Dissura episcopus*, and *Abdimia sphenorhyncha*, where the ambiens is absent, and *Leptoptilus crumeniferus* and *argala*, where there is no femorocaudal. The tendency to an ardeine structure in the syrinx of these birds has already been remarked upon, and may possibly be correlated with the absence of the ambiens. Storks have no biceps slip, but a typical expansor secundariorum. The humeral head of the anconæus is generally present. The patagial tendons are usually of a somewhat complicated form.

The *tensor patagii brevis* is constituted upon a similar plan in all storks, though there are naturally some little differences in detail.

In *Leptoptilus*, according to WELDON,[3] there is but one tendon which, widening out just before its insertion on to the fore arm, gives off a recurrent slip to the tendon of the *longus*.

In *Ciconia nigra*, according to FÜRBRINGER, the tendons

---

[1] 'On the Trachea of *Tantalus*,' &c., *P. Z. S.* 1878, p. 625 ; 'On the Form of the Trachea in certain Species of Storks and Spoonbills,' *ibid.* 1875, p. 297.

[2] A. H. GARROD, 'Note on an Anatomical Peculiarity in certain Storks,' *ibid.* 1877, p. 711. In a specimen of *Xenorhynchus australis* a few fibres corresponding to the accessory femorocaudal were found.

[3] *Loc. cit.* (on p. 419).

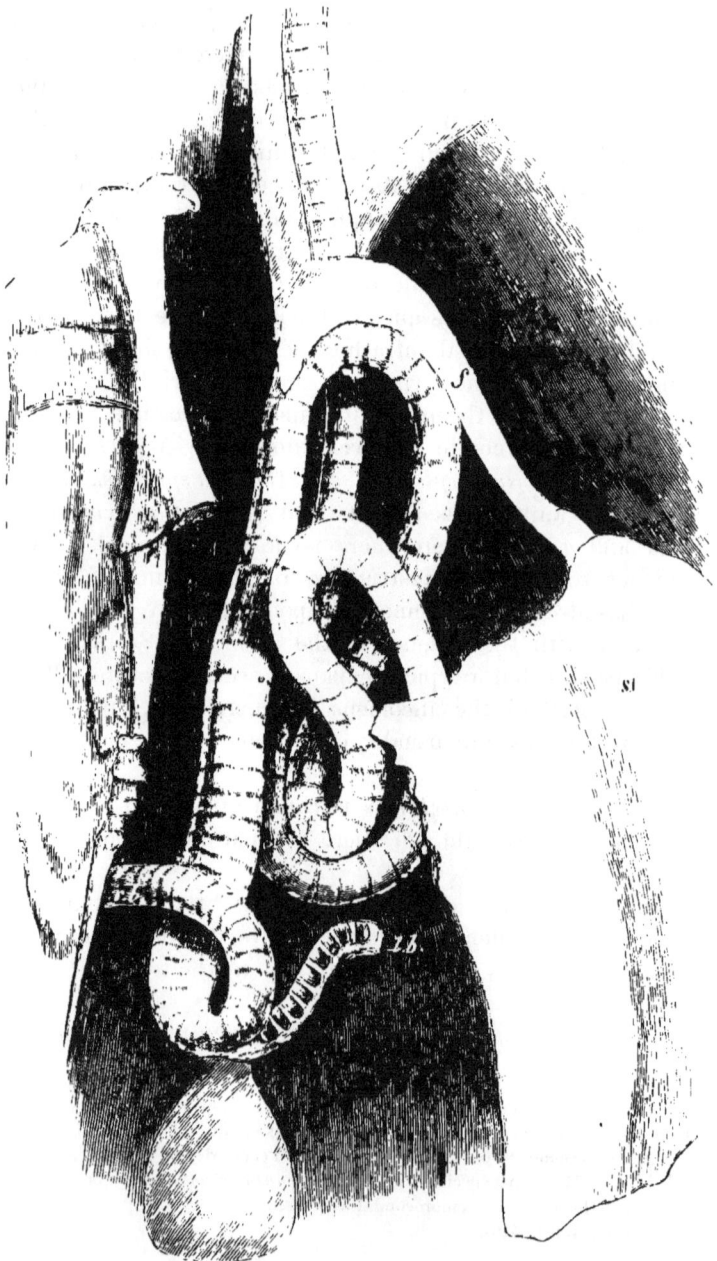

FIG. 201.— CONVOLUTED WINDPIPE OF *Tantalus ibis* (AFTER GARROD).
c, coracoid ; f, furcula ; st, sternum ; r.b, l.b, bronchi.

are a little more complicated. The tendon of the brevis is obscurely divided into two from nearly its commencement ; the more anterior of these again divides into two, one of which runs forward to be inserted on to the fore arm separately from the hinder part, which remains continuous with the rest of the tendon ; there is a recurrent slip to longus.

In *Abdimia* the tendons are much the same ; in both these genera the *propatagialis pectoralis* of FÜRBRINGER is muscular.

In *Tantalus leucocephalus* the broad fascia-like tendon of the brevis gives off a wristward slip, from the junction of which with tendon of *extensor matacarpi radialis* a patagial fan arises.

The pterylosis of the Ciconiidæ has been studied by NITZSCH. The neck is continuously feathered down to about the middle, where the spinal and ventral tracts respectively become divided into two. The two spinal tracts are narrow but strongly feathered, and cease abruptly at about the end of the scapula ; after a short space they recommence as a bifid but feebly feathered tract, the limbs of which unite a little way in front of the oil gland. The ventral tracts are broad upon the pectoral region but narrow towards the vent.

In *Pseudotantalus leucocephalus* NITZSCH could not find the aftershaft, but nevertheless one appears to exist.

In *Mycteria* and *Leptoptilus argala* the dorsal tract has posteriorly no spinal space.

In *L. argala* NITZSCH states the aftershaft to be absent.

The oil gland has two apertures on each side in *Ciconia alba*, five in *C. nigra* ; *L. argala* has no less than six. *Anastomus coromandelicus* has three ; *Tantalus leucocephalus* has the same number.

The deep flexor tendons of the Ciconiidæ are constructed on the plan of type I. In *Tantalus leucocephalus* there is a slight variation ; a small vinculum runs to flexor communis before the latter divides into three, and then a broader vinculum, chiefly going to tendon of digit II., but also slightly to III. and IV., binds together the two tendons. In *C.*

nigra there is the ordinary vinculum and a special slip to digit II.

The lungs in the Ciconiidæ are at least often distinguishable from those of the Ardeidæ by the great deficiency of the muscles arising from the ribs and attached to the pulmonary aponeurosis. In *Cancroma* there were four pairs of such muscles, arising from a corresponding number of ribs, in *Nycticorax* five; but in *Ciconia alba* 1 only found one pair inserted on to the aponeurosis in front of the septum bounding the anterior intermediate air sac anteriorly. Professor WELDON found no such muscles in a considerable number of storks.

The tongue is always small. The proventriculus is zonary. The liver is nearly equilobed, and there is always a gall bladder. The intestinal measurements of a number of species are as follows :—

—	S. I.	L. I.	C.
	Ft. In.	Inches	Inches
Ciconia nigra .	4·9	3	·5
„ alba .	3 6	2	·5
„ maguari . .	7·6		·2
„ boyciana . .	5 7		·3
Dissura episcopus . .	3 3	2·5	·3
Abdimia sphenorhynchus .	2 6	3·5	·15
Mycteria americana . .	6 4	4	·25
Xenorhynchus australis .	5	3·5	·25
„ senegalensis	5·5	5	·4
Leptoptilus crumeniferus .	8·5		·25
„ argala . .	6·5		·25
Tantalus ibis . . .	5		
„ loculator . .	6·6	2	·5

The number of cervical vertebræ is seventeen (*Leptoptilus, Tantalus*) or eighteen (*Xenorhynchus*): the hypapophyses are feeble; there is a ventral canal formed by union of catapophyses of C7–C11 (*Xenorhynchus*), C7–C12 (*Tantalus*). Four ribs reach the sternum in *Xenorhynchus*, five in the others. The sternum has one pair of notches, and the spina externa is absent or small. The procoracoid is of fair size, but does not reach the clavicle; the coracoids meet

at their sternal insertion in *Leptoptilus*.[1] The hypocleidium articulates with carina sterni. The skull is desmognathous, holorhinal, and without basipterygoid processes; there
are in *Tantalus* rudiments of these processes in the shape of
a minute spine. The holorhinal character of the nostrils is
largely marked by ossifications of the alinasals; the nostrils
are thus much reduced in size, in a fashion suggestive of the
Steganopodes and possibly significant, but there is no bony
septum between them. This distinguishes the storks from
the herons, as does also the form of the palatines. These
bones are in the first place not cut off at right angles behind,
as in the herons, while the internal lamina only bounds the
interparietal space, and is at most (*Xenorhynchus*) carried
back to the end of the bones as a slight median keel. This
is absent in other storks. In *Xenorhynchus* and *Leptoptilus*
the palatines again approach each other, and are only separated by the vomers just behind the maxillo-palatines. Opposite to this point each palatine is produced into a strong
outwardly directed snag, large in *Xenorhynchus*, hardly indicated in other storks. The interorbital septum is entire.
The large lacrymal is perforated or deeply notched for duct
of gland.

Family **Ardeidæ.**—The herons contrast with the storks in
(1) the tracheo-bronchial syrinx always furnished with a
pair of intrinsic muscles, (2) non-division of *pectoralis primus*
into two layers, (3) *invariable* absence of ambiens, (4) presence of powder-down patches, (5) absence or weakness of
vinculum, (6) presence of a single cæcum only.

On the other hand the two families agree in (1) absence
of biceps slip to patagium, (2) presence of expansor secundariorum, in addition, of course, to the points enumerated in
the definition of the group.

In *Cancroma* the oil gland is nude. The herons have
four or six *powder-down tracts*. Six are found in *Cancroma*,
Butorides atricapillus, *Ardea cocoi*, and other species;

---

[1] Not in *Xenorhynchus* and *Dissura*; there is a trace of an overlap in
Abdimia, *Tantalus*, and *Platalea*.

four in *Ardetta* and *Botaurus.* The number of rectrices differs
much ; there are only eight in *Ardetta exilis* and *A. involucris,*
ten in *Botaurus* (not invariably), twelve in *Ardea cocoi* and
Cancroma cochlearia.

The *pterylosis* differs from that of the storks by the
narrowness of the tracts. The spinal space begins with the
commencement of the neck, and only terminates a little way
in front of the tufted oil gland. The anterior part of the
spinal tract is not always more strongly feathered than the
posterior part, and there is (according to the figures of
NITZSCH) hardly a break between them. The ventral tracts
also are separate early on the neck. In *Cancroma* they divide
on the breast into a broader, stronger outer tract, which ceases
just below the metapatagium, and a narrow inner tract. The
anterior ventral powder-down patches constantly interrupt
the continuity of the anterior and posterior sections of the
ventral tracts.

	Small Intestine	Large Intestine	Cæcum
Ardea Goliath .	90		·25
„ *sumatrana*	57		·15
„ *purpurea*	46	1·5	·15
„ *cocoi* .	61	3	·4
„ *egretta* .	70	1·5	·25
„ *garzetta* . .	45		·15
Butorides atricapillus	25	3·6	
„ *cyanurus* .	30	3·5	·15
Ardetta involucris .	22·5	2·25	
Nycticorax caledonicus	40	3·5	·15
„ *violaceus* .	60		·2
Tigrisoma brasiliense	33	3	·14
Cancroma cochlearia	33	3	·15
Botaurus stellaris .	51	4	·15

A curious absence of any apparent relationship between
the relative length of the sections of the alimentary canal,
and of the alimentary canal as a whole, and of the food, is
shown by the above table, drawn up by Mr. GARROD,
which I reproduce here from his notes.

They seem, however, to ally the last four genera, which

have a long large intestine. As a rule the left lobe of the liver is the smaller, while in the storks the lobes are equal. A gall bladder is always present except in *Botaurus stellaris*. The cæcum is single, and, as will be seen from above measurements, rudimentary.

The carotids are, as a rule, two, and separate : but in *Botaurus* they fuse, and in *Ardetta involucris* the right only is present.

Of the leg muscles the *ambiens* is always absent, and the formula is typically AXY −. The *femorocaudal* is never strong, and is particularly slender in *Ardea Goliath*.[1] In *A. sumatrana*, *A. ludoviciana*, *Nycticorax Gardeni*, *Cancroma cochlearia*, and *Tigrisoma brasiliense* it is totally absent.

The *deep plantar tendons* are characteristic ; there is almost always a very slender vinculum between the two, which is totally wanting in *Botaurus stellaris*, *Ardetta involucris*, and *A. exilis*.

The *tensores patagii* are stork-like. The tendon of the *brevis* bifurcates, and from the point where the anterior limb is inserted on to tendon of extensor of fore arm a recurrent slip is given off to *longus*. This arrangement holds good for *Ardea purpurea*, *A. Goliath*, *Cancroma cochlearia*, and *Nycticorax griseus*; but in *Cancroma* the recurrent slip is sometimes absent. The *pectoralis abdominalis* is present, and thus serves to differentiate them from the storks, ibises, and spoonbills, in which the muscle is absent.

The *skull* of the Ardeidæ has been chiefly studied by SHUFELDT.[2] In the more normal forms (e.g. *Ardea cinerea*, *Butorides cyanurus*) the skull is holorhinal, the holorhiny not being obscured -as it is often among the storks—by the irregular ossification of alinasals. The vomer is well developed, much compressed laterally, and largely double. The maxillo-palatines are spongy bones, largely free from each other posteriorly. The palatines (see fig. 202)

---

[1] Sometimes absent in this species.

[2] 'Osteological Studies of the Subfamily Ardeinæ,' *Journ. Comp. Med. and Surg.* 1889.

are very straight bones, usually cut very squarely behind, but notched postero-laterally in *Ardea bubulcus, A. minuta, A. comata,* with well-marked internal laminæ, which extend quite to the posterior end of the bone. The interorbital septum is largely fenestrate. The bony ectethmoids are but little developed; *Cancroma* is in several ways rather anomalous. It is heron-like in the fenestrated interorbital septum, and in the fact that the internal lamina of each palatine is continuous to the posterior end of that bone. The nostrils are continued forward by a deeper groove than that which is formed in the more normal herons—a point of likeness this to *Scopus* and *Balæniceps* (*qq.v.*) The very broad palatines join the vomer again in front of their posterior junction with each other, thus dividing the interpalatine vacuity into two areas. It is thus to a certain degree 'doubly desmognathous,' and is so far like *Xenorhynchus* (see p. 429). There is a well-marked lateral process of the palatines, as in *Scopus* and storks.

As in *Scopus* the *procoracoid* is very small and the *coracoids* overlap each other at their insertion. Like the storks, and unlike *Scopus,* the hypocleidium articulates with the end of the carina sterni. The hypocleidium, moreover, projects backwards between the two clavicles as a narrowish piece.

The hæmapophyses of the *cervical* and *dorsal vertebræ* are small, those of the latter being sometimes quite absent. There is a catapophysial canal (in *Cancroma* as well as *Ardea*), formed in the two types mentioned by cervical vertebræ 7–12.

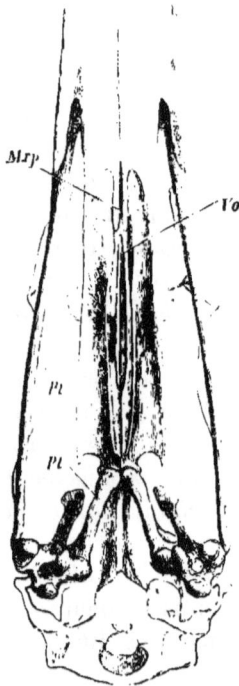

FIG. 202.—VENTRAL SURFACE OF SKULL OF *Ardea cinerea* (AFTER HUXLEY).

Pt, pterygoids; *Pl,* palatines; *Vo,* vomer; *Mxp,* maxillo-palatines.

Family **Balænicepidæ.**—The great ' whale head ' of Africa, *Balæniceps rex*, requires further study before its exact position can be determined. It is admittedly a member of the present group, though its original describer, GOULD, regarded it as a pelican.[1] We know the skeleton through the labours of PARKER,[2] while our at present scanty know-

FIG. 203.—SYRINX OF *Balæniceps*. FRONT VIEW. (AFTER BEDDARD.)
b, free margin of bronchidesmus.

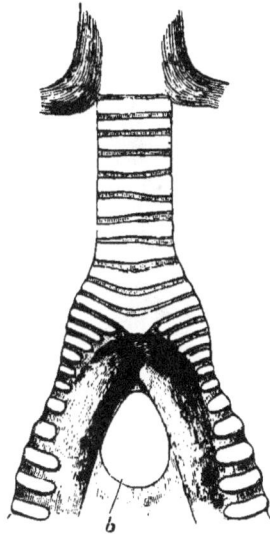

FIG. 204.—THE SAME. BACK VIEW. (AFTER BEDDARD.)

ledge of the soft parts is due to myself.[3] Its powder-down patches were discovered by BARTLETT.[4] As in the herons also the right lobe of the liver is the largest, and the cæcum is single. The *syrinx* (see figs. 203–205) is ardeine in form, but lacks the intrinsic muscles. These, however, are not en-

[1] Not a serious mistake in view of the admitted relationships between the **Steganopodes** and **Herodiones.**

[2] 'On the Osteology of *Balæniceps rex*,' *Trans. Zool. Soc.* iv. p. 269 (abstr. in *P. Z. S.* 1860, p. 324).

[3] 'On certain Points in the Visceral Anatomy of *Balæniceps* bearing upon its Affinities,' *P. Z. S.* 1888, p. 284.

[4] 'On the Affinities of *Balæniceps*,' *ibid.* 1861, p. 131. See also REINHARDT, 'On the Affinities of *Balæniceps*,' *ibid.* 1860, p. 377, and GIEBEL, 'Ueber *Balæniceps rex*,' *Zeitschr. f. d. ges. Nat.* lxi. 1873, p. 350.

F F

tirely absent; their former presence is indicated by a narrow ligament on each side (fig. 42, p. 62), which occupies the place that a muscle should, and is attached precisely where the intrinsic muscles are attached in other Ardeidæ. The membrana tympaniformis is well formed, the bronchidesmus is incomplete, while the general form of the organ is purely tracheo-bronchial and thoroughly ardeine. This will be apparent from the annexed woodcuts.

'The nearest relations of *Balæniceps*,' said PARKER, 'are the South American boatbill (*Cancroma cochlearia*) and the little South African umbre (*Scopus umbretta*).' The interorbital septum is stork-like in its completeness. The lacrymal, as in *Scopus* alone among **Herodiones**, reaches as far down as the quadrate jugal bar, but it is fused anteriorly with the walls of the skull. The nostrils are continued forward by a groove precisely like that of *Scopus* and *Cancroma*. In the palatine bones the fusion of the internal laminæ to form a median keel behind the interparietal space is precisely like *Scopus*; so, too, is

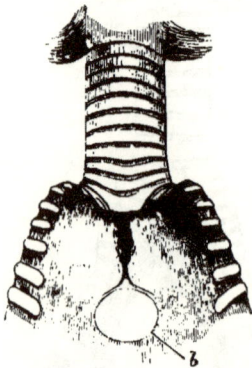

FIG. 205.—SYRINX OF *Balæniceps*, ARRANGED TO DISPLAY PESSULUS AND MEMBRANA TYMPANIFORMIS (AFTER BEDDARD).

the lateral angle of these bones (see p. 422). There is a firm synostosis between the furcula and the carina sterni.

Cervical vertebræ 7–13 have, as in most other **Herodiones** (excluding, however, the supposed ally of *Balæniceps*, *Scopus*), a ventral catapophysial canal.

The family **Plataleidæ** includes not only the spoonbills but the ibises. The name Hemiglottides was applied by NITZSCH to the group 'on account of the surprising smallness of their tongues.'

The *pterylosis* is exactly as in the storks.[1] The rectrices

---

[1] According to NITZSCH It appeared to me (in *Platalea rosea*) to be more like that of *Tantalus loculator*, in that the hinder part of the spinal tract was not bifid, but continuously though sparsely feathered.

are twelve. The oil gland of *Platalea leucorodia* has three
distinct orifices on each half, that of *Ibis* only one. The
long downwardly bent bill of the ibises distinguishes them
from the storks and suggests *Numenius*. NITZSCH, indeed,
regarded the birds as intermediate between the two groups
represented by these types.

The *tensores patagii* have always a *biceps slip* running to
the tendon of the longus, and there is a patagial fan.

In *Ibis æthiopica* the tendon of the tensor brevis is
simple and rather diffuse. In *Eudocimus ruber* and *G. melanopis* the tendon gives off a distinct wristward slip, while the
patagial fan is formed of two rather separate strands, with
the posterior of which, rather high up, the wristward slip
fuses in *Geronticus melanopis*.

In *Platalea* the muscle and its tendons are much the
same, but the *brevis* is very broad and fascia-like.

The *muscle formula* of the leg is complete (*i.e.* ABXY +)
in all Plataleidæ.

The *plantar tendons* are connected by a vinculum which
in *Eudocimus ruber* extends on to the special slip to digit II.
By the division of this vinculum may have arisen the two
vincula of *Ciconia nigra* (see above, p. 427).

The *liver* is equilobed (*I. æthiopica*), or the right is a
little larger (*E. ruber*). A *gall bladder* is present. The
following are intestinal measurements :—

	s. l.	l. I.	C.
Eudocimus ruber	31	1·6	·25
Nipponia Temminckii	62		·35
Plegadis falcinellus .	42		·25 [1]
Ibis æthiopica .	40		·12
„ *strictipennis* .	36		·2
Platalea leucorodia .	70		·12
„ *ajaja* . .	52		

The intestinal convolutions of *Platalea leucorodia* are
shown in fig. 207. The greater part of the gut has preserved

[1] HUNTER, in *Essays and Observations* (ed. Owen, London), 1861, writes of
this species : ' The cæca are about four inches long and very small, attached to
the ileum their whole length.'

the primitive arrangement. The duodenal loop is curved, and in other storks (*s.s.*) this (cf. fig. 208) is converted into a spiral. There is a tendency, in fact, among the Ciconiidæ

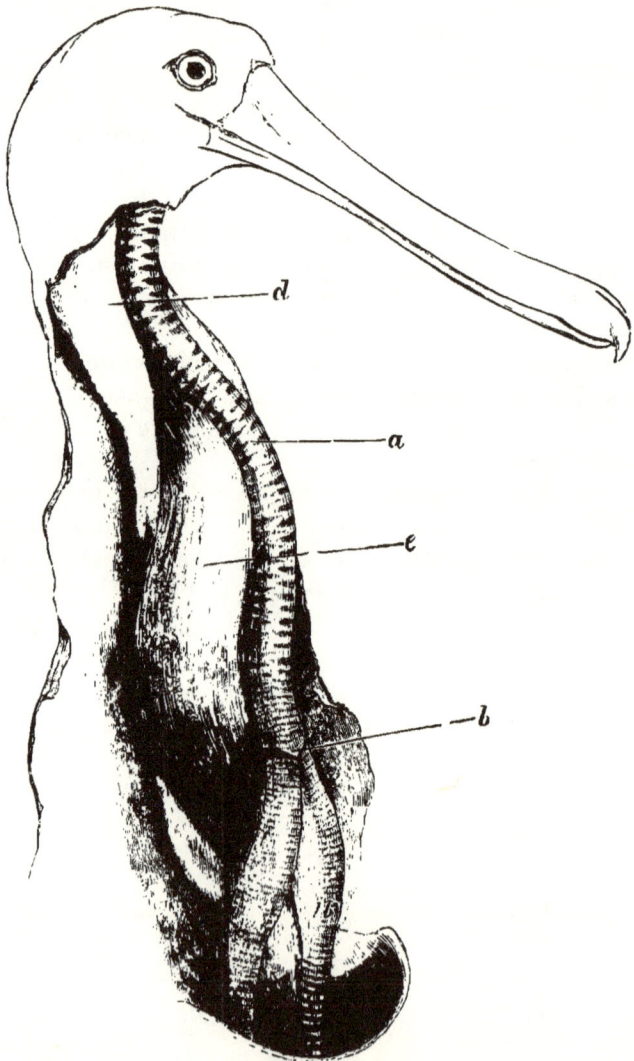

FIG. 206.—WINDPIPE OF *Platalea ajaja* (AFTER GARROD).

a, trachea ; *b*, bifurcation of bronchi in front of sternum ; *r.b*, *l.b*, bronchi ; *d*, œsophagus *e*, cervical muscles.

to the formation of these spirals, which are also found in the
Accipitres.

The *windpipe* in the ibises is simple, not convoluted;

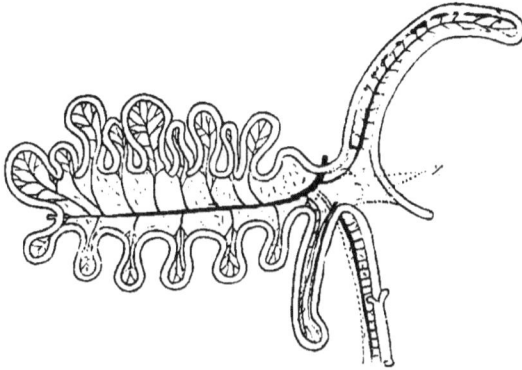

FIG. 207.— INTESTINES OF *Platalea leucorodia* (AFTER MITCHELL).
x, short-circuiting vessel divided.

but in *Platalea leucorodia* it is convoluted.[1] In this bird
the windpipe runs in a straight or slightly sinuous course

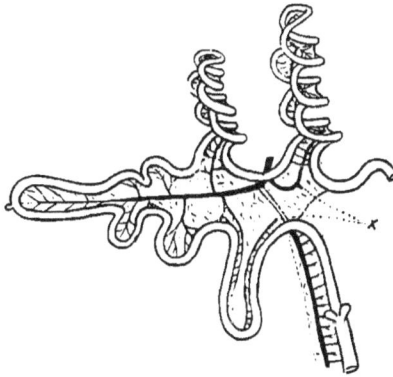

FIG. 208.—INTESTINES OF *Ciconia nigra* (AFTER MITCHELL).
x, as in fig. 207.

into the thorax; it then bends upon itself, and after an inch
and a half or so reverts to its original direction, and divides
within the thorax into the two bronchi.

[1] I have a windpipe of *P. leucorodia* which is not convoluted, but which
bifurcates within the thorax.

In *Platalea ajaja* (see fig. 206) the trachea is peculiar in that its bifurcation takes place about the middle of the neck, some distance at any rate before the entry of the bronchi into the thoracic cavity.

In a female which I measured the bifurcation was three inches in front of sternum, and the distance from the bifurcation to the end of the larynx was six and a quarter inches. Hudson has stated [1] that there are two species of spoonbill in the Argentine : one has a windpipe like that figured above, and may be considered to be the true *ajaja*, while the other has a windpipe which is not modified in any way. The characters of the latter—apart from the question of the windpipe—have been considered by some to be those of immaturity ; but this is denied by Hudson.

I have in my possession seven windpipes of *P. ajaja*, of which two are males and the rest females. In the two males the lengths of the trachea are practically identical. In the females the position of the syrinx varies considerably, being more than an inch lower in some than in others ; this may possibly account for the differences observed by Mr. Hudson.

The extrinsic muscles of this species arise from inner side of coracoid, not far from the middle line of sternum. The intrinsic muscles of the syrinx are present ; they cease, however, some little way in front of the syrinx. These muscles are also inconspicuous, and form a very thin though rather wide band. The syrinx is not especially stork-like, except in the fact that the bronchi are for the greater part of their extent tubes with complete rings. The first dozen or so, however, are incomplete, there being thus a membrana tympaniformis.

There is a cartilaginous three-way piece, and generally there is but little ossification in the windpipe.

The windpipe of *P. alba* is different from both the species already described.

It is not convoluted, but the bifurcation does not take place outside of the thoracic cavity, as in the last species. It

---

[1] Sclater and Hudson, *Argentine Ornithology*, vol. ii. p. 115.

is also stork-like in the fact that there is no membrana tympaniformis; the bronchial rings are complete rings from the very first.

The main peculiarity of the windpipe, however, lies in the fact that from a point about three inches from the larynx it appears to bifurcate and to consist of two closely applied tubes. By cutting windows it was ascertained that this appearance corresponded to the reality—that the windpipe did consist of two tubes. This arrangement, which seems to characterise the present species only, is, of course, suggestive of the median tracheal septum of the penguins, &c.[1]

Of the ibises I have examined the windpipe of *Eudocimus ruber*, *E. alba*, and of *Ibis æthiopica*. There are differences between the species similar to those which exist between the species of *Platalea*. In *Ibis æthiopica* there is no membrana tympaniformis, the bronchial rings being complete. In the two other species this structure is present, and moreover the last few tracheal rings are defective in the middle line posteriorly.

The extrinsic pair of muscles are attached in *E. ruber* to the inner surface of sternum not far from the middle line. The intrinsic muscles stop short some way in front of the syrinx.

In *Platalea ajaja* cervical vertebræ 7–10 have a catapophysial canal. Some of the dorsal vertebræ are fused.

The number of *cervical vertebræ* is seventeen or eighteen (*Platalea ajaja*). The sternum is two-notched (*Platalea*) and has a small spina externa, but no spina interna.

The *skull* has, as in charadriiform birds,[2] a pair of occipital vacuities.

It is schizorhinal and has rudimentary basipterygoid processes, which in *Platalea ajaja* have the form of small sharp thorns. Supra-orbital impressions are feebly developed. The lacrymals with the ectethmoids very nearly form a complete ring of bone, the aspect of this part of the skull being

---

[1] GADOW, in NEWTON'S *Dict. Birds*, speaks of the trachea (*sub voce*) of *Platalea* as divided by a cartilaginous septum. I have not found this to be the case with *P. ajaja* or *P. leucorodia*.

[2] Also, however, in geese.

distinctly charadriine.   A groove runs forward from the
nostrils, as in *Scopus*, &c. (*q.v.*)

Family **Phœnicopteridæ.**[1]—*Phœnicopterus* agrees with
the storks in the subdivision of the prebronchial air sacs by
many septa into smaller chambers.   But, as this also occurs

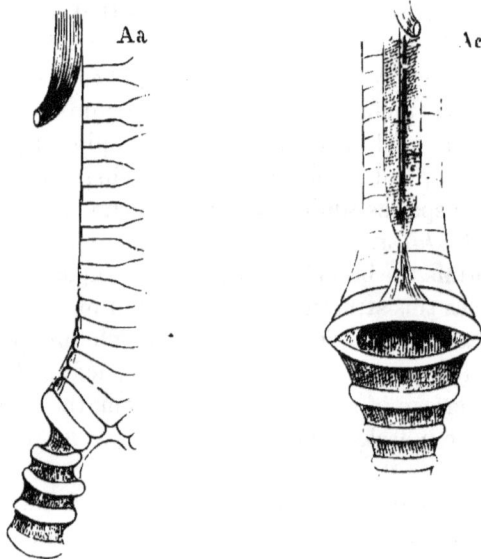

FIG. 209.—SYRINX OF *Phœnicopterus* (AFTER WELDON).
Aa, front view ; *Ac*, lateral view.

in *Chauna* and to a much greater extent, less weight must
be laid upon it than upon some of the muscular characters.

In the storks, as in *Scopus*, but not the Anatidæ, the pectoralis major is divided into two distinct layers.   This is also
the case with *Phœnicopterus*.   The *tensores patagii* are
closely similar in the birds under comparison and diverge
from those of the duck.   The ducks are peculiar in the origin
of the smallest head of the gastrocnemius from the biceps ;

[1] Our knowledge of the anatomy of the 'soft parts' of *Phœnicopterus* is
mainly due to WELDON (*P. Z. S.* 1883) and to GADOW (*Journ. f. Orn.* xxv.
p. 382.)

no such connection occurs in either stork or flamingo. That the deep flexor tendons of *Phœnicopterus* are not stork-like is surely related to the diminutive hallux of that genus.

On the other hand the *accessory femorocaudal*, though small, is present in the flamingo, though absent in the storks,[1] while the *syrinx* (fig. 209) is not stork-like. The *cæca* are long (three inches), but the *intestines* are not duck-like.

The *atlas* is notched by the odontoid process, and the notch is very nearly converted into a foramen. There are nineteen cervical vertebræ. None of the catapophyses fuse to form a canal. The transition between catapophyses and hæmapophyses is more complete than in most birds ; on the last cervical the two catapophyses are raised on a common platform, and on the first dorsal is the first (and last) hæma-pophysis, which is flattened and obscurely bifid. The last cervical and the first three dorsals are fused. Five ribs reach the single-notched sternum. The coracoids overlap at their insertion.

The *skull* is desmognathous, with basipterygoid processes, to which the anterior ends of the pterygoids are attached. It is holorhinal with pervious nostrils. There are lateral occipital fontanelles. The lacrymals are large and rather duck-like, notched externally ; they nearly reach the jugal bar. There are no ossified ectethmoids.

In including the Plataleidæ and Ibididæ with the **Herodiones** I shall have the assent of most ornithologists. Alone among recent observers who have occupied themselves with the structure of the group, GARROD and FORBES placed the spoonbills and ibises apart. The latter included them in his group Pluviales with the Charadrii, *Rhinochetus*, &c. These diverse opinions about the Plataleidæ appear to me to be largely due to the primitive position which they occupy among the **Herodiones**. They are to my thinking not far from a basal ' gralline ' stock. The Plataleidæ have the complete muscle formula, and the biceps slip to the patagium. Some of these muscles have been lost in the other **Herodiones**, in which

[1] With the partial exception noted on p. 425, footnote.

there is traceable a gradual series of modifications. In the
storks the accessory femorocaudal is always absent, and in
some the ambiens and femorocaudal also ; in the Ardeidæ both
ambiens and accessory femorocaudal are gone, and the
modification of these leg muscles culminates in the aberrant
Cancroma, where the formula is merely XY−. In the
Plataleidæ (as in *Tantalus*) there are just faint traces of the
basipterygoid processes, missing elsewhere in the group.
The ibises are schizorhinal, and in *Platalea* the ends of the
nasal grooves are rounded, thus tending towards the holorhinal, while the lower part of the bony nostril is wider than
the upper, appearing thus to show a commencing occlusion
of the schizorhinal nostril into a holorhinal one ; in *Ardea
cinerea* a faint trace of a former schizorhinal condition is
seen in a slight groove which runs back from the end of the
holorhinal nostril, as in *Chunga* (see p. 144). The schizorhinal condition, as has been before pointed out, is probably
the more archaic. Finally the Plataleidæ have, according
to MITCHELL, the most primitive form of gut among the
Herodiones. Significant points of likeness between the
Plataleidæ and **Grues** are not wanting ; the occipital foramina
and the impressions of the supra-orbital glands were among
the facts that led FORBES to associate them together. The
convoluted windpipe is common to *Platalea*, *Tantalus*, and
Grus. Both *Platalea* and *Grus* have the complete muscle
formula and schizorhinal nostrils.

It has been often asserted that there are likenesses
between the **Herodiones** and the accipitrine birds. This
largely reduces itself to a comparison between the **Herodiones**
on the one hand and the Cathartidæ and *Serpentarius* on
the other ; for if the two latter forms of accipitrine birds are
rightly so placed the falcons must, on account of the various
reductions in their structure, be derived from some form near
to these, and cannot be their ancestors. There are two at
first sight rather striking likenesses between the **Herodiones**
and these lower accipitrines. GADOW has commented upon
resemblances in the lie of the intestines, and MITCHELL has
still further emphasised this likeness. In both groups (the

falcons only among the **Accipitres** were examined) there is a
tendency for the intestine to be thrown into spirally twisted
loops; but MITCHELL is of opinion that this is really no
more than a convergent resemblance, for as the simpler
types are considered the spiral arrangement becomes less
and less obvious, thus indicating its special development, and
independent development, in both.

The second point concerns the syrinx, which as a specialised organ is wanting in both storks and Cathartidæ. In
neither group are there intrinsic muscles. This point of
resemblance rests, however, upon mere negativity. The
details of the conformation of the lower part of the larynx
are, as may be inferred from the preceding explanations and
figures, extremely different in both; it seems as if the syrinx
has degenerated in both, but along quite different lines, the
loss of the intrinsic muscles being about the only point in
common. And we know from other groups that this muscle
may be independently lost. There is, to my mind, as much
to be said for a derivation of the **Accipitres** from the crane as
from the pelargine stock, the fact being that we must
probably seek for the origin of both from a low branch,
perhaps common to all. The matter is further dealt with
under the section devoted to the **Grues**.

The skull of the flamingo is not duck-like. The back
view of it, with the occipital fontanelles, might, it is true, be
mistaken for that of an anserine bird; but there are no
salient likenesses elsewhere. There are but rudimentary
basipterygoid facets, and the palatines have (which they
have not in the duck tribe) a well-developed internal lamina,
which, as is so often the case, is sharply bent downwards at
its edge. In front of the fused maxillo-palatines there is no
palatal vacuity, as in nearly all **Anseres**. The lacrymals,
like those of the **Anseres**, are certainly large; but their size is
not so conspicuously marked in the length of their base of
attachment to the margin of the orbit as in the length and
great breadth of the descending process, not a feature of the
anserine skull. The ectethmoids seem hardly ossified. The
interorbital septum is largely deficient in front, as in *Chauna*,

but is also more vacuolated behind, and in a different way from what is found among the few **Anseres** in which the interorbital septum is not solid. The length of the angular process of the mandible, though a duck character, is also found in the Plataleidæ. There are so many characteristic duck-like characters wanting in the skull of *Phœnicopterus* that we cannot place it with that group.

Of fossil **Herodiones** a considerable number of genera and species have been described. These range from the Cretaceous to cave deposits, and have been found in Europe, India, Mauritius, and Rodriguez. If *Scaniornis*, described lately by DAMES,[1] from the Cretaceous of Sweden, be really an ally of the flamingo-like bird *Palælodus*, it is important to note that this family goes back further into the past than any other living family, so far as our information allows us to say. HUXLEY's name of ' Amphimorphæ ' for the group, and his remark to the effect that they are so thoroughly intermediate between the storks and ducks, will occur to the mind in this connection. This bird is known by a scapula, coracoid, and humerus.

The other forms upon which new genera have been founded are not known by even so much as the scanty remains of *Scaniornis*. Thus *Palæociconia*, *Propelargus*, *Ibidopodia* are only known by the tarso-metatarsus (incomplete), while the other genera are founded upon equally fragmentary remains. *Tantalus Milne-Edwardsii*[2] (nearly perfect tibio-tarsus), from middle Miocene, France. *Palælodus*, however, is known by coracoid, scapula, and some of the ' long bones,' as well as the sternum, ' scarcely to be distinguished from the somewhat larger sternum of *Phœnicopterus roseus*,' furcula, one or two vertebræ, and metacarpal bones. The bird seems not to have had such long legs as the modern flamingo, but longer toes. *Elornis* (Eocene and Miocene) was also a flamingo, but intermediate in the length of its legs between *Phœnicopterus* and *Palælodus*. *Agnopterus* (Miocene) is reckoned a flamingo by FÜRBRINGER, placed ' incertæ sedis' by LYDEKKER.

---

[1] ' Über Vogelreste aus dem Saltholmskalk von Limhamn bei Malmö,' *Bih. K. Svensk. Ak. Handl.* xvi. 1891.

[2] SHUFELDT. ' Fossil Bones of Birds,' &c., *P. Acad. Nat. Sci. Philad.* 1896, p. 507.

TUBINARES

Definition.—Nostrils produced into tubes. Aftershaft present, aquinto-cubital. Oil gland tufted. Skull schizognathous, holorhinal. Two carotids present. Large supra-orbital glands.

The peculiar form of the external nares, which has given to the group its name, characterises it. It is a character which is found in no other group,[1] but in all members of the present. The **Tubinares** are nearly the only large group of birds which can thus be diagnosed by a single character. The tube itself is, according to FORBES, whose ' Challenger ' memoir upon the group forms the chief classic upon the subject, in *Majaqueus*, and probably in other genera, caused by a growth of the catilaginous walls of the nasal sacs. The degree of fusion between the two tubes varies in different genera. In *Procellaria*, for example, they quite coalesce and the external aperture is single. In *Pelecanoides* there is a distinct and not broad septum ; while in *Bulweria* and others the septum is so broad that the tubes almost appear double.

The petrels are web-footed birds with a small hallux, which in *Pelecanoides* is quite absent. The web does not take in the rudimentary hallux. The general number of the *rectrices* is twelve, but *Ossifraga* has as many as sixteen. The *aftershaft* is always present. The *pterylosis* does not vary greatly ; the dorsal and ventral tracts are well separated upon the neck. The ventral pteryla is divided on the neck ; in the pectoral region each branch again divides. The dorsal tract divides at the middle of the scapulæ into two, which unite later, thus enclosing a space.

The *tongue* varies much in size and in the amount of its spiny bordering or covering. In *Diomedea*, for instance, it is much covered superiorly with spines. In *Œstrelata* and others the tongue has a bordering of spines which are lateral

---

[1] It has been pointed out that the complex and somewhat protuberant nostrils of *Chionis* bear a little resemblance to those of the Tubinares, but are differently developed.

as well as posterior; in *Daption capensis*, and in many others, the spines are confined to the hind margin of the organ.   The more usual condition is the intermediate form.

There is a well-developed but small *gizzard* present. The *cæca* are absent in the Oceanitidæ, but present, as a rule, in the Procellariidæ; they are small and nipple-like, and in *Cymochorea* appear to be reduced to a single cæcum.

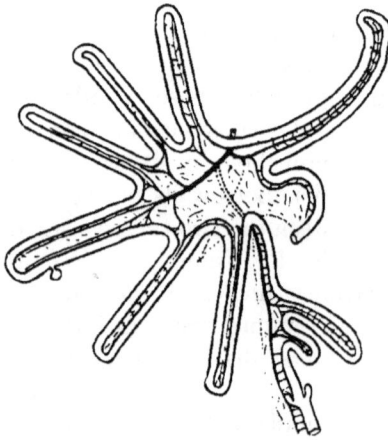

Fig. 210.—Intestines of *Fulmarus glacialis* (after Mitchell).
r, short-circuiting vessel divided.

The *gall bladder* is always present, and the lobes of the *liver* are equal or nearly so. The arrangement of the intestine is shown in fig. 210.   The duodenal loop is double; the greater part of the intestine is drawn out into a considerable number of straightish loops.

As to *muscular anatomy*, the *great pectoral* is divisible into two layers by an interposed tendon, as in storks and Steganopodes.   But in *Larus*, according to FÜRBRINGER (and in *Podica* also), the same division occurs, which tends to lessen the differences between the **Tubinares** and the Laridæ, so insisted upon by GARROD, FORBES, and some others.   The *tensores patagii* in the **Tubinares** are complicated, but not in all the genera.   In the Oceanitidæ they are simplest.   In these petrels the tensor brevis is a simple tendon.   In *Pelecanoides*, and in some others, there is the additional complication that the tendon bifurcates near to its attachment on the extensor tendon, and gives off an anterior slip inserted more wristwards.   In *Prion* affairs are still further complicated by the metamorphosis into tendon of the whole of the *extensor metacarpi radialis*

superficialis. In *Œstrelata brevirostris* we first meet with
a recurrent slip going to the longus tendon and arising
from the brevis in front of the termination of its anterior
branch. The more typical tubinarian arrangement is seen
in *Œstrelata Lessoni.* We find here the characteristic
ossicle of the tendon of the brevis which is found in
so many **Tubinares**, and which has been held by some to
be a character of much systematic importance in differen-
tiating the group—a character which FÜRBRINGER thinks
has been 'overvalued.'[1] From this ossicle spring some
of the fibres of the extensor muscle of the fore arm ; it is
also the starting point of the recurrent tendinous fibres,
which unite the brevis and longus tendons ; these tendons
are, moreover, in close apposition for nearly the whole of
their course—itself a characteristic feature of the **Tubinares**.

In *Ossifraga* and in some other genera there are no
ossicles, but the tendons are highly complicated. In some
petrels—for instance, in *Diomedea exsulans*—the wing sesamoid is double, and in this bird also there are considerable
complications of the various tendons. While, therefore, we
cannot define the petrels by the arrangement of the tendons
of the tensores, as is sometimes possible, it is evident that
we have what might be expected in a large and important
group—a very considerable series of modifications of these
organs. No petrel has, strictly speaking, a *biceps slip*, and
the *biceps* itself, though perfectly normal in origin and
insertion, has much more largely degenerated into tendon
than is usual for this muscle. There is, however, a curious
modification of this muscle in *Pelecanoides* and in a few
others. Here the coracoidal head alone forms the muscle :
the humeral head goes entirely to the *tensor patagii longus* ;
this slip is, therefore, as FORBES remarks, 'functionally a
biceps slip.' Something apparently representing the true
biceps slip is occasionally found in the **Tubinares**. In a
few there is a tendon derivable from the humeral head of the

[1] Justly, as it is found not only among the **Steganopodes**, which may be
fairly regarded as allies of the **Tubinares**, but also in so remote a type as
Merops !

biceps, which appears to end on the fascia of the wing.[1]
The *expansor secundariorum*, found only among the
Oceanitidæ, is peculiar in that it arises from the surface of
the pectoral ; its muscular belly is, as usual, at the elbow,
and the tendon is joined by a branch from the scapularies.
The *anconæus* has a well-marked tendinous attachment to
the humerus.

The muscles of the hind limb, to which GARROD attached
so much classificatory importance, vary much in the group.
The *ambiens* is present in all except *Fregetta* and *Pelecanoides*. In *Garrodia* and some others the tendon does
not cross the knee.[2]   All have a femorocaudal, but the
accessory is absent in *Bulweria* and *Pelecanoides*. The
semitendinosus has an accessory in the Oceanitidæ, but not
in the others. It is inserted separately from the semimembranosus. The *deep flexors* blend about halfway down
the leg ; but when a hallux is present it receives no slip
from the conjoined tendons. The *syrinx* of the **Tubinares**
shows an interesting series of gradations, from a quite
ordinary tracheo-bronchial type to what is very much like the
bronchial syrinx of the Caprimulgidæ, though FÜRBRINGER,
while admitting the 'bronchophone tendency' of the syrinx
of the Strigidæ, as a point of similarity between that group
and the Caprimulgidæ, considers that only 'artificially' can
the **Tubinares** and the Spheniscidæ be brought into the same
line. Nevertheless in the series of **Tubinares** the bronchial
rings to which the intrinsic muscles are attached seem to
move further down. FORBES, however, regards this as a
splitting of the trachea, and holds that the intrinsic muscles
are invariably fastened to the fifth semi-ring. 'It is in the
genus *Pelecanoides*,' remarks FORBES, 'that the typical
construction of the syrinx of the **Tubinares** is seen in its
simplest form.' In this bird all the bronchial rings are
semi-rings, and there is a three-way piece of the usual

---

[1] In a specimen of *Nycticorax griseus* on one side of the body I found a
tendon from the biceps running to the tensor brevis tendon, which may be comparable to the above-described slip.

[2] Cf. as to this *Opisthocomus*, *Œdicnemus*, and *Casuarius*.

structure. In *Garrodia* the first three bronchial rings are complete. In *Thalassœca glacialoides* the last four tracheal rings are incomplete behind, and are quite like the four succeeding bronchial semi-rings, being, moreover, like them,

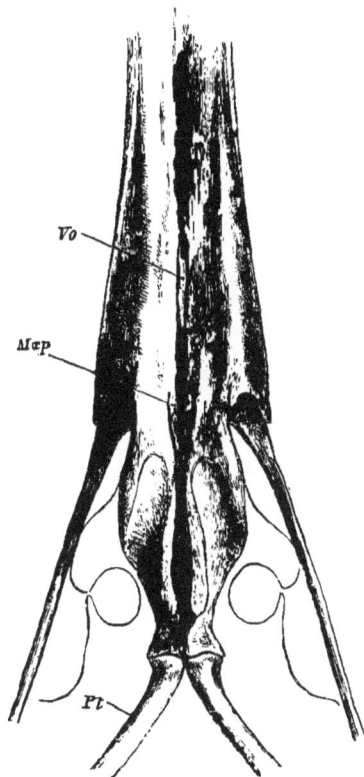

FIG. 211.—SKULL OF *Diomedea exulans* (AFTER HUXLEY).

Vo, vomer; *Mxp*, maxillo palatines ; *Pt*, pterygoid.

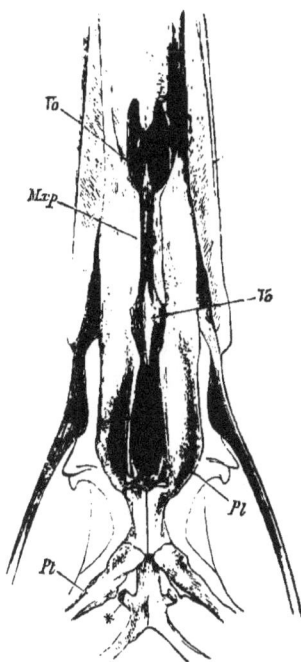

FIG. 212.—SKULL OF *Procellaria gigantea* (AFTER HUXLEY).

Pl, palatines ; *, basipterygoid process. Other letters as in fig. 211.

ossified. This modification is carried to its furthest extreme in *Ossifraga*. In this petrel the last nine or ten tracheal rings are incomplete in front, and the last twelve or so are incomplete behind. Thus the membrana tympaniformis does not commence for some distance away from the bifurcation of the respiratory tube, the rings being double and

G G

complete for the same way in front of it. The division of the trachea is carried upwards above the point where it is externally divided into two tubes by an internal septum. And in *Acipeter*, though there is no external division of the trachea, there is an incomplete internal septum. In *Diomedea*, according to SWINHOE, at any rate in two species, the bronchi are long and convoluted, as in certain storks.

The petrels are schizognathous and holorhinal birds, some with basipterygoid processes. The *skull* possesses a small peculiar bone, which has been termed the 'ossiculum lacrymo-palatinum,'[1] whose relations are sufficiently indicated by its name. This bone occurs in many genera; but it is said to be of no great classificatory significance,[2] since it also occurs in birds so remote from the petrels and from each other as Cuculidae, *Cariama*, and Laridae. The skull is markedly excavate above the orbits for the supra-orbital glands. The skull of an albatross (*Diomedea melanophrys*) is in more than one particular like that of the steganopods. It is, it is true, schizognathous; but the interval separating the maxillo-palatines and the palatine expansions of the maxillae in front of these is very slight; and, as FORBES has observed, the downwardly curved extremity of the vomer partly fills up this gap, though there is no actual fusion between it and the maxillo-palatines. The hooked bill is *Fregata*-like; and the closely approximated and downwardly produced internal laminae of the palatines are highly suggestive of the pelican. There are always fifteen *cervical vertebrae*. The *sternum* varies so much, from having a plain hinder contour to the presence of two notches on each side, that its characters need not be given in detail.

The petrels can be divided into two families, the Oceanitidae and the Procellariidae, which may be thus defined:—

[1] By BRANDT. PARKER has termed it 'uncinate.'
[2] See, however, under **Steganopodes**, p. 416.

Oceanitidæ.

> *Tarsi ocreate or covered by large transversely oblique scutes anteriorly. No cæca. Expansor secundariorum present. Accessory semitendinosus present. Ambiens (when present) does not pass over knee. Wing shorter than leg.*

Procellariidæ.

> *Tarsi covered with hexagonal scutella. Cæca present (except in Haloeyptena). No expansor secundariorum or accessory semitendinosus. Ambiens (absent in Pelecanoides) crosses knee. Leg shorter than wing.*

Not many fossil **Tubinares** are known. If *Hydrornis natator* of MILNE-EDWARDS, from the lower Miocene of France, be referable to this group, it has a considerable antiquity. *Diomedea anglica* has been lately described by LYDEKKER.[1]

The nearest allies of the petrels appear to be the **Steganopodes** (*q.v.*)

PALAMEDEÆ

Definition.—**Aftershaft rudimentary. Aquincubital. Oil gland tufted. Muscle formula, A(B)XY +. No biceps slip. Expansor secundariorum present. Carotids, two. Cæca large. Skull desmognathous, holorhinal, with basipterygoid processes. Ribs without uncinate processes.**

This well-marked group of birds has only two genera at most, which collectively contain but three species. It has been chiefly investigated by PARKER,[2] GARROD,[3] MITCHELL,[4] and myself.[5]

---

[1] In *Quart. Journ. Geol. Soc.* xlii. p. 366.

[2] 'On the Systematic Position of the Crested Screamer,' *P. Z. S.* 1863, p. 511.

[3] 'On the Anatomy of *Chauna derbiana*,' &c., ibid. 1876, p. 189.

[4] 'On the Anatomy of *Chauna chavaria*,' ibid. 1895, p. 350.

[5] 'On some Points in the Anatomy of *Chauna chavaria*,' ibid. 1886, p. 178, and BEDDARD and MITCHELL, 'On the Anatomy of *Palamedea*,' ibid. 1894, p. 536.

The *skin* in these birds is excessively emphysematous, save only on the shoulder, and in *Palamedea* and *Chauna derbiana*—on the tibia. The feathers, however, do not perforate the cutaneous air cells, 'but cause the skin to be indented where they are inserted.' The *aftershaft*, though present, is confined to the feathers on the nape of the neck. The *rectrices* are twelve in *Chauna* and fourteen in *Palamedea*. The *oil gland* in *Ch. derbiana* has a single orifice on each side; it is encircled by feathers which constitute the tuft.

In *Ch. chavaria* the summit of the oil gland is covered by feathers, a line of which separates the two orifices of the gland. In *Palamedea* the same is the case, but the encircling ring of feathers is not complete on the ventral side. The *pterylosis* is almost unique in the fact that there are no apteria, except, indeed, a space in the axillary cavities, and these are covered with down feathers. The strong horny spur borne on a bony core, an outgrowth of the first metacarpal, is comparable to a thickened featherless patch of skin in a corresponding situation in *Sarcidiornis*.[1]

The *patagialis* muscle is not reinforced by a *biceps slip*; the brevis tendon is single, but broad, and without a patagial fan. The *expansor secundariorum* is present and 'ciconiine.' The insertion of the *deltoides posterior* is extensive in *Palamedea*—for about three inches down the humerus. The *anconæus* has a well-marked humeral head. The division of the *biceps* commences in the fleshy belly of the muscle. In all the members of this family a very peculiar muscle exists, to which Mr. MITCHELL and I have given the name of *costo-sternalis externus*. It arises from the third, fourth, and fifth ribs by a tendinous head, and is inserted on to the costal edge of the sternum half an inch from the posterior end.

The muscles of the leg are complete, as regards those upon which GARROD laid stress in his classification, in *Chauna*. In *Palamedea*, however, the *accessory femorocaudal* is absent. The *biceps* has not an anserine insertion, but

---

[1] Not, of course, to the *carpal* spur of *Plectropterus*.

passes through a loop in the ordinary way. Both *peroneal* muscles are present ; the *tibialis anticus* is split just at its insertion. The *deep flexor tendons* differ somewhat in the

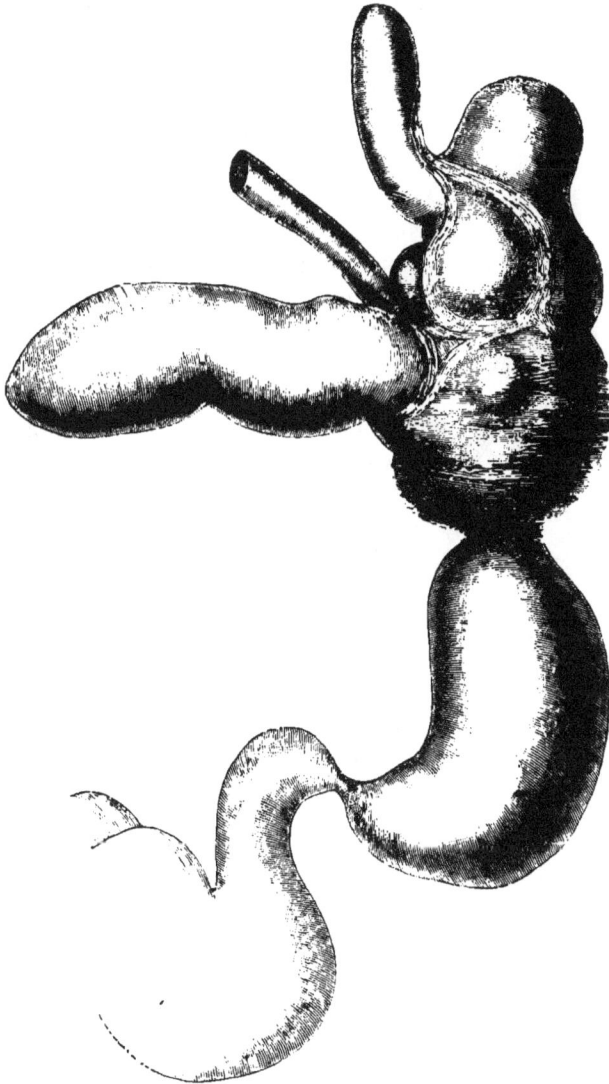

Fig. 213.—CÆCA OF *Chauna chavaria* (AFTER BEDDARD).

three species. In *Palamedea* the *flexor longus hallucis* is slender, and gives off a vinculum to the *flexor communis* tendon, before supplying the first digit ; in *Chauna derbiana* there is no branch to the first digit at all ; *Chauna chavaria*

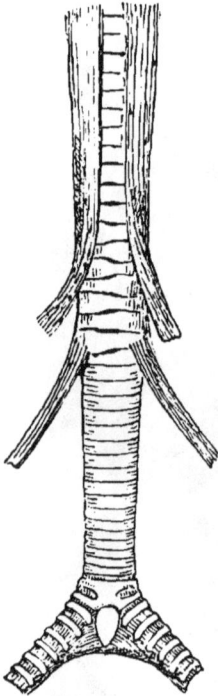

Fig. 214.—WINDPIPE OF *Palamedea* (AFTER BEDDARD AND MITCHELL).

is like *Palamedea*, but there are two distinct vincula.

The *palate* (of *Ch. chavaria* at any rate) is provided with three longitudinal rows of papillæ ; the *tongue* is just over an inch long, and its base is edged with spines. There is no transverse constriction or oblique groove, such as is found in many anatiform birds. The *proventriculus* is peculiar ; there is a narrow zone of glands round the œsophageal aperture, from which a broad triangular patch extends down one side of the cavity. The gizzard is decidedly small. The lobes of the *liver* are more nearly equal in size in *Chauna* than in *Palamedea* ; in all there is a conspicuous *gall bladder*. The *cæca* (fig. 213) are in some respects unique in structure ; they are in the first place large, measuring three inches or so in length ; they are sacculated by a single band.

The *windpipe* agrees with that of some of the **Anseres** in having two pairs of extrinsic muscles :[1] the upper pair is inserted into the middle of the membrane, which runs between the coracoid and the corresponding limb of the furcula ; the lower pair close to the costal processes of the sternum. The intrinsic muscles cease some little way in front of the syrinx.

The prebronchial and subbronchial air sacs are, in *Ch. chavaria* at least, much divided up, as in the storks. In

[1] Apparently first noted by CRISP. ' On the Visceral Anatomy of the Screamer,' *P. Z. S.* 1864, p. 14.

Chauna the lower pair of extrinsic muscles fan out upon the aponeurosis of the lungs.

There are eighteen *cervical vertebræ* in *Palamedea* and in *Ch. chavaria*, nineteen in *Ch. derbiana*. There are seven complete ribs in *Palamedea*, eight in *Chauna*. The *sternum*, which has one notch on either side, has neither external nor internal spina. There are in neither genus any traces of uncinate processes on the ribs, a character which is unique among living birds.[1]

The *skull* of the Palamedeidæ has many anserine characteristics, which have been emphasised by PARKER, and perhaps rather too lightly touched upon by GARROD. In their desmognathism (which is a complete fusion) they are, of course, anserine, the form of the maxillo-palatines most recalling those of *Mergus*. The form of the palatines is duck-like or gallinaceous in the rudimentary character of the internal lamina, which is a mere ridge. The pterygoids are articulated to the large oval duck-like basipterygoid processes, nearer to their middle than is the case with the **Anseres**. The lacrymals are small, quite contrary to what is found among the **Anseres**,

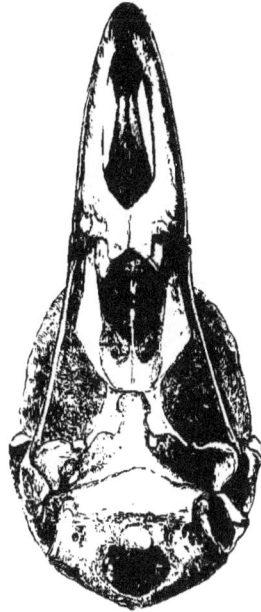

Fig. 215. Skull of *Chauna derbiana*. Ventral Aspect. (After Garrod.)

though the ectethmoids are not unlike those of the latter group. The anterior part of the face, with the clear-cut holorhinal nostrils and the hooked bill, is suggestive of *Cariama* or a gallinaceous bird; it does not at least recall the duck or goose. The interorbital septum is deficient in front,

[1] Not even a rudiment of these characteristically avian structures has, so far as I am aware, been detected.

but not fenestrate posteriorly.   Occipital fontanelles are absent.

Though perhaps rightly placed in the neighbourhood of the geese, it is obvious, from what has been said, that the Palamedeidæ are distinguished from them by many differences, of which the most important are perhaps—

(1) Continuous feathering.

(2) Absence of biceps slip to patagium.

(3) Peculiar form of intestinal cæca.

(4) Normal character of biceps cruris.

(5) Emphysematous character of skin and breaking up of cervical air sacs.

On the other hand there is nothing in the skull which forbids an association with the **Anseres**, and the windpipe, with its two pairs of extrinsic muscles, is decidedly goose-like. But it must be remembered that this feature is also found among **Galli**, and in a few other forms.

ANSERES [1]

Definition.—Oil gland tufted. Aftershaft small or absent. Aquincubital.   Two carotids.   Trachea with two pairs of extrinsic muscles.   Cæca long.   Gall bladder present.   Biceps slip present.   Glutæus maximus large.   Muscle formula of leg, $ABX+$.   Skull desmognathous, with basipterygoid facets.

The swans, geese, ducks, and mergansers, which make up this large assemblage of birds, are all aquatic, or semi-aquatic, in habit, and in correspondence have webbed feet, with the exception only of *Anseranas*.   They are also for the most part strong flyers, excepting only the living *Tachyeres cinereus* and the extinct *Cnemiornis calcitrans*.

Tachyeres cinereus, the ' steamer duck,' from the shores of Patagonia, has been investigated by R. O. CUNNINGHAM.[2] It does not appear from his memoir, which relates chiefly to osteology, but in the course of which he describes and

---

[1] H. SEEBOHM, ' An Attempt to diagnose the Sub-Orders of the Ancient Ardeo-Anserine Assemblage,' &c., *Ibis*, 1889, p. 92.

[2] ' On the Steamer Duck,' *Tr. Z. S.* vii. p. 493.

figures the gizzard and windpipe, that there are noteworthy modifications of structure. The bird flies strongly when young, but swims only when adult. CUNNINGHAM suggests a greater density of the bones; but when weighed and compared with the skeleton of some other **Anseres** no reliable differences were apparent.

Cnemiornis calcitrans is a large anserine bird from the Pleistocene deposits of New Zealand, which, according to Sir JAMES HECTOR,[1] stood over two feet high, and was at least three feet in length. *Cnemiornis* is characterised by the great weight of its bones; the following comparative table from HECTOR's paper brings out this fact :—

		Bulk	Weight
Cnemiornis		10	244
Ocydromus	Non-volant	10	210
Stringops	,,	10	187
Nestor .	Volant	10	131
Hieracidea	,,	10	126

It seems clear from these comparisons that *Cnemiornis* could not have been a flying bird. Moreover the sternum has a keel whose highest elevation is under three lines. This bone appears to have possessed no lateral notches, but only a slight median concavity. On the other hand the ' rough tubercular surface ' of the keel is perhaps a little suggestive of a missing cartilaginous piece.

The suggestion of FÜRBRINGER to separate *Cnemiornis* into a sub-family, Cnemiornithinae, seems perhaps, in the light of the above facts, to be justifiable. But the remainder of the members of the family cannot be divided with precision. It is above all impossible to divide swans, geese, and ducks from each other into three such groups. Nevertheless, as will be gathered from the following account of their structure, the **Anseres** vary very much among themselves—more especially in the structure of the windpipe.

All the **Anseres**, as noted in the definition, have a tufted *oil gland.*

[1] ' On *Cnemiornis calcitrans*,' &c., P. Z. S. 1873, p. 768, and 1874, p. 307.

The *aftershaft* is always small, and sometimes absent.

In the *pterylosis* (according to NITZSCH) the pectoral tract divides about halfway down the neck ; it is broad, and gives off on each side a stronger outer band. The dorsal tract also divides about halfway down the neck, and encloses a long narrow space.

The number of *rectrices* in various genera is shown in the annexed table.

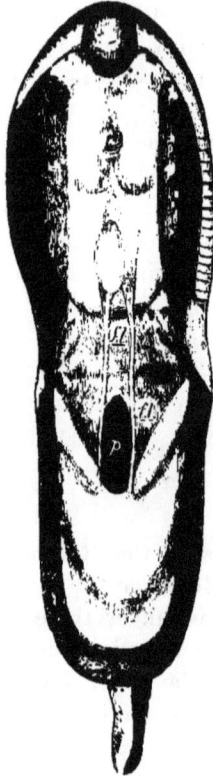

FIG. 216.—MOUTH OF *Biziura lobata* (AFTER FORBES).
p, pouch : *t*, tongue : *f.l.* frænum linguæ.

	Rectrices
Biziura lobata . . .	24
Anas specularis . .	14
Anas boschas . . .	14
Aix sponsa . . .	16
Erismatura rubida . .	18
Cercopsis Novæ Hollandiæ	14 [1]
Plectropterus gambensis .	16
Cygnus musicus . .	24
Fuligula rufina .	16
Dafila acuta . .	16
Bernicla canadensis .	18
Mergus merganser .	18
„ *albicillus* .	16

The *tongue* is strong and bordered with spines. In *Mergus* only is it thin and pointed.

The *liver* has a large *gall bladder*.

Biziura lobata is unique among **Anseres** [2] in having a subgular pouch (fig. 216), formed by a duplicature of the frænum linguæ.

The most noteworthy point that appears from the facts given on the next page is the gradual reduction of the cæca in the mergansers and smews, and the great range of variation in size which they exhibit in *Cygnus nigricollis*, a variation

[1] NITZSCH says 16.   [2] Cf. **Otides** for a similar structure.

which does not, it will be observed, bear any relation to the variation of the total length of the intestine.

The following are some *intestinal measurements* :—

	S. I.		L. I.	Cœc.
	Ft.	In.	Inches	Inches
Mergus merganser	5	4	2·25	1·25, 1·5
„ *castor*	4	8	3	1·8
„ *albicillus*	4	3	3	·2
Biziura lobata	8	6·5	4·5	6·75
	6	11·25	6·5	7·75
Anas specularis	5	2·5	2·75	5
Rhodonessa caryophyllacea	4		—	2·2, 1·75
Aix sponsa	4	1·2	3·5	5
Sarcidiornis melanota ♂	4	7	3	2·75, 3
„ *carunculata* ♂	3	6	4	3·25
Bernicla canadensis	7	6	—	9
„ *jubata* ♂	3	4	3·5	2·5, 2·75
Chenalopex jubata	4		2·5	3·75, 4
	11		—	15
Cygnus nigricollis	11	4	—	8
	9	6	—	9, 10
Dendrocygna autumnalis	3	6	2·25	3·2, 3·4
Metopiana peposaca	4	10·5	2	3·5
	4	3	—	5·5, 6
Plectropterus gambensis	6	9	7	6·75, 7·25
„ *niger* ♂	6	1	3	6
Cereopsis Novæ Hollandiæ	6		3·5	11·5
Anser albifrons ♂	5	8	5	7·75
„ *indicus* ♀	5	1	3·5	5·5
Chloephaga magellanica ♀	6	8	4	12, 13
Cygnus olor ♂	11	9	8	14, 15
„ *ferus* ♂	8	6	9	9·25
	8	4	8	9
„ *buccinator* ♂	8	6	7	7·5
„ *atratus* ♂	9	2	6	11
Tadorna tadornoides ♀	5	6	3	3·5, 4·5
Dafila acuta ♀	4	6	4	4, 4·5
Pœcilonetta bahamensis ♂	2	10	2·5	4, 4·25
„ *bahamensis* ♂	3	4	3	3, 3·3
Querquedula circia ♀	4		—	3
Hymenolæmus malacorhynchus ♀	4		4	3·8, 4·25
Spatula clypeata ♂	8	9	4	5
	6	6	3·5	3·5, 3·75
Fuligula ferina ♂	4	2	4	4·5
„ *rufina* ♂	4	5	4	6·75
	4	2	3·5	6·5, 7·5
Nyroca leucophthalma ♀	4	8		1·5

¹ Only one present.

The *muscular anatomy* of the **Anseres** is very uniform, which coincides with their uniformity of life habit. Even presumed aberrant types, such as *Mergus, Biziura,* and

Tachyeres, are hardly if at all to be distinguished anatomically
from the typical geese and ducks. Our knowledge of the
muscular structure of this group of birds is mainly due to
FÜRBRINGER, who has illustrated the fore half of the body
by two double plates referring to *Anser cinereus*.

An interesting duck character (also, however, found in
Colymbus, **Tinami**, and some **Galli**) is the meeting of the two
great pectoral muscles over the carina sterni. In *Mergus*

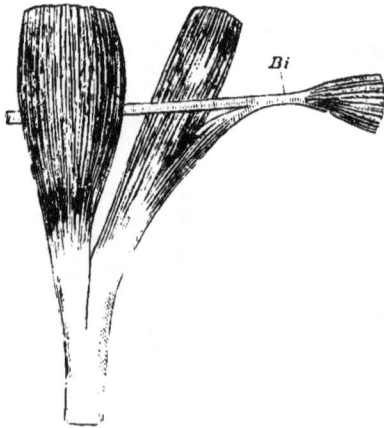

Fig. 217.—Biceps Femoris of Duck (*Bi*), to show its
Relations to Gastrocnemius (after Weldon).

merganser, for example, they blend for a space of half
an inch.

There is a *biceps slip* present,[1] and this has at least
sometimes a peculiar arrangement, which is remarkably like
that of the **Colymbi**. In *Anser cinereus* FÜRBRINGER
figures the biceps slip as attached in the ordinary way to
the tendon of the tensor patagii longus; but before it is thus
attached it gives off a slender tendon, which—exactly as in
the **Colymbi**—runs over the patagium and is inserted on to
the fore arm in front of the broad and diffuse tensor patagii
brevis tendon.

In *Anas* FÜRBRINGER figures this tendon as joining the

[1] Not in *Cygnus Bewicki*.

patagial fan, a state of affairs which is exactly paralleled in
Æchmophorus (see p. 387). In ducks there is a peculiarity in
the biceps to which attention appears to have been first
called by FÜRBRINGER. There is a tendinous sheath
partly covering the patagial side of the muscle and derived

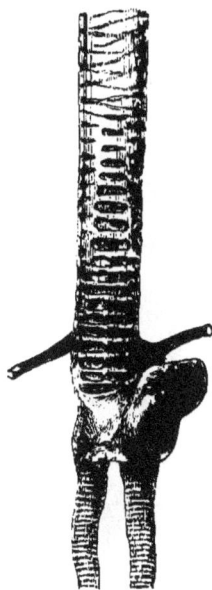

FIG. 218.—WINDPIPE OF
Sarcidiornis melanota ♂
(AFTER GARROD).

FIG. 219.—SAME OF
S. melanota ♀
(AFTER GARROD).

FIG. 220.—SAME OF
*Rhodonessa caryo-
phyllacea* ♀ (AFTER
GARROD).

from the pectoralis primus. A similar structure occurs in
Colymbl and storks, and in *Casuarius* its homologue is more
independent of the biceps. The latter at any rate, if not
the former, seems to me to correspond to a peculiar muscle
found in the tinamous (cf. p. 489).

The *expansor secundariorum* shows some variations. In
Biziura lobata it is entirely absent. In *Aix sponsa* the
tendons thin off and are lost on the interthoracic septa. In
others, as, for instance, *Bernicla canadensis*, the expanded
tendons end upon the œsophagus ; on the way thither they

blend with the sheath of the carotids, which when pulled
upon they compress, as also the jugulars.   This may conceivably be a provision for increasing the blood supply of the
wings by interrupting that of the head.   Finally in *Cygnus*,
Mergus, and *Sarcidiornis* the expansor secundariorum is
ciconiiform.

The *deltoid* has a strong scapular slip.   The *anconæus*

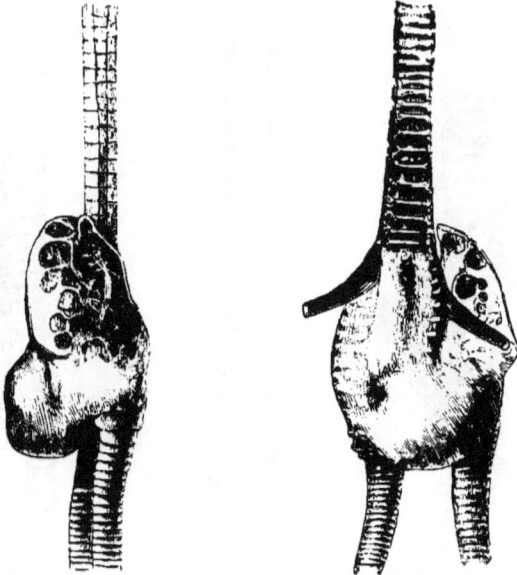

FIG. 221.   SYRINX OF *R. caryophyllacea* ♂.   RIGHT-HAND FIG., FRONT
VIEW; LEFT-HAND, SIDE VIEW.   (AFTER GARROD.)

has a humeral slip.   In the hind limb all **Anseres** have the
formula ABX+.   The *femorocaudal* is slender, the accessory very large.   In *Biziura lobata* the *ambiens* has a peculiarity found also in *Phalacrocorax* and the extinct *Hesperornis*—that its tendon perforates the patella.   The *glutæus
maximus* is very large, and its origin descends below the acetabulum.   A marked peculiarity of the **Anseres**, found elsewhere in *Struthio*, is that the *biceps femoris* gives off a slip
to the gastrocnemius.   In most **Anseres** the *flexor longus
hallucis* gives off a slip to the hallux before fusing with the

flexor profundus ; in *Biziura lobata* the same slip is given off, but it becomes lost on one of the annular masses of fibrocartilage surrounding the other flexors.

The *windpipe* in the **Anseres** is nearly always straight, the only exception among the ducks and geese as yet recorded being *Anseranas melanoleuca*, in the males of which the trachea forms a double loop, extending to quite the end of the pectoral muscles. A second peculiarity of the trachea is seen in the males of *Metopiana peposaca*, *Strictonetta nævosa*,[1] *Melanitta fusca*, *Nyroca*, *Mergus*, *Somateria*, and *Clangula*. In these (see fig. 222) there is a bulbous enlargement of the trachea some little way in front of the syrinx. It is present, but very slightly developed, in *Fuligula rufina*. In nearly all the ducks the syrinx in the male has a remarkable asymmetrical enlargement, which is as a rule entirely bony, but is sometimes (*Mergus*, *Clangula*, *Nyroca*, *Fuligula*) mainly formed of membrane. The accompanying figures will give some idea of the form of this structure, which shows differences in different species. The figures are taken from a memoir upon the subject by Professor GARROD.[2] YARRELL[3] and EYTON[4] have

[1] E. P. RAMSAY, ' Note on the Tracheæ of certain Australian Ducks,' *Proc. Linn. Soc. N. S. W.* iii. 1879, p. 154.
[2] ' On the Form of the Lower Larynx in certain Species of Ducks,' *P. Z. S.* 1875, p. 151.
[3] *British Birds.*
[4] *Monograph of the Anatidæ or Duck Tribe.*

FIG. 222.—WINDPIPE OF *Metopiana peposaca* ♂ (AFTER GARROD).

also figured a good many of the syringes of the ducks in illustration of this matter. The only ducks in which there is certainly no modification of the syrinx of this kind are *Biziura lobata, Œdemia nigra,* and *Melanitta fusca* ; in the former bird FORBES[1] has described a plain syrinx (fig. 223) with a box at the bifurcation of the bronchi formed of the last tracheal and of a few of the anterior bronchial semi-rings.

FIG. 223.— SYRINX OF *Biziura* ♂ (AFTER FORBES).

He suspects, however, that the genus *Erismatura*[2] will be found to have a similar syrinx minus a lateral outgrowth, as MACGILLI-VRAY appears to have described something of the kind. *Somateria mollissima* has a very slight symmetrical enlargement of the syrinx. A very marked characteristic of the **Anseres** is the possession of two pairs of extrinsic tracheal muscles. In this they agree with Palamedeidæ. The single pair of intrinsic muscles are as a rule attached to the third or fourth tracheal ring in front of the syrinx in the ducks.

Among the *Cygninæ* (swans) there is frequently a looped trachea, the coils being intrasternal. This is so with both sexes of *C. ferus, C. buccinator, C. americanus,* and *C. Bewicki*; there appears to be a trace of the looping in *C. atratus.* In *C. olor, C. immutabilis, C. nigricollis,* and *C. coscoroba* the windpipe is straight in both sexes. It is interesting to note that in *C. buccinator,* at any rate, the intrinsic muscles do not follow the coils that bridge across the loop. This species of swan is also remarkable for the extraordinary dilatation of the middle of each bronchus, which is, again, characteristic of both sexes. These dilatations would be almost spherical were it not for the irregular crumpling

[1] 'A Note on some Points in the Anatomy of an Australian Duck (*Biziura lobata*),' *P. Z. S.* 1882, p. 155.

[2] 'There is no expansion or tympanum, as in other ducks' (*Orn. Biogr.* iv. 1838, p. 331).

here and there; their diameter is about an inch. The bronchial rings which cover them have largely lost their individuality, and form an irregular network of partly cartilaginous and partly osseous bars.

The males (? as to females) of *Cygnus ferus* and *C. Bewicki* at any rate show less marked traces of the same peculiarity.

In the former there is a distinct fusiform dilatation, but further down the bronchi than in *C. buccinator*, between the rings of which there is some slight formation of anastomoses. In *C. Bewicki* both features are still less marked.

Cygnus olor, *C. atratus*, *C. nigricollis*, and *C. coscoroba* have no trace of this remarkable structure. I have examined males of all and females of two.

As to the geese, *Bernicla canadensis*, *Anser indicus*, and *Cercopsis Novæ Hollandiæ* (and doubtless many others) have a syrinx without the anatine bulbus. It is present in *Plectropterus* (*gambensis*, *Rüppeli*, *niger*) and *Sarcidiornis*, and present, though small and solid, in *Chenalopex jubata*. It is fenestrated in *Plectropterus*, not so in *Sarcidiornis*.

Dendrocygna appears to illustrate the commencement of the syringeal enlargement. In the male of *D. arcuata* the last twelve tracheal rings are widened and enclose a spacious chamber about twice the diameter of the rest of the trachea. The intrinsic muscles are attached to the beginning of this thin-walled box. In the female there is an indication of this in the fact that eleven of the tracheal rings in front of the last three are imperfect posteriorly, being closed by membrane. The intrinsic muscles also are attached opposite to the commencement of this modified region of the trachea.

In *D. autumnalis* there is the same box, which is strengthened posteriorly by a strong bony bar. The windpipe of the female has no such modification as has been described above in *D. arcuata*.

It is possible that this single enlargement in *Dendrocygna* is the beginning of both the tracheal swelling and the syringeal bulbus in *Mergus*, &c.

H H

The number of *cervical vertebræ* [1] among the **Anseres** varies
considerably. The smallest number is found, for example,
in *Plectropterus gambensis*, *Biziura lobata*, and *Tachyeres
cinereus*, where there are only sixteen. In *Œdemia nigra*
there are seventeen ; among the swans, twenty to twenty-
four. The number of true ribs also varies considerably.
The smallest number is to be seen in *Cereopsis*, where there
are but five. *Tachyeres*, *Plectropterus*, and a number of
other genera have seven ; there are eight in *Tadorna rulp-
anser*, and as many as nine in certain swans and geese.

The *sternum*, which has a moderate spina externa, but
no spina interna, is whole in *Cnemiornis*, but has two
notches or foramina in other **Anseres**. The *coracoids* come
into contact, but do not overlap at their sternal articulation.
The procoracoid is small, and does not reach the clavicle,
which, however, reaches the scapula.

The *skull* has large oval sessile basipterygoid processes.
It is holorhinal and desmognathous. There are frequently
lateral occipital fontanelles, as in many 'pluvialine' birds.[2]
The palatines are remarkable for the rudimentary character
of their inner laminæ, which brings about a resemblance to
the gallinaceous birds, as has been pointed out by HUXLEY,
and, it may be added, to the parrots. That part of the
palatine is only indicated by a not well marked ridge which
is totally absent in *Berniela leucopsis*, *Chen cærulescens*.
The general direction of the bone, therefore, is oblique ; it is
only near to the attachment with the pterygoids that it
becomes feathered out in a horizontal direction. That, at
least, is the more normal arrangement ; for in *Mergus* the
greater part of the bone has its upper and lower surface coincident with the horizontal axis. This, too, is the case with
Biziura lobata.

The oval basipterygoid facets for articulation with the

---

[1] For osteology see PARKER, ' On the Morphology of the Duck and the Auk
Tribes,' *Cunningham Memoirs R. Irish Ac.* No. 6, 1890, and SHUFELDT, ' On N.
American Anseres,' *P. U. S. Nat. Mus.* xi. p. 215.

[2] Absent in *Cereopsis*, *Biziura*, *Cygnus*, and *Cnemiornis*. See OWEN, *Tr. Z.
S.* ix. pt iii.

pterygoids are placed so far forwards that the anterior ends
of those bones articulate with them.

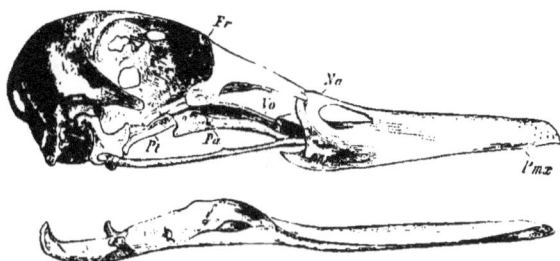

FIG. 224.—SKULL OF *Querquedula crecca*. LATERAL VIEW.
(AFTER HUXLEY.)

Fr, frontal ; *Na*, nasal ; *Pmx*, premaxilla ; *Vo*, vomer ; *Pa*, palatine ; *Pt*, pterygoid.

The vomer is a thin deepish plate of bone which is more
or less intimately connected with the
median septum and maxillo-palatines in
front. The maxillo-palatines are completely fused across the middle line in
many **Anseres** (e.g. *Chen, Hymenolæmus*) ;
in *Biziura* and *Mergus* they come into
contact but are not fused. The latter
has very un-ducklike palatine in that the
somewhat delicate maxillo-palatines diverge
from each other after their junction posteriorly as well as anteriorly, the palatine
vacuity in front being (for a duck) unusually extensive. In *Chen cærulescens*,
indeed, secondary bony growths have
almost completely obliterated this vacuity,
a kind of 'false' palate having been
formed.

The lacrymal bones are large, having
a considerable length of line of union with
the skull ; they are sometimes (e.g. *Chloë-
phaga*) and sometimes not (e.g. *Mergus*)
ankylosed with the orbital margin. *Cere-
opsis* has (among the genera which I have examined) a skull

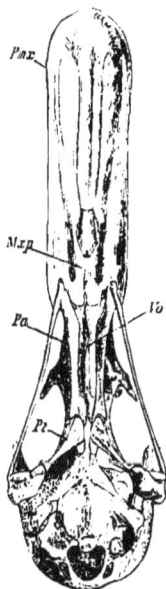

FIG. 225.—VENTRAL
VIEW OF SAME.
Mxp, maxillo-palatines.
Other letters as in fig. 224.

H H 2

which is peculiar in that posteriorly the lacrymal is free from the orbital wall, but is fused with a process of the frontal at the anterior end of the supra-orbital impression, leaving (as in *Chionis* and some other birds) a foramen.[1] In *Cereopsis* the descending process of the lacrymal curves backwards and comes near to the zygoma;[2] the junction is completely effected, and there is ankylosis, in *Dendrocygna*.

The ectethmoids of the **Anseres** are often largely deficient as ossifications; when present they are thin-walled bubble-like structures, coming into relation with the lacrymals. The interorbital septum is generally very complete; but it is largely vacuolate in *Mergus* and *Biziura*.

The following table shows the number of *cervical vertebræ* and the condition of the hæmapophyses and catapophyses in a series of anserine birds :—

---	C. V.	Hyp.	Cat. nearly unite	Posterior Catapoph.
Plectropterus gambensis	16	C13–D3	C12, 13	C16–D1
Metopiana peposaca	17	C14–D5	C12, 13	C17–D3
Sarcidiornis carunculata	16	C13–D4	C11, 12	C15–D2
Hymenolæmus	16	C13–D4	C11, 12	C15–D2
Tachyeres cinereus	16	C13–D5	C11, 12[1]	C15–D3
Œdemia nigra	16	C12–D5	C10, 11	C15–D3
Biziura lobata	16	C13–D5	C11, 12	C15–D3
Dendrocygna autumnalis	17	C13–D3	C11, 12	C16–D1
Bernicla brenta	19	C15–D4	C12–14	C18, 19[2]

[1] Fused to form a solid bifid 'hæmapophysis,' as in some other birds.

[2] They are rudimentary and do not mount upon the hypapophysis.

As a possible appendix to the **Anseres** must be mentioned three or four species of an extinct genus of birds, *Gastornis*, flightless and larger than an ostrich. It has been found only in Europe and from Eocene beds. It is placed among the 'ratites' by LYDEKKER and some others; this is largely on account of the coracoid, which is imperfect above, and appeared to LEMOINE and others as probably 'platycoracoidal.' FÜRBRINGER, however, considers that the tuberosity

[1] This is figured by OWEN in his paper on *Cnemiornis*, *Tr. Z. S.* ix. pl. 35, fig. 8.

[2] Sometimes joining. Cf. OWEN, *loc. cit.* pl. 35, fig. 6.

described by the latter as ' tubérosité préglénoïdienne ' is really the broken end of the scapula, which would be thus, as in *Didus*, ankylosed with the coracoid, and would also form with it an angle approaching to a right angle. The supposed remains of the scapula, on the other hand, are for FÜRBRINGER the acrocoracoid. On this interpretation the shoulder girdle of *Gastornis* would be a nearly typical carinate shoulder girdle. The length and slenderness of the coracoid too is not a ratite character, but it does ally *Gastornis* with *Cnemiornis* (and also for that matter with *Phororhacos*). In spite of the freedom of the metacarpals (a character only known elsewhere in *Archæopteryx*), the complete furcula and various points in the bones of the lower limb, pointed out by NEWTON, together with the facts already referred to, seem to point to a greater likeness to *Cnemiornis* than to any other known type. The skull had basipterygoid processes and seemingly teeth in sockets. The pygostyle appears to have been at most very small and probably absent.

ICHTHYORNITHES

Definition.—**Small toothed birds with carinate shoulder girdle and sternum. Bones of pelvis not united. Quadrate single-headed. Vertebræ amphicœlous.**

This group of birds, from the Cretaceous of North America, has been investigated by MARSH, who, in his great work upon the toothed birds, placed it in a group Odontotormæ, as opposed to Odontolcæ (*Hesperornis*), on account of the fact that the teeth are implanted each in a distinct socket. Its relationship to other groups is doubtful ; but it is probably not greatly misplaced if we consider it in the neighbourhood of the stork and plover tribe, as has been done by FÜRBRINGER. The group contains two genera—*Ichthyornis* and *Apatornis*. Of the former MARSH refers to several species, viz. *I. dispar*, *I. victor*, *I. validus*, *I. tener*, *I. agilis*, and *I. anceps*.

Of *Apatornis* there is but one species known, *A. celer*.

These two genera comprise a number of 'small birds, scarcely larger than a pigeon. In their powerful wings and small legs and feet they remind one of the terns, and according to present evidence they were aquatic birds of similar life and habits.'

The restoration of *Ichthyornis* given by MARSH has been extensively copied in various works, in some of which it would appear as if our knowledge of the osteology of the species selected were greater than is really the case. It has been made, for example, to show schizorhinal nostrils and a pelvis constructed after the carinate type, with the ischia and ilia fused. It is not known whether the *skull* was schizorhinal, as only the calvarium and the lower jaw and a fragment of the upper jaw have been discovered. The skull has well-marked grooves for the supra-orbital glands ; the quadrate, as stated in the definition, is single-headed, as in *Hesperornis* and many **Struthiones**. The *brain*, like that of *Hesperornis*, is small, and the cerebellum is remarkably large as compared with the hemispheres. The *teeth* of *Ichthyornis* are implanted in distinct sockets.

MARSH has remarked upon the close resemblance between the lower jaw, with its teeth, and that of the smaller mosasauroid reptiles. In *Ichthyornis dispar* there are twenty-one distinct sockets in each ramus of the jaw. *I. victor* had the same number of teeth : in *I. anceps* the teeth were more numerous, and at the same time more slender. The jaws were united, as in *Hesperornis*, by cartilage or ligament.

The *vertebræ*, as already mentioned, are amphicœlous ; but an approach to the typical saddle-shaped vertebræ is seen in some of them. The *atlas* is notched for the odontoid process of the axis. None of the dorsal vertebræ appear to have coalesced, and there is a pygostyle quite typical in form, but rather small. The *shoulder girdle* of both *Ichthyornis* and *Apatornis* is constructed upon the carinate plan.[1] There is the same angle between the scapula and the coracoid, and the clavicles are well developed. There are, however, differ-

---

[1] SHUFELDT (' Notes on the Extinct Bird *Ichthyornis*,' *J. Anat. Phys.* xxvii. p. 336) especially compares *Ichthyornis* with *Rhynchops* and *Sterna*.

ences in detail. In *Apatornis* there is a very long acromial
process. The coracoids overlap at their articulation with the
sternum, more so in *Ichthyornis* than in *Apatornis*. The
clavicles are generally figured as typically carinate ; but the
only part of this bone known is ' a fragment from the upper
end of that bone in *Apatornis*.' The *sternum* is deeply
keeled.

The bones of the fore limb are well developed ; the
humerus has a very large crest, surpassing in comparative
size that of any recent bird ; this clearly indicates a powerful
flyer, and the rest of the bones of the limb bear out this view.
Though nothing is known of the structure of the feathers,
there are upon the ulna impressions for the quill feathers.

In the *pelvis* all the bones are free posteriorly, as in *Hesperornis*, *Apteryx*, &c. The acetabulum is perforate, as in
most recent birds ; but the perforation is of moderate size, as
in the tinamous.

In the neighbourhood of *Ichthyornis* are possibly to be
placed MARSH's genera *Baptornis*, *Telmatornis*, and SEELEY'S
Enaliornis. These birds, however, Cretaceous, like *Hesperornis* and *Ichthyornis*, are known by such limited material
that their position is absolutely uncertain.

The affinities of *Ichthyornis* to *Hesperornis* have been
dwelt upon by some ; but it appears that LYDEKKER's
remark, that ' the Odontornithes are a *series* of birds ancestral to the modern series of toothless carinates,' expresses the
truth. He has furthermore added that this series ' differs
from the Euornithes (STEJNEGER's name for carinates) by the
absence of union between the rami of the mandible and
between the distal ends of the ischium and ilium,' likenesses
which do not mean a near relationship, but express the
degree of development of bird structure at that period.

ACCIPITRES

Definition.—**Aquincubital. Oil gland present. Two carotids. Skull desmognathous and holorhinal. Cæca rudimentary or absent. Ambiens present. Biceps slip absent.**

This large group of birds admits of but a scanty definition, if we are to include in it, as is here done, the secretary
bird and the American vultures; for it then shows a considerable amount of structural variation. The *oil gland,*

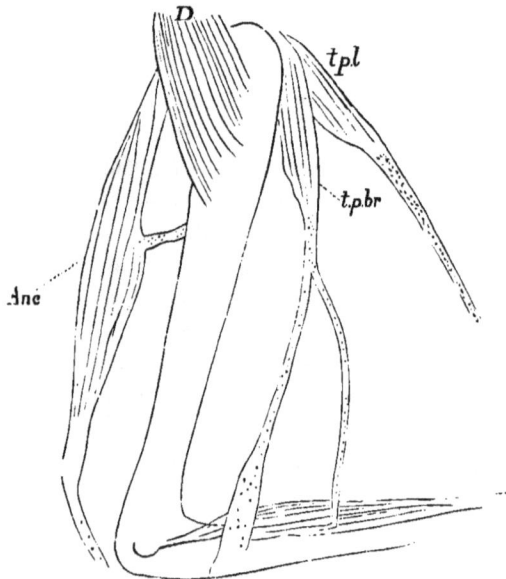

Fig. 226.—Tensores Patagii of *Polyboroides* (after Beddard).

t.p.l, tensor patagii longus; *t.p.br*, tensor brevis; *Anc*, ancouæus; *D*, deltoid.

invariably present, is generally feathered, but nude in the
Cathartidæ. In *Serpentarius* this gland varies in size; in
one specimen it was found to be very small and to have a
very minute tuft.

The *aftershaft* is absent in the Cathartidæ and in *Pandion,*
present in other **Accipitres**. Twelve *rectrices* is the usual

number, but fourteen occur in *Neophron percnopterus* and *Rhinogryphus californianus.*

There are *powder-down patches* in *Elanus, Circus,* and *Gypaetus.* The *pterylosis* is described for a variety of types by NITZSCH.

The ventral tract broadens out on the breast, where it is even sometimes (*Gyps fulvus*) divided into an outer and inner branch. The dorsal tract forks upon the shoulders; in *Gypaetus barbatus* each limb of the fork is connected by a single row of feathers with the long single median posterior portion of the tract.

In *Pernis apivorus* these latter slender forks are figured as being much longer, and in *Falco peregrinus* they dilate into four or five rows of feathers before uniting. In *Falco brachypterus* there is the usual dorsal fork, but between its extremities lies the beginning of the very broad posterior part of the tract. In all these birds there are lateral neck spaces. The lumbar tract is but little marked, or is entirely deficient. A large amount of detail is given in NITZSCH's account of this family, which is treated more fully than many others.

The *tensor patagii brevis* is simple in all accipitrines, and there is never a *biceps slip.* There is, however, a certain amount of variation in the tendon. The simplest form of the tendon is seen in

Vultur monachus	*Falco melanogenys*
„ *auricularis*	„ *subbuteo*
Gyps fulvus	„ *æsalon*
Tinnunculus alaudarius	*Thrasaetus harpyia*
Microhierax cœrulescens	

where it is a simple tendon without branches, as in many picarian birds. On the other hand in

Buteo vulgaris	*Spilornis bacha*
Circus maurus	„ *cheela*
„ *Gouldi*	*Neophron percnopterus*
Helotarsus ecaudatus	*Milvago chima-chima*

Milvago chimango · *Melierax polyzonus*
Gypaetus barbatus · *Polyborus brasiliensis*
Dryotriorchis spectabilis · *Polyboroides typicus*
Aquila imperialis · *Haliaetus albicilla*
Lophoaetus occipitalis · *Milvus ictinus*
Melierax monogrammicus · *Astur approximans*

a wristward branch (as shown in fig. 226) is given off, the tendon, in fact, bifurcating. FÜRBRINGER, however, while figuring the two types, distinguishes in the apparently single tendon of *Tinnunculus* two separate tendons in close contact for their entire length.

A further complication is seen in *Gypaetus* and *Gypohierax*,[1] where a small recurrent tendon (patagial fan) joins the anterior branch of the *brevis* with the *longus*, a state of affairs found to characterise *Serpentarius*, as will be pointed out later.

The *expansor secundariorum*[2] is present in *Milvago chimango, Harpyhaliaetus coronatus, Falco, Polyborus, Tinnunculus,* and *Microhierax cærulescens* ; in others it is absent.

The *anconæus* arises in *Polyboroides* and in some other types by a single head from the scapula, which is partly fleshy and partly tendinous. In *Vultur auricularis*, on the other hand, the muscle arises by two completely tendinous heads, so that the muscle has not that value in the classification of the **Accipitres** that I at one time thought.[3]   In all

---

[1] In *Geranoaetus melanoleucus*, which has, as have all the last-mentioned genera, a bifurcate tendon of the *brevis*, a small muscular belly ending in a tendon which becomes lost upon the patagium arose on the right side, in a specimen which I dissected, from the extensor muscle near to the end of the anterior branch of the brevis tendon.   I am uncertain of the exact homology of this structure in *Geranoaetus*.

[2] For muscular anatomy, &c., of **Accipitres** see GIEBEL, ' Bemerkungen über *Cathartes aura*,' &c., *Zeitschr. f. d. ges. Naturw.* ix. (1857), p. 426 ; ' Zur Anatomie von *Vultur fulvus*,' *ibid.* xxi. (1863), p. 131 ; ' Zur Anatomie des Lämmergeiers,' *ibid.* xxviii. (1866), p. 149 ; S. HAUGHTON, ' On the Comparative Myology of certain Birds,' *P. R. Irish Ac.* ix. (1867), p. 524 (crane and goose as well as hawks).   He deals with weight only.   E. NEANDER, *Undersökningar af Muskulaturen hos slägtet* Buteo, Lund, 1875 ; MILNE-EDWARDS in *Recherches Anatomiques, &c., des Oiseaux Fossiles de la France*, Paris, 1867.

[3] ' On certain Points in the Anatomy of the Accipitres,' *P. Z. S.* 1889, p. 81.

there appears to be an accessory tendinous origin from the humerus. The *pectoralis primus* is commonly divisible into two layers, but not in *Milvago* and *Dryotriorchis*.

The *deep flexor tendons* of the foot belong in the majority of species to type described above on p. 101. But there are a few variations of the typical arrangement. In *Astur tibialis* the slip to digit II. is present, but it is very small. In *Baza*, on the other hand, the vinculum is alone present, there being no special slip to the tendon supplying digit II.

In *Dryotriorchis spectabilis*, *Vultur auricularis*, and *Milvus ictinus*, the fibres of the vinculum are perfectly continuous with the slip to digit II., and form with it one single band of connection.

Both *peroneal muscles* appear to be present in the **Accipitres**.

All genera have the *ambiens* and the *femorocaudal*. In *Falco* and *Circus maurus* there is also a slender *semitendinosus*.[1] *Glutæus I.* is generally absent, *glutæus V.* commonly but not always present.

The *syrinx* of the **Accipitres** is of the ordinary tracheo-bronchial form.

In *Falco peregrinus* the intrinsic muscles are inserted on to a transversely elongate fibro-cartilaginous bar which runs across the interannular membrane of bronchial semi-rings 1 and 2. This membrane is very wide, owing to the fact that the first bronchial semi-ring is much arched, the concavity being downwards, while the second semi-ring is equally arched, but the concavity is upwards. None of the tracheal rings are fused, and the last gives rise to a pessulus. *F. candicans*, *F. lanarius*, *F. biarmicus*, *F. Feldeggi*, *F. æsalon*, *F. sacer* are perfectly similar, and the bronchidesmus (in those specimens in which it had been preserved) is complete.

Much like the syrinx of *Falco* is that of *Hieraeidea berigora*; I can, indeed, detect no differences. So too *Tinnunculus alaudarius* and *Erythropus vespertinus*. In

---

[1] FORBES in a MS. note records what I call 'semitendinosus' in *Microhierax cærulescens* as a product of the division of the semimembranosus on account of its origin from ischium and pubis.

Milvago chimango and *M. chima-chima* the syrinx is at first
sight perfectly similar, but the intrinsic muscles only just
get beyond the first bronchial semi-ring. In *Herpetotheres
cachinnans* this divergence from the normal falconine
syrinx is carried still further, the intrinsic muscle being
attached to the first semi-ring.

The syrinx of *Polyborus brasiliensis* is an exaggeration of
the falconine type. The first and second bronchial semi-rings
are very prominent and wide apart, thus leaving a very
spacious interannular membrane, to which the intrinsic
muscles are attached. The last few tracheal rings are fused
mesially in front and behind. The remaining forms, so far
as I have studied them (comprising the genera *Melierax,
Nisaetus, Gypaetus, Thrasaetus, Buteo, Milvus, Spizaetus,
Urubitinga, Haliaetus, Vultur, Spilornis, Morphnus, Helo-
tarsus, Leucopternis, Circus, Aquila, Circaetus, Gyps, Archi-
buteo, Geranoaetus,* and *Asturina*), differ from each other
in details—such as the completeness or incompleteness of
the bronchidesmus, the degree of ossification of the rings
and semi-rings, the number of the last tracheal rings which
are fused, and the attachment of the intrinsic muscles
(semi-rings 1, 2, or 3)—but they agree to differ from the
falcons in the absence of a pronounced oval gap between
the first and second bronchial semi-rings, which gives to
the syrinx of the falcons so characteristic an appearance.

The lobes of the liver are subequal, and a gall bladder
is present. The cæca of the Falconidæ are minute.

Haliaetus albicilla is a fish-eating bird, and for some
reason birds with such habits are furnished with a long intestine, as will be seen from the measurements in the table
on p. 477. The duodenal loop in this bird, exceptionally,
is thrown into a series of subsidiary loops, a state of affairs
which, as it occurs in the remote penguin, may have some
relation to habits and may not be a character upon which
stress is to be laid. The greater part of the intestine pre-
serves the simple archaic form of a number of irregular
coils ; but near to the cæca are two spirally twisted, elongated
loops. In other **Accipitres** it is more usual for the upper

loops to be long and twisted, a circumstance which recalls the structure of the loops in the stork (see fig. 208, p. 437).

The following are a few intestinal measurements :—

	Ft. In.	Inches	Inches
Serpentarius reptilivorus ♂	6 6	3	·15
" " ♂	7 9	4·5	·25
" " ♂	6 8·5	3	·25
" " ?	7 6	4·5	·25
Gypagus papa ♂ . .	5 1		
" " . . .	4 10		
Cathartes atratus . .	4 1		
Polyborus brasiliensis ♂ .	5 3	3	·5
Spizaetus coronatus ♀ .	3	4	·12
" *caligatus* ♂	2 6	1·25	·12
Falco biarmicus ♀ .	2 3	2	·15
Milvus ictinus ? .	3 11	2	·12
Circus Gouldi ? .	4 7·75	2	·15
" *æruginosus* .	3 8·5	2·75	·25
Haliaetus albicilla ? .	11 2		·25
" *vocifer* ♂ .	8 7	2·5	·15
Aquila nævioides ? . .	4 4	2	·09

The *skull* of the Falconidæ is described and figured by HUXLEY,[1] PARKER,[2] and SHUFELDT.[3] The palate is described as desmognathous; but it is always the case that a large portion of the maxillo-palatines—the posterior region—are not in contact. In two skulls of *Lophoaetus occipitalis* the palatal surfaces of the bones were *nowhere in contact*, and were only in contact for a minute space in a skull of *Vultur calvus*. Neither is *Elanus* desmognathous, according to SHUFELDT. The maxillo-palatines are large and swollen. The vomer is long and knifeblade-shaped;[4] there is often a medio-palatine, for instance in *Haliaetus albicilla*, where it is embraced by the bifurcate posterior extremity of the vomer. The lacrymal is large and has a separate ossification, the so-called infraorbital, attached to its posterior extremity in many

[1] In *P. Z. S.* 1867.   [2] *Linn. Trans.* (2) i.

[3] 'Some Comparative Osteological Notes on the N. American Kites,' *Ibis*, 1891, p. 228; 'Osteology of *Circus hudsonianus*,' *J. Comp. Med.* 1889.

[4] It has been found to be bifid in front, after the charadriiform plan, in young of *Tinnunculus* (cf. SUSCHKIN, 'Zur Anat. u. Entwicklungsgesch. d. Schädel d. Raubvögeln,' *Anat. Anz.* xi. p. 767).

hawks, for example in *Haliaetus albicilla, Lophoaetus occipitalis, Circus Gouldi, Asturina Natteri, Astur Novæ Hollandiæ*; in others, such as *Herpetotheres cachinnans, Vultur calvus*, the large size of the lacrymal suggests that such a bone is present, but ankylosed with the lacrymal. The bony nostrils of the Falconidæ are holorhinal, sometimes (e.g. *Herpetotheres cachinnans*) reduced in extent by alinasal ossifications; the long septum between them is more or less perfect. The number of cervical vertebræ, ribs, hæmapophyses, and uncinate processes of a few types is shown in the following table :—

—	Cerv. Vert.	Ribs	Hæmapophyses	Uncinate Processes
Herpetotheres cachinnans .	13	(8) r + r' + 6	C10- D3	On 3-7
Dryotriorchis spectabilis .	14	(8) r + r' + 5 + r	C10–D2	„ 3-6
Circus Gouldi . . .	14	(9) v + v' + 7	C10–D4	„ 3-8
Asturina Natteri .	14	(9) r + v' + 6 + r	C10–D4	„ 3-8
Accipiter nisus . . .	14	(10) r + v' + 7 + v'	C11–D3	„ 4-8
Astur Novæ Hollandiæ .	14	(9) r + 8	C10–D4	„ 3-8
Lophoaetus occipitalis .	14	(9) v + v' + 6 + r	C10–D4	„ 3-8
Melierax monogrammicus	13	(9) v + r' + 6 + v	C10–D4	„ 3-8

The sternum is whole or with one pair of foramina, sometimes notches, and often only developed on one side. The coracoids slightly overlap in *Dryotriorchis, Herpetotheres,* and *Melierax*; they do not quite meet in *Accipiter, Lophoaetus,* &c.

Pandion is undoubtedly an aberrant genus, which is by several (*e.g.* GADOW) made the type of a separate family, and is thought by some to lead towards the owls. It differs from other falcons in having no aftershaft, in its somewhat peculiar tensores patagii and deep plantar tendons.

The *tensor patagii brevis* has the additional 'aquiline' wristward slip, from the middle of which rises a short recurrent slip which joins the insertion of the main tendon. The tendon of the *biceps* muscle is split for nearly its whole length.

The *deep plantar tendons* are not accipitrine; they blend completely, as in owls, hornbills, &c., the area of fusion being ossified.

The *syrinx* is not remarkable in form. Anteriorly the last three tracheals are fused medianly ; posteriorly the fusion is more extensive, and includes the first bronchial semi-ring. The second bronchial semi-ring is in front close to the first ; behind it is united with the third, upon which latter are inserted the intrinsic muscles.

The *skull* is accipitrine and not strigine. The descending process of the lacrymals, however, is firmly and entirely blended with the ectethmoid, but the former bone has no backwardly projecting frontal portion, let alone a separate ossification at the end of it, such as is met with in some **Accipitres**. The vomer is long and ends in front in an olive-shaped swelling which fits in between, but is not attached to the diverging limbs of the anteriorly fused maxillo-palatines. ·

The ring of the *atlas* is incomplete in the middle line above ; there are fifteen *cervical vertebræ*. The hæmapophyses are very feeble on the earlier cervical vertebræ ; they commence on C10, where they are double ; they are strong over the last cervical and the first three dorsals, where they end. Six ribs reach the sternum, of which the first four have uncinate processes. Both the tibio-tarsus and the tarso-metatarsus have a bony bridge for tendons ; the latter has one behind as well as in front.

This bird possesses a scapula accessoria in the glenoid capsule, the significance of which as a point of affinity with the owls is marred by its occurrence in toucans, &c. (see p. 192). The coracoids slightly overlap, as in some **Accipitres**.

Whatever may be thought about *Pandion*, it is clear that the separation of the secretary bird to form a distinct family, Serpentariidæ, is perfectly justifiable.[1]

Serpentarius has basipterygoid processes, and its muscle formula is BXY+.

The *tensor patagii brevis* is more stork or crane like than accipitrine, and indeed resembles *Cathartes* in the presence of

[1] The claims of *Polyboroides* to be a member of this family have been dismissed by MILNE-EDWARDS (*Hist. Nat. Madagascar*) and myself (*loc. cit.* on p. 474).

a slip (see fig. 227) uniting the brevis and longus tendon.   It
must be remembered, however, that this also exists in some
eagles.

In *Serpentarius* there is a longer attachment of the
deltoid to the humerus than in other birds of prey : and there
is an accessory biceps muscle (see fig. 227).[1]

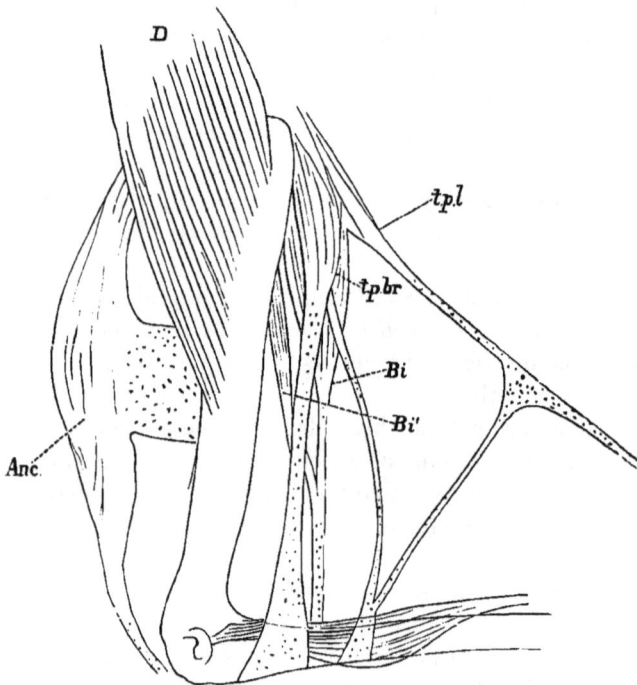

FIG. 227.—TENSORES PATAGII OF *Serpentarius* (AFTER BEDDARD).
tp.l, tensor longus ; *tp.br*, tensor brevis ; *Bi*, biceps ; *Bi'*, accessory biceps ;
Anc, anconæus ; *D*, deltoid.

The *anconæus* has a very broad tendon of origin from
humerus.

In the syrinx a strong box is formed by the last tracheal
ring, and the intrinsic muscles are attached to bronchial
semi-ring 2.

There is a powerful *expansor secundariorum*.

[1] Cf. *Rhinochetus*, p. 371.

The *skull* has strong basipterygoid processes. The lacrymals are large and extend backwards in close connection with the skull wall; they are not ankylosed to it. The descending process is thin and articulates with the slight ectethmoid. There is a small knife-shaped vomer.

The family **Cathartidæ** consists of the genera *Sarcorhamphus* (condor), *Gyparchus* (or *Gypagus*, king vulture), *Cathartes* (turkey vulture), and *Rhinogryphus*. They all have, so far as is known, the *oil gland* nude, twelve *rectrices*, no *aftershaft*, and are aquincubital.

The *tongue* is large and fleshy, with denticulations of its upturned lateral margins.

The *stomach* is not a gizzard. There are no intestinal *cæca*. The intestines are 61-inch in *Gyparchus*, 49-inch in *Cathartes atratus*. Of the heart of the condor some observations will be found above (p. 50); both *carotids* are present. The *liver* is equilobed, with a *gall bladder*. There are in *Gyparchus* traces of a crop.

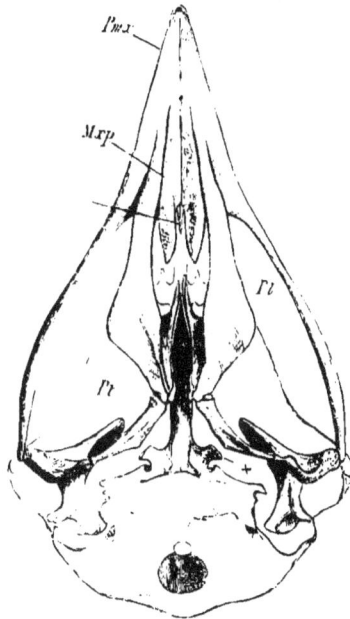

Fig. 228. - Skull of *Serpentarius* (after Huxley).

Pmx, premaxilla : *Mxp*. maxillo-palatines : *Pl*. palatine ; *Pt*, pterygoid ; *x*, basipterygoid process.

The most distinctive feature of the Cathartidæ, however, is the *windpipe*, from which a proper syrinx may be really said to be absent. The only muscles upon the trachea are the sterno-tracheales, which (in *G. papa*) are very short and broad, and arise from the sternum in the middle line, close together between the inner ends of the coracoids. Intrinsic syringeal muscles are entirely absent in the Cathartidæ,

I I

unless, indeed, their homologues exist[1] in the form of a muscular covering to the terminal purely membranous section of the bronchus (fig. 229) which, dividing into three slips, runs from thence to the parietes.   There are in any case no muscles at the actual bifurcation.   Nor is there any change in the character of the rings themselves such as to suggest even the rudiment of a syrinx.   In *Cathartes* the rings at the bifurcation are extraordinarily thin, leaving wide membranous intervals, which are occasionally traversed by bridges putting successive rings into communication.   In *Sarcorhamphus* and *Gyparchus*, which also agree (see below) in their 'muscle formula,' the rings are thicker and closer together (see fig. 229).   And in these two genera the bronchi are incomplete internally, giving rise to what may be termed a membrana tympaniformis.

In *Gyparchus papa* the tendons of the patagium are somewhat complicated.   The brevis consists of a separate anterior and posterior section, of which the latter is thinner and more diffuse.   The anterior tendon divides into two, of which the foremost gives off a slip to the longus.   There is no *biceps slip*.   The tendons, in fact, are thoroughly stork-like, as are those of the condor (*Sarcorhamphus*) and *Cathartes*.   In this character the family is very uniform.

The *expansor secundariorum* is present in all.

The *pectoralis primus* is well divided into two parts, of which the lower (in *G. papa*) is inserted by a thin round tendon altogether below insertion of superficial layer.   The head of the *anconæus* is distinctly bifid and entirely tendinous, arising from scapula and from supinator muscle.   There is a humeral slip of moderate size.   (This muscle is described from *Cathartes*.)

The *ambiens* is present in all Cathartidæ ; so too the *semitendinosus* and its accessory.   *Cathartes* has in addition the *femorocaudal*, which is absent in the other genera.   In *Gyparchus* (? as to the others) the *semitendinosus* and *semi-*

---

[1] BEDDARD, 'Notes on the Anatomy of the Condor,' *P. Z. S.* 1890, p. 146.

membranosus are inserted in common. There are two *peroneals* (at any rate in *Cathartes*). The deep flexors are fused before origin of four slips to four toes.

The *glutæus I.* covers over biceps, and glutæus V. is present.

The *skull* of the Cathartidæ has basipterygoid processes. It is desmognathous, but the desmognathism is totally

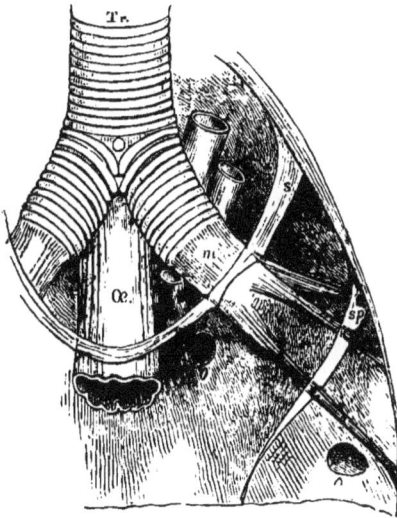

FIG. 229.—WINDPIPE OF CONDOR (AFTER BEDDARD).

Tr, trachea; *œ*, œsophagus; *O*. ostia of lungs; *s, sp*, septa between air sacs; *m*, muscles ensheathing end of bronchi.

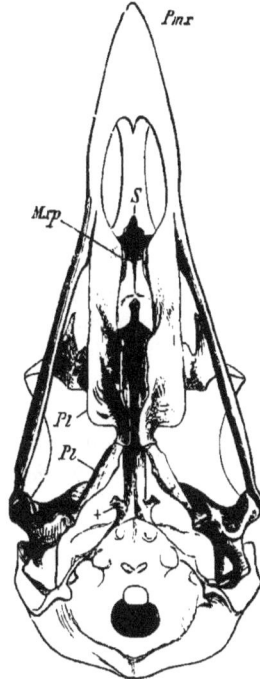

FIG. 230.—SKULL OF *Cathartes aura* (AFTER HUXLEY).

Pmx, premaxilla; *Mxp*, maxillo-palatines; *Pl*, palatines; *Pt*, pterygoid; *S*, ossified septum; *s*, basipterygoid processes.

different from that of other vultures and hawks. The maxillo-palatines proper (see fig. 230) are very far indeed from meeting in the middle line; indeed, they only just get beyond the shelter of the palatines. But a flat dorsal process of each of these bones (*S*) meets and is co-ossified with the nasal septum in the middle line. In *Gyparchus papa*, at any rate, there is a small medio-palatine. The lacrymal (at least in *Cathartes atratus*) is a smallish bone completely filling a notch in the frontal margin ; its descending process ankyloses with the ectethmoid, forming the usual ring.   In *Gyparchus papa* the orbital portion of the lacrymal is greatly reduced ; the nostrils are not so elongated as in *Cathartes* ; the palatal bridge is more plainly an alinasal fold.   The bony nostrils are holorhinal, but much more elongated than in the Falconidæ ; there is no trace of an ossified internarial septum.

Of fossil Accipitres the remains of a number of different species have been found.   The most interesting of these, on account of its age, is the *Lithornis vulturinus* of OWEN,[1] from the London clay. It had been held to come nearest to *Cathartes*, an interesting fact in view of its occurrence in this country ; but LYDEKKER regards it as clearly accipitrine and allied to *Accipiter* and *Circus*.   *Harpagornis*,[2] from the Pleistocene of New Zealand, was a large bird, one and a half time the bulk of a golden eagle, also belonging to the same division of Accipitres.   *Teracus* and *Palæohierax* are extinct genera from the lower Miocene of France, known only by femur and tarso-metatarsus respectively.   They are also probably true falcons.   *Serpentarius* is known by an extinct form, *S. robustus*, from the lower Miocene of the same country.

The following table shows the main differences between the several families of the **Accipitres** :—

—	Falconidæ	Serpentariidæ	Cathartidæ
Aftershaft . . .	+ (exc.Pandion)	+	—
Oil gland . . .	Tufted	Tufted	Nude
Muscle formula .	A +	BXY +	(A)XY +
Accessory semimemb. .	+	—	—
Cæca . . .	+ (rud.)	+ (rud.)	—
Syrinx . . .	Trach.-bronch.	Trach.-bronch.	—
Basipt. proc. . .	—	+	+
Desmognathism .	Of maxillo-pal.	Of max.-pal.	Of alinasals

[1] ' Description of the Fossil Remains of . . . a Bird (*Lithornis vulturinus*) from the London Clay,' *Trans. Geol. Soc.* (2), vi. 1841, p. 206.

[2] Cf. HAAST in *Trans. N. Zealand Inst.* iv. 1871, p. 192, and *ibid.* vi. 1874, p. 64, and OWEN in *Extinct Birds of New Zealand.*

It is clear from the few characters—the principal ones, however—given in the above list that the Cathartidæ are more aberrant (considering the Falconidæ to be the typical birds of prey) than are the Serpentariidæ; for the Cathartidæ diverge in all eight characters from the Falconidæ, while the secretary vulture only diverges in three. What reason is there, it might be asked, to retain the American vultures within this order at all, particularly if the owls are to be—as I think they should—excluded? The only group which has the distinctive characters of the Cathartidæ (besides, of course, the present group) is that of **Herodiones**. There only do we find birds with ambiens and expansor secundariorum, without biceps slip, holorhinal, and with rudimentary or absent cæca. The **Steganopodes** also are not far off. It really comes to the beak and claws, the ceroma, and to the presence of various structures (*e.g.* the peculiar palate, the basipterygoid processes) which forbid their association with the **Herodiones**. The several groups are not far off, but on the whole the American vultures are more like the remaining birds of prey than like the stork tribe (see also under the discussion of the affinities of the **Grues,** p. 382).

TINAMI

Definition.—**Oil gland tufted. Quintocubital. Muscle formula of thigh, ABXY+. Expansor secundariorum present. Biceps slip absent. Both carotids present. Large cæca and crop. Skull dromæognathous. Tail short without ploughshare bones. Bones of pelvis free distally.**

The tinamous are purely South American birds, of which in his recent catalogue Count SALVADORI allows nine genera.

The tinamous have a tufted *oil gland*, but the tuft is often very minute, and in *Calodromas elegans* consists of only four feathers, two larger and two smaller, the larger ones being uppermost.

I take my account of the *pterylosis* of the tinamous from

PYECRAFT's careful description [1] of *Calodromas* (which I can confirm) and *Rhynchotus rufescens*. In the former bird the body is fairly covered with feathers, the apteria being narrow. There is no down save on the wings. The spinal tract soon divides into two ; but they rejoin near the base of the neck. These tracts again divide and reunite some way in front of the oil gland, enclosing thus a dorsal apterion. The ventral tracts also divide early upon the neck, and each of them again divides on the pectoral region into a stronger, outer, and a somewhat weaker, inner, tract. Until about halfway down the neck the dorsal and ventral tracts are in contact. In *Rhynchotus rufescens* there is no spinal apterion. The *aftershaft* is much more rudimentary than in *Calodromas*, where it is well developed. Both birds have ten feebly developed rectrices. *Rh. perdicarius* has eight.

The aftershaft is apparently in the process of disappearance among the tinamous. In *Nothocercus*, writes Mr. PYECRAFT, ' it is evidently degenerating, inasmuch as the shaft is almost, if not quite, obsolete, only the rami remaining.' In *Tinamus solitarius* the aftershaft is absent. *Powder-down patches* exist in a few tinamous. They occur, for example, in *Tinamus major*. In *Crypturus tataupa* the powder-down patches extend down on each side of dorsal tract from a little in front of humerus nearly to oil gland. After the end of the scapula they thicken and spread outwards as far as the head of the femur, and are in contact for nearly two inches along mid-line ; they then narrow again and terminate half an inch in front, and slightly to the side, of the oil gland.

Rhynchotus perdicarius has apparently no powder-downs.

The *tongue* of the tinamous is small and triangular in form. The *crop* is present and large. The *proventriculus* is zonary : the *liver* subequilobed, with a *gall bladder*.

The following are measurements of the *alimentary canal* :—

[1] *Ibis*, 1895, p. 1.

—	S. I.	L. I.	C.
Crypturus tataupa .	19·75	3·75	2·5
„ obsoletus	34		3·3
„ salleri .	25	2·5	4
Rhynchotus rufescens	42		8·5, 9·5
Nothura maculosa .	19	3	5
Tinamus solitarius	73		4·5

The *cæca*, it will be observed (fig. 231, p. 488), are well developed, particularly in *Rhynchotus rufescens*; they are also large and very peculiar in form in *Calodromas*.[1] The cæca of this bird are not merely much wider than is customary, but they are beset with numerous small diverticula, which diminish in size towards the apex of the cæcum. These peculiar cæca are absolutely unique among birds, and nothing at all like them has been described in any other tinamou.

A curious feature of at any rate some tinamous (shared, however, by the **Anseres**, Palamedeidæ, some gallinaceous birds, and perhaps *Toccus*) is the existence of two pairs of extrinsic muscles upon the *trachea*. In *Crypturus tataupa* one of these pairs is stouter than the other, and they are both lost on the fascia covering lungs. This genus has no intrinsic muscles.

In other tinamous intrinsic muscles are present.

In *Calodromas elegans* the anterior face of the lower part of the trachea (about an inch in length) is covered with a sheet of muscle, which is the extrinsic muscle, and probably (judging from the conditions which obtain in the female) is attached to the long fascia. The very broad intrinsic muscles underlie this, and are inserted a long way down the bronchus to the four or five rings following the third. When viewed laterally the curvature of the last four tracheal rings is seen to gradually increase; there is thus a considerable membranous interval left between the last tracheal and the first bronchial, which is straight. The membrana tympaniformis is narrow. In the hen bird the extrinsic muscles are

[1] BEDDARD, 'On the Cæca of *Calodromas*,' *Ibis*, 1890, p. 61.

small, and do not form a sheet of muscle covering the end of
the trachea.

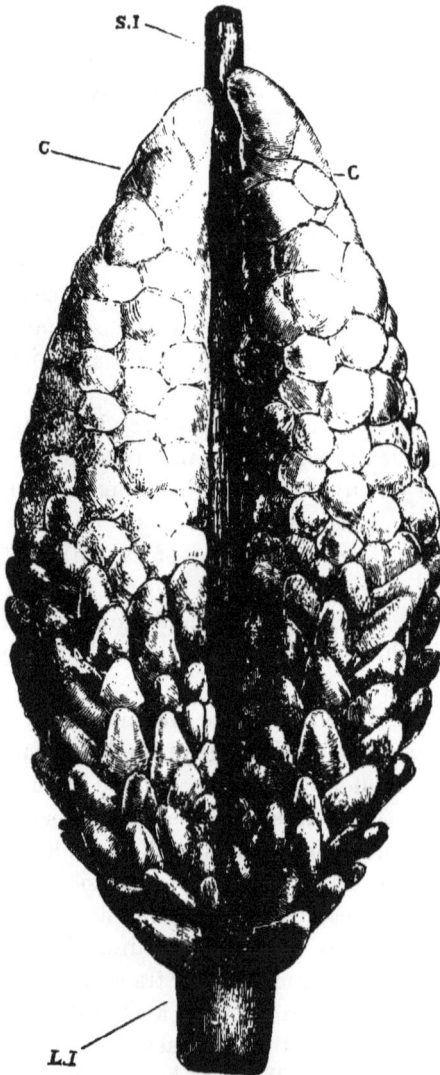

FIG. 231.—CÆCA OF *Calodromas elegans* (AFTER
BEDDARD).

S.I, small intestine ; *L.I*, large intestine ; *C*, cæca.

FIG. 232.—CÆCA OF *No-
thura maculosa* (AFTER
BEDDARD). LETTERS AS
IN FIG. 231.

In *Tinamus solitarius* the intrinsic muscles are large and are inserted upon the fifth or sixth bronchial semi-ring. In *Rhynchotus rufescens* they are also large, but attached higher up.

The *muscles*[1] of the tinamous are remarkable on account of their soft texture and pale colour. The *pectoralis I.* is very large and meets its fellow of the opposite side and for a considerable portion of its extent. The actual junction of the fibres is prevented by a fibrous septum, which is a continuation of the carina sterni. The second pectoral is also large. The *tensor patagii* tendon is a broad diffuse band, as in gallinaceous birds ; it has no biceps slip unless a muscle that will be referred to in connection with the *biceps* immediately really represents this. The *biceps* itself presents no remarkable features, but a kind of accessory biceps runs along the front of the humerus, which is quite distinct from the biceps proper ; I have found this both in *Calodromas* and *Rhynchotus*. This is regarded by FÜRBRINGER as part of the coracobrachialis externus.

In *Crypturus* (as also in gallinaceous birds and sand grouse—a noteworthy fact, perhaps) the *pectoralis abdominalis* has a remarkable ending.

Instead of being inserted directly upon the humeral crest it ends upon a tendinous bridge, to which are also attached the pectoralis, the latissimus dorsi posterior, and the expansor secundariorum.

The tinamous have the *gallinaceous muscle* in the fore arm and the *expansor secundariorum*. The *anconæus* has a humeral slip.

In *Crypturus tataupa* and *Nothura maculosa* at any rate the *expansor secundariorum* ends in a tendon which is inserted on to the scapula on the one hand and the manubrium sterni on the other.

The tinamous have the complete muscular formula of the leg, *i.e.* ABXY +.

---

[1] ALIX, 'Sur la Myologie du *Rhynchotus rufescens*,' *Journ. de Zool.* v. (1876), p. 411. 'Mémoire sur l'Ostéologie et Myologie du *Nothura major*,' *Journ. Zool.* iii. 1874, pp. 167, 252.

The *glutæi I.-V.* are well developed. A very interesting feature of the thigh muscles (referred to by GARROD [1]) is the existence of a small 'suprasciatic' slip of muscle arising behind the acetabulum, which reinforces the accessory femorocaudal. The interest of this small muscle lies in the fact that it has its precise counterpart in the struthious birds (*q.v.*) This muscle was found by FORBES to be absent in a male *Cryptarus tataupa*; it was present in a female of the same species.

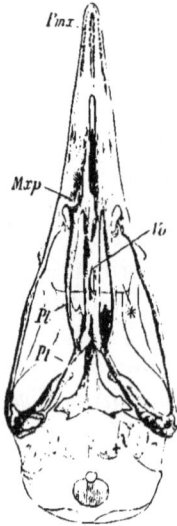

The two *deep flexor tendons* fuse and then supply digits II.-IV. ; before uniting the flexor hallucis gives off a slender slip to hallux, which is wanting in *Crypturus undulatus.*

The *skull* of the tinamous, as was first pointed out by PARKER,[2] is completely 'struthious' so far as concerns the palate. As will be seen from the annexed cut (fig. 233) the vomer is broad and unites in front with the maxillo-palatines, as in *Dromæus.* Its ends receive behind the pterygoid and palatines, which are thus prevented from articulation with the basisphenoidal rostrum. There are large basipterygoid processes and the head of the quadrate is single, as in *struthious* birds. The supraorbital chain of bones figured by PARKER in *Tinamus robustus* is another archaic skull character of these birds.[3] The nasals, lacrymals, and adjoining bones are very much like those of *Rhea* and not at all like those of gallinaceous birds. Between the nasals posteriorly is a considerable tract of ethmoid,[4] appearing

FIG. 233.—SKULL OF *Tinamus robustus* (AFTER HUXLEY).

Pmx, premaxillæ; *Mxp*, maxillo-palatines; *V*, vomer; *Pl*, palatines; *Pt*, pterygoids.

---

[1] 'On certain Muscles of the Thigh of Birds,' &c., *P. Z. S.* 1873, p. 642.

[2] 'On the Osteology of Gallinaceous Birds and Tinamous,' *Zool. Trans.* v.

[3] Absent, according to LUCAS ('Notes on the Osteology of the Spotted Tinamou,' *Proc. U. S. Nat. Mus.* x. 1887, p. 157), in *Nothura maculosa.*

[4] Prof. PARKER wrote, in 1862 (*loc. cit.* p. 213): 'I suppose that in the tinamou, as in other ostriches, the broad top of the ethmoid is separately developed

upon the surface of the skull. The outer descending part of the nasal reaches the maxilla, and with the upper part of the bone encircles the holorhinal nasal foramen; it is not ankylosed with the lacrymal. The latter descends and articulates with the jugal by a very distinct facet, especially distinct in *Rhynchotus rufescens*. The bone also becomes fused with the lateral wing-like process of the ethmoid, forming a complete ring of bone round a relatively very wide foramen. A special point of resemblance to *Rhea* and *Dromæus* is the perforation of the descending process of the lacrymal itself. This is best seen in *Rhynchotus rufescens*; in *Calodromas elegans*, *Nothura maculosa*, and *Crypturus tataupa*, there is merely a notch which in the fresh skull may possibly be converted into a foramen by a ligament.[1]

FÜRBRINGER gives 16-18 as the number of *cervical vertebræ*. The cup of the *atlas* is perforated for the odontoid process in *Crypturus*, *Rhynchotus*, and *Nothura*. As in the gallinaceous birds, &c., some of the dorsal vertebræ are fused together. In *Tinamus solitarius* this was the case with the first three and to a less extent with the last cervical. In *Nothura maculosa* five vertebræ were thus fused, and a strongish longitudinal piece of bone, formed of ossified ligament, connected their transverse processes. In *Crypturus tataupa* there were four vertebræ fused and one in front partially so. Four ribs reach the sternum in *Tinamus solitarius*. The *sternum* of the tinamous is very remarkable in form. The manubrium is slightly bifurcate; the middle portion of the sternum, which bears the keel, is exceedingly narrow, and a wide space is left on each side between it and the lateral processes, which are thin and as long as the middle piece. The anterolateral processes are well developed.

The *pelvis* is so far on the struthious pattern that the

by a long piece growing from above downwards between the anterior ends of the frontals. No suture remains to tell me that; but if it be so all is perfectly *struthious*, for those birds differ in this from all others examined by me.'

[1] This, however, is not a unique feature of the birds in question. It occurs, for example, in *Tantalus* and *Xenorhynchus*.

three bones do not fuse posteriorly; the pectineal process is large.

There are at most faint traces of a ploughshare bone.

In *Crypturus* the clavicles come into contact with the acrocoracoid and the scapula, but not with the small procoracoid.

The two coracoids are not nearly in contact at their articulation with the sternum.

The only birds with which the tinamous have been compared are the ostrich tribe, the gallinaceous birds, rails, bustards, and some of the **Limicolæ**. PARKER saw in the tinamou a ' cock ostrich mule ; ' and perhaps the prevalent opinion is that they lie on the confines of these two groups. It is unquestionably to the **Struthiones** that they show the greatest number of important likenesses,[1] so much so, indeed, that their inclusion in one great group with them would be by no means an unreasonable way of disposing of them. The salient points of resemblance are by no means confined to the skeleton, but the most numerous resemblances are in that part of the body of the birds. The skull, with its ' dromæognathous ' palate, is strikingly like. The appearance of the ethmoid as a median ossification of the skull roof is struthious, but it also occurs, though not so markedly, in *Gallus* and (according to SELENKA) in the Caprimulgidæ. It is very conspicuous in the tinamou and the **Struthiones**. The open pelvis is especially like that of *Apteryx*. The single-headed quadrate is struthious; but, as already mentioned, the struthious birds are not uniform in this character.

The sternum, with its antero-lateral and postero-lateral processes, recalls that of *Apteryx*, in spite of the enormous length of these parts and a consequent superficial dissimilarity. The absent, or rudimentary, ploughshare bone may perhaps be passed over as correlated with the imperfect flight. As to the soft parts, the peculiar additional accessory

---

[1] A singular if less important likeness than some mentioned above has been referred to by Mr. BARTLETT ('Notes on the Breeding of several Species of Birds,' &c., *P. Z. S.* 1868, p. 114). The male *Rhynchotus rufescens* incubates, as in the *Struthiones*, while the chick ' much resembles the young of a *Rhea*.'

femorocaudal muscle is a striking resemblance to the **Struthiones**; the peculiar accessory biceps muscle of the arm may have its degenerate counterpart in a sheet of strong tendinous tissue which runs along the humerus in certain ratites.

The following table shows some of the more striking likenesses of the **Tinami** to the **Struthiones** and **Galli** :—

	Tinami	Struthiones	Galli
Skull .	Dromæognathous	Drom.	Schizognathous
Pelvis .	Bones free	Bones free or but little united	Ischia and ilia fused posteriorly
Oil gland .	Tufted	0	Tufted, or nude, or 0
Remex V. . .	+	+ ?	+
Leg muscles . .	ABXY +	(A)BXY(+)	(A)BXY +
Suprasciatic muscle .	+	+	0
Entepicondylo-ulnaris	+	In Apteryx	+
Trachea . . .	Sometimes with two pairs extrins. muscles	Only one pair	Sometimes two pairs

STRUTHIONES

Definition.—**Flightless birds without stiff contour feathers. Oil gland absent. Wing small. Expansor secundariorum and biceps slip absent. Semitendinosus and its accessory always present. An additional slip to accessory femorocaudal present. Skull dromæognathous with basipterygoid processes; holorhinal. Sternum without a well-developed carina. Coraco-scapular angle wide. Coracoid fused with scapula. Cæca large.**

As will be seen from the above definition, the characters of this group are to a considerable extent negative characters. They are for the most part such characters as are correlated with the loss of the power of flight. We need not, therefore, lay too much stress upon them as indicative of the naturalness of the group. But even when these characters (as, for instance, the absence of the carina sterni, the open angle between the coracoid and scapula, the absence of a ploughshare bone, which, moreover, is occasionally and exceptionally present. There is a skeleton of an old *Struthio* in the Cambridge Museum in which several of the last vertebræ

are fused) are set aside as comparatively valueless as marks of near relationship, there remain enough anatomical resemblances to justify the older view that all these birds are but members of one and the same group. FÜRBRINGER has denied this in his 'Untersuchungen,' and places *Apteryx*, together with the Dinornithidæ, apart from the other struthious birds, and has again separated Struthiiformes from Rheiformes and Casuariiformes, deriving all from different levels of the ornithic tree. There is no doubt that the various types of struthious birds do require separating into at least six families; but the likenesses among them appear to me to forbid any wider separation. The close resemblance of the palate throughout the group, so far as we know it (*Æpyornis* is not known), is a strong reason for associating them together; perhaps even the osteological and other characters, which, as already suggested, are but evidence of the loss of the flight power, may be of more importance as an argument for affinity than is generally admitted; it may show that they are allied, because the degeneration has proceeded along the same lines. There is, it is true, not a great deal of evidence in favour of this view; but we have the penguins also with a degenerate wing, in which the modifications of structure [1] have progressed along different paths. They have, for example, lost the biceps, which is present in all **Struthiones**, while the feathers of the wing are equally inefficient as aids to flight with those of the **Struthiones**, but are quite unlike them. The peculiar muscle of the thigh, which will be found described as an adjunct of the accessory femorocaudal, is one of those apparently small facts of structure which, on account of their very minuteness, seem of importance as a mark of true relationship.

The fact that all of the struthious birds have large or moderately developed cæca is further evidence of affinity. It might be thought that the usual absence of the oil gland was one of those characters affording clear evidence of degeneration; but its capricious appearance and disappearance in

---

[1] See, however, the qualifying remarks with regard to the wings of *Apteryx* on p. 499.

other birds forbid us to assume this without any further argument.

While the **Struthiones** present collectively and individually a larger number of important differences from other birds, their organisation is essentially on the plan of that of the remaining members of the class. To take only one—at the same time one of the most striking—of these correspondences in anatomical structure, the respiratory organs may be considered. It is hardly too much to say that there are not even differences of detail in the arrangement of the lungs and air sacs among the struthious birds. Professor HUXLEY exploded some years ago the idea that the oblique septa of the Apteryx were more like the mammalian diaphragm than the corresponding structures of other carinate or ratite birds. It is inconceivable that there should be this minute correspondence of detail with detail, if we are to assume with some that the struthious birds have arisen from a totally different stock from that which produced the carinates. They would derive the former from the dinosaurs and the latter from the pterodactyles.

The existing struthious birds are the genera *Struthio*, Afro-Arabian in range; *Rhea*, South American; *Dromæus*, Australia; *Apteryx*, New Zealand; and *Casuarius*, Australian region. The structure of these living members of the group will be considered first, after which some account will be given of the Dinornithidæ and other extinct and undoubted members of the group, as well as of a few dubious forms which have been placed here—rather because they do not definitely fit in anywhere in particular than from their obvious affinities with the **Struthiones**.

The genus *Struthio* appears to contain two species, the more common *Struthio camelus* and the Somaliland *S. molybdophanes*. The ostrich has two toes, Nos. III. and IV. There is no oil gland. The pterylosis, continuous in the adult bird, shows two distinct apteria in the embryo, as has been shown by Miss LINDSAY.[1] In the young chick

[1] 'On the Avian Sternum,' *P. Z. S.* 1885, p. 684. See also W. MARSHALL, 'Beobachtungen über das Verhältniss der Federn,' &c., *Zool. Gart.* xvi. (1875),

there is a ventral apterion in the sternal region and a lateral apterion outside each half of the ventral tract. The adult ostrich has a claw on each of digits I. and II. The arrangement of the wing feathers has been carefully worked out by the late Mr. WRAY.[1] He finds the remiges to be quite distinct, as well as the tectrices majores; the tectrices mediæ are but scantily represented, and there is an incomplete row of tectrices minores. The number of remiges upon the hand, including one upon the carpus, is sixteen. There are four to the ala spuria. The number of cubitals is about twenty. It has, therefore, more primaries than any bird except the penguin.

The genus *Apteryx*, entirely confined to New Zealand, consists of three or four species, viz. *A. australis*, *A. Mantelli*, *A. Oweni*, *A. Haasti*, and *A. Bulleri*.

It has been described as possessing a continuous, uninterrupted plumage; but this, according to T. J. PARKER,[2] is far from the truth. ' In a fresh specimen of *A. Bulleri*,' he remarks, ' I find the lateral apterium to be fully 2 cm. wide, and to extend about 5 cm. cephalad and 9 cm. caudad from the axilla, its total length being, therefore, about 14 cm. In the same specimen the ventral or inferior space was of about equal width (2 cm.), and extended about 11 or 12 cm. caudad from between the origins of the wings. Moreover the inner (ventral) surface of the wing is always nearly devoid of feathers and so constitutes a well-marked lower wing-space.'

The oil gland is present and the feathers have no aftershaft.

The relatively minute wing of the Apteryx has a true alar membrane, which, as PARKER has justly pointed out, is further evidence for regarding this bird as the derivative of a flying form.

p. 121, and ZANDER, ' Über das Gefieder des afrikanischen Strausses,' *Schr. phys.-ök. Ges. Königsb.* xxix. 1889, SB. p. 31.

[1] ' On some Points in the Morphology of the Wings of Birds,' *P. Z. S.* 1887, p. 343.

[2] ' Observations on the Anatomy and Development of *Apteryx*,' *Phil. Trans.* 1891.

Though no rectrices can be distinguished, there are recognisable remiges. PARKER counted nine or ten cubitals and two or three metacarpals and a single mid-digital ; there are also tectrices majores. An extraordinary peculiarity of *Apteryx* is the situation of the nostrils near the very end of the beak.

Of *Casuarius* there are some ten species which are found in several of the islands lying to the north of the continent of Australia, such as New Britain, Ceram, &c., as well as— one species, *Casuarius australis*—in the north of Australia itself. They are remarkable externally for their black coloration, brown in the young, and for the horny casque upon the head. The neck is naked and adorned with bright colours, in which blue is especially prominent, and there are often dependent folds of bright-coloured skin in this region. The feathers have an aftershaft as large as the feather itself; the rectrices are unrecognisable, but the remiges are present in the shape of long spines which correspond to the stems of the feathers. The claw of the inner of three toes is very elongate.

The emu, *Dromæus*,[1] is entirely Australian in range, and contains two species. This genus, agreeing with the cassowary in laying a green egg, has no helmet or wattle, or stiff spines upon the wing. It has, however, like the cassowary, a large aftershaft.

The fourth genus of **Struthiones** is the South American *Rhea*, of which three species are recognised. These have been carefully compared by GADOW.[2] The genus is characterised, so far as external characters are concerned, by the want of an aftershaft and by the feathered neck—not naked, as in the ostrich ; it lays a yellowish white egg. The *Rhea* is three-toed. There is a distinct ventral apterion running from sternal callosity to vent.[3]

---

[1] G. DUCHAMP, 'Observations sur l'Anatomie du *Dromæus*,' *Ann. Sci. Nat.* (5), xvii. 1873.

[2] 'On the Anatomical Differences in the Three Species of *Rhea*,' *P. Z. S.* 1885, p. 308.

[3] 'A. BOECKING, *De Rhea Americana*, Diss. Inaug. Bonn, 1863 ; J. F.

Struthio has been found fossil in the Siwalik Hills, in South Russia and Samos. *Rhea* is found fossil in America (South). LYDEKKER considers *Hypselornis sivalensis*, whose place of interment is indicated by the name, to be an emu. It is only known from the second phalanx of the third digit of the pes.

Genyornis Newtoni, from Australia,[1] with a skull a foot long, seems to have been a gigantic emu. But it has not as yet been fully described.

Dasornis londinensis (from the Eocene clay of Sheppey) is placed by FÜRBRINGER among the **Ratites**, rather in deference to the opinion of Sir R. OWEN [2] than from conviction. GADOW, on the other hand, places it among **Stereornithes**. It is only known by a water-worn skull fragment, indicating a skull as large as that of the Dinornithidæ. It seems useless to speculate upon the affinities of this fragment.

Macrornis of SEELEY must remain for the present a name.

In surveying the *muscular system* of the **Struthiones** [3] it is clear that, so far as concerns the muscles of the manus, *Apteryx* is, in accordance with other reductions in the bones of that limb, the most degenerate type. On the other hand (assuming, of course, the derivation of the Struthiones from some carinate form) the shoulder girdle of *Apteryx* has retained more of the primitive musculature than the other genera.

In all the genera the following muscles have disappeared : the *pectoralis propatagialis, biceps propatagialis, deltoides propatagialis,*[4] *deltoides minor, scapulo-humeralis anterior, expansor secundariorum.*[5]

The *pectoralis major* is in all very reduced.

All the struthious birds except *Apteryx* have also lost the

VAN BEMMELEN, 'Onderzoek van een *Rhea*-Embryo,' *Tijd. Ned. Dierk. Ver.* 1888, p. ccv.

[1] STIRLING and ZIETZ, 'Preliminary Notes on *Genyornis*.' &c., *Tr. Roy. Soc. S. Australia*, xx.

[2] On *Dinornis* (part xiv.), *Tr. Zool. Soc.* vii. p. 145, pl. xvi.

[3] GADOW, *Zur vergleichenden Anatomie der Muskulatur des Beckens und der hinteren Gliedmässe der Ratiten.* Jena, 1880.

[4] In *Apteryx* some elastic tissue in the patagium possibly represents this.

[5] Traces have been asserted to exist in *Apteryx* and *Dromæus*, but require confirmation.

serratus metapatagialis, the *latissimus dorsi metapatagialis*, and the *pectoralis abdominalis*.

On the other hand *Apteryx* has lost what the other struthious birds have retained, the *latissimus dorsi anterior* and the *rhomboideus profundus*; the latter muscle, however, is not distinguishable in the cassowary.

It must be admitted, therefore, that *Apteryx*, so far as concerns the anterior extremity, has diverged from the hypothetical ancestral condition in slightly different lines from other **Struthiones**.

In the cassowary [1] both *rhomboidei* are present, but they originate from ribs and not from the cervical vertebrae. The *rhomboideus profundus* is parallel with and hardly distinguishable from a portion of the *serratus profundus*; hence FÜRBRINGER is indisposed to admit the existence of a separate *rhomboideus profundus*.

The *serratus superficialis* consists of two separate fan-shaped bands of muscle. The *coraco-brachialis internus* is entirely converted into tendon. The *biceps* originates only from the coracoid, and ends without being definitely split into two tendons upon both radius and ulna. There is only one *scapulo-humeralis* muscle, which is, however, of fair size.

The *subscapularis* is a single-headed muscle arising from the scapula only.

The *anconæus* has a single origin from the scapula, and has no attachment to the humerus.

In the hind limb all five *glutæi* are present; they are all large, especially gl. I. and gl. V. The *ambiens* is absent as a rule; it is occasionally present, but is then imperfect, reaching only as far as the knee. The *semitendinosus* and its accessory are well developed. The *femorocaudal* is a small slender muscle; it is inserted in common with the accessory, which is enormous in size. In *Casuarius Bennettii* at any rate there is an additional adductor of peculiar origin; the muscle is two-headed, one head being a tendon which

---

[1] J. F. MECKEL, 'Beiträge zur Anatomie des indischen Casuars,' *Arch. f. Anat. u. Phys.* 1830, p. 200, 1832, p. 273.

springs from the muscular fibres of the accessory femoro-
caudal, the other fleshy and springing from the pelvis
just behind the acetabulum.   It is inserted along the femur
below the vastus internus and over the conjoined femoro-
caudals.

In *Struthio* the *rhomboideus superficialis* arises, as in
carinate birds, from the spinous processes of the vertebræ
(1–3 cervicals) ; it is inserted only on to the scapula.   The
rhomboideus profundus arises from the spinous processes of
the last cervical and first dorsal vertebræ ; it is inserted on to
the end of the scapula.   The *serratus superficialis* of *Struthio* is a single muscle arising as two or three bands, either
from the last cervical and first dorsal rib or, in addition,
from the second dorsal rib.   It is attached to the ventral
border of the scapula.   The *serratus profundus* is divisible
into a more superficial and a deeper layer ; the former is
the less extensive, and arises either by a slip from the rib
of cervical vertebra 19, and by two slips from the last
cervical rib, or by two larger slips and one very small one
from between the last cervical and the first dorsal rib, and
from the latter ; they are inserted on to the inner border of
the scapula.   The deeper layer also varies, but arises in
several slips from the last two cervical ribs.   It is also
inserted on to the inner border of the scapula.   The *coraco-
brachialis externus* is very large as compared with the same
muscle in the carinate birds ; it is not quite so large as in
Rhea.   The *coraco-brachialis internus* is larger in *Struthio*
than in any other ratite.   The *biceps* arises from the spina
coracoidea ; its muscular belly is not well developed ; it is
inserted on to the radius and ulna, and on to the membrane
between them.   As with *Rhea* the *deltoid* arises from the
scapula and neighbouring region of coracoid.   The *teres
major*, again, as in *Rhea*, is a comparatively large muscle.

On p. 87 *et seq.* will be found an account of the muscles of
the hand in *Palamedea*, which I have taken to illustrate that
of the carinate birds in general.

The differences which are to be noticed in *Struthio* are,
apart from minor divergences, the following :—

The *extensor metacarpi radialis* is single.

The *ectepicondylo-ulnaris* is absent or fused with the *extensor metacarpi ulnaris*.

The *extensor digitorum communis* supplies only the index.

The two *pronators* form only one muscle.

The *flexor digitorum sublimis* and the *fl. dig. profundus* arise by a single head from the flexor condyle of the humerus. The two muscles immediately divide; the upper part (= sublimis) ends in two tendons, of which one is inserted on to radiale, the other fuses with the upper tendon of profundus, and also gives off two slips which surround that tendon and, reuniting, fuse with the lower tendon of the profundus. The lower part of the muscle (= profundus) gives off two tendons, of which the upper ends on the first metacarpal, while the lower runs to the base of the last phalanx of the index.

The *flexor metacarpi ulnaris* ends fleshily on ulnare, but is prolonged beyond this bone, receiving also some fibres from it, to the metacarpal.

The *radio-metacarpalis ventralis*—or at least a muscle which, if it be not this, is not found in *Palamedea*—arises from the ulna and not from the radius.

The total number of muscles in the hand of the ostrich is twenty-three, allowing for the absent ectepicondylo-ulnaris. The additional muscle is a small *pronator quadratus*, running from the ulna to the radius.

It appears, therefore, that, in spite of the small size of the manus of the ostrich relatively to that of flying birds, there is but little if any evidence of degeneration in its musculature. On the contrary, indeed, for it might be said that the wing muscles of *Struthio* are less degenerate, or at any rate less modified, than those of carinates in that amount of muscle as compared with tendon is greater. The complication of the conjoined flexores digitorum is highly suggestive of a walking or climbing animal. It seems to be conceivable that the ostrich branched off from the avian stem before the power of flight was perfectly established.

The ostrich [1] has the complete leg muscle formula ABXY +.

The *femorocaudal* is fleshy, but not large, and has no distinct tendon of its own. It blends above with the accessory. The *accessory femorocaudal* is an enormous muscle ending in a broad thin tendon which distally is lost in a fibrous expansion round the great vessels and nerves of the thigh. The *accessory semitendinosus* is small. The tendons of the *semimembranosus* and the semitendinosus become united just after the attachment to the latter of the accessory; they soon, however, diverge, the semimembranosus being continued as a long thin tendon down the leg to join the tendon of the *gastrocnemius*. The *obturator externus* and the *adductors* are small; the *obturator internus* is very large. The *ambiens* does not arise from the pectineal process, or even from the pubis, but from the ilium. An additional adductor muscle which has been referred to in the cassowary also exists in the ostrich; it has, however, but one (tendinous) head, arising from the femorocaudal muscle.

The *rhomboideus superficialis* of *Rhea* [2] springs, like that of *Struthio*, from the spinous processes of the cervical vertebræ, but from a larger number (four). It is inserted on to the coracoid as well as the scapula. The *rhomboideus profundus* arises from the spinous processes of the first three dorsals. As in carinate birds the *serratus superficialis* is composed of an anterior and a posterior section; the former arises as a single band from the last cervical rib, and is attached to the front part of the scapula; the latter is large and consists of three broad slips springing from the first three dorsal ribs and their uncinate processes; it is attached to the hinder end of the scapula. It may be, FÜRBRINGER thinks, that a portion of this is really the *pars superficialis*

---

[1] S. HAUGHTON, 'On the Muscular Mechanism of the Leg of the Ostrich,' *P. R. Irish Ac.* ix. (1866), p. 50; A. MACALISTER, 'On the Anatomy of the Ostrich,' *ibid.* 1867, p. 1; ALIX, 'Sur l'Appareil Locom. de l'Autruche de l'Afrique,' *Bull. Soc. Philom.* 1868.

[2] S. HAUGHTON, 'Muscular Anatomy of the Rhea,' *P. R. Irish Ac.* ix. (1867), p. 497.

of the *serratus profundus*; otherwise that muscle only consists of the deeper portion which arises as two slips from the last two cervical ribs and runs directly backwards to be inserted on to the lower border of the scapula.

The *coraco-brachialis internus* is largely tendinous: its origin, contrary to what is found in other **Struthiones**, just extends on to the sternum. The origin of the *biceps* is peculiar; it arises not only from the coracoid spine by a rounded tendon, but also by a sheet of tendon edged with muscle from the whole, of the coracoid and from just an adjacent bit of the sternum. It is inserted on to both radius and ulna.

In the manus of *Rhea*, on the other hand, we have more evidence of degeneration than in *Struthio*. There are, in the first place, only twenty-one muscles at most, and some of these are much simplified.

The muscles that appear to be totally wanting are (1) the *extensor digitorum communis*, (2) the *pronator profundus*. The *extensor indicis* is only represented by the belly arising from the wrist. The *flexor sublimis* may possibly be represented by a slip of muscle arising from the tendinous edge of the *flexor metacarpi ulnaris*, which goes to be inserted, partly by tendon, partly by fleshy fibres, on to the ulnare and base of metacarpals 2 and 3.

As in the ostrich the *radio-metacarpalis ventralis* arises from the ulna. In *Rhea* there is a special peculiarity in the presence of a muscular slip running from the tendon of the *extensor metacarpi ulnaris* near to its insertion to the *extensor indicis*. Finally the *ectepicondylo-ulnaris* is distinct.

In the leg there is no *femorocaudal*, the formula being BXY+. The *accessory femorocaudal* is enormous, and there is a good struthious accessory adductor. *Glutæus primus* is very large and overlaps biceps; *glutæus V.* is present and large.

In *Dromæus* [1] the *rhomboideus superficialis* and pro-

---

[1] S. HAUGHTON, 'Muscular Anatomy of the Emu,' *P. R. Irish Ac.* ix (1867), p. 487; G. ROLLESTON, 'On the Homologies of certain Muscles connected with the Shoulder Joint,' *Trans. Linn. Soc.* xxvi. 1870, p. 609.

fundus arise from ribs, the latter from only one, the former from three.

The *biceps* apparently arises like that of *Rhea*.

In the leg the *ambiens* and the *femorocaudal* are wanting, the formula being, therefore, BXY – . All the *gluteals* are present, and the first covers the biceps. The *accessory femorocaudal* is very large, but it does not appear to possess the struthious accessory muscle.

The muscles of the wing of *Apteryx* are, of course, described by OWEN in his account of the anatomy of the bird. But a fuller and later description, with illustrations, is to be found in T. J. PARKER's paper upon *Apteryx*, and in FÜRBRINGER.

There is no *rhomboideus profundus*. The *serratus superficialis* is one muscle arising from the first two cervical ribs ; it has the *pars metapatagialis* wanting in the other ratites. There is also a *pectoralis abdominalis* wanting elsewhere, but *no latiss. dorsi anterior*. The *latiss. dorsi metapatagialis* is well developed.

The muscles running from the shoulder girdle to the humerus are reduced to six ; these are the two *pectorals*, the *deltoides* (single), the *teres major*, the *coraco-brachialis longus*, and the *c. br. brevis*.

The *biceps* is single-headed, arising from the coracoid ; it has long tendons at each end and a small belly in the middle ; it is inserted only on to the radius. The *anconæus longus* fuses early with the single-headed *triceps*.

As might be expected from the presence of but a single finger, the muscles of the hand are much reduced. Perhaps the most noteworthy peculiarity is the presence of the gallinaceous and tinamine muscle, the *entepicondylo-ulnaris*. There is a peculiar *accessory brachialis anticus*, seemingly only met with in *Apteryx*. The remaining ten muscles are *brachialis anticus*, *ectepicondylo-ulnaris*, *ectepicondylo-radialis*, *pronator* (? *sublimis* or *profundus*), *extensor metacarpi ulnaris*, *extensor indicis longus*, with one head from contiguous surfaces of radius and ulna inserted in *A. australis* on to carpo-metacarpus, in *A. Bulleri* on to base of distal phalanx, *extensor longus pollicis* [1] inserted on to the thumb side of the carpo-metacarpus, *flexor digitorum profundus, ulni-metacarpalis ventralis.* [2] 'In one specimen (*A. australis*) a minute tendon was seen preaxial of that of the deep flexor and passing to the preaxial side of the carpo-metacarpus,' a rudimentary *interosseous dorsalis* in one specimen of *A. Bulleri*.

In the leg the muscle formula is complete.

The *tongue* in all these birds is a flat triangular organ, relatively small in size. The accompanying cut (fig. 234) shows its characters in *Rhea*: it does not show any differences of importance in the other genera.

The course of the intestine in *Struthio camelus* is shown in fig. 12, p. 28. It will be observed that it is in many respects exceedingly simple: thus the greater part of the small intestine is thrown into a series of short folds, with none of the longer and more specialised folds found in many other birds. The duodenal loop is the only part of the small intestine which shows a special fold, and this is a loop with a short lateral diverticulum—Y-shaped, in fact.

FIG. 234. TONGUE AND WINDPIPE OF *Rhea Darwini* (AFTER GADOW).
N, hypoglossal nerve.

The next most characteristic feature of the intestines of this bird is the enormous large intestine, which is for the greater part of its extent thrown into folds like those of the small intestine. In *Casuarius* (see fig. 11, p. 28) the small intestine is quite as simple as that of the ostrich; the large intestine is short and straight. The emu is practically identical with the

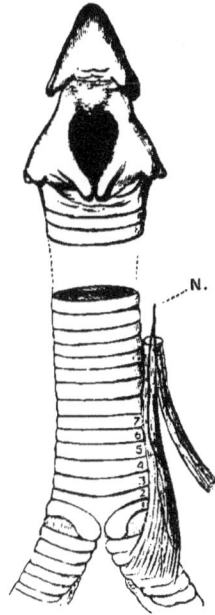

[1] PARKER terms it *extensor metacarpi radialis brevis*. I identify it as above.

[2] *Flexor carpi radialis* of PARKER.

cassowary, but the duodenal loop was strong, longer, and narrower. It is interesting to find from MITCHELL's paper that the intestine of *Rhea* is somewhat intermediate between that of *Struthio* and that of *Casuarius*. 'The anterior portion resembled *Casuarius*; the rectum had an expansion recalling that in the ostrich, but much less strongly marked.'

The following table contains measurements of the alimentary tract [1] in the **Struthiones** :—

—	Small Int.		Large Int.		Cæc.			
	Ft.	In.	Ft.	In.	Ft.	In.	Ft.	In.
Rhea macrorhyncha ♀	4	2	1		1	4½,	1	9
„ „ ♂	4	5¾	1	11	2	9		
„ „	5		1	4	2			
Rhea americana	6	3	1	8	2	4,	2	0
„ „ ?	9	8	2	2	4	8		
„ „ (young)	5	10	1	4	2	6,	2	8
Struthio camelus ?	23	1	32	0	2	8,	2	11
„ „	23	4	30	8	2	10½		
„ „ ♀	23		29	8	2	7		
„ „ ♂	24	6	31	8	2	8,	3	1
„ „ ♂	23		24	0	1	10		
„ „ ?	28	6	33	2	2	7		
Casuarius uniappendiculatus	3	8		10		4½,		5
„ *Beccarii* ♀	4	8	1			4½		
„ *picticollis* ♂	4			10		4		
„ *Bennettii* ♂	3	10½		11½		3¼,		3¾
Apteryx australis	4	4		4½		7		
„ *Oweni* ♂	3	4¼		4¼		7		
Dromæus Novæ Hollandiæ	10	6	1			5		
Casuarius bicarunculatus	5		1			7		

All the **Struthiones** have cæca, which are especially developed in the ostrich, where they have been described by Sir EVERARD HOME as well as by GADOW. Apart from their length and structure the most remarkable fact about them is that, in contradistinction to what we find in other birds, they are inserted by a common orifice.

These cæca dwindle gradually in diameter towards the tip, and are provided internally with a spiral valve of about twenty turns. I have attempted to compare these cæca with those of the Martineta tinamou (see p. 488). In the latter bird the cæca are furnished with numerous short diverticula,

---

[1] E. REMOUCHAMPS, 'Sur la Glande Gastrique du Nandou,' *Bull. Ac. Belg.* l. (1880), p. 114.

which are the outward expression of a reticulate internal
structure, like a ruminant's stomach. Towards the extremity
of the cæcum the folds cease to be so definitely arranged in
a network, and some to present an indication of a spiral disposition. *Rhea* has also very large cæca, and there are
traces of the spiral valve of the ostrich. The cæca of the
cassowaries and the emu, as will be seen from the table of
measurement, are very much smaller than those of the
ostrich and *Rhea*. GADOW mentions an obscure formation of

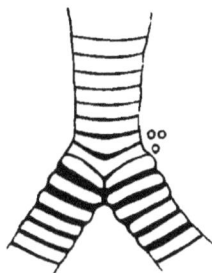

FIG. 235.—SYRINX OF *Apteryx*
Mantelli. FRONT VIEW.
(AFTER FORBES.)
a.o, in this and following
figs., tracheal rings.

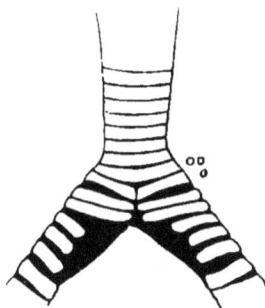

FIG. 236.—THE SAME, FROM BEHIND.

an internal network by the presence of folds which may be
compared with the structure of the cæca of *Calodromus*,
already referred to. The cæca of *Apteryx* are long and narrow, like those of the tinamous (excluding *Calodromus*).

Struthio has the most remarkable liver of all the **Struthiones**. The two lobes are intimately fused into one heart-
shaped lobe. There is an indication of a spigelian lobe, as
with other **Struthiones**, and the single bile duct (there is no
gall bladder) opens, exceptionally, only 4 cm. from the
pylorus. *Rhea* has, as a rule, no gall bladder, but GADOW
found traces of one in a specimen dissected. This occurred
(see figs. 19, 20, p. 34) both in *Rh. Darwini* and *Rh. americana*.
Casuarius and *Dromæus* have a well-developed gall bladder,

and MITCHELL noted in the former genus that the gall duct and the pancreatic duct opened into a distinct diverticulum of the duodenum.

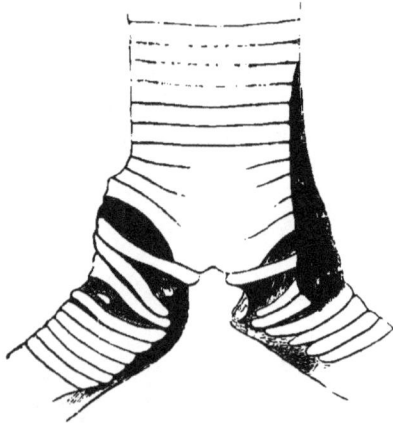

FIG. 237.—SYRINX OF *Rhea americana.* FRONT VIEW. (AFTER FORBES.)

In *Apteryx* there is a gall bladder.

The chief source of information about the *windpipe* and

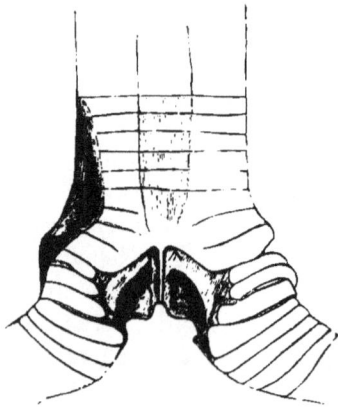

FIG. 238.—THE SAME, FROM BEHIND.

syrinx of the Struthionidæ is contained in a memoir by FORBES.[1] These birds, as a rule, possess no specially modified

---

[1] 'On the . . . . Trachea in the Ratite Birds,' *P. Z. S.* 1881, p. 778.

syrinx, a statement, however, which does not, as was first
shown by ALIX,[1] apply to *Rhea*. In that bird (see woodcut,

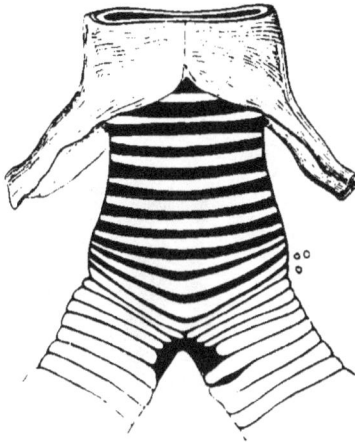

FIG. 239. SYRINX OF *Casuarius galeatus*. FRONT VIEW. (AFTER FORBES.)

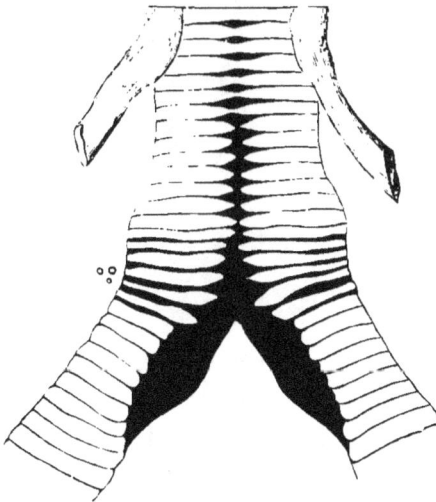

FIG. 240.—THE SAME, FROM BEHIND.

figs. 237, 238) there is not merely a pair of intrinsic muscles,[2]

[1] *Bull. Soc. Phil.* 1874, p. 38.
[2] First noted, apparently, by PARKER in *Tr. Z. S.* vol. v. p. 238, foot note.

but some considerable modification of the last tracheal and
early bronchial rings.   The last four tracheal rings are
soldered together to form a cartilaginous box, which behind
and in front shows no lines of demarcation between the
several rings of which it is composed.   There is a membrana

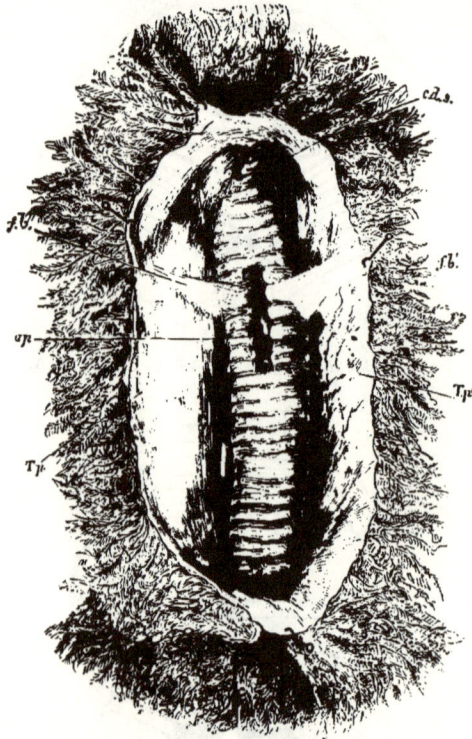

FIG. 241.— TRACHEAL POUCH OF EMU CUT OPEN (AFTER MURIE).
Tp, Tp', the pouch ; *op*, opening into trachea ; *c.d.s*, prolongation of upper end of pouch ;
fb, fb', fibrous glands.

tympaniformis and a pessulus ; the first three or four
bronchial semi-rings are different from those which follow.
The remaining Ratitæ have no distinct syrinx.   In *Struthio*,
for example (see figs. 43, 44, p. 64), although there is a membrana tympaniformis completing the bronchi internally, there
is neither pessulus nor intrinsic muscles.   The syrinx of
Apteryx is about on the same level.   *Casuarius* is rather

different. The last few tracheal rings are incomplete posteriorly; the space left between them is continuous with a membrana tympaniformis. There is no pessulus or intrinsic musculature; in the division of the last tracheal rings there is a suggestion, faint perhaps, of the tracheal syrinx. *Dromæus*, as might be imagined, closely resembles *Casuarius*. It has, however, a peculiarity which has been fully gone into by MURIE,[1] who quotes the pre-existing literature upon the matter. In front of the trachea some way down the neck a certain number of the tracheal rings are deficient in front; and the lining membrane of the tube here projects as a sac, which can be inflated, and has, no doubt, something to do with the drumming voice of the bird. The accompanying illustration shows this peculiarity, which is not met with in the cassowary.[2]

As to *osteology*,[3] in *Rhea* (fig. 77, p. 140) the vomer tends to be bifurcate posteriorly where it is much widened out, and articulates both with the palatines and the pterygoids. The palatines are posteriorly flat and fenestrated. The maxillo-palatines are thin plates which meet the anterior bifurcation of the vomer. The descending process of each lacrymal has a large foramen, the presence of which led R. O. CUNNING-HAM to distinguish *Rh. Darwini* from *Rh. americana*, where the foramen is simply a notch. GADOW, however, showed that the question of notch or foramen is simply individual variability, and I am in a position to assert the same of *Rh. macrorhyncha*. The existence of a complete descending process of the nasal has been denied. PARKER, however, has figured an ascending pillar of bone from the maxillary, and in a specimen of *Rh. macrorhyncha* this was joined by a suture to the anterior margin of the lacrymal. It has, it is true, no connection with the premaxillary part of the nasal; but this can scarcely interfere with a comparison of the

[1] 'On the Tracheal Pouch of the Emu,' *P. Z. S.* 1867, p. 405.

[2] For the lungs and air sacs of Struthiones see *ante*, p. 495. Those of *Rhea* have been described by W. N. PARKER, 'Note on the Respiratory Organs of *Rhea*,' *P. Z. S.* 1883, p. 141; of *Dromæus* by MALM, 'Om Luftrör-säcken,' &c., *Öfv. K. Vet. Ak. Förh.* 1880, p. 33.

[3] PANDER and D'ALTON, *Die Skelete der straussartigen Vögel.* Bonn, 1827.

bone to the outer part of a nasal and to the naso-maxillary of the Dinornithidæ (see below).

There is a well-developed, though thin and curved, ectethmoid lamina, which joins the maxillo-palatine below and the descending process of the lacrymal above. This has been also stated to be absent.

Rhea has seventeen *cervical vertebræ.* The *atlas* is notched, as in *Struthio*, but not so widely. In the shoulder girdle the *procoracoid* is short, but is continued down to the articulations of the coracoid by the membrana coracoidea, of which, in a specimen of *Rhea macrorhyncha* before me, a

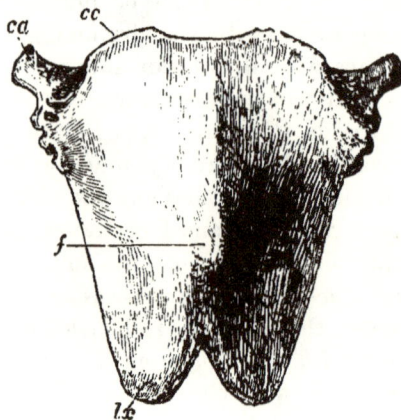

FIG. 242.—STERNUM OF *Rhea* (AFTER MIVART).
cc, coracoid grooves ; *ca*, anterior lateral process ; *f*, keel (?) ; *l.x*, posterior lateral process.

portion is ossified as a thin spicule of bone shutting in the foramen coracoideum. The *sternum* (see fig. 242) has a median ventral prominence and two lateral thin rings of the bone, which may be indications of foramina. Three (sometimes four) pairs of ribs reach the sternum. The *pelvis* (fig. 243) has a small pectineal process. The pubes join the ischia posteriorly, and anteriorly an interobturator process, of which there are faint indications in *Struthio*, unite the two bones. Posteriorly the ilia are attached to the ischia.

The structure of the *skull* of the emu [1] is not widely different from that of the skull of *Rhea*. The vomer is widely bifurcate behind, where it articulates both with pterygoids and palatines. The basipterygoids articulate with the pterygoids at the extreme posterior end of the

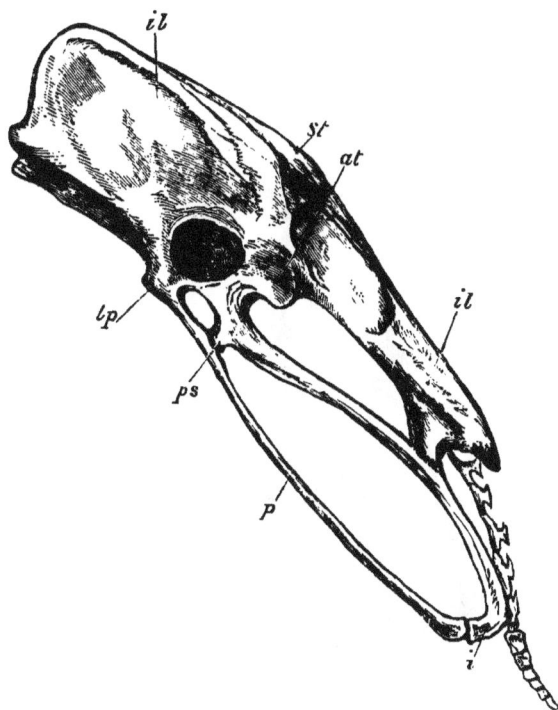

FIG. 243.—PELVIS OF *Rhea* (AFTER MIVART).

il, ilium ; *lp*, pectineal process ; *at*, antitrochanteric process ; *st*, supratrochanteric process ; *ps*, interobturator process ; *i*, ischium ; *p*, pubis.

latter, instead of nearly halfway along them, as in other birds. The maxillo-palatines are hollow swollen plates, which unite with the vomer and premaxillaries, but come apart in the dried skull. The descending process of the lacrymal has a foramen, as in *Rhea*. It joins the thin lamina of the ectethmoid. The descending process of the

---

[1] W. K. PARKER, 'On the Structure and Development of the Skull in the Ostrich Tribe,' *Phil. Trans.* 1868, p. 113.

nasal is only represented in the specimen before me by a minute pointed bit of bone attached above to a point corresponding to that whence the 'naso-maxillary' arises in *Rhea*.

Dromæus has twenty *cervical vertebræ*. The *atlas* is

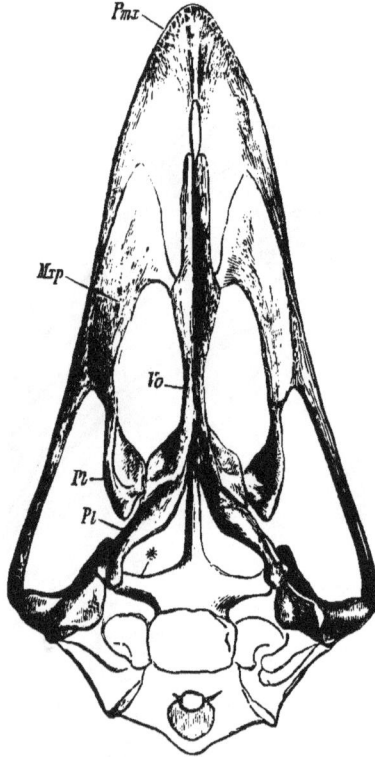

FIG. 244.—SKULL OF EMU (AFTER HUXLEY).

Pmx, premaxilla ; *Mxp*, maxillo-palatine ; *Vo*, vomer ; *Pl*, palatine ; *Pt*, pterygoid : *, basipterygoid process.

notched, very nearly perforated (fig. 66, p. 118). The *procoracoid* is not quite so well developed as in *Rhea*, but there are a pair of rudimentary *clavicles*.

The *sternum* (fig. 74, p. 129) ' much resembles that of *Rhea* ;' it is not notched and is rather pointed at its extremity. Three or four ribs reach it.

The *pelvis* (fig. 245) has the pubes and ischia quite free posteriorly in the dry skeleton; but they are united by cartilage, as in the latter with the ilium. The interobturator process is present, and shuts off an anterior portion of the obturator foramen.

The *skull* of *Casuarius* [1] is very like that of *Dromæus*.

The number of *cervical verterbræ* in *Casuarius* varies from eighteen to nineteen. In the *atlas* the indication of the

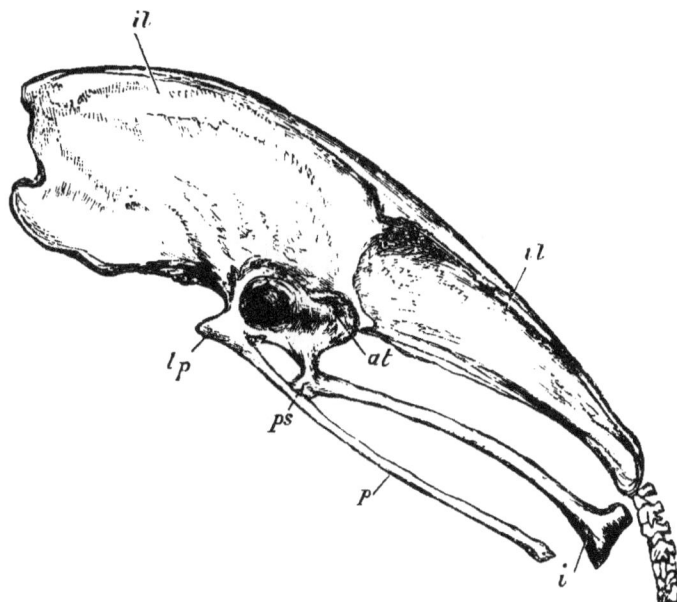

FIG. 245.—PELVIS OF EMU (AFTER MIVART). LETTERS AS IN FIG. 243.

closing of the notch for the odontoid process may be completed, as is shown in the cut (fig. 68, p. 118). The shoulder girdle is very like that of the emu, possessing also rudimentary *clavicles*. The membrana coracoidea may, however, be ossified, and there are two foramina. One of these lies between the membrana coracoidea and the coracoid, and is therefore apparently the homologue of the foramen in

[1] W. H. FLOWER, 'On the Skeleton of the Australian Cassowary,' *P. Z. S.* 1871, p. 32.

Dromæus; the other is smaller and further back in the substance of the coracoid.

The *sternum* (fig. 246) is an exaggeration of that of *Dromæus*, being longer and more pointed posteriorly. Four or five ribs articulate with it.

The *pelvis* too, though very like that of the emu, is (fig. 247) an advance upon it in structure. There may be (*C.*

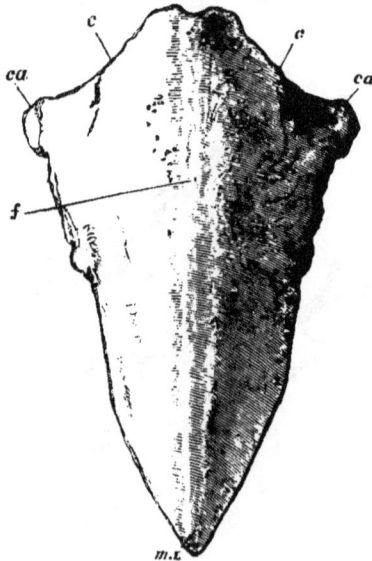

FIG. 246.—STERNUM OF CASSOWARY (AFTER MIVART).
c, coracoid groove; *mx*, posterior end. Other letters as in fig. 242.

galeatus) or may not be an osseous union between pubis and ischium and between ischium and ilium.

The ostrich *skull* is rather unlike that of the other two.

The vomer is very short [1] and does extend back as far as the articulation of the palatines and pterygoids. The latter bones articulate not only with the basipterygoid processes but with the basisphenoid; they bear off the palatines, which

[1] 'W. GRUBER, 'Ueber das Thränenbein der straussartigen Vögel,' &c., *Bull. Ac. Sci. St.-Pétersb.* 1855, p. 161. It varies somewhat in length according to FÜRBRINGER, and was found in one case to be not unlike the vomer of an ægithognathous bird.

run forward in a straight course. The maxillo-palatines articulate with the vomer.

The axial skeleton has been described in the greatest detail by MIVART.[1]

There are twenty *cervical vertebræ*, and then five which have ribs articulating with the sternum. The *atlas* is more simply ring-like than in other birds; it has a very wide notch

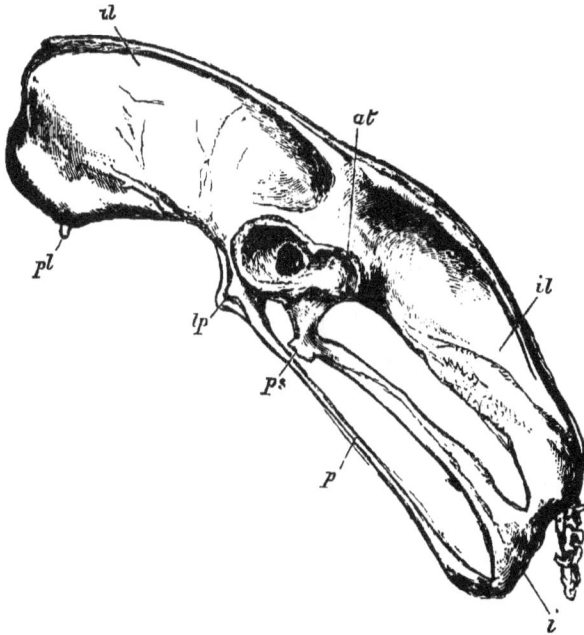

FIG. 247.—PELVIS OF CASSOWARY (AFTER MIVART).
pl, pelvic rib. Other letters as in fig. 243.

for the odontoid process of the axis. The catapophyses of the seventeenth cervical unite.

The *sternum* has a raised and flattened tract posteriorly, which may be the equivalent of the keel; it has two posterior lateral processes, which extend beyond the median portion of the bone.

[1] 'On the Axial Skeleton of the Ostrich,' *Tr. Z. S.* viii. p. 385.

The *pelvis* is remarkable for the symphysis of the pubes, which is shown in the accompanying figure (fig. 250). The ischia also unite each with its corresponding pubis. There is a well-developed pectineal process. GARROD [1] and F. DARWIN described a small ossification attached to the front

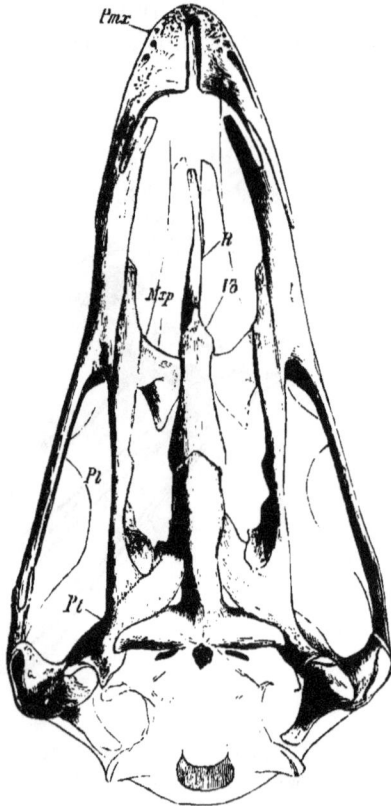

FIG. 248.—SKULL OF OSTRICH (AFTER HUXLEY).
R, rostrum.   Other letters as in fig. 241.

margin of the pubis which may conceivably be the homologue of the marsupial bone of the marsupialia.

The *skull* of *Apteryx* has been described by OWEN [2] as

---

[1] ' Notes on an Ostrich lately living in the Society's Gardens,' *P. Z. S.* 1872, p. 356.

[2] ' On the *Apteryx australis*,' *Trans. Z. S.* ii. p. 57, and iii. p. 277.

regards its adult structure, and the development has been
lately treated of by T. J. PARKER.[1]

Apart from the very elongated anterior part of the skull

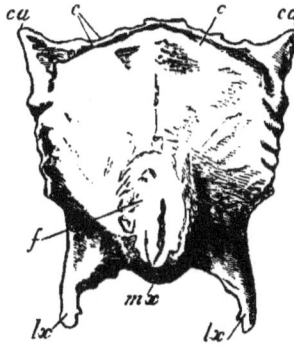

FIG. 249.—STERNUM OF OSTRICH (AFTER MIVART).
LETTERS AS IN FIG. 74.

the characteristics are those of the struthious birds generally.
The Y-shaped posterior end of the vomer bears off from

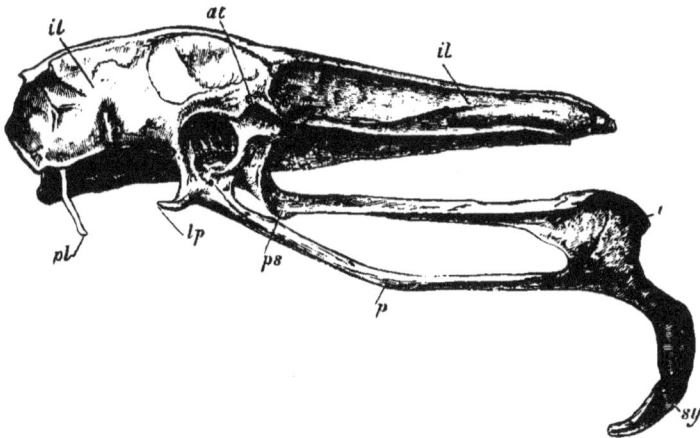

FIG. 250.—PELVIS OF OSTRICH (AFTER MIVART).
sy, symphysis pubis. Other letters as in fig. 243.

articulation with the rostrum the palatines and pterygoids.
The basipterygoid processes are large. As in all other

[1] 'Observations on the Anatomy and Development of *Apteryx*,' *Phil. Tran*
clxxxii. (1891), p. 25.

struthious birds, with the exception of the adult cassowary and the Dinornithidæ, the ossified ethmoid appears on the dorsal surface of the skull between the nasals. In the adult the sutures disappear, and the bones are so firmly united that the quadrate and the columella are the only movable bones in the skull. The quadrate has a two-headed otic process, differing so far from other struthious birds and agreeing with the Carinatæ.

PARKER's statements as to this matter are opposed to those of FÜRBRINGER in the table of differential characters which he gives in his great work. The vertebral column is described by the authors already quoted as well as by MIVART.[1] There are sixteen *cervical vertebræ*, and four ribs articulate with the sternum. The *atlas* is either perforated or only notched by the odontoid process, and it is imperfectly joined above, not always, but in many cases, at the summit of the neural arch. The tenth and eleventh vertebræ have sometimes a ventral hypapophysial canal, as in **Herodiones.** MIVART found this to be the case with *A. Oweni.* I found the catapophyses to approach each other very closely in that species and in *A. australis,* but not to fuse.

The *sternum* is somewhat variable in form. Occasionally the posterior lateral processes exceed the middle process in length ; sometimes they are less or subequal. As a rule the sternum appears to be broader than long, but this is not invariably the case. The varying proportions of the sternum and the lengths of its several processes seem to offer characters diagnostic of the species. In two specimens of *A. Bulleri* PARKER found a ' low ridge nearly as well marked as the vestigial keel of *Stringops.*' The *shoulder girdle,* like the sternum, is subject to great individual variation. The relative lengths of the scapula and coracoid vary ; the curve of the scapula varies, but it is in the coracoid that the most interesting variations occur. The coracoid notch, converted by a ligament into a foramen, but being in the embryo a distinct foramen in the cartilage, is sometimes absent, its place being

---

[1] 'On the Axial Skeleton of the Struthionidæ,' *Trans. Z. S.* vol. x. p. 1. See also ALLIS, ' On the Skeleton of the Apteryx,' *J. Linn. Soc.* 1873, p. 523.

indicated by a thinning of the bone. The rudiments of tuberosities for the attachment of the missing furcula are often fairly evident. The coraco-scapular angle oscillated between 150 and 122. There is a supracoracoid foramen.

In the *pelvis* there is no fusion between ilium, ischium, and pubis. The pectineal process is long and appears to be ossified equally by pubis and ilium. In the skeleton of the adult foot two of the tarsals are present as free bones not fused with either the tibia or the metatarsus ; these are, according to T. J. PARKER, two centralia.

The bones of the wing in the struthious birds are especially reduced in the emu, cassowary, and *Apteryx*.

The wing of the adult emu (*Dromæus ater*) has been figured and described by PARKER. There is no trace of a separate carpus either in young or adult. In a six-weeks-old chick the first metacarpal is half the length of the second, but in the adult it is reduced to a small prominence not a third of its length. There is no trace of a third metacarpal. The single finger (the index) has three phalanges and a long strong claw. ' The wings of an adult are about the size of those of a jay or a bower bird ; in the young chick, with legs the size of those of a turkey, the wings are no longer than a wren's.'

In *Apteryx* the wing is in some respects further reduced than that of the emu ; in others less so. In the adult *A. australis* (T. J. PARKER) there are no distinct carpals, but a broad flattened carpo-metacarpus, with traces of being composed of three metacarpals. There are sometimes two and sometimes three phalanges—the last clawed—to the single finger (index) ; where one is atrophied it is the second. In *A. Oweni* there appears to be invariably a distinct radiale ; the third metacarpal is more distinct than in the last species, and in one case was entirely free. The clawed index has two or three phalanges. The single example of *A. Haasti* which PARKER examined had an ulnare as well as a radiale in the carpus, a fairly distinct metacarpale III., and three phalanges to the index.

In *A. Bulleri* the manus shows much greater variations ;

in one specimen a radiale is present in the carpus, in another
a bone which appears to represent radiale and distal carpals ;
this specimen had a free third metacarpal. In two other
instances there is a carpo-metacarpus, as in *A. australis*.
There are two or three phalanges and, as always, a claw to
the index.

The development of the manus of *Apteryx* shows plainly
what is also apparent from its adult structure, that it is in a
condition of degeneration. Traces of three distal carpals, as
well as of radiale and ulnare, are visible ; all the metacarpals
are distinct, the third being as long as the second and having
a rudimentary phalanx.

The **Æpyornithidæ**, containing the type genus *Æpyornis*
and a recently established new genus, *Mullerornis*,[1] was for
some time only known by the subfossil egg and by the
bones of the hind limb. More recently Messrs. MILNE-
EDWARDS and GRANDIDIER, and more recently again Mr.
C. W. ANDREWS, have described other parts of the skeleton,
so that now, though there are still many lacunæ, we have a
fair knowledge of several important parts of the skeleton.
This family is limited to Madagascar, where its remains have
been found chiefly in marshes.

The *skull* is only incompletely known—the palate, for
instance, so important in determining its affinities, is quite
unknown—being only represented by what is little more
than a calvaria, and by an imperfect mandible. The occipital
condyle is pedunculate, as in the moas. The frontal region
of the skull is covered by many pits, which are arranged in
a fairly regular fashion ; it is suggested that these may be
the marks of the inplantation of feathers, of which, therefore,
the *Æpyornis* may have possessed a frontal crest—a feature
which has also been observed in certain moas. There are
also, as in the moas, a prominent basi-temporal platform,

[1] 'Observations sur les *Æpyornis* de Madagascar,' *Comptes Rend.* cxviii.
1894, p. 122 ; 'Sur les Ossements d'Oiseaux,' &c., *Bull. Mus. Nat. Hist.* 1895,
p. 9 ; 'On the Skull, Sternum, and Shoulder Girdle of *Æpyornis*,' *Ibis* (7), ii.
p. 376 ; 'On some Remains of *Æpyornis* in the British Museum,' *P. Z. S.* 1894,
p. 108.

an open Eustachian groove, and a similar structure of the
articular facet for the quadrate.

The *sternum* is singular by its extraordinary breadth
and great shortness; the length in the middle line is only
one-fifth of the greatest breadth. The hinder border is not
notched, but forms a ' gently concave curve.' The antero-
lateral processes are stout. There is, of course, no keel.
The coraco-scapula is typically ratite, the angle between

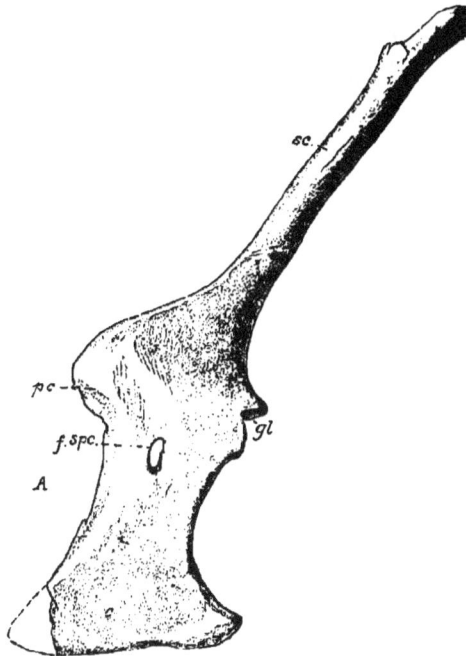

Fig. 251.— Shoulder Girdle of *Æpyornis* (after Andrews).
sc, scapula; *pc*, procoracoid; *f.spc*, foramen supracoracoideum; *gl*, glenoid cavity.

the two being very slight. As will be seen from the figure,
it most resembles that of *Casuarius*. The bird had a
rudimentary humerus.

Dinornithidæ.—This family consists of a number of
genera, all New Zealand in habitat; their remains are so
abundant in various parts of the country that they must
have existed in countless numbers. That there should have

been within so limited an area at least twenty-five distinct species is explained by Captain Hutton by the view that at one time the two islands of New Zealand were divided up into a greater number—an archipelago, in fact—the result of this being what we now see among the cassowaries, where each of the islands inhabited by them has its own peculiar species, isolation, indeed, permitting of the specialisation. All the moas, however, became extinct at a period not less than three or four hundred years ago. T. J. PARKER,[1] whose work on the cranial osteology of the group is the most recent, allows the genera *Dinornis*, *Pachyornis*, *Mesopteryx*, *Anomalopteryx*, *Emeus*, and probably *Megalapteryx*, distributed among three subfamilies. The moas—a general term applied to all these genera—were birds of fair, often large, size. The smaller species ranged from $2\frac{1}{2}$ to 4 feet in height; the largest were at least thirteen feet high.

The skull of the moas had a short and wide beak. The occipital condyle is remarkable on account of its 'more or less pedunculate character,' a circumstance which is of importance in considering the relationship to the moas of the Madagascar *Æpyornis* (see p. 522).

The orbit is smaller than in other struthious birds. The nasals are peculiar in that they meet behind above the ethmoid, so that no part of the latter bone appears on the upper surface of the skull. It is only in the adult cassowary among recent struthious birds that the ethmoid is entirely hidden on a superficial view, a state of affairs which is brought about by the development of the crest, and does not exist in the young bird. The palate is like that of the emu and cassowary, but is most like that of *Apteryx*.

The nasal bone is furnished with a slender maxillary process, or, as in emus, there is a corresponding bone separately ossified. The lacrymal is firmly ankylosed to frontal; its descending process joins ectethmoid. T. J. PARKER has figured and described a peculiar thin scroll-like bone which appears on a lateral view of the skull and pro-

[1] 'On the Cranial Osteology, Classification, and Phylogeny of the Dinornithidae,' *Tr. Z. S.* xiii. p. 373.

jects beyond the anterior margin of the maxillo-nasal ; this he has termed the alinasal.

The number of cervical vertebræ is large, at any rate in *Anomalopteryx parva*, the only species in which they are all without doubt preserved.   There are in this bird twenty-one.   The sternum is longish and rather narrow in *Anomalopteryx casuarina* ; it is short and broad in *Dinornis maximus*. In all it has a pair of lateral notches strongly marked; the lateral processes are strongly divergent.   There is also a median posterior notch.

The pectoral girdle is but little known, and appears sometimes to have been completely absent.

In the pelvis the bones are separate and the pectineal process but little marked.

That the feathers have large aftershafts, like the emu, &c., was first discovered by the late Mr. DALLAS.[1]   Sir R. OWEN has figured the ossified rings of the trachea ; but they present no special features of interest.

As to their relationships with other ratites, T. J. PARKER is of opinion that they form, together with the *Apteryx* and cassowaries, a definite branch of the struthious tree, as in the annexed diagram, which is from his paper.   FÜRBRINGER comes to conclusions which are not greatly different.   The relations of the Dinornithidæ to *Struthio* and *Rhea* are 'ganz entfernt,' to *Dromæus* and *Casuarius* 'fern,' but to *Apteryx* 'nahe.'

There is no doubt that *Struthio* is removed far from the Dinornithidæ, as well as from other ratites, by the structure of its palate, which diverges much.   But it not clear that *Rhea* is so remote ; the existence of an apparent homologue of the maxillo-nasal bone, to which I have referred in the description of the skull of *Rhea*, is a point of somewhat striking likeness to *Emeus*, while the conformation of the skull generally in *Rhea* does not seem to divide it very deeply from *Casuarius*, &c.   Though no doubt T. J. PARKER is right in directing attention to the special resemblances in the skulls of *Apteryx* and the Dinornithidæ, it must not be

[1] ' On the Feathers of *Dinornis robustus*,' OWEN, *P. Z. S.* 1865, p. 265.

forgotten that *Dinornis*, like other ratites, *except Apteryx*, has a single head to the quadrate. In the characters of the pelvis *Dinornis* is near to *Apteryx* and the Casuariidæ and remote from *Rhea* (as well as from *Struthio*). The large aftershaft allies it to the Casuariidæ. NATHUSIUS has commented upon the practical identity in egg-shell structure which *Rhea* shows to *Dinornis*, a likeness which impressed him so greatly that he proposed to place them in the same genus. A considerable number of the special relations between *Apteryx* and the Dinornithidæ, upon which FÜR-BRINGER writes, such as failing pneumaticity, absence of clavicle, mutual distance between coracoids, and even the form of the sternum, may largely depend upon the loss of flight, which is more complete in these birds than in the ostrich, for example. In no less than three footnotes FÜRBRINGER comments upon the supposed absence of uncinate processes to the ribs of the Dinornithidæ; but this difference from other ratites does not exist, as T. J. PARKER has definitely asserted their presence. It is significant that in his tables of differential characters FÜRBRINGER refers little to those of the fore limb girdle (including sternum) as distinguishing Dinornithidæ from *Rhea*. A detailed account of the *pros* and *cons* will be found in the systematic part of FÜRBRINGER's work, and as regards the skull in PARKER's paper already referred to.

The **Struthiones** have been often held to be more primitive than any of the existing groups of birds.

There are really, however, not a large series of characters in which they may be fairly said to be more primitive than some other groups, and most of these are shared by some others.

The form of the palate and the single-headed quadrate appears to be a low character; but the former is shared with the tinamous, the latter with some other groups. The incompleteness of the fusion of the cranial bones may be looked at in the same way; but the penguin is on the same level as the **Struthiones**. The absence of any fusion distally between the bones of the pelvis in *Apteryx* and *Dinornis* is

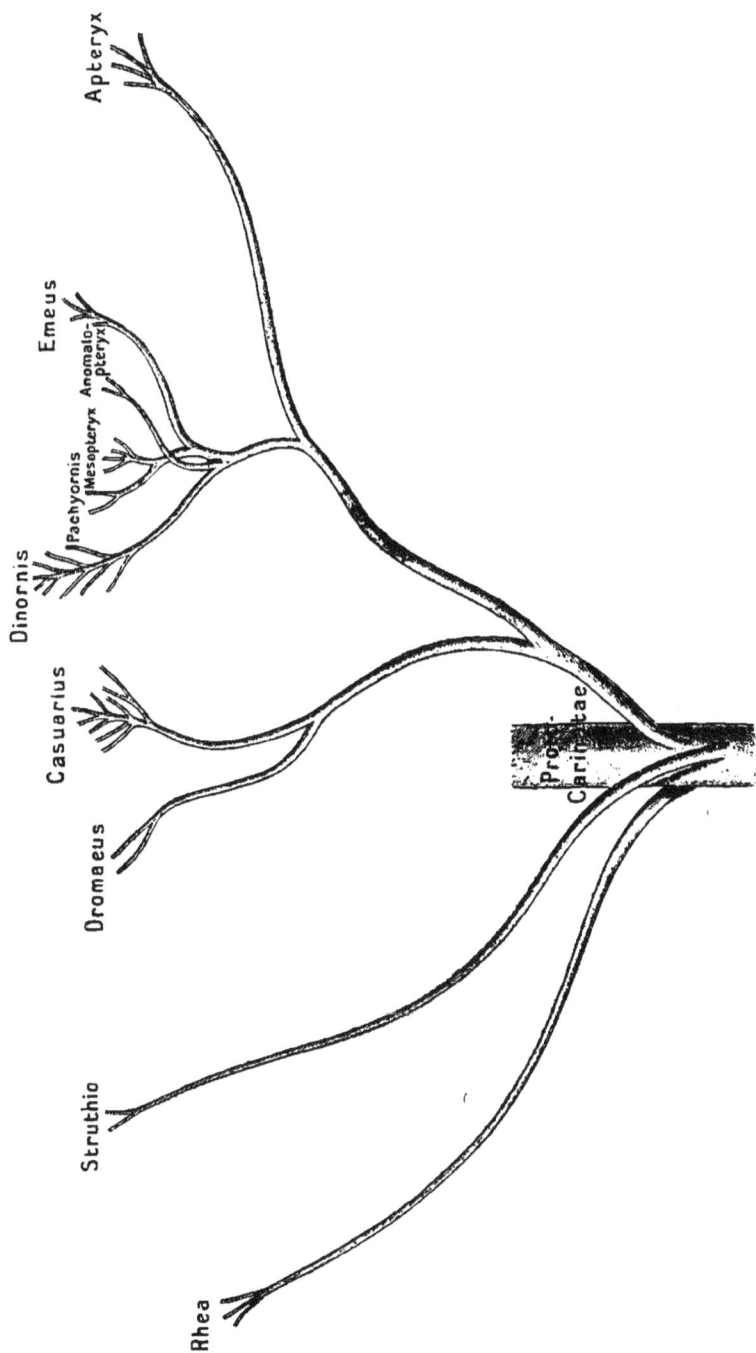

Apteryx
Emeus
Dinornis
Pachyornis
Mesopteryx
Anomalopteryx
Casuarius
Dromaeus
Struthio
Rhea

Pro...
Carinatae

Fig. 252.

dinosaurian ; but the tinamous are like *Apteryx* in this. The complete procoracoid of *Struthio* seems to be an archaic character, as do the two free centralia in the foot of *Apteryx*. As to negative characters, the most important of those that are possibly, but not certainly, to be regarded as primitive appears to be the usual absence of the oil gland.

The long rectum of *Struthio* is probably an ancient character ; but whether the absence of a bird-like syrinx in all ratites except *Rhea* is a similar feature seems to be doubtful. The large size of the blood corpuscles in the ratites is noteworthy in this connection.

The following table gives the principal characters of the existing genera. From it may be inferred the somewhat less modified condition of *Apteryx* and the very isolated position of *Struthio* among the members of the group :—

—	Struthio	Rhea	Dromæus	Casuarius	Apteryx
Aftershaft . . .	0	0	+	+	0
Rhomboideus profundus	+	+	+	0	0
Lat. dorsi anterior .	+	+	+	+	0
Lat. dorsi metapat. .	0	0	0	0	+
Serratus metapat. .	0	0	0	0	+
Pectoralis abdom. .	0	0	0	0	+
Muscle formula of leg	ABXY+	BXY+	BXY−	ABXY+ or −	ABXY+
Carotids . . .	2	L.	2	2	L.
Syrinx . . .	R.	+	R.	R.	R.
Large intestine .	Very long	Not long	Not long	Not long	Short
Cœca . . .	Long	Long	Short	Short	Moderate
Manus . . .	3 digits	3 digits	1 digit	1 digit	1 digit
Cervical vertebræ .	20	17	20	18, 19	16
Atlas . . .	Notched by odontoid	Notched	Notched	Notched or perforated	Notched
Pelvis .	A symphysis pubis : ischia join pubes	Ischia and ilia join	Bones free	Bones free	Bones free
Clavicles .	0	0	R.	R.	0

Group SAURURÆ

SAURORNITHES

As there is but a single genus, and in all probability but a single species,[1] in this group, it is useless to attempt any formal definitions of family or other characters. I shall merely give the more important facts in its structure, as I have with the foregoing groups. As to *external characters*, the *Archæopteryx* has an anisodactyle foot, like that of the **Passeres**. The feet and the digits of the manus have been stated to have been covered with scales. That scales may have been present, at least on the foot, is very probable, but there is not the faintest evidence of their having been there. Of *feathers* the remiges and rectrices are plain, while of the general body feathering there is not so much evidence. With the exception of a circle of feathers upon the neck, suggestive of those of the condor, and similar rings of feathers upon the ankle, it is thought by some that the *Archæopteryx* was naked. Most of the restorations, however, admit a general feathering. The chief criticism to be offered is the extreme perfection of the remains of such feathers as are visible in the slab of stone in which the dead bird was originally imbedded. This being the case, the apparent absence of feathers over the general body surface gains more weight. That they may have been present and of the nature of down feathers is believed by reason of certain faint indications of something to round the contours of the body; the group of contour feathers upon the leg are plainly visible even in photographs of the Berlin example. This example is much better than the specimen in London, which is the only other skeleton in existence. The *rectrices* are quite obvious, a pair to each of the separate vertebræ of the tail. There appear to have been not fewer than thirty

---

[1] It has been argued that specific and even generic differences exist between the London and Berlin examples.

of these : FÜRBRINGER places the number between that
figure and forty.   Most of the restorations allow thirty-
two or thirty-four.   This number is important ; it is in
excess of that generally found in living birds, although the
tail itself is not composed of actually more vertebræ.
Among recent birds it is perhaps a significant fact that
the penguins alone have this number.   Of *remiges* seventeen appears to have been the number, six or seven
primaries and ten secondaries.   No existing bird has
so few primaries, the nearest approach being nearly all the
Anomalogonatæ (and some other birds too), which have ten.
There is some difference of opinion as to how these remiges
were attached to the arm and hand. DAMES, in his elaborate
monograph upon *Archæopteryx*, puts forward the view that
they were attached to the metacarpal and down to the
claw of digit II.   MENZBIER limits the attachment of the
primaries to the basal phalanx of the third, not second
digit.   FÜRBRINGER thinks that the greater number of the
primaries were attached to metacarpal III. and the third
finger, only a few being inserted upon the phalanges of
digit II., where the latter is overlapped by the last-mentioned
digit. HURST has adopted the revolutionary view that there
are missing, and probably cartilaginous, digits IV. and V., to
which the primaries were attached.   As to the presumed
additional fingers, if they were really present, where did
they articulate ?   The entire available space appears to be
taken up with the digits which are already known.   In its
primaries *Archæopteryx* is the very reverse of the penguin,
which it appears to resemble in its rectrices.   The exceptional number to be found in that bird is not in the least
explained by the conditions observable in *Archæopteryx*.
A *beak* seems to have been absent in *Archæopteryx*, owing
to the fact that the teeth extend to the end of the jaws.
The *vertebral column* of this bird has some fifty vertebræ, of
which ten or eleven are reckoned cervical ; the smallness of
the number, which probably belongs to this category, is only
approached among the parrots and the Pico-Passeres and
some of their nearest allies, where, however, thirteen is the

lowest number, fourteen being more general. This fact
may, however, have some significance, especially when it is
remembered that fourteen is also found in the penguins, and
when the remarks on p. 164 are taken into consideration as to
the possible low position among birds of the Pico-Passeres.

The vertebræ were apparently amphicœlous. There are
only two sacral vertebræ, and, as already stated, the tail is
long and composed of a long series of elongated vertebræ, to
each of which a pair of rectrices are attached.

The *ribs* seem to have had no *uncinate processes*, but these
may have been present and cartilaginous; another remarkable feature about them is the fact that they had, as in many
reptiles, but one articulation. There are also a number of
abdominal ribs which might be supposed to be merely the
sternal parts of the vertebral ribs, were it not for the close
approximation of the **V**-shaped pairs.

The *skull* is toothed to the very end of the jaws, thus
rendering improbable the presence of a beak. The nostrils
are definitely holorhinal, and are divided into two holes by an
alinasal growth, as in some living birds. Or else the supposed posterior part of the nasals is really the antorbital
space present in so many birds. It does not seem certain
whether in the latter event the nostrils are bounded behind
by the posterior division of the nasal bone or whether, as in
pterodactyles, a process of the maxilla rises up to join the
nasal. The space for the eye, which has a ring of bones
in the sclerotic, is completed below, as in certain parrots,
by a bony arch.

Concerning the *sternum* we must, I suppose, agree with
HURST, who has observed that 'nothing is known, though
much has been written.'

The *scapula* is eminently bird-like, as is the *furcula*, with
its **U**-shaped meeting of the two ankylosed bones. The
coracoid is imperfectly known. The large size of the deltoid
crest of the humerus and the apparent absence of the crest
for the insertion of the pectoralis are the two most salient
facts in its structure: the latter fact supports those who
hold that the sternum, if present, must have been small or

cartilaginous. It is usual to consider that the *carpus* of *Archæopteryx* contained but one carpal ; HURST, however, asserts that there are two, a radiale and an ulnare. As in modern birds, *Archæopteryx* is generally held to have possessed three fingers, and, again as in modern birds, the second is the longest. HURST holds that the bird had five, and bases his view upon both fact and theory. As to the former he sees differences in the supposed second and third metacarpals in the Berlin and London specimens ; it is possible, therefore, that the bones are not the same in the two cases ; hence those of the one may be metacarpals four and five. In the second place he considers that (for reasons which will be referred to more fully immediately) the bird used its fingers for grasping purposes, and that those fingers which were thus used could not have been hampered with feathers of the stiff kind shown in the fossils as apparently attached to them ; hence there were missing digits to which these feathers were attached. In support of this he recalls the young *Opisthocomus*, which uses the fore limb as a grasping organ before the remiges are developed, and is unable to do so afterwards.

In any case the metacarpals of the three digits are movable, and the number of phalanges progressively increases from two to four.

The *pelvis* is ornithic and has a perforated acetabulum ; but the bones, as in no other bird, are not fused but separated by sutures.

It has been held that the supposed furcula is composed of two ventrally united prepubes, as in dinosaurs (and pterodactyles ?)

The hind limb is avian, with nothing remarkable about it.

The *Archæopteryx* differs from all birds in the following characters :—

(1) The tail is as long as the body, with a pair of rectrices fastened to each vertebra.

(2) The cervical vertebræ (nine) are fewer than in any other bird.

(3) There are apparently free cervical ribs, and the thoracic ribs have but one head.

(4) The sternum is absent or weak (?).

(5) There are abdominal ribs.

(6) The number of phalanges to the fingers of the hand is as in reptiles.

(7) The constituent bones of the pelvis are separate.

(8) There was no beak.

The *Archæopteryx* also differs from all birds excepting those specially mentioned in the following characters :—

(1) The jaws are toothed (also *Hesperornis*, *Ichthyornis*, *Laopteryx* ?).

(2) The ribs have no uncinate processes (so in *Chauna*, *Palamedea*).

(3) The metacarpals are free (also *Gastornis*).

It is, furthermore, supposed that the bones were not aërated, no pneumatic foramina having been discovered. This would militate against flight, and there are other facts of structure that indicate at most a feeble power of flight. It must, however, be observed that the series of rectrices was apparently continued along the sides of the body, and that the tibiæ seem to have borne strong quill feathers. From this HURST infers that the *Archæopteryx* was 'fitted for flight, if not for prolonged flight.' But though it has an 'insessorial' foot it seems doubtful whether the attitude when at rest was not quadrupedal. The heavy head and neck and the slenderness of the hind limbs would tend to throw the centre of gravity further forwards than in recent birds, HURST thinks.[1]

As an appendix to the present group may be mentioned the very imperfectly known *Laopteryx*—not on account of any definitely ascertained resemblances, but merely by reason

[1] DAMES, 'Über *Archæopteryx*,' *Paläont. Abhandl.* ii. 1884, is the principal memoir upon the bird. BARN, in *Zool. Anz.* ix. p. 106, has summed up the literature down to 1886. Since then HURST and PYCRAFT have written upon *Archæopteryx* in *Natural Science*, vols. v. vi.

of the fact that it existed at about the same period. *Lao-pteryx priscus* is known from a skull fragment from the upper Jurassic of Wyoming at about the same horizon as the ' *Atlantosaurus* beds.' It was about the size of the heron (*Ardea herodias*). The back part alone of the skull has been found, and the remains show that the head of the quadrate was undivided, as in ratites (except *Apteryx*). Close by was found a single tooth which may or may not have belonged to it. MARSH [1] considers the bird to have been ratite in its characters.

[1] ' Discovery of a Fossil Bird in the Jurassic of Wyoming,' *Amer. Journ. Sci.* xxi. (1881), p. 341.

INDEX

PRINTED BY
SPOTTISWOODE AND CO., NEW STREET SQUARE
LONDON

www.ingramcontent.com/pod-product-compliance
Lightning Source LLC
Chambersburg PA
CBHW020854210326
41598CB00018B/1663